Nanoparticles: Synthesis and Applications

Nanoparticles: Synthesis and Applications

Editor: Andrew Green

NYRESEARCH
P R E S S

New York

Published by NY Research Press
118-35 Queens Blvd., Suite 400,
Forest Hills, NY 11375, USA
www.nyresearchpress.com

Nanoparticles: Synthesis and Applications
Edited by Andrew Green

International Standard Book Number: 978-1-63238-637-3 (Hardback)

Cataloging-in-Publication Data

Nanoparticles : synthesis and applications / edited by Andrew Green.
    p. cm.
Includes bibliographical references and index.
ISBN 978-1-63238-637-3
1. Nanoparticles. 2. Nanostructured materials--Synthesis. 3. Materials science. I. Green, Andrew.
TA418.9.N35 N36 2019
620.5--dc23

Contents

Permissions

List of Contributors

Index

Preface

Every book is initially just a concept; it takes months of research and hard work to give it the final shape in which the readers receive it. In its early stages, this book also went through rigorous reviewing. The notable contributions made by experts from across the globe were first molded into patterned chapters and then arranged in a sensibly sequential manner to bring out the best results.

Nanoparticles are a result of rapid scientific progress. Nanoparticles refer to particles that are sized between 1-100 nm in at least one dimension. Nanoparticles are surrounded by an interfacial layer which affects its properties. This layer is made up of organic molecules, ions and inorganic molecules. Due to the on-going technological advancements, nanoparticles are gaining prominence across all scientific fields such as medicine, manufacturing, materials science, chemistry, etc. This book covers in detail some existing theories and innovative concepts revolving around nanoparticles. Comprehensive language and content ranging from the basic to the most complex theories and practices in the field of nanotechnology makes this book an ideal reference guide for students, researchers and experts.

It has been my immense pleasure to be a part of this project and to contribute my years of learning in such a meaningful form. I would like to take this opportunity to thank all the people who have been associated with the completion of this book at any step.

Editor

Green synthesis of gold nanoparticles using marine algae and evaluation of their catalytic activity

. Ramakrishna[1] · Dandamudi Rajesh Babu[1] · R. M. Gengan[2] · S. Chandra[3] · G. Nageswara Rao[1]

Abstract The hazardous effects of current nanoparticle synthesis methods have steered researchers to focus on developing newer eco-friendly methods for synthesizing nanoparticles using non-toxic chemicals. Owing to the diverse applications of nanoparticles in various fields such as catalysis, medicine, diagnostics, and sensors, several novel green approaches have been explored for synthesizing nanoparticles using different natural sources such as plants, algae, bacteria, and fungi. Hence, in the present work, a green method for the synthesis of gold nanoparticles (AuNPs) under ambient conditions using aqueous extracts of marine brown algae is reported and the synthesized AuNPs were evaluated for their catalytic efficiency. The aqueous extracts of algae comprise reducing as well as capping agents required for the formation of AuNPs. The Fourier transform infrared spectra of the extracts revealed the presence of compounds having hydroxyl groups that are largely responsible for the reduction of auric chloride to AuNPs at room temperature. Results from high-resolution transmission electron microscopy and dynamic light scattering studies suggested that most of the biosynthesized AuNPs are nearly spherical in shape with an average size in the range of 27–35 nm. High negative values of zeta potential measurement confirmed the stability of AuNPs. Moreover, the reduction kinetics of AuNPs studied by UV–visible spectrophotometry showed that they have good catalytic efficiency in the degradation of dyes as well as reduction of nitro compounds in the presence of sodium borohydride as reducing agent. This simple process for the biosynthesis of gold nanoparticles is rapid, cost-effective and eco-friendly. The formation of AuNPs was observed with the change of pale yellow gold solution to ruby red color of gold nanoparticles and confirmed by surface plasmon spectra using UV–visible spectroscopy. Nanoparticles synthesized through such environmentally benign routes can be used for synthesizing many other metal nanoparticles as well as for a wide range of biomedical applications, for commercial production on a large scale and also can be used as efficient catalysts for different organic reactions.

✉ G. Nageswara Rao
gnageswararao@sssihl.edu.in

M. Ramakrishna
mramakrishna@sssihl.edu.in

Dandamudi Rajesh Babu
drajeshbabu@sssihl.edu.in

R. M. Gengan
genganrm@dut.ac.za

S. Chandra
chan1958@gmail.com

[1] Department of Chemistry, Sri Sathya Sai Institute of Higher Learning, Prasanthinilayam, Puttaparthi 515134, Andhra Pradesh, India

[2] Chemistry Department, Durban University of Technology, Durban 4001, South Africa

[3] LN Government College, Ponneri 601204, Tamil Nadu, India

Graphical abstract

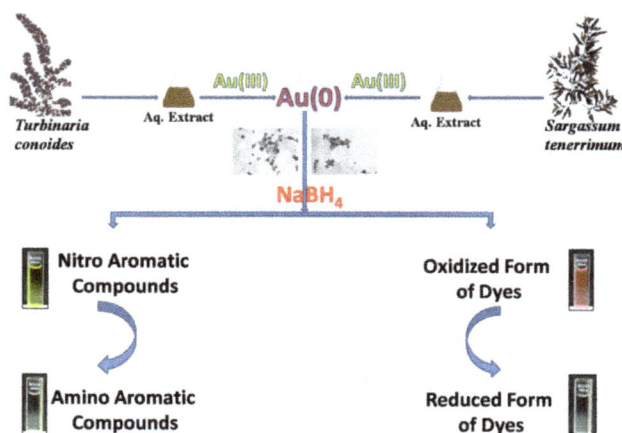

Keywords Green synthesis · *Turbinaria conoides* · *Sargassum tenerrimum* · Gold nanoparticles · Catalysis

Introduction

Nanoparticles are gaining enormous research attention in various fields such as chemistry, physics, materials science, life sciences and engineering. This high interest in nanoparticles is because of their unique optical, magnetic, electronic and catalytic properties with their distinctive feature of size and shape [1, 2]. Many of the existing physical and chemical methods suffer from few drawbacks such as high cost, use of environmentally hazardous chemicals and non-availability for medical applications due to presence of toxic capping agents [3, 4]. These factors contributed towards exploration of new methods and materials for the production of nanoparticles based on the principles of "Green Chemistry". The emphasis in this approach is on the synthesis and application of the nanoparticles for a maximum societal benefit, with minimal impact on the ecosystem [5]. As both the synthesis and applications of nanoparticles are important, many researchers from both academic and industry are focusing towards biological systems such as plants, marine algae, fungi and bacteria for the production of nanoparticles [6–9]. The compounds present in the extracts can act as reducing as well as stabilizing agents and render more biocompatibility to biosynthesized nanoparticles [10].

Oceans are a rich source for many varieties of natural products. Seaweeds are one such variety that belongs to a group of plants known as marine algae [11]. Seaweeds are considered as a source of bioactive compounds such as proteins, lipids, carbohydrates, carotenoids, vitamins and many other secondary metabolites with a wide range of biological activities [12–14]. Few algae such as

Acanthophora spicifera [15], *Chlorella pyrenoidusa* [16], *Kappaphycus alvarezii* [17], *Sargassum wightii* [18], *Sargassum myriocystum* [19], *Stoechospermum marginatum* [20] and *Laminaria japonica* [21] have been used for the synthesis of stable and polydispersed gold nanoparticles. The biosynthesized nanoparticles are even tested for their bacterial pathogenicity and other biological activities. Thus, algae stand as a prospective material for green synthesis of nanoparticles. One of the promising applications of gold nanoparticles (AuNPs) is in the area of catalysis and AuNPs can act as excellent catalysts for many organic reactions even at ambient temperatures. This unique property led many researchers to study the catalytic efficiency of gold nanoparticles. Aromatic nitrocompounds are one of the most commonly used chemical groups in the manufacturing industry and are extremely hazardous if released in environment. However, the reduced products of these aromatic nitro compounds are largely used in preparation of polymers, rubber products, hair dyes, as intermediates for drugs in pharmaceutical industry, etc. [22]. AuNPs act as efficient catalysts for this reduction process. Owing to the toxic effects as well as undesirable by product formation in the reaction medium, the chemically synthesized AuNPs are not suitable as catalysts for many industrial applications despite exhibiting high catalytic activity. Hence, there exists a boundless necessity to develop efficient catalysts through alternate methods for this chemical reduction which is very essential for beneficial applications. One of the major sources of environmental contamination is through dyes from the textile, paint, and paper manufacturing industries [23, 24]. Dyes belonging to the family of xanthines such as Rhodamine B, and Sulforhodamine are extremely harmful because of their severe ecological impact and are non-biodegradable. These dyes should be treated before discharging into the environment. Many physical, chemical treatment methods and biotransformation agents are being used for the reduction of these pollutants. But these are highly ineffective, do not totally eliminate and also involve high operational costs [25]. Hence, there is a need to develop inexpensive methods which degrade the dye molecules into non-toxic compounds.

In this regard, we report the synthesis of AuNPs using aqueous extracts of two brown algae *Turbinaria conoides* and *Sargassum tenerrimum*. The reported nanoparticles synthesis is very simple, efficient and economical. Though there are many reports on biosynthesis of AuNPs from natural sources, application-based reports are few and so we evaluated the biosynthesized AuNPs for their catalytic efficiency using nitro compounds and dyes as model substrates for decolorization. The AuNPs are characterized by UV–visible spectroscopy, high-resolution transmission electron microscopy (HRTEM), Fourier transform infrared

(FTIR) and dynamic light scattering (DLS) with Zeta potential measurements.

Results and discussion

Biosynthesis of gold nanoparticles and UV–vis spectroscopy

Aqueous extracts of *T. conoides* and *S. tenerrimum* were added into two separate flasks containing gold solution. The formation of AuNPs was confirmed by the development of ruby red or pinkish color which is a characteristic of AuNPs [26]. The color change from light yellow to ruby red or pinkish red is due to the excitation of surface plasmon resonance (SPR) in the gold nanoparticles induced by passing light and this observation was confirmed by UV–vis spectral analysis [11]. Change in the SPR of AuNPs with respect to reaction time can be seen in Fig. 1. Intensity of the color as well as the absorbance values increased gradually in the reaction medium along with the reaction time which implied an increasing AuNPs concentration and indicated continuous reduction of gold ions. AuNPs from *T. conoides* (Fig. 1a) showed a weak SPR band at 536 nm and after 90 min, a red shift was observed in SPR at 540 nm due to slight modification in the size and shape of AuNPs. Absorption spectra of AuNPs synthesized from *S. tenerrimum* are shown in Fig. 1b. The bioreduction of Au(III) ions and formation of AuNPs is confirmed by the gradual appearance of pinkish red color. A weak SPR band at 525 nm can be seen in Fig. 1b and after 60 min, a red shift to 547 nm was observed due to the formation of AuNPs. A gradual increase in the absorption intensity and saturation at absorbance value 2.069 indicate the complete reduction of gold ions in the reaction medium. The

incubation period of 15 days did not show any considerable change in SPR intensity of *T. conoides* thereby ascertaining the stability of biosynthesized nanoparticles. However, with respect to *S. tenerrimum*, there was a slight decrease in the absorption intensity of AuNPs and broadening of SPR after an incubation period of 15 days. In Fig. 1, the absorbance around 520–560 nm is due to the SPR exhibited by spherical nanoparticles. However, the spectral pattern in the near infrared (NIR) region shows a shoulder band around 750 nm which increased along with reaction time, thereby indicating that the biosynthesized AuNPs are also anisotropic in nature which is confirmed by TEM images. The biosynthesized AuNPs relatively show a sharper absorbance peak around 540 nm and a broader peak between 700 and 800 nm in Fig. 1b. Brown algal cell walls are rich in polysaccharides and so have abundant hydroxyl groups. Results from Mata et al. [27] confirmed the participation of hydroxyl groups during biosynthesis of AuNPs using *Fucus vesiculosus*. Fucoxanthins which are carotenoids rich in hydroxyl groups and are algal pigments can also contribute with respect to gold reduction as these have good reducing properties [28]. Fucoidans refer to a type of polysaccharide which contains considerable percentages of L-fucose and sulfate ester groups [29]. Results from [[30], [31]] show that *T. conoides* has higher fucose content than *S. tenerrimum*. Hence, this may be the result for a faster formation of AuNPs in *T. conoides*. The plasmon absorption of AuNPs is dependent on particle size and nature of the nanoparticles can be qualitatively related to the shape of the resonance peak [32]. Sharp absorbance peaks are exhibited by small and uniform sized nanoparticles, whereas broad absorbance peaks imply a wider size distribution or aggregation of nanoparticles [33]. According to Mie's theory, spherical AuNPs exhibit strong absorption at 520 nm without any additional band;

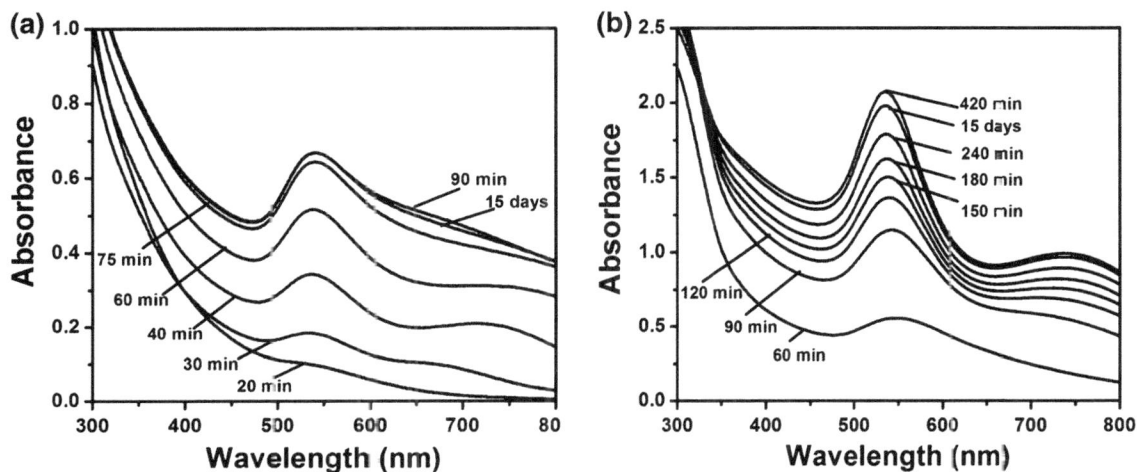

Fig. 1 UV–vis absorption spectra of time-dependent formation of AuNPs synthesized from **a** *Turbinaria conoides* and **b** *Sargassum tenerrimum*

however, the triangular shape AuNPs absorb at 540 nm along with additional absorption in near infrared region. Anisotropic AuNPs give rise to more than one SPR band depending on the shape of the particle. The number of SPR peaks and the symmetry of nanoparticles are inversely proportional to each other i.e., as the number of SPR peaks increases, the symmetry of nanoparticles decreases [34]. Thus, spherical, triangular and other shaped AuNPs show one, two or more SPR peaks. Results from the present study show the presence of isotropic and anisotropic AuNPs as can be seen in Fig. 1 with two SPR bands. Anisotropic AuNPs have their characteristic absorbance in the NIR region and also they are more compatible for usage in many biological applications [35].

Fourier transform infrared (FTIR) studies

FTIR spectral measurements on aqueous extracts of marine algae and the obtained AuNPs are carried out to identify the possible biomolecules that are responsible for the reduction of Au(III) ions as well as capping of the AuNPs. Figure 2a shows IR spectra of the aqueous extract of *T. conoides* with many prominent peaks containing diverse

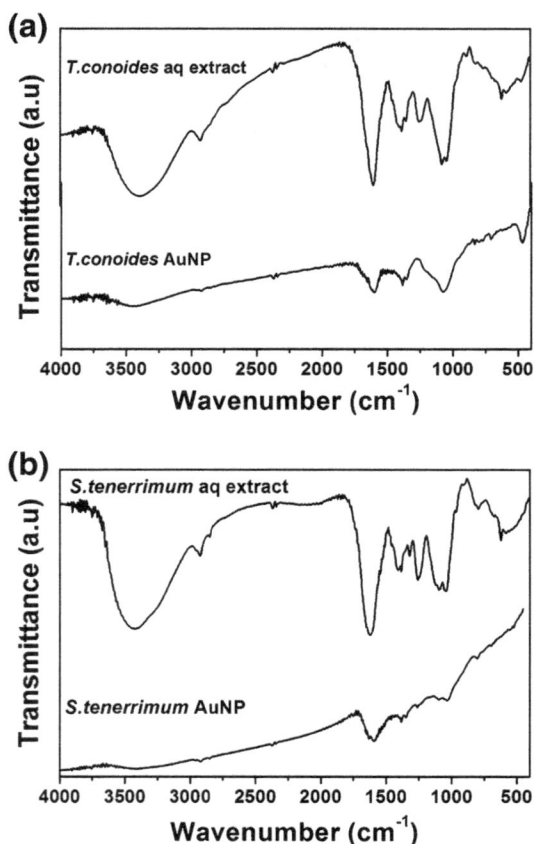

Fig. 2 FTIR spectra of **a** *T. conoides* aqueous extract and its AuNPs, **b** *S. tenerrimum* aqueous extract and its AuNPs

functional groups. Polysaccharides are rich in brown algae and so they have abundant hydroxyl groups. The broad peak around 3385 cm^{-1} in the IR spectra of aqueous extract of *T. conoides* corresponds to the existence of both $-NH_2$ and $-OH$ groups. This broadness is due to the overlap of both O–H bond stretching of high concentration of alcohols or phenols and also the N–H stretch of 1° amines. *T. conoides* is rich in polyphenolic substances [36, 37] and phytochemical investigation led to the isolation of sulfated polysaccharides [30, 38]. The presence of carboxyl groups on the surface of *T. conoides* as reported earlier [39] is confirmed by the presence of a strong band at 1602 cm^{-1} and a weak band around 1400 cm^{-1} due to the asymmetrical and symmetrical stretching of carboxylate ions. The peak at 1253 cm^{-1} denotes C–N stretching of aliphatic amines. The peaks at 1078 and 1043 cm^{-1} are due to the C–N stretching vibrations of aliphatic amines present in the algae. IR spectra of AuNPs in Fig. 2a showed weaker bands around 3385, 1602 cm^{-1} and suppressed band at 1253 cm^{-1} than compared with the extract. This observation revealed that compounds having amine and hydroxyl functional groups could possibly be associated in the reduction as well as stabilization of AuNPs which is in agreement with studies reported earlier [28, 40].

IR spectra of the aqueous extract of *S. tenerrimum* are shown in Fig. 2b and previously reported studies showed the presence of various secondary metabolites such as amino acids, alkaloids, carbohydrates, flavonoids, saponins, sterols, tannins, proteins and phenolic acids from *S. tenerrimum* [41]. The presence of an absorption band at 3446 cm^{-1} is due to the N–H stretching vibrations of peptide linkages and O–H stretch vibrations of phenolic OH groups indicating the existence of polyphenols and phenolic acids. Peaks at 1610 and 1420 cm^{-1} are characteristic of COO^- stretching of alginates [42]. The peak at 1321 cm^{-1} is due to C–N stretch of amine groups. It can be observed from Fig. 2 that FTIR spectra of aqueous extracts of both the species of algae appear almost similar and there exist some common absorption peaks. IR spectra for AuNPs synthesized from *T. conoides* look similar to that of the crude extract with a slight variation in the intensities and wavelengths. The peak at 3447 cm^{-1} has been shifted from 3385 cm^{-1} after reducing the chloroauric acid which indicates the capping with gold nanoparticles. Similarly, IR spectra in case of *S. tenerrimum* too show a large variation of intensity at 3446 cm^{-1} and almost similar absorption bands with that of the extract. Previous studies have shown that the hydroxyl groups have a stronger ability to interact with nanoparticles and therefore the secondary metabolites containing hydroxyl group may act as capping agents for the formation of stable gold nanoparticles [28, 43]. Both the brown algae studied are rich in polysaccharides [30, 31, 44] and these may be responsible for the formation of stable gold nanoparticles. Also, it is apparent from Fig. 2 that the presence of different secondary metabolites with various functional groups in the aqueous extracts may act as reducing as well as capping agents for AuNPs.

High-resolution transmission electron microscopy

Figure 3 shows representative TEM images and the corresponding size distribution histograms of AuNPs synthesized from the aqueous extracts of *T. conoides* and *S. tenerrimum*. As shown in Fig. 3a, AuNPs synthesized from *T. conoides* are anisotropic. The AuNPs formed are polydisperse and shape disparity can be seen. Inset of Fig. 3a shows that the diameters of AuNPs vary from approximately 12 to 57 nm with average size of around 27.5 nm. Figure 3b shows the TEM image of AuNPs synthesized from *S. tenerrimum*. Inset of Fig. 3b indicates the AuNPs that have diameters ranging from approximately 5 to 45 nm and the average particles size is around 35 nm. From Fig. 3, we observe a variation in particle size and AuNPs that are anisotropic. The effect of reducing agent ratio has been varied and the ratio of 4.5:0.5 (extract:gold solution) was found to produce stable AuNPs than other variations which gave fewer as well as unstable AuNPs.

Dynamic light scattering (DLS) and zeta potential studies

DLS has been used to measure the particle sizes in colloidal solution and the size distribution data of AuNPs obtained at pH 7 from *T. conoides* and *S. tenerrimum* are shown in Fig. 4a, b. DLS analysis indicated that the hydrodynamic radius is 28.60 ± 20.65 (Z-average ± (SD) d nm) for the AuNPs synthesized from *T. conoides* with a polydispersity index of 0.521. The corresponding zeta potential and zeta deviation values are −26.3 and 27.5 mV. The average particle size distribution is 82.30 ± 52.82 (Z-

average ± (SD) d nm) for nanoparticles synthesized from *S. tenerrimum* with a polydispersity index of 0.412. The zeta potential and zeta deviation values obtained are −29.1 and 82.8 mV, respectively. Both the zeta potential values obtained imply a stable dispersion of the biosynthesized AuNPs. A zeta potential higher than 30 mV or lesser than −30 mV is indicative of a stable system [3]. The large negative potential value suggests the presence of negatively charged moieties in the extracts that confer electrostatic stability to the nanoparticles.

Catalytic activity of the biosynthesized gold nanoparticles

Reduction of nitro compounds (4-nitrophenol and p-nitroaniline)

To study the catalytic activity of the biosynthesized gold nanoparticles, we have chosen the reduction of nitroarenes (4-nitrophenol and *p*-nitroaniline) to their corresponding aminoarenes (4-aminophenol and *p*-phenylenediamine). Progress of the reduction reaction was visualized with discoloration of characteristic yellow color of the nitro compounds and the reaction was monitored with UV–vis spectroscopy as shown in Fig. 5.

The maximum absorbance of aqueous solution of 4-nitrophenol (4-NP) was found to be at 317 nm and the addition of $NaBH_4$ caused a red shift in the absorbance from 317 to 400 nm. This is due to the formation of 4-nitrophenolate ion and is indicative due to the color change from light yellow to intense yellow. The absorption peak at 400 nm remained unaltered with time in the absence of nanoparticles, suggesting that the reduction did not take place. But with the addition of nanoparticles synthesized from the aqueous extract of *T. conoides*, the yellow color faded gradually to a colorless solution. The intensity of the absorption peak at 400 nm decreased

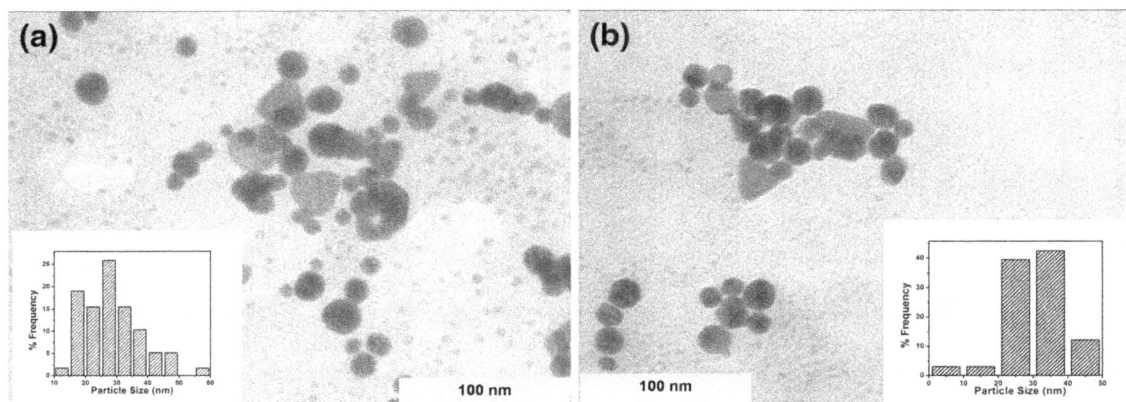

Fig. 3 HRTEM and particle size distribution histogram (*inset*) of AuNPs synthesized from **a** *T. conoides* and **b** *S. tenerrimum*

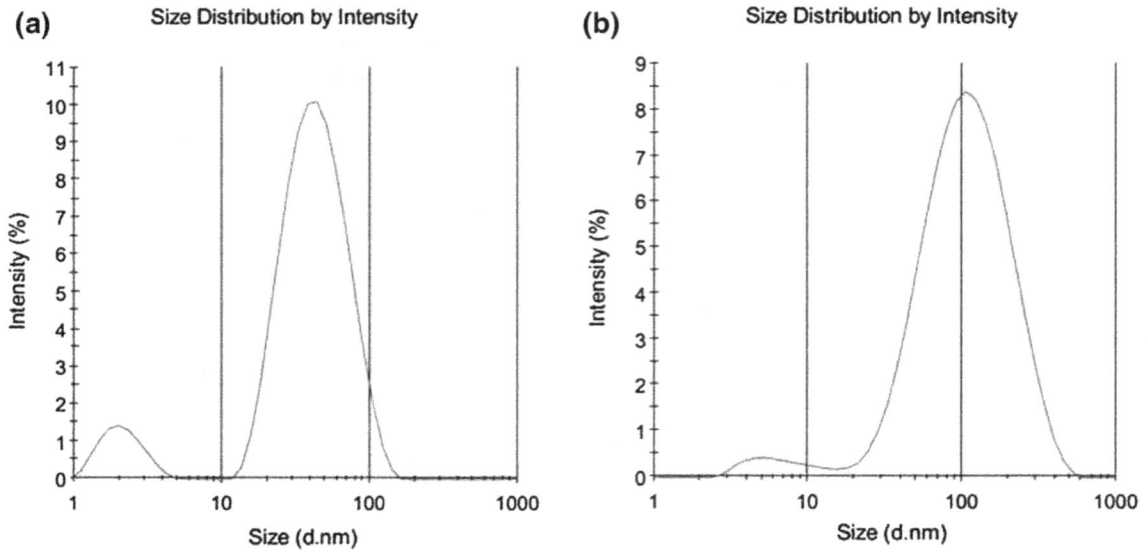

Fig. 4 Size distribution of AuNPs synthesized from **a** *T. conoides* and **b** *S. tenerrimum*

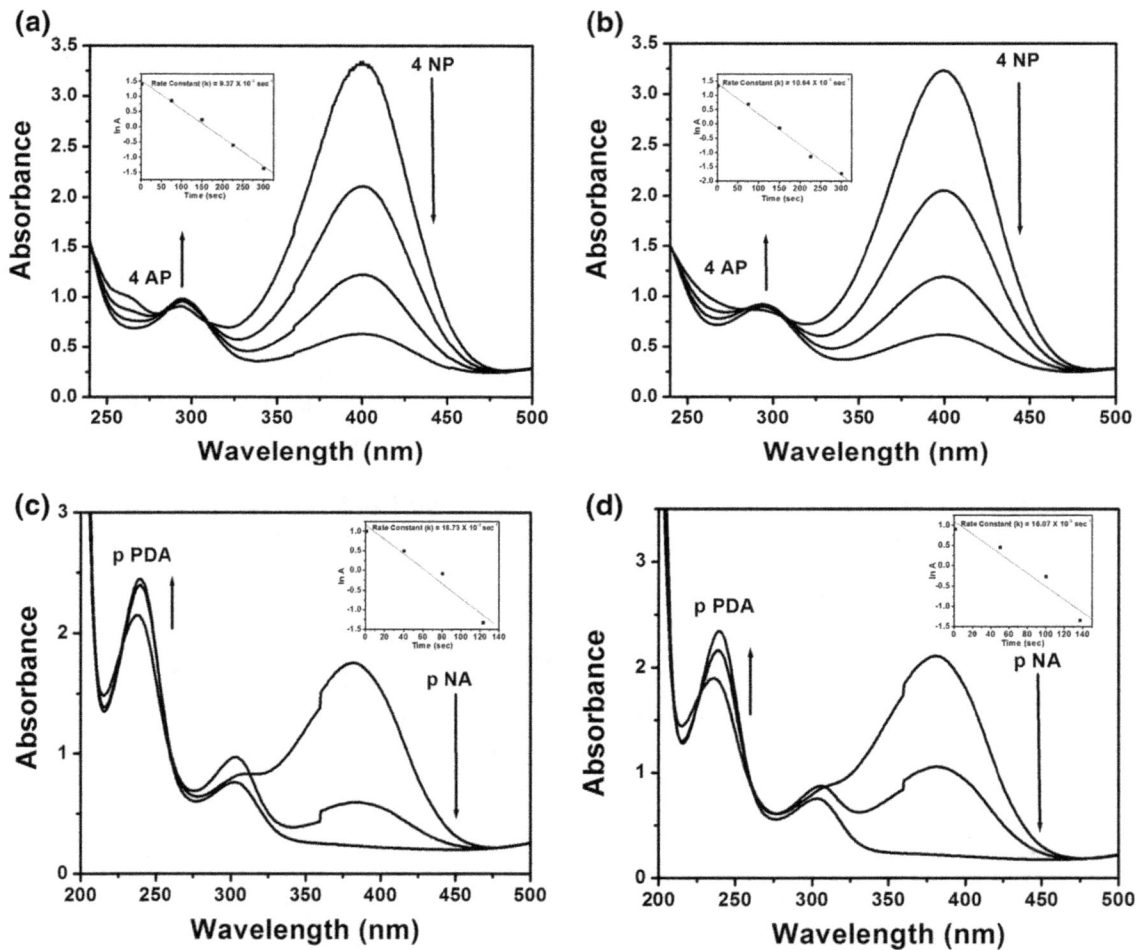

Fig. 5 Time-dependent UV–vis spectra of the borohydride reduction of 4-NP and *p*-NA catalyzed by AuNPs obtained from *T. conoides* (**a**, **c**) and *S. tenerrimum* (**b**, **d**). *Inset graphs* are the corresponding kinetic plots of ln *A* (*A* = absorbance of 4-nitrophenolate ion at 400 nm) versus time

gradually with the concomitant appearance of a new peak at 298 nm, corresponding to the formation of 4-aminophenol (4-AP) (Fig. 5a) [45]. In this process of catalytic reduction, the nanoparticles transfer the electrons from BH_4^- ions to nitro compound, which was qualitatively monitored by UV–vis spectrophotometer. As seen in Fig. 5a, the presence of isosbestic points indicates that 4-NP is fully converted to only 4-AP and no side reaction took place [46]. The reaction was complete within 300 s at room temperature. Inset of Fig. 5a shows that the logarithm of the absorbance of 4-nitrophenolate at 400 nm (ln A) versus time showed a good linear correlation and the rate constant (k) was calculated to be 9.37×10^{-3} s^{-1}. Furthermore, the same method was followed for evaluating the catalytic effect of AuNPs obtained from S. tenerrimum. The reaction was complete within 300 s and k value was determined to be 10.64×10^{-3} s^{-1} (Fig. 5b).

The catalytic efficiency of the AuNPs obtained from both the species of algae was also examined for the reduction of p-nitroaniline (p-NA) in the presence of NaBH$_4$. The absorption spectrum of a mixture of p-NA and NaBH$_4$ has shown a band at 380 nm. However, with the addition of AuNPs to the reaction mixture, the band at 380 nm decreased gradually, whereas a new band at 238 nm evolved gradually due to the formation of the reaction product p-phenylenediamine (p-PDA) in the solution (Fig. 5c, d). Initially, the p-NA molecules are adsorbed on the surface of the gold nanoparticles which play an important role in the electron transfer process. The electron transfer occurs from the negatively charged BH$_4^-$ to the p-NA via the AuNPs. The p-NA is then reduced to p-PDA. This has been observed by the slow disappearance of the characteristic yellow color of p-NA into colorless solution due to successful reduction reaction. The complete decolorization has been observed within 123 s in case of AuNPs synthesized from T. conoides and within 140 s with AuNPs synthesized from S. tenerrimum. The reaction rates have been calculated by taking the logarithm values (ln A) of absorbance at 380 nm with respect to time and the plots showed a good linear correlation. The k values for this catalytic conversion of p-NA to p-PDA were calculated to be 18.73×10^{-3} s^{-1} for AuNPs synthesized from T. conoides (Fig. 5c) and 16.07×10^{-3} s^{-1} for AuNPs synthesized from S. tenerrimum (Fig. 5d).

For the above reduction reactions, control experiments have been performed by taking water instead of nanoparticles solution. The yellow color of the nitro compounds solution has not changed which confirmed the catalytic role of nanoparticles. The concentration of NaBH$_4$ used in the reaction mixture is very much higher than the concentration of the nitro compounds. It is assumed that the concentration of NaBH$_4$ remains constant during the reaction and the reduction rate can be assumed to be independent of NaBH$_4$ concentration. In this context, the order of the reactions is considered to be a pseudo-first-order reaction. Results from present study showed better catalytic efficiency than previously reported studies under similar conditions. Biogenic AuNPs obtained from Breynia rhamnoides were evaluated for reduction of 4-NP [47] and the k value obtained was 7.66×10^{-3} s^{-1} which is lesser than the values obtained for 4-NP reduction in the present study. The k values obtained for 4-NP reduction using poly(amidoamine) dendrimer–metal nanocomposites (silver, platinum, palladium) of different generations exhibited catalytic rates in the range 0.0263×10^{-3} s^{-1} to 3.59×10^{-3} s^{-1} [48] were lower than the results obtained in the present study with respect to reduction of 4-NP. Results from the present study when compared with previous works are more promising for further application of AuNPs. Both 4-AP and p-PDA are very useful, and important as they are widely used as intermediates in organic synthesis. They are also used for preparing dyes such as azo dyes, sulfur dyes and fur dyes. 4-AP is used in the production of medicines such as paracetamol and clofibrate. p-PDA is also used in the manufacture of polymers, rubber antioxidants and as photo developer material [49]. The conventional methods for synthesizing aminobenzenes such as 4-AP and p-PDA suffer from the limitations of rigorous reaction conditions, high costs and tedious procedures. Thus, the above method is a green, simple and highly efficient one for preparing gold nanoparticles as the results are encouraging compared to those reported from the literature.

Reduction of organic dye molecules (Rhodamine B and Sulforhodamine 101)

The characteristic absorption peak occurs at 553 nm for Rhodamine B (RhB) in the UV–vis spectrum [50] and Fig. 6 shows the discoloration of RhB. In the absence of sodium borohydride, the reduction of dyes by AuNPs did not take place. The reduction process started spontaneously after mixing the dye solution with NaBH$_4$ along with catalytic amounts of AuNPs and the color of the dye faded gradually. The absorption spectra of RhB (λ_{max} 553 nm) degradation at different time intervals in the presence of AuNPs obtained from both T. conoides and S. tenerrimum are shown in Fig. 6. The reaction rates have been calculated by taking the logarithm values of absorbance (ln A) at 553 nm wrt time and the plots showed a good linear correlation. Inset graphs in Fig. 6 show the corresponding ln A versus time (s) plot for the same and the rate constant value for the reduction. The same procedure has been followed for the degradation studies of Sulforhodamine 101(SRh and λ_{max} 586 nm).

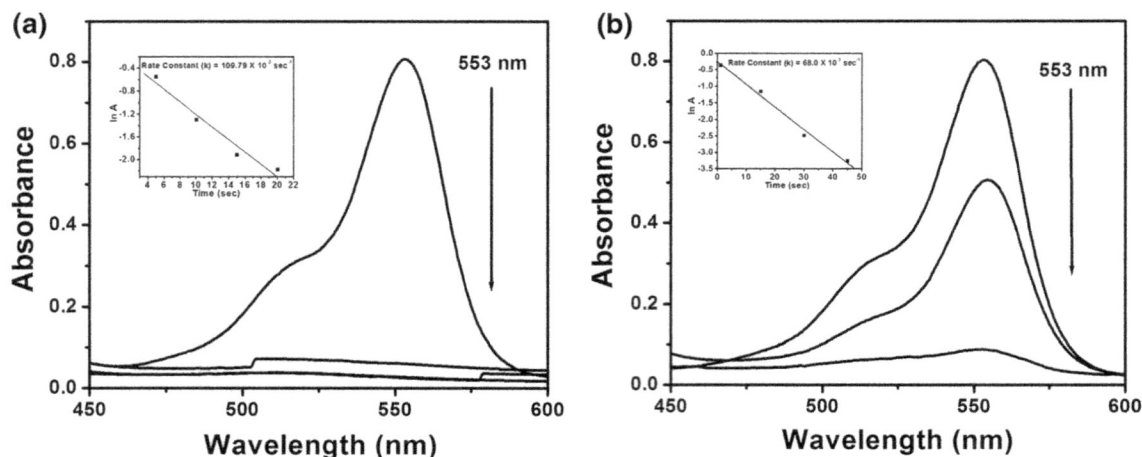

Fig. 6 UV–vis absorption spectra for catalytic reduction of Rhodamine B at 553 nm using AuNPs obtained from **a** *T. conoides*, **b** *S. tenerrimum*. *Inset graphs* are the corresponding kinetic plots ln A (A = absorbance at 553 nm) versus time

The catalytic rates of the studied dye reductions are dependent on the nanoparticle concentration. For three catalytic runs, AuNPs concentration was varied while other conditions were kept constant. With increasing AuNP concentration, k increased (Figs. 6, 7, 8, 9; Table 1) due to increase in number of reaction sites. Table 1 shows that both the reduction time as well as the k values changed with varying amount of catalyst. The reaction rates are higher for 50 μl, medium for 25 μl and lower for 10 μl of catalyst added, thereby showing a dose dependency. The spectra obtained in Figs. 6, 7, 8 and 9 show that with decreasing absorbance at all wavelengths the dye molecules are getting decolorized completely without the formation of any side products. Earlier investigations reported that carboxyl, amino and hydroxyl functional groups act as efficient source for adsorbing dye molecules [51]. Results of FTIR spectra show the presence of hydroxyl, amino and carboxylate groups in the extracts as well as the synthesized gold nanoparticles. An efficient electron transfer occurs when the dye molecules are in contact with the surface of catalyst. During the reduction process, NaBH$_4$ molecules transfer electrons to the dye molecules via the gold nanoparticles and the dye molecules get reduced to give colorless solution.

Methods

Chemicals and materials

Chloroauric acid was obtained from SRL Chemicals, India. 4-nitrophenol, *p*-nitroaniline, Rhodamine B and Sulforhodamine 101 were purchased from Sigma Aldrich. The two species of brown algae *S. tenerrimum* and *T. conoides* were collected from Mandapam, South Coast of Tamilnadu,

India. The algal materials were washed thoroughly with distilled water to remove debris and other associated biota. The samples were shade dried, powdered and stored at 4 °C for further use.

Preparation of extracts

The powdered sample (1 g) of *T. conoides* was mixed with distilled water (20 ml), boiled for 5 min and filtered hot through Whatman No. 1 filter paper. The filtered extract was centrifuged at 5000 rpm for 10 min and the supernatant was used both as a reducing agent and as a stabilizer for preparing gold nanoparticles. The same procedure was adopted for preparing the aqueous extract of *S. tenerrimum*. When not in use, the extracts were refrigerated and stored at 4 °C.

Synthesis of gold nanoparticles

Aqueous extracts (5 ml) of the algae were added to 1 mM aqueous AuCl$_4$ solution (45 ml) in a 250-ml Erlenmeyer flask. The flasks were kept on a magnetic stirrer at room temperature and the solutions changed to ruby red indicating the formation of gold nanoparticles. The reduction of Au^{3+} ions in the solution was monitored at periodic intervals with the help of UV–vis spectrophotometer (Shimadzu 2450). After the reaction reached saturation, the gold nanoparticles solution was centrifuged at 10,000 rpm (Beckman Coulter Avanti J-26SXPI) for 15 min and the obtained pellet was redispersed in distilled water to remove any uninteracted biomass. This process of centrifugation and redispersion was carried out twice to get a better separation of nanoparticles. The obtained nanoparticles were lyophilized using a MiniLyodel lyophilizer.

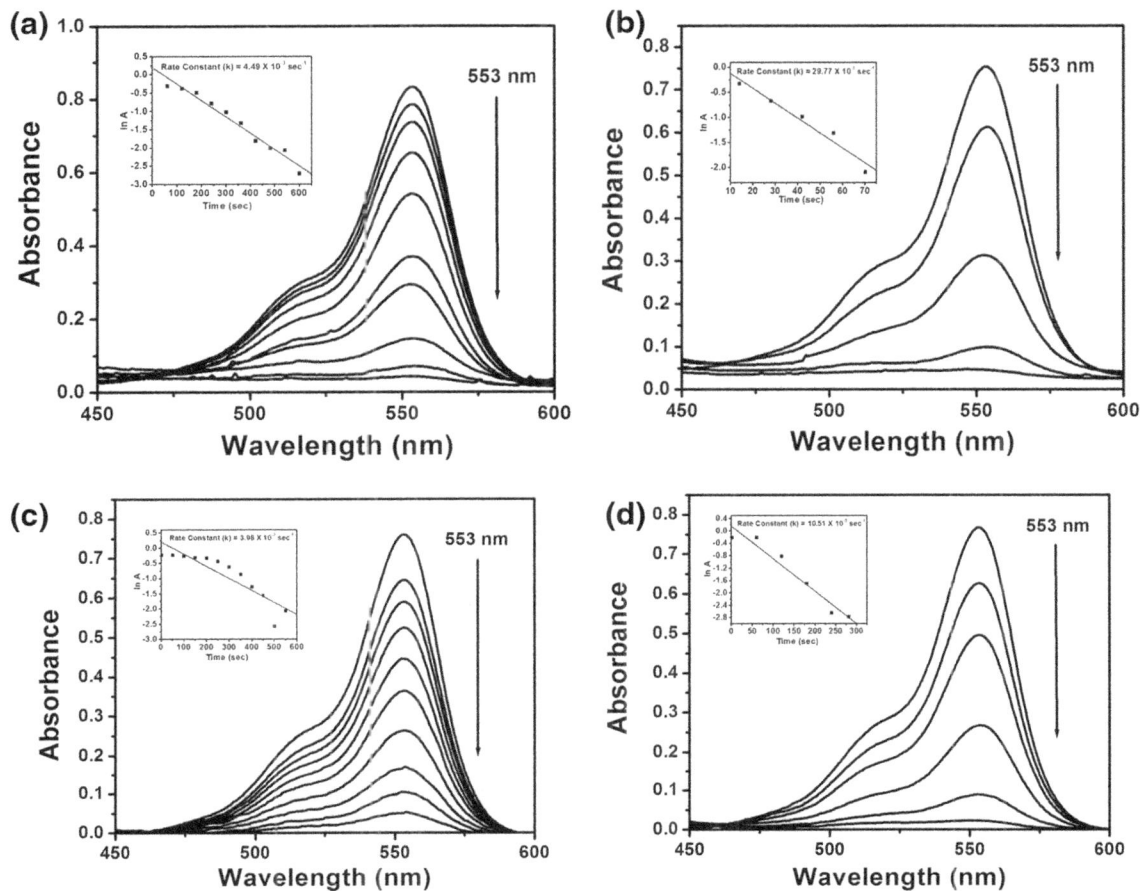

Fig. 7 UV–vis absorption spectra for catalytic reduction of Rhodamine B at 553 nm with NaBH$_4$ using **a** 10 μl of AuNPs and **b** 25 μl of AuNPs as catalyst obtained from *T. conoides*. Below them are the successive reduction spectra of Rhodamine B catalyzed by **c** 10 μl of AuNPs and **d** 25 μl of AuNPs obtained from *S. tenerrimum*. *Inset graphs* are the corresponding kinetic plots ln *A* (*A* = absorbance at 553 nm) versus time

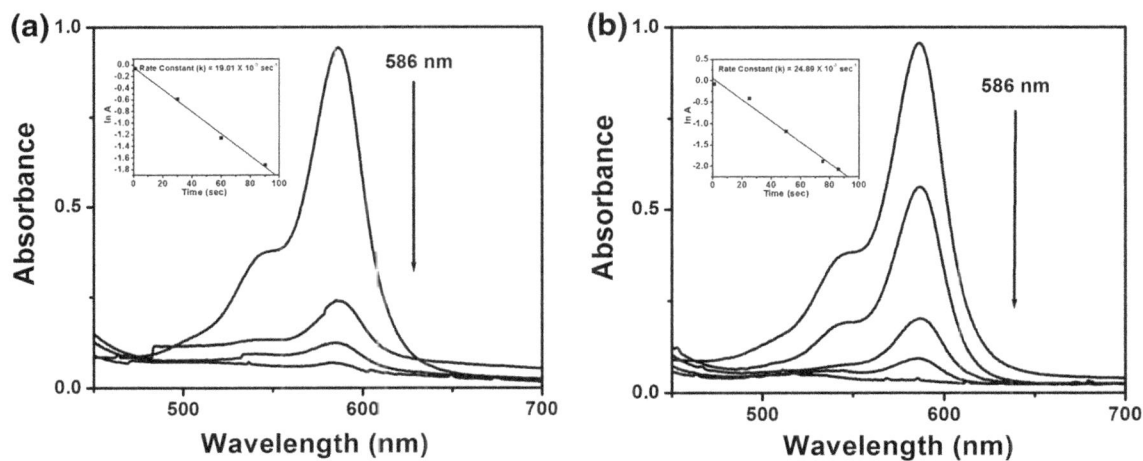

Fig. 8 UV–vis absorption spectra for catalytic reduction of Sulforhodamine 101 at 586 nm using AuNPs obtained from **a** *T. conoides*, **b** *S. tenerrimum*. *Inset graphs* are the corresponding kinetic plots ln *A* (*A* = absorbance at 586 nm) versus time

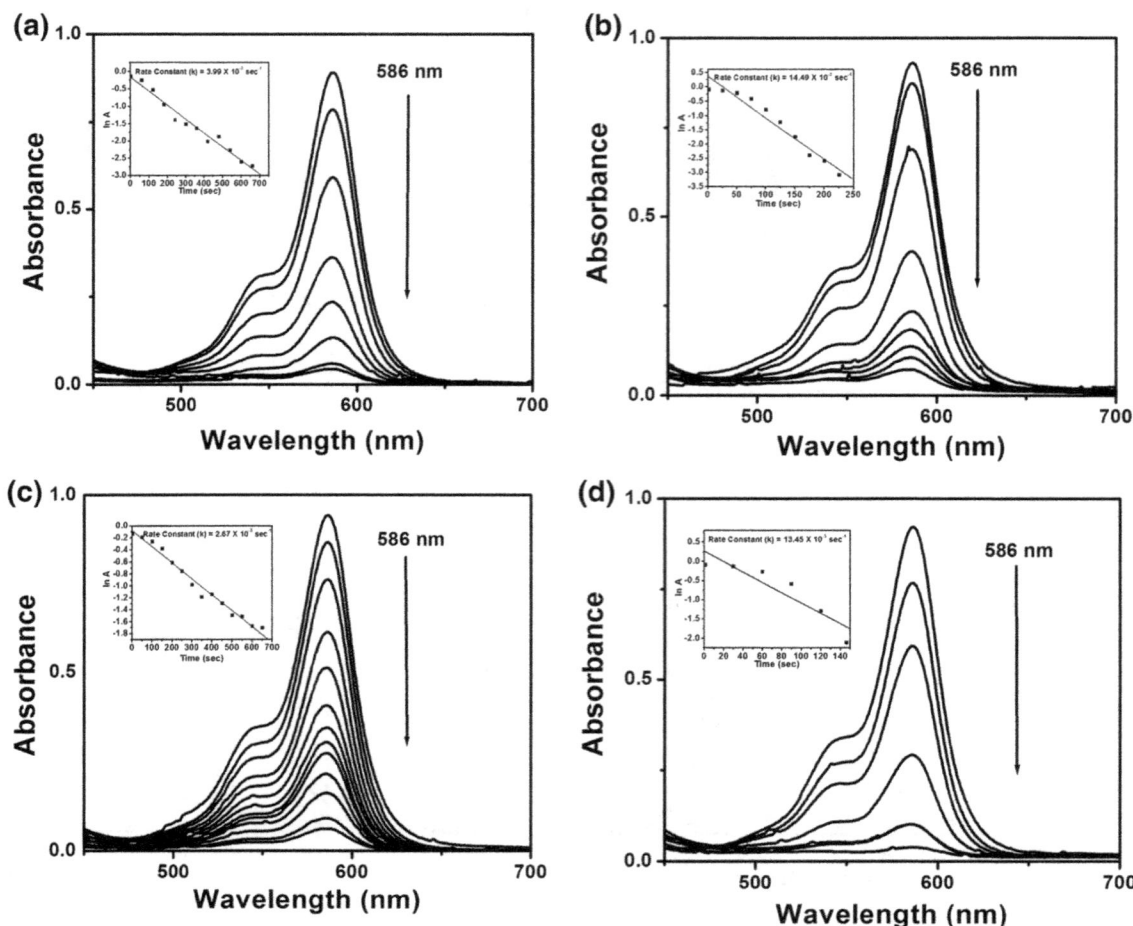

Fig. 9 The UV–vis absorption spectrum for the successive reduction of Sulforhodamine 101 at 586 nm with NaBH₄ using **a** 10 µl of AuNPs and **b** 25 µl of AuNPs as catalyst obtained from *T. conoides*. Below them are the successive reduction spectra of Sulforhodamine 101 catalyzed by **c** 10 µl of AuNPs and **d** 25 µl of AuNPs obtained from *S. tenerrimum*. *Inset graphs* are the corresponding kinetic plots ln *A* (*A* = absorbance at 553 nm) versus time

Table 1 The volume of AuNPs added as catalyst, reduction time and rate constant values for the reduction of dye molecules in the presence of NaBH₄

Name of the dye molecule	Volume of AuNPs added (µl)	Rate constant (k) (s^{-1}) (time taken for full reduction)	
		T. conoides	*S. tenerrimum*
Rhodamine B	10	4.49×10^{-3} (600 s)	3.98×10^{-3} (550 s)
	25	29.77×10^{-3} (70 s)	10.51×10^{-3} (281 s)
	50	109.79×10^{-3} (20 s)	68.0×10^{-3} (45 s)
Sulforhodamine 101	10	3.99×10^{-3} (660 s)	2.67×10^{-3} (650 s)
	25	14.49×10^{-3} (225 s)	13.45×10^{-3} (146 s)
	50	19.01×10^{-3} (90 s)	24.89×10^{-3} (86 s)

Characterization studies

The biosynthesized nanoparticles were characterized by recording the UV–vis spectra at periodic time intervals until the absorption maxima reached saturation. Millipore water was used as blank and the UV–vis spectra were recorded from 300 to 800 nm operated at a resolution of

1 nm. The spectral data were plotted using Origin 6.0 version. To obtain the particle size and shape, the gold nanoparticles solution (1 µl) was placed on Formvar-coated grids, air dried and viewed at 100 kV to carry out transmission electron microscopy (JEOL 1010 TEM using a Megaview III camera and iTEM software) studies. The nanoparticle size distribution was determined using *ImageJ*

software and the resultant data were plotted in histograms. The FTIR spectra of the lyophilized powders of both the aqueous extracts and AuNPs dispersion were recorded using KBr pellet method with Shimadzu IRAffinity-1 spectrophotometer. The spectra were recorded in the range of 400–4000 cm^{-1} and with a resolution of 4 cm^{-1}. The FTIR spectra revealed information about possible functional groups involved in the formation of AuNPs. The stability and size distribution of AuNPs were measured using DLS and zeta potential studies that were carried out using Zetasizer Nano S90 (Malvern).

Evaluation of the catalytic effect of synthesized gold nanoparticles

The catalytic decolorization reactions were carried out in a 3-ml quartz cuvette with a path length of 1 cm and were monitored by UV–vis spectrophotometer by following the method reported by Gangula et al. [47]. An aqueous stock solution of p-nitrophenol or p-nitroaniline (0.3 ml of 2 mM) was mixed with distilled water (1.4 ml) and then an ice cold solution of sodium borohydride (1 ml of 0.03 M) was added. To this, the biosynthesized nanoparticles dispersion solution (0.3 ml) was added and the contents were mixed well. The reaction progress was monitored by recording the time-dependent absorption spectra in the range of 200–500 nm at room temperature.

The obtained nanoparticles were also used as a catalyst for the reduction of two different organic dye molecules Rhodamine B (RhB) and Sulforhodamine 101 hydrate (SRh) by following method reported by Siddhardha et al. [52]. An aqueous stock solution of RhB or SRh (2.5 ml of 10^{-5} M) was added to ice cold solution of sodium borohydride (0.5 ml of 0.1 M) and to this, nanoparticles dispersions of different volumes (50, 25 and 10 µl) were added as catalyst. The mixture was shaken well and the reaction was monitored by UV–vis spectrophotometer in the range of 450–700 nm at room temperature.

Conclusions

The studies revealed that the aqueous extracts of brown algae (T. conoides and S. tenerrimum) can reduce Au(III) ions to gold nanoparticles (AuNPs) and also have the potential to stabilize them. The method followed for synthesis is very simple, cost-effective, efficient, non-toxic and eco-friendly. The biosynthesized AuNPs are of sizes ranging from 5 to 57 nm. The synthesized AuNPs act as efficient catalysts for the reduction of aromatic nitro compounds and organic dye molecules. AuNPs synthesized

from T. conoides exhibited greater catalytic potential than S. tenerrimum. The present biosynthesis method can be extended for preparation of other metal nanoparticles which can be explored in future for quick organic synthesis and for a wide range of catalysis reactions. According to our knowledge, this is the first report dealing with the catalytic potential of biosynthesized AuNPs from T. conoides and S. tenerrimum.

Acknowledgments The authors would like to express their gratitude to Bhagawan Sri Sathya Sai Baba, Founder Chancellor, Sri Sathya Sai Institute of Higher Learning, Puttaparthi, India for his constant support, inspiration and guidance.

Compliance with ethical standards

Conflict of interest The authors declare that they have no competing interests.

Authors' contributions MR synthesized and evaluated the catalytic efficiency of nanoparticles. DRB and RMG characterized the nanoparticles. SC helped in sample collection and identification. GNR supervised the research work. All the authors read, corrected and approved the final manuscript.

References

1. Dubey, S.P., Lahtinen, M., Särkkä, H., Sillanpää, M.: Bioprospective of *Sorbus aucuparia* leaf extract in development of silver and gold nanocolloids. Colloids Surf. B **80**, 26–33 (2010)
2. Das, S., Marsili, E.: A green chemical approach for the synthesis of gold nanoparticles: characterization and mechanistic aspect. Rev. Environ. Sci. Biotechnol. **9**, 199–204 (2010)
3. Edison, T.J.I., Sethuraman, M.: Instant green synthesis of silver nanoparticles using *Terminalia chebula* fruit extract and evaluation of their catalytic activity on reduction of methylene blue. Process Biochem. **47**, 1351–1357 (2012)
4. Ghosh, S., Patil, S., Ahire, M., Kitture, R., Gurav, D.D., Jabgunde, A.M., Kale, S., Pardesi, K., Shinde, V., Bellare, J.: *Gnidia glauca* flower extract mediated synthesis of gold nanoparticles and evaluation of its chemocatalytic potential. J. Nanobiotechnol. **10**, 17 (2012)
5. Dahl, J.A., Maddux, B.L., Hutchison, J.E.: Toward greener nanosynthesis. Chem. Rev. **107**, 2228–2269 (2007)
6. Doane, T.L., Burda, C.: The unique role of nanoparticles in nanomedicine: imaging, drug delivery and therapy. Chem. Soc. Rev. **41**, 2885–2911 (2012)
7. Narayanan, K.B., Sakthivel, N.: Green synthesis of biogenic metal nanoparticles by terrestrial and aquatic phototrophic and heterotrophic eukaryotes and biocompatible agents. Adv. Colloid Interface Sci. **169**, 59–79 (2011)
8. Sadhasivam, S., Shanmugam, P., Veerapandian, M., Subbiah, R., Yun, K.: Biogenic synthesis of multidimensional gold nanoparticles assisted by *Streptomyces hygroscopicus* and its electro-

chemical and antibacterial properties. Biometals **25**, 351–360 (2012)

9. Lengke, M.F., Sanpawanitchakit, C., Southam, G.: Biosynthesis of gold nanoparticles: a review. In: Rai, M., Duran, N. (eds.) Metal nanoparticles in microbiology, pp. 37–74. Springer, Berlin (2011)

10. Dumur, F., Guerlin, A., Dumas, E., Bertin, D., Gigmes, D., Mayer, C.R.: Controlled spontaneous generation of gold nanoparticles assisted by dual reducing and capping agents. Gold Bull. **44**, 119–137 (2011)

11. Inbakandan, D., Venkatesan, R., Khan, S.A.: Biosynthesis of gold nanoparticles utilizing marine sponge *Acanthella elongata* (Dendy, 1905). Colloids Surf. B **81**, 634–639 (2010)

12. Kim, S.-K.: Handbook of Marine Macroalgae: Biotechnology and Applied Phycology. Wiley, Chichester (2011)

13. Chanda, S., Dave, R., Kaneria, M., Nagani, K.: Seaweeds: a novel, untapped source of drugs from sea to combat infectious diseases. In: Méndez-Vilas, A. (ed.) Current Research, Technology and Education Topics in Applied Microbiology and Microbial Biotechnology, pp. 473–480. Formatex Research Center, Badajoz, Spain (2010)

14. Mohamed, S., Hashim, S.N., Rahman, H.A.: Seaweeds: a sustainable functional food for complementary and alternative therapy. Trends Food Sci. Technol. **23**, 83–96 (2012)

15. Swaminathan, S., Murugesan, S., Damodarkumar, S., Dhamotharan, R., Bhuvaneshwari, S.: Synthesis and characterization of gold nanoparticles from alga *Acanthophora spicifera* (VAHL) Boergesen. Int. J. Nanosci. Nanotechnol. **2**, 85–94 (2011)

16. Oza, G., Pandey, S., Mewada, A., Kalita, G., Sharon, M., Phata, J., Ambernath, W., Sharon, M.: Facile biosynthesis of gold nanoparticles exploiting optimum pH and temperature of fresh water algae *Chlorella pyrenoidusa*. Adv. Appl. Sci. Res. **3**, 1405–1412 (2012)

17. Rajasulochana, P., Krishnamoorthy, P., Dhamotharan, R.: Potential application of *Kappaphycus alvarezii* in agricultural and pharmaceutical industry. J. Chem. Pharm. Res. **4**, 33–37 (2012)

18. Singaravelu, G., Arockiamary, J., Kumar, V.G., Govindaraju, K.: A novel extracellular synthesis of monodisperse gold nanoparticles using marine alga, *Sargassum wightii* Greville. Colloids Surf. B **57**, 97–101 (2007)

19. Dhas, T.S., Kumar, V.G., Abraham, L.S., Karthick, V., Govindaraju, K.: *Sargassum myriocystum* mediated biosynthesis of gold nanoparticles. Spectrochim. Acta, Part A **99**, 97–101 (2012)

20. Rajathi, F.A.A., Parthiban, C., Kumar, V.G., Anantharaman, P.: Biosynthesis of antibacterial gold nanoparticles using brown alga, *Stoechospermum marginatum* (Kützing). Spectrochim. Acta, Part A **99**, 166–173 (2012)

21. Ghodake, G., Lee, D.S.: Biological synthesis of gold nanoparticles using the aqueous extract of the brown algae *Laminaria japonica*. J. Nanoelectron. Optoelectron. **6**, 268–271 (2011)

22. Orendorff, C.J., Sau, T.K., Murphy, C.J.: Shape-dependent plasmon-resonant gold nanoparticles. Small **2**, 636–639 (2006)

23. Wong, Y., Yu, J.: Laccase-catalyzed decolorization of synthetic dyes. Water Res. **33**, 3512–3520 (1999)

24. Banat, I.M., Nigam, P., Singh, D., Marchant, R.: Microbial decolorization of textile-dye containing effluents: a review. Bioresour. Technol. **58**, 217–227 (1996)

25. Srinivasan, A., Viraraghavan, T.: Decolorization of dye wastewaters by biosorbents: a review. J. Environ. Manag. **91**, 1915–1929 (2010)

26. MubarakAli, D., Thajuddin, N., Jeganathan, K., Gunasekaran, M.: Plant extract mediated synthesis of silver and gold nanoparticles and its antibacterial activity against clinically isolated pathogens. Colloids Surf. B **85**, 360–365 (2011)

27. Mata, Y., Torres, E., Blazquez, M., Ballester, A., González, F., Munoz, J.: Gold (III) biosorption and bioreduction with the brown alga *Fucus vesiculosus*. J. Hazard. Mater. **166**, 612–618 (2009)

28. Vijayaraghavan, K., Mahadevan, A., Sathishkumar, M., Pavagadhi, S., Balasubramanian, R.: Biosynthesis of Au (0) from Au(III) via biosorption and bioreduction using brown marine alga *Turbinaria conoides*. Chem. Eng. J. **167**, 223–227 (2011)

29. Li, B., Lu, F., Wei, X., Zhao, R.: Fucoidan: structure and bioactivity. Molecules **13**, 1671–1695 (2008)

30. Chattopadhyay, N., Ghosh, T., Sinha, S., Chattopadhyay, K., Karmakar, P., Ray, B.: Polysaccharides from *Turbinaria conoides*: structural features and antioxidant capacity. Food Chem. **118**, 823–829 (2010)

31. Sinha, S., Astani, A., Ghosh, T., Schnitzler, P., Ray, B.: Polysaccharides from *Sargassum tenerrimum*: structural features, chemical modification and anti-viral activity. Phytochemistry **71**, 235–242 (2010)

32. Nellore, J., Pauline, P.C., Amarnath, K.: Biogenic synthesis by *Sphearanthus amaranthoids*; towards the efficient production of the biocompatible gold nanoparticles. Dig. J. Nanomater. Biostruct. **7**, 123–133 (2012)

33. Bakshi, M.S., Sachar, S., Kaur, G., Bhandari, P., Kaur, G., Biesinger, M.C., Possmayer, F., Petersen, N.O.: Dependence of crystal growth of gold nanoparticles on the capping behavior of surfactant at ambient conditions. Cryst. Growth Des. **8**, 1713–1719 (2008)

34. Shankar, S.S., Ahmad, A., Pasricha, R., Sastry, M.: Bioreduction of chloroaurate ions by geranium leaves and its endophytic fungus yields gold nanoparticles of different shapes. J. Mater. Chem. **13**, 1822–1826 (2003)

35. Treguer-Delapierre, M., Majimel, J., Mornet, S., Duguet, E., Ravaine, S.: Synthesis of non-spherical gold nanoparticles. Gold Bull. **41**, 195–207 (2008)

36. Chandini, S.K., Ganesan, P., Bhaskar, N.: *In vitro* antioxidant activities of three selected brown seaweeds of India. Food Chem. **107**, 707–713 (2008)

37. Devi, G.K., Manivannan, K., Thirumaran, G., Rajathi, F.A.A., Anantharaman, P.: *In vitro* antioxidant activities of selected seaweeds from Southeast coast of India. Asian Pac. J. Trop. Med. **4**, 205–211 (2011)

38. Sokhi, G., Vijayaraghavan, M.: Extracellular polysaccharides in *Turbinaria conoides*-structure and ultrastructure. Curr. Sci. **54**, 1192–1193 (1985)

39. Vijayaraghavan, K., Sathishkumar, M., Balasubramanian, R.: Biosorption of lanthanum, cerium, europium, and ytterbium by a brown marine alga, *Turbinaria conoides*. Ind. Eng. Chem. Res. **49**, 4405–4411 (2010)

40. Rajeshkumar, S., Malarkodi, C., Gnanajobitha, G., Paulkumar, K., Vanaja, M., Kannan, C., Annadurai, G.: Seaweed-mediated synthesis of gold nanoparticles using *Turbinaria conoides* and its characterization. J. Nanostruc. Chem. **3**, 1–7 (2013)

41. Kumar, P., Senthamil Selvi, S., Lakshmi Prabha, A., Prem Kumar, K., Ganeshkumar, R., Govindaraju, K.: Synthesis of silver nanoparticles from *Sargassum tenerrimum* and screening phytochemicals for its antibacterial activity. Nano Biomed. Eng. **4**, 12–16 (2012)

42. Arivuselvan, N., Radhiga, M., Anantharaman, P.: *In vitro* antioxidant and anticoagulant activities of sulphated polysaccharides from brown seaweed (*Turbinaria ornata*) (Turner) J. Agardh. Asian J. Pharm. Biol. Res. **1**, 232–238 (2011)

43. Aravindhan, R., Madhan, B., Rao, J.R., Nair, B.U., Ramasami, T.: Bioaccumulation of chromium from tannery wastewater: an approach for chrome recovery and reuse. Environ. Sci. Technol. **38**, 300–306 (2004)

44. Mohankumar, K., Meenakshi, S., Balasubramanian, T., Manivasagam, T.: Sulfated polysaccharides of *Turbinaria conoides* dose-dependently mitigate oxidative stress by ameliorating

antioxidants in isoproterenol induced myocardial injured rats: evidence from histopathological study. Egypt. Heart J. **64**, 147–153 (2012)

45. Lee, K.Y., Hwang, J., Lee, Y.W., Kim, J., Han, S.W.: One-step synthesis of gold nanoparticles using azacryptand and their applications in SERS and catalysis. J. Colloid Interface Sci. **316**, 476–481 (2007)

46. Chen, L.-J., Ma, H., Chen, K., Cha, H.-R., Lee, Y.-I., Qian, D.-J., Hao, J., Liu, H.-G.: Synthesis and assembly of gold nanoparticle-doped polymer solid foam films at the liquid/liquid interface and their catalytic properties. J. Colloid Interface Sci. **362**, 81–88 (2011)

47. Gangula, A., Podila, R., Ramakrishna, M., Karanam, L., Janardhana, C., Rao, A.M.: Catalytic reduction of 4-nitrophenol using biogenic gold and silver nanoparticles derived from *Breynia rhamnoides*. Langmuir **27**, 15268–15274 (2011)

48. Esumi, K., Isono, R., Yoshimura, T.: Preparation of PAMAM-and PPI-metal (silver, platinum, and palladium) nanocomposites and their catalytic activities for reduction of 4-nitrophenol. Langmuir **20**, 237–243 (2004)

49. Bai, X., Gao, Y., Liu, H.-G., Zheng, L.: Synthesis of amphiphilic ionic liquids terminated gold nanorods and their superior catalytic activity for the reduction of nitro compounds. J. Phys. Chem. C **113**, 17730–17736 (2009)

50. Chen, Y., Yin, R.-H., Wu, Q.-S.: Solvothermal synthesis of well-disperse ZnS nanorods with efficient photocatalytic properties. J. Nanomater. **2012**, 34 (2012)

51. Das, S.K., Ghosh, P., Ghosh, I., Guha, A.K.: Adsorption of rhodamine B on *Rhizopus oryzae*: role of functional groups and cell wall components. Colloids Surf. B **65**, 30–34 (2008)

52. Siddhardha, R.S., Kumar, V.L., Kaniyoor, A., Muthukumar, V.S., Ramaprabhu, S., Podila, R., Rao, A., Ramamurthy, S.S.: Synthesis and characterization of gold graphene composite with dyes as model substrates for decolorization: a surfactant free laser ablation approach. Spectrochim. Acta, Part A **133**, 365–371 (2014)

2

Cocoa pod husk extract-mediated biosynthesis of silver nanoparticles: its antimicrobial, antioxidant and larvicidal activities

Agbaje Lateef[1] · Musibau A. Azeez[1] · Tesleem B. Asafa[2] · Taofeek A. Yekeen[1] · Akeem Akinboro[1] · Iyabo C. Oladipo[3] · Luqmon Azeez[4] · Sunday A. Ojo[1] · Evariste B. Gueguim-Kana[5] · Lorika S. Beukes[6]

Abstract The present investigation reports utility of cocoa pod husk extract (CPHE), an agro-waste in the biosynthesis of silver nanoparticles (AgNPs) under ambient condition. The synthesized CPHE-AgNPs were characterized by UV–visible spectroscopy, Fourier-transform infrared spectroscopy, Energy dispersive X-ray (EDX) spectroscopy and transmission electron microscopy. The feasibility of the CPHE-AgNPs as antimicrobial agent against some multidrug-resistant clinical isolates, paint additive, and their antioxidant and larvicidal activities were evaluated. CPHE-AgNPs were predominantly spherical (size range of 4–32 nm) with face-centered cubic phase and crystalline conformation pattern revealed by selected area electron diffraction, while EDX analysis showed the presence of silver as a prominent metal. The synthesized nanoparticles effectively inhibited multidrug-resistant isolates of *Klebsiella pneumonia* and *Escherichia coli* at a concentration of 40 µg/ml, and enhanced the activities of cefuroxime and ampicillin in synergistic manner at 42.9–100 % concentration, while it completely inhibited the growth of *E. coli*, *K. pneumoniae*, *Streptococcus pyogenes*, *Staphylococcus aureus*, *Pseudomonas aeruginosa*, *Aspergillus flavus*, *Aspergillus fumigatus* and *Aspergillus niger* as additive in emulsion paint. The antioxidant activities of the CPHE-AgNPs were found to be excellent, while highly potent larvicidal activities against the larvae of *Anopheles* mosquito at 10–100 µg/ml concentration were observed. Our study demonstrated for the first time the utility of CPHE in the biosynthesis of CPHE-AgNPs with potential applications as antimicrobial and larvicidal agents, and paint additives for coating material surfaces to protect them against microbial growth while improving their shelf life.

Keywords CPHE-AgNPs · Antimicrobial activity · Multidrug resistance · Paint additive · Antioxidant · Larvicidal

A. Lateef, M. A. Azeez, T. B. Asafa, T. A. Yekeen, A. Akinboro, I. C. Oladipo: Nanotechnology Research Group (NANO⁺).

✉ Musibau A. Azeez
maazeez@lautech.edu.ng

[1] Department of Pure and Applied Biology, Ladoke Akintola University of Technology, PMB 4000, Ogbomoso, Nigeria

[2] Department of Mechanical Engineering, Ladoke Akintola University of Technology, PMB 4000, Ogbomoso, Nigeria

[3] Department of Science Laboratory Technology, Ladoke Akintola University of Technology, PMB 4000, Ogbomoso, Nigeria

[4] Department of Chemical Sciences, Osun State University, Osogbo, Nigeria

[5] Department of Microbiology, University of KwaZulu-Natal, Private Bag X01, Scottsville, Pietermaritzburg 3209, South Africa

[6] Microscopy and Microanalysis Unit, School of Life Sciences, University of KwaZulu-Natal, Private Bag X01, Scottsville, Pietermaritzburg 3209, South Africa

Introduction

Biosynthesis of nanoparticles of metals such as gold, silver, zinc, copper, platinum and palladium has received great attention from researchers all over the world. This is due to their wide application in medical and pharmaceutical fields, especially as ingredients of most consumer products like shampoo, soaps, detergents, shoes, cosmetics and toothpastes [1]. Green chemistry has become an expanded area of nanotechnology for a number of reasons which include its low cost, less demanding and eco-friendly nature that is devoid of use of hazardous chemicals and procedures. The advent of green technology into the synthesis of nanoparticles has greatly revolutionized the field of nanotechnology. Firstly, it has opened up the possibility of using biomolecules/substances of diverse origin in its synthesis and secondly, it has widened its applicability in different areas of human endeavors. Utilization of green synthesized nanoparticles transverses medical and biomedical applications to solving environmental problems such as land and water pollution, through material engineering to applications in agriculture.

Quite a number of biological macromolecules/substances have been employed as capping and stabilizing agents for the green synthesis of nanoparticles. For instance, metabolites of some arthropods have been used for green synthesis of nanoparticles [2–5]. In several studies, metabolites and enzymes of microbial origin (fungi and bacteria), even whole microbes have been used in the synthesis of nanoparticles [6–11], while many others have reported the synthesis of nanoparticles by employing extracts from various parts of the plants such as seeds, fruits, flowers, leaves, stem and roots [12–19]. Kaviya et al. [20], Njagi et al. [21] and Roopan et al. [22] synthesized AgNPs using *Sorghum* spp bran extract, *Citrus cinensis* peel extract and *Cocos nucifera* coir, respectively, suggesting the usefulness of agro-wastes in green nanotechnology. Most recently, the usefulness of *Cola nitida* pod, seed and seed shell for the green synthesis of AgNPs has been demonstrated in our laboratory [23, 24].

Cocoa (*Theobroma cacao*) is one of the key economic crops cultivated in Nigeria. As the third largest producer of cocoa in Africa and one of the highest cocoa producer in the world, the production capacity of Nigeria was reported to have reached about 385,000 tonnes per annum on the cultivable area of 966,000 ha with an appreciable increase of about 215,000 from year 2000 production level [25, 26]. The implication of this is that as the cocoa production industry is expanding, so also the cocoa pod husk (CPH) which is the by-product of cocoa processing that account for 52 to 76 % of the cocoa pod wet weight [27]. Cocoa is the principal ingredient of chocolate and other derived products such as cocoa liquor, cocoa butter, cocoa cake and cocoa powder, whereas the abundantly produced cocoa pod husks as waste in cocoa plantations during the extraction of cocoa beans are often discarded as of no market value. However, to create valuable products, cocoa pod husks have been explored as food antioxidants [28], dietary fibers [29], animal feed [30], as a precursor in the activated carbon production [31], fertilizer [32] and thermoplastic polyurethane composites [33]. It has also been investigated for use in the production of enzyme, with resultant improvement of the nutritional quality of the husk through fungal solid substrate fermentation [34].

The need to find alternative usage for agro-wastes motivated the present investigation into the biotechnological potential of cocoa pod husk extract (CPHE) in green chemistry for the synthesis of CPHE-AgNPs. Therefore, the present study was designed to explore the utility of CPHE in the synthesis of CPHE-AgNPs, evaluation of the antioxidant activities, mosquito larvicidal activities, and antimicrobial potentials of the synthesized nanoparticles against multidrug-resistant clinical bacteria. In addition, the usefulness of the synthesized CPHE-AgNPs as antimicrobial additive in emulsion paint was demonstrated. To the best of our knowledge, this report is the first of its kind on the use of cocoa pod husk extract (CPHE) for the synthesis of CPHE-AgNPs.

Materials and methods

Collection and processing of cocoa pods

Fresh cocoa fruits were obtained from Ipetumodu, Osun State, Nigeria. They were brought to the laboratory and thoroughly washed to remove dirt and other extraneous substances. The pods were split opened with the pod husk and beans separated. The pod husks were first transformed into chips and then air dried for seven days at room temperature (30 ± 2 °C). The pod husk chips were milled into powder with the aid of electric blender (Fig. 1).

Preparation of cocoa pod husk extract

Cocoa pod husk extract (CPHE) was obtained following the methods of Lateef et al. [23] by weighing 0.1 g of the powder and suspended in 10 ml of distilled water, and heated in water bath at 60 °C for 1 h. The extract was filtered using Whatman No. 1 filter paper and then centrifuged at 4000 rpm before the collection of the final clear extract (CPHE), which was stored at 4 °C for further use.

Green synthesis of CPHE-AgNPs and characterization

The CPHE prepared was used to synthesize CPHE-AgNPs using the protocol previously described [23, 24]. To 40 ml of 1 mM silver nitrate ($AgNO_3$), 1 ml of the extract was added at room temperature (30 ± 2 °C) and the reaction mixture was allowed to stand for some minutes. A change in color of the reaction mixture was visually observed, followed by the measurement of its absorbance spectrum using UV–visible spectrophotometer (Cecil, USA) operated at the range of 200–800 nm. FTIR spectroscopy analysis was carried out on the synthesized CPHE-AgNPs

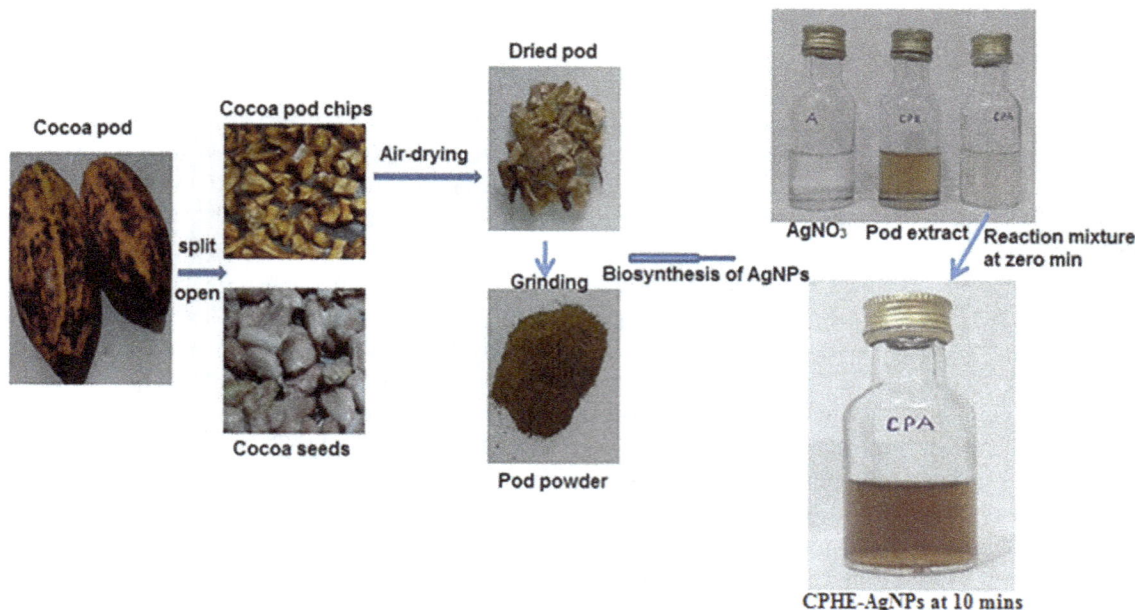

Fig. 1 Green biosynthesis of CPHE-AgNPs using the cocoa pod husk extract

using IRAffinity-1S Spectrometer to identify the functional groups of the various biomolecules that took part in the green synthesis. Transmission electron microscopy (TEM) and EDX of the synthesized CPHE-AgNPs were conducted to determine the size, morphology, nature and their elemental composition. For TEM analysis, the colloidal sample was placed on a 200 mesh hexagonal copper grid (3.05 mm) (Agar Scientific, Essex, UK) coated with 0.3 % formvar dissolved in chloroform. The grids were dried before viewing under TEM model JEM-1400 (JEOL, USA) which was operated at 200 kV to obtain the micrographs.

Antimicrobial activities of the synthesized CPHE-AgNPs

CPHE-AgNPs' antibacterial properties against clinical isolates of *Escherichia coli* and *Klebsiella pneumonia* obtained from LAUTECH Teaching Hospital, Ogbomoso were investigated using agar diffusion method as previously described [5, 10, 23, 24]. The culture broth was obtained through overnight growth in peptone water and this was used to seed the freshly prepared plates of Mueller–Hinton Agar (Lab M Ltd., UK). Thereafter, plates were bored with the aid of cork borer, and 100 µl of the graded concentrations of CPHE-AgNPs was introduced into the wells. This was subsequently followed by incubation at 37 °C for 24 h, after which zones of inhibition were measured.

Antibacterial susceptibility test

The drug susceptibility of test isolates was carried out as previously demonstrated [35, 36]. The isolates were tested on the discs (Abtek Biologicals Ltd., UK) impregnated with antibiotics containing (µg): ceftazidime (Caz), 30; cefuroxime (Crx), 30; gentamicin (Gen), 10; ceftriaxone (Ctr), 30; ofloxacin (Ofl), 5; augmentin (Aug), 30; erythromycin (Ery), 30; and cloxacillin (Cxc), 5 for Gram-positive isolates. The Gram-negative isolates were tested against antibiotics (µg): ceftazidime (Caz), 30; cefuroxime (Crx), 30; gentamicin (Gen), 10; ampicillin (Amp), 10; ofloxacin (Ofl), 5; augmentin (Aug), 30; nitrofurantoin (Nit), 300; and ciprofloxacin (Cpr), 5. After incubation at 37 °C for 48 h, the zones of inhibition were measured and interpreted [36] taking into cognizance of the recommended breakpoints [37].

Antimicrobial properties of synthesized CPHE-AgNPs as additive in paint

The antimicrobial usefulness of CPHE-AgNPs as additive in paint was investigated as described earlier [5, 10, 23], by inoculating 19 ml of sterilized commercially procured white emulsion paint with 1 ml ($\sim 10^6$ cfu/ml) of 18-h broth cultures of *E. coli*, *K. pneumoniae*, *S. pyogenes*, *Staphylococcus aureus* and *Pseudomonas aeruginosa*. In the case of antifungal assay, 48-h broth cultures of *A. flavus*, *A. fumigatus* and *A. niger* were used as inocula.

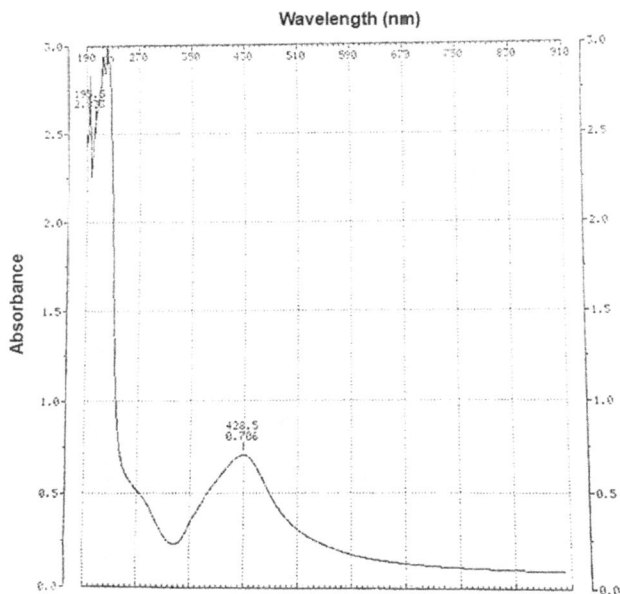

Fig. 2 The UV–vis absorption spectrum of the biosynthesized CPHE-AgNPs

The test experiment consisted of the paint and test organism, which was supplemented with 1 ml of 100 μg/ml of CPHE-AgNPs. These were incubated at 37 and 30 ± 2 °C for 48 h for bacteria and fungi, respectively. Thereafter, 1 ml of the contents of each bottle was inoculated on nutrient agar for bacteria and potato dextrose agar for fungi, and incubated appropriately for 48 h before examination for growth.

Antioxidant activities of CPHE-AgNPs

DPPH radical scavenging activity

This was carried out using the methods of William et al. [38] by reacting one milliliter of graded concentrations of CPHE-AgNPs prepared in methanol with 4.0 ml methanolic solution of 0.1 mM DPPH. The mixture was shaken and left in a dark box to stand for 30 min at room temperature (30 ± 2 °C). One milliliter of absolute methanol mixed with 4.0 ml of 0.1 mM methanolic DPPH was also prepared and used as blank. The absorbance of the resulting solution was measured at 517 nm on a UV/Vis spectrophotometer (model 6405, Jenway Ltd. Essex, UK). The inhibitory percentage of DPPH was determined accordingly [39].

$$\% \text{ inhibition} = \frac{A_{\text{blank}} - A_{\text{sample}}}{A_{\text{blank}}} \times 100$$

The efficient concentration of CPHE-AgNPs that decreased the initial concentration of DPPH radical by 50 % (IC_{50}) was obtained by interpolation from linear regression analysis [34]. This same procedure was used for standard antioxidant compounds such as quercetin and β-carotene.

Ferric reducing activity

This was investigated using the methods of Tan et al. [40], which involved addition of 250 μl of phosphate buffer

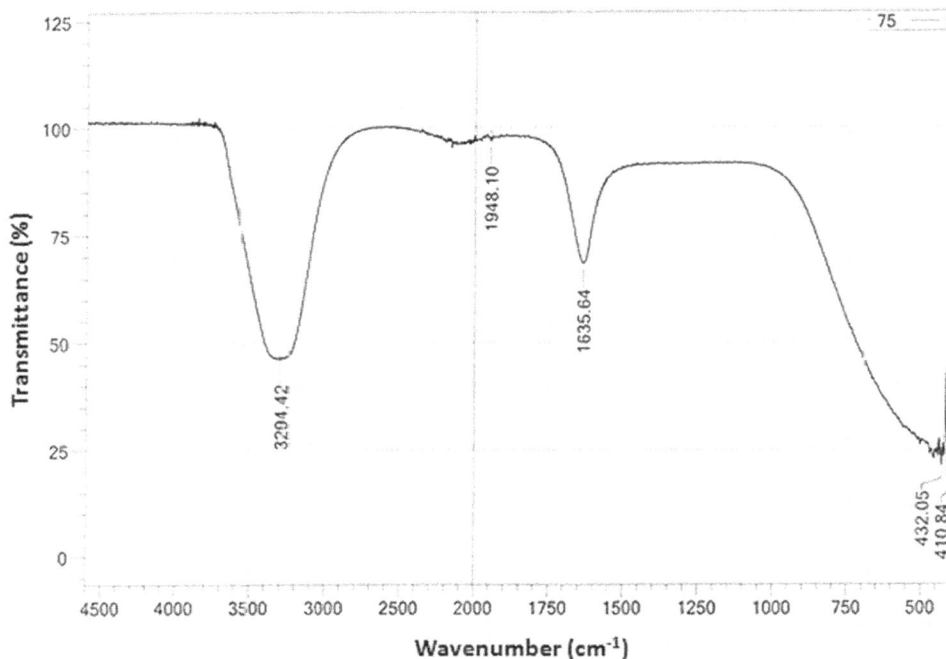

Fig. 3 The FTIR spectrum of the biosynthesized CPHE-AgNPs

(pH 6.6) and 2.5 ml of potassium ferricyanide to 1 ml of different concentrations of CPHE-AGNPs prepared using distilled water. This was followed by incubation of the resulting solution at 50 °C for 20 min. After cooling, 250 µl of trichloroacetic acid was added and centrifuged for 10 min at 500 rpm. Then, 250 µl of the supernatant with 250 µl of deionized distilled water and 500 µl of ferric (II) chloride were mixed thoroughly and absorbance was read at 700 nm. The blank was prepared with all reagents without the CPHE-AgNPs.

Larvicidal activity

This was evaluated in a dose–response bioassay against the first instar anopheline larvae as previously described [10], by exposing ten *Anopheles* mosquito larvae to 10 ml of graded concentrations of CPHE-AgNPs (10–100 µg/ml) in triplicate at room temperature (30 ± 2 °C).The number of dead larvae was recorded at specific intervals after exposure until total death was obtained. In the control experiment, the larvae were exposed to sterile distilled water

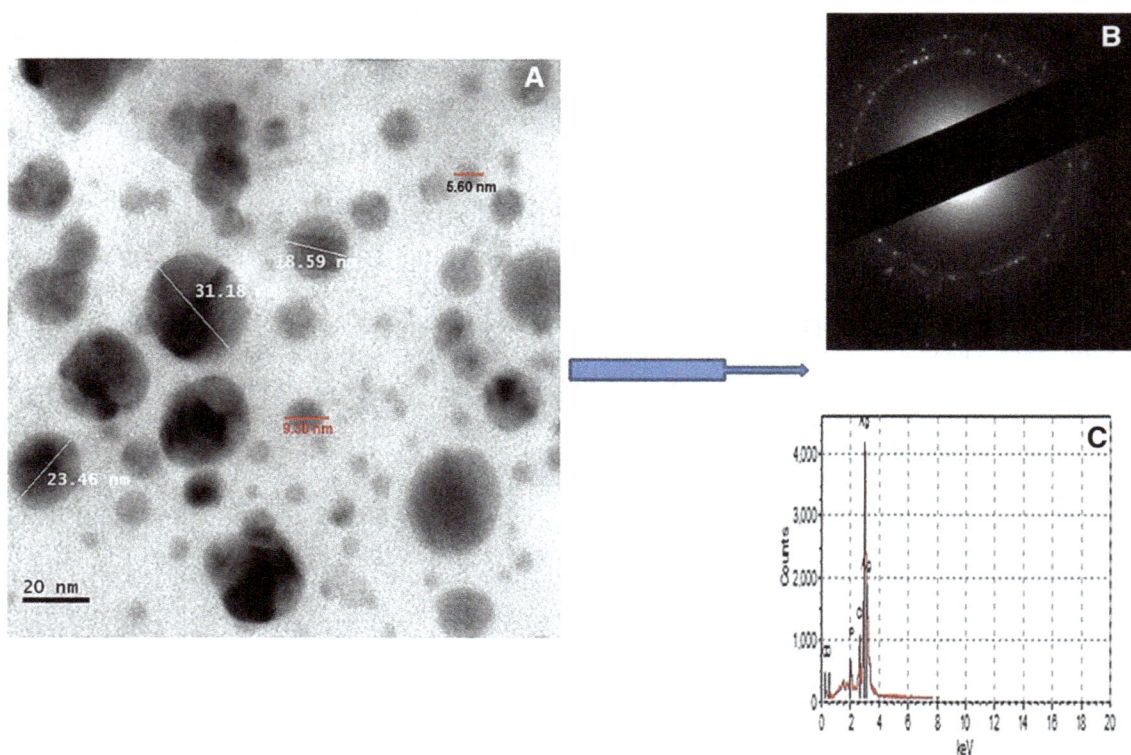

Fig. 4 Transmission electron micrograph (**a**), selected area electron diffraction pattern (**b**), and energy dispersive X-ray signal (**c**) of the biosynthesized CPHE-AgNPs

Table 1 The antibiotic resistance pattern of the test bacterial isolates

No of antibiotics	Isolates*	Source	Resistance pattern
3	KU	Urine	Crx, Gen, Amp
5	PA	Sputum	Caz, Crx, Aug, Nit, Amp
7	KW	Wound	Caz, Crx, Gen, Cpr, Ofl, Aug, Amp
8	EU	Urine	Caz, Crx, Gen, Cpr, Ofl, Aug, Nit, Amp
	SP	Sputum	Caz, Crx, Gen, Ctr, Ery, Cxc, Ofl, Aug
	SA	Ear	Caz, Crx, Gen, Ctr, Ery, Cxc, Ofl, Aug

* K, *K. pneumoniae*; PA, *P. aeruginosa*; EU, *E. coli*; SP, *S. pyogenes*; SA, *S. aureus*; antibiotics abbreviations are as defined under Experimental Details

K. pneumoniae (urine)   E. coli (urine)   K. pneumoniae (wound)

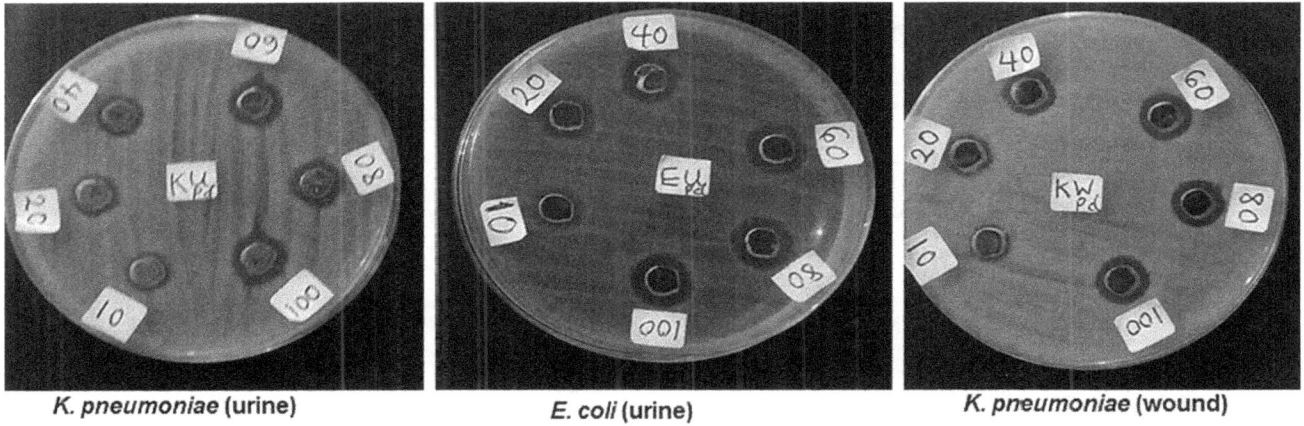

Fig. 5 The antibacterial activities of synthesized CPHE-AgNPs against some clinical bacterial isolates

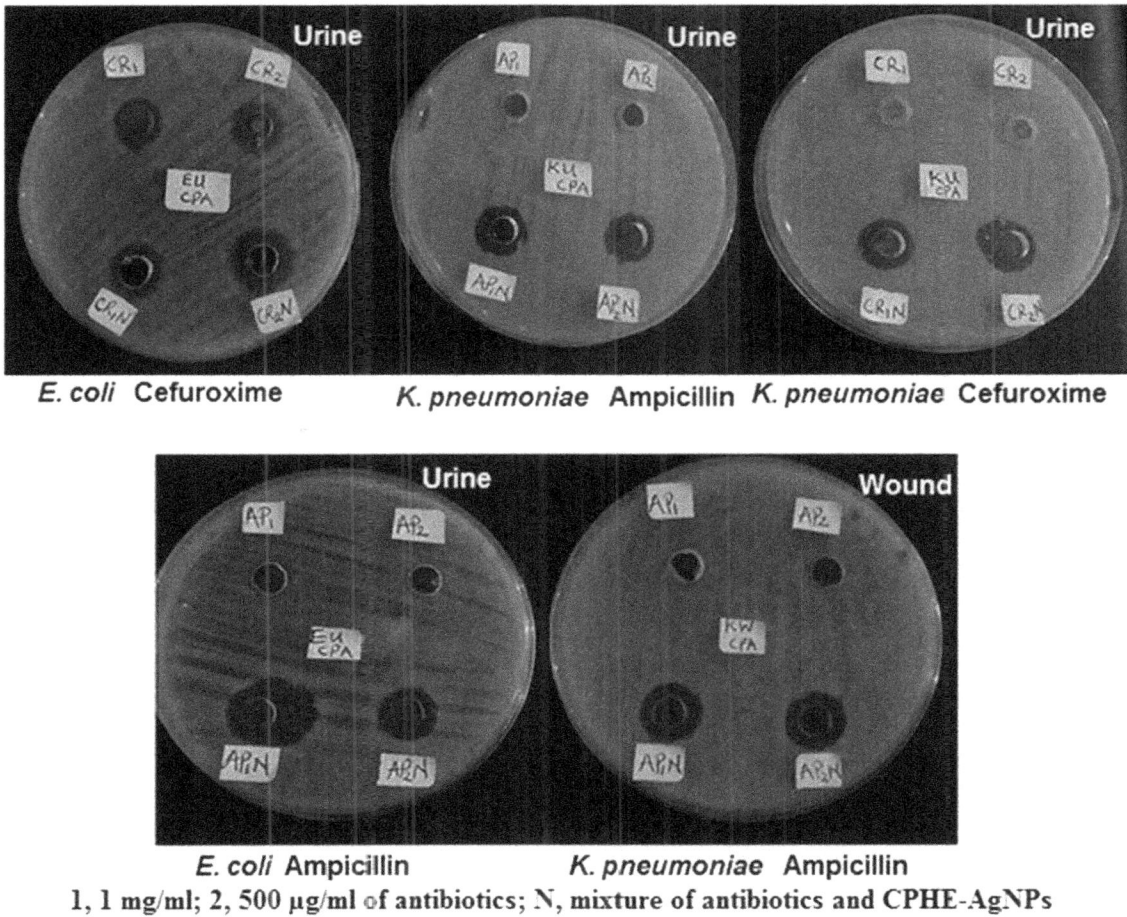

E. coli Cefuroxime   K. pneumoniae Ampicillin   K. pneumoniae Cefuroxime

E. coli Ampicillin   K. pneumoniae Ampicillin
1, 1 mg/ml; 2, 500 µg/ml of antibiotics; N, mixture of antibiotics and CPHE-AgNPs

Fig. 6 The synergistic activities of synthesized CPHE-AgNPs with ampicillin and cefuroxime on some clinical bacterial isolates

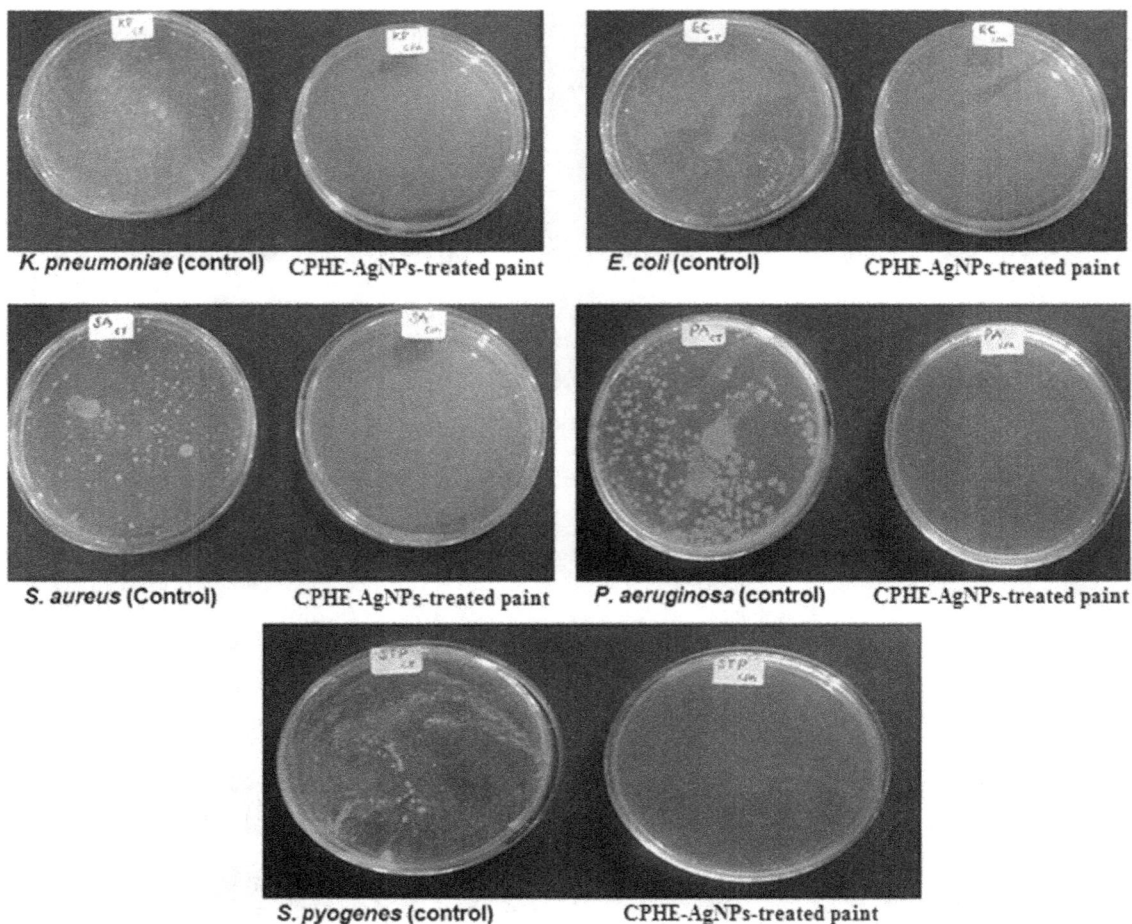

Fig. 7 Antibacterial activities of synthesized CPHE-AgNPs on bacteria inoculated into emulsion paint

under the same conditions. The percentage mortality was plotted against the concentration of the CPHE-AgNPs.

Results and discussion

Biogenic formation of CPHE-AgNPs

In the present study, CPHE mediated the formation of CPHE-AgNPs within a period of 10 min, producing brown color (Fig. 1). Colloidal green-synthesized AgNP solutions exhibiting shades of color from yellowish through brown to dark brown have been reported by several authors [5, 9, 10, 23, 24, 41, 42], suggesting the presence of different macromolecules in the extracts that played catalytic and stabilization roles in the formation of the particles. The maximum absorbance readings for the biosynthesized CPHE-AgNPs occurred at wavelength of 428.5 nm (Fig. 2). The value obtained is within the range reported for AgNPs [10, 23, 43–46]. The CPHE-AgNPs exhibited a high level of stability, devoid of aggregation or deterioration.

FTIR spectrum for CPHE mediated CPHE-AgNPs (Fig. 3) manifested strong peaks at 3294.42, and 1635.64 cm^{-1}, implicating proteins and phenolic compounds as the capping and stabilization molecules involved in the biotransformation process that produced the CPHE-AgNPs. The band 3294 is typical of N–H bond of amines, while that of 1635 is indicative of C=C stretch of alkenes or C=O stretch of amides [47]. In cocoa, total phenolic (determined at 45.6–46.4 mg gallic acid equivalent of soluble phenolic), 32.3 % carbohydrate, 21.44 % lignin, 19.2 % sugars, 8.6 % protein and 27.7 % minerals were previously reported [27], and they are biomolecules known to be very rich in the identified chemical bonds. Specifically, compounds such as citric acid, malic acid protocatechuic acid, p-hydroxybenzoic acid, salicyclic acid, kaempferol, linarin, resveratrol, apigenin, luteoin, crysoplenol, linoleic acid and oleic acid have been identified in cocoa pod extract [48].

The CPHE-AgNPs were fairly spherical in shape with sizes ranging from 4 to 32 nm (Fig. 4a), which is in agreement with those earlier reported [5, 9, 10, 23, 24, 44, 45]. The particles were well dispersed within the organic

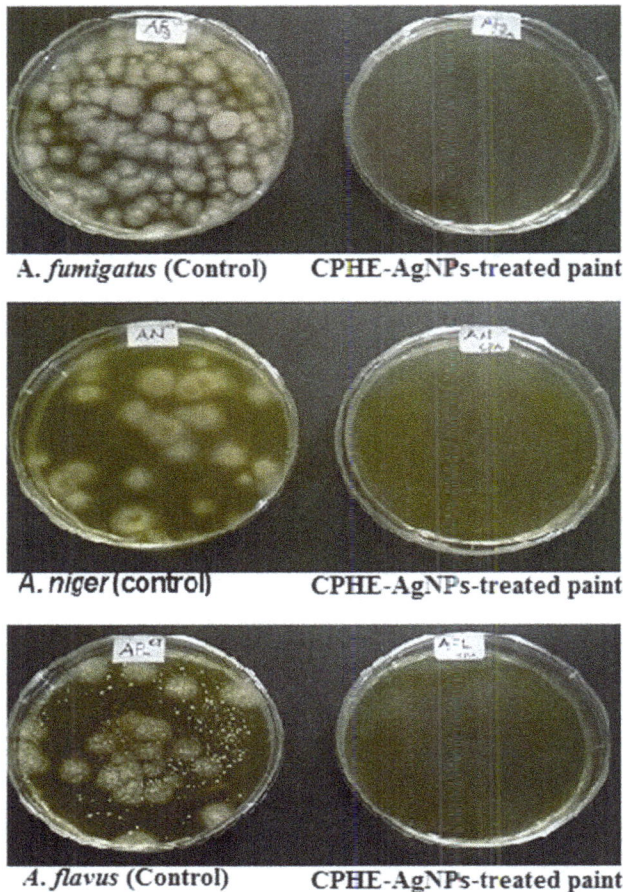

Fig. 8 Antifungal activities of synthesized CPHE-AgNPs on fungi inoculated into emulsion paint

matrix, indicating good stability against aggregation. The EDX patterns (Figs. 4c) showed the intense presence of silver in the CPHE-AgNPs colloidal solution [5, 10, 23, 24, 49, 50] to the tune of 95 %, having the ring-like SAED pattern (Fig. 4B) associated with the face-centered cubic crystalline structure of silver [47]. It can, therefore, be inferred from these results that cocoa pod husk extract is a veritable source of biomolecules in the biogenic and eco-friendly synthesis of CPHE-AgNPs that could be of wide application in the expanding field of nanobiotechnology. This report adds to bioresource utilization of agro-wastes in the synthesis of nanoparticles.

Antibacterial activities of CPHE-AgNPs against multidrug-resistant bacteria isolates

CPHE-AgNPs strongly inhibited the growth of multidrug-resistant *K. pneumoniae* and *E. coli* (Table 1) with the zones of inhibition of 10–14 mm (Fig. 5) at concentrations of 40–100 µg/ml. The activities shown by the particles against these resistant isolates are of considerable importance, indicating that the particles can be deployed to reduce the growth of drug-resistant isolates that abound in the environment [35, 36, 51–57]. Similar results on the tremendous antibacterial activities of green synthesized AgNPs have been reported [5, 9, 10, 23, 24, 47, 49].

Furthermore, the CPHE-AgNPs contributed to improvement (42.9–100 %) in the antibacterial activities of cefuroxime and ampicillin through synergy (Fig. 6). It is interesting to note that in several cases where the resistant isolates were not inhibited by the antibiotics, the CPHE-AgNPs-antibiotic treatments produced outstanding growth inhibitions against strains of *K. pneumoniae* and *E. coli*. These results concurred with those reported in similar studies [5, 10]. This pronounced activity is a further testimony to the potentials of CPHE-AgNPs in combating multidrug-resistant isolates, which would be of immense application in biomedical industry.

Antimicrobial activities of CPHE-AgNPs in paint

The incorporation of CPHE-AgNPs into emulsion paint led to effective inhibition of the growth of *E. coli*, *K. pneumoniae*, *S. pyogenes*, *S. aureus* *P. aeruginosa* (Fig. 7), *A. flavus*, *A. fumigatus* and *A. niger* (Fig. 8) as against the abundant growth on the control plates We have previously shown that biosynthesized AgNPs can protect paint from microbial deterioration through antimicrobial activities [5, 10, 23], thus reiterating the relevance of AgNPs as antimicrobial additives in paint for applications in the built environment [58].

Antioxidant activities of CPHE-AgNPs

The DPPH-free radical scavenging activities of the biosynthesized CPHE-AgNPs were in the range of

Table 2 The ferric ion reducing activity of the biosynthesized CPHE-AgNPs

Test material (µg/ml)*	Ferric ion reducing power activity (%)				
	20	40	60	80	100
CPHE-AgNPs	0.1614.44	0.3733.39	0.4641.52	0.9182.13	0.9383.94
Standards (µg/ml)*	200	400	600	800	1000
Quercetin	12.050.13	20.390.22	28.730.31	63 950.69	1001.12
β-carotene	11.530.03	26.920.07	19.230.05	42 310.11	65.380.17

* Concentration; each value is an average of three readings

32.62–84.50 % at the investigated concentration of 20–100 μg/ml, while those of β-carotene and quercetin were in the range of 11.11–66.67, and 43.87–74.62 % inhibitions, respectively, at the concentrations of 0.2–1 mg/ml. The IC_{50} obtained were 49.70, 430 and 710 μg/ml for CPHE-AgNPs, quercetin and β-carotene, respectively. In the same vein, the ferric ion reducing activities of the CPHE-AgNPs were in the range of 14.44–83.94 % for concentrations of 20–100 μg/ml (Table 2), whereas β-carotene had activities in the range of 11.53–65.38 %, and quercetin displayed ferric ion reduction in the range of 12.05–100 % at concentrations of 0.2–1.0 mg/ml. These antioxidant activities shown by the CPHE-AgNPs are similar to those previously reported [10, 23, 59, 60], with the particles manifesting greater potencies than the antioxidant standards. Previous studies have established the antioxidant capabilities of extracts obtained from cocoa pod husk [28, 34, 48]. Generally, the free radical scavenging activities of AgNPs have been attributed to the bioreductant molecules present on the surface of the nanoparticles which increase the surface areas for antioxidant activity [60]. The presence of compounds such as citric acid, malic acid, terpenoid and resveratrol has been previously attributed to antioxidant activities of cocoa pod extracts [48].

Larvicidal activity of CPHE-AgNPs

CPHE-AgNPs showed potent larvicidal activities (70–100 %) against the larvae of Anopheles mosquito at concentrations of 10–100 μg/ml within 2 h (Fig. 9) with the LC_{50} of 43.52 μg/ml. The larvicidal activity of CPHE-AgNPs is similar to those previously reported on the larvicidal activities of some plant and bacterial extract-mediated AgNPs on Anopheles larvae [10, 22, 61, 62]. However, in the present study, the potency of the CPHE-AgNPs was found to be greater within 2 h of exposure than those previously reported. It can, therefore, be inferred from these results that the biosynthesized CPHE-AgNPs can find useful application in the malaria control programme by killing the larvae of the vector of Plasmodium parasites.

Conclusion

The present study has clearly demonstrated the usefulness of cocoa pod husk extract (CPHE) as a cost-effective and eco-friendly bio-resource in the green synthesis of CPHE-AgNPs. The synthesized particles were fairly spherical with size ranging from 4 to 32 nm. The highly remarkable antibacterial activities of CPHE-AgNPs and enhanced activities in synergy with antibiotics against multidrug-resistant clinical isolates of bacteria, including excellent larvicidal activities against larvae of the vector of Plasmodium parasites indicated the possibility of their exploitation in biomedical industry. Furthermore, the successful inhibition of microbial growth when used as additive in paint and very strong antioxidant activities suggests the biotechnological potential of the synthesized nanoparticles in the biomedical industry and as a coating for surfaces of materials to protect them against microbial growth while improving their shelf life. To the best of our knowledge, this report represents the first reference to cocoa pod husk extract in the green synthesis of CPHE-AgNPs.

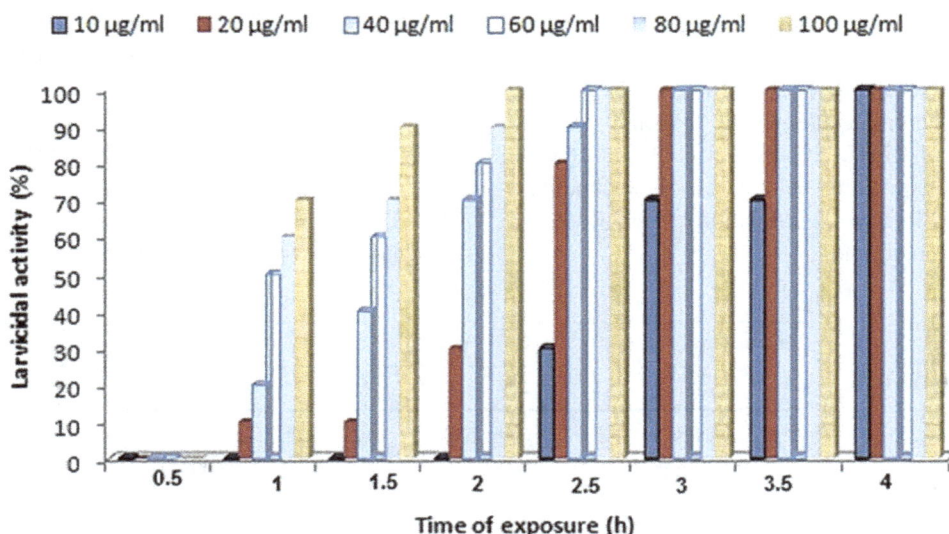

Fig. 9 Larvicidal activity of the biosynthesized CPHE-AgNPs on *Anopheles* mosquito larvae

Acknowledgments AL thanked the authority of LAUTECH, Ogbomoso, Nigeria for the provision of some of the facilities used in this study. MAA gratefully thanked DBT-TWAS for Postdoctoral opportunity and visit to the Department of Chemistry, University of Pune, India (Sept, 2013–Aug, 2014).

References

1. Kim, B.S., Song, J.Y.: Biological synthesis of gold and silver nanoparticles using plant Leaf extracts and antimicrobial applications. In: Hou, C.T., Shaw, J.F. (eds.) Biocatalysis and Biomolecular Engineering, pp. 447–457. Wiley, New Jersey (2010)

2. Philip, D.: Honey mediated green synthesis of silver nanoparticles. Spectrochim. Acta Part A **75**, 1078–1031 (2010)

3. Sreelakshmi, C., Datta, K.K.R., Yadav, J.S., Reddy, B.V.: Honey derivatized Au and Ag nanoparticles and evaluation of its antimicrobial activity. J. Nanosci. Nanotechnol. **11**, 6995–7000 (2011)

4. Obot, I.B., Umoren, S.A., Johnson, A.S.: Sunlight- mediated synthesis of silver nanoparticles using honey and its promising anticorrosion potentials for mild steel in acidic environments. J. Mater. Environ. Sci. **4**, 1013–1018 (2013)

5. Lateef, A., Ojo, S.A., Azeez, M.A., Asafa, T.B., Yekeen, T.A., Akinboro, A., Oladipo, I.C., Gueguim-Kana, E.B., Beukes, L.S.: Cobweb as novel biomaterial for the green and ecofriendly synthesis of silver nanoparticles. Appl. Nanosci. (2015). doi:10.1007/s13204-015-0492-9

6. Shivaji, S., Madhu, S., Singh, S.: Extracellular synthesis of antibacterial silver nanoparticles using psychrophilic bacteria. Process Biochem. **46**, 1800–1807 (2011)

7. Rajeshkumar, S., Ponnanikajamideen, M., Malarkodi, C., Malini, M., Annadurai, G.: Microbe-mediated synthesis of antimicrobial semiconductor nanoparticles by marine bacteria. J. Nanostruct. Chem. **4**, 96–102 (2014)

8. Sarsar, V., Selwal, M.K., Selwal, K.K.: Biofabrication, characterization and antibacterial efficacy of extracellular silver nanoparticles using novel fungal strain of *Penicillium atramentosum* KM. J. Saudi Chem. Soc. **19**, 682–688 (2015)

9. Lateef, A., Adelere, I.A., Gueguim-Kana, E.B., Asafa, T.B., Beukes, L.S.: Green synthesis of silver nanoparticles using keratinase obtained from a strain of *Bacillus safensis* LAU 13. Int. Nano Lett. **5**, 29–35 (2015)

10. Lateef, A., Ojo, S.A., Akinwale, A.S., Azeez, L., Gueguim-Kana, E.B., Beukes, L.S.: Biogenic synthesis of silver nanoparticles using cell-free extract of *Bacillus safensis* LAU 13: antimicrobial, free radical scavenging and larvicidal activities. Biologia **70**, 1295–1306 (2015)

11. Jena, J., Pradhan, N., Dash, B.P., Panda, P.K., Mishra, B.K.: Pigment mediated biogenic synthesis of silver nanoparticles using diatom *Amphora* sp. and its antimicrobial activity. J. Saudi Chem. Soc. **19**, 661–666 (2015)

12. Dhand, V., Soumya, L., Bharadwaj, S., Chakra, S., Bhatt, D., Sreedhar, B.: Green synthesis of silver nanoparticles using *Coffea arabica* seed extract and its antibacterial activity. Mater. Sci. Eng., C **58**, 36–43 (2016)

13. Reddy, N.J., Vali, D.N., Rani, M., Rani, S.S.: Evaluation of antioxidant, antibacterial and cytotoxic effects of green synthesized silver nanoparticles by *Piper longum* fruit Mater. Sci. Eng. C **34**, 115–122 (2014)

14. Gogoi, N., Babu, P.J., Mahanta, C., Bora, U.: Green synthesis and characterization of silver nanoparticles using alcoholic flower extract of *Nyctanthes arbortristis* and in vitro investigation of their antibacterial and cytotoxic activities. Mater. Sci. Eng., C **46**, 463–469 (2015)

15. Anwar, M.F., Yadav, D., Kapoor, S., Chander, J., Samim, M.: Comparison of antibacterial activity of Ag nanoparticles synthesized from leaf extract of *Parthenium hystrophorus* L. in aqueous media and gentamicin sulphate: in-vitro. Drug Dev. Ind. Pharm. **41**, 43–50 (2015)

16. Nayak, D., Ashe, S., Rauta, P.R., Kumari, M., Nayak, B.: Bark extract mediated green synthesis of silver nanoparticles: evaluation of antimicrobial activity and antiproliferative response against osteosarcoma. Mater. Sci. Eng., C **58**, 44–52 (2016)

17. Dare, E.O., Oseghale, C.O., Labulo, A.H., Adesuji, E.T., Elemike, E.E., Onwuka, J.C., Bamgbose, J.T.: Green synthesis and growth kinetics of nanosilver under bio-diversified plant extracts influence. J. Nanostruct. Chem. **5**, 85–94 (2015)

18. Logeswari, P., Silambarasan, S., Abraham, J.: Synthesis of silver nanoparticles using plants extract and analysis of their antimicrobial property. J. Saudi Chem. Soc. **19**, 311–317 (2015)

19. Agharkar, M., Kochrekar, S., Hidouri, S., Azeez, M.A.: Trends in green reduction of graphene oxides, issues and challenges: a review. Mater. Res. Bull. **59**, 323–328 (2014)

20. Kaviya, S., Santhanalakshmi, J., Viswanathan, B., Muthumary, J., Srinivasan, K.: Biosynthesis of silver nanoparticles using *Citrus sinensis* peel extract and its antibacterial activity. Spectrochim. Acta A **79**, 594–598 (2011)

21. Njagi, E.C., Huang, H., Stafford, L.: Biosynthesis of iron and silver nanoparticles at room temperature using aqueous *Sorghum* Bran extracts. Langmuir **27**, 264–271 (2011)

22. Roopan, S.M., Madhumitha, G.R., Abdul Rahuman, A., Kamaraj, C., Bharathi, A., Surendra, T.V.: Low-cost and eco-friendly phyto-synthesis of silver nanoparticles using *Cocos nucifera* coir extract and its larvicidal activity. Ind. Crop Prod. **43**, 631–635 (2013)

23. Lateef, A., Azeez, M.A., Asafa, T.B., Yekeen, T.A., Akinboro, A., Oladipo, I.C., Azeez, L., Ajibade, S.E., Ojo, S.A., Gueguim-Kana, E.B., Beukes, L.S.: Biogenic synthesis of silver nanoparticles using pod extract of *Cola nitida*: antibacterial, antioxidant activities and application as additive in paint. J. Taibah Univ. Sci. (2016). doi:10.1016/j.jtusci.2015.10.010

24. Lateef, A., Azeez, M.A., Asafa, T.B., Yekeen, T.A., Akinboro, A., Oladipo, I.C., Ajetomobi, F.E., Gueguim-Kana, E.B., Beukes, L.S.: *Cola nitida*-mediated biogenic synthesis of silver nanoparticles using seed and seed shell extracts and evaluation of antibacterial activities. BioNanoSci. **5**, 196–205 (2015)

25. Franzen, M., Mulder, M.B.: Ecological, economic and social perspectives on cocoa production worldwide. Biodivers. Conserv. **16**, 3835–3849 (2007)

26. Amao, O.D., Oni, O., Adeoye, I.: Competitiveness of cocoa-based farming household in Nigeria. J. Dev. Agric. Econ. **7**, 80–84 (2015)

27. Vriesmann, L.C., Amboni, R.D.D.M.C., de Oliveira Petkowicz, C.L.: Cacao pod husks (*Theobroma cacao* L.): Composite and hot-water-soluble pectins. Ind. Crop Prod. **34**, 1173–1181 (2001)

28. Azizah, A., Nikruslawati, N., Tee, T.S.: Extraction and characterization of antioxidant from cocoa by-products. Food Chem. **64**, 199–202 (1999)

29. Redgwell, R., Trovato, V., Merinat, S., Curti, D., Hediger, S., Manez, A.: Dietary fibre in cocoa shell: characterisation of component polysaccharides. Food Chem. **81**, 103–112 (2003)

30. Aregheore, E.: Chemical evaluation and digestibility of Cocoa (*Theobroma cacao*) byproducts fed to goats. Trop. Anim. Health Pro. **34**, 339–348 (2002)

31. Adeyi, O.: Proximate composition of some agricultural wastes in Nigeria and their potential use in activated carbon production. J. Appl. Sci. Environ. Manag. **14**, 55–58 (2010)

32. Agbeniyi, S.O., Oluyole, K.A., Ogunlade, M.O.: Impact of Cocoa Pod Husk Fertilizer on Cocoa Production in Nigeria. World J. Agric. Sci. **7**, 113–116 (2011)

33. El-Shekeil, Y.A., Sapuan, S.M., Algrafi, M.W.: Effect of fiber loading on mechanical and morphological properties of cocoa pod husk fibers reinforced thermoplastic polyurethane. Composites. Mat. Design **64**, 330–333 (2014)

34. Lateef, A., Oloke, J.K., Gueguim-Kana, E.B., Oyeniyi, S.O., Onifade, O.R., Oyeleye, A.O., Oladosu, O.C., Oyelami, A.O.: Improving the quality of agro-wastes by solid state fermentation: enhanced antioxidant activities and nutritional qualities. World J. Microbiol. Biotechnol. **24**, 2369–2374 (2008)

35. Lateef, A., Davies, T.E., Adelekan, A., Adelere, I.A., Adedeji, A.A., Fadahunsi, A.H.: *Akara* Ogbomoso: microbiological examination and identification of hazards and critical control points. Fd Sci. Technol. Int. **16**, 389–400 (2010)

36. Lateef, A., Ojo, M.O.: Public health issues in the processing of cassava *(Manihot esculenta)* for the production of 'lafun' and the application of hazard analysis control measures. Qual. Assur. Saf. Crops Foods. **8**, 165–177 (2016)

37. Andrews, J.M.: BSAC Standardized disc susceptibility testing method (version 4). J. Antimicrob. Chemother. **56**, 60–76 (2005)

38. Williams, B.W., Cuverlier, M.E., Berset, C.: Use of free radical method to evaluate antioxidant activity. Food Sci. Technol. LWT **28**, 25–30 (1995)

39. Olajire, A.A., Azeez, L.: Total antioxidant activity, phenolic, flavonoid and ascorbic acid contents of Nigerian vegetables. Afr. J. Food Sci. Technol. **2**, 22–29 (2011)

40. Tan, C., Xue, J., Abbas, S., Feng, B., Zhang, X., Xia, S.: Liposome as a delivery system for carotenoids: comparative antioxidant activity of carotenoids as measured by ferric reducing antioxidant power, DPPH assay and lipid peroxidation. J. Agric. Food Chem. **62**, 6726–6735 (2014)

41. Kalishwaralal, K., Deepak, V., Ramkumarpandian, S., Nellaiah, H., Sangiliyandi, G.: Extracellular biosynthesis of silver nanoparticles by the culture supernatant of *Bacillus licheniformis*. Mater. Lett. **62**, 4411–4413 (2008)

42. Shaligram, N.S., Bule, M., Bhambure, R., Singhal, R.S., Singh, S.K., Szakacs, G., Pandey, A.: Biosynthesis of silver nanoparticles using aqueous extract from the compactin producing fungal strain. Process Biochem. **44**, 939–943 (2009)

43. Thirumurugan, A., Tomy, N.A., Kumar, H.P., Prakash, P.: Biological synthesis of silver nanoparticles by *Lantana camara* leaf extracts. Int. J. Nanomater. Biostruct. **1**, 22–24 (2011)

44. Zaki, S., El-Kady, M.F., Abd-El-Haleem, D.: Biosynthesis and structural characterization of silver nanoparticles from bacterial isolates. Mater. Res. Bull. **46**, 1571–1576 (2011)

45. Kannan, R.R.R., Arumugam, R., Ramya, D., Manivannan, K., Anantharaman, P.: Green synthesis of silver nanoparticles using marine macroalga *Chaetomorpha linum*. App. Nanosci. **3**, 229–233 (2013)

46. Priyadarshini, S., Gopinath, V., Priyadharsshini, N.M., Ali, D.M., Velusamy, P.: Synthesis of anisotropic silver nanoparticles using novel strain, *Bacillus flexus* and its application. Colloids Surf. B: Biointerf. **102**, 232–237 (2013)

47. Shankar, S., Jaiswal, L., Aparna, R.S.L., Prasad, R.G.S.V.: Synthesis, characterization, in vitro biocompatibility, and antimicrobial activity of gold, silver and gold silver alloy nanoparticles prepared from *Lansium domesticum* fruit peel extract. Mater. Letters **137**, 75–78 (2014)

48. Karim, A.A., Azlan, A., Ismail, A., Hashim, P., Gani, S.S.A., Zainudin, B.H., Abdullah, N.A.: Phenolic composition, antioxidant, anti-wrinkles and tyrosinase inhibitory activities of cocoa pod extract. BMC complementary and Alternative Med. **14**, 381 (2014)

49. Salem, W.M., Haridy, M., Sayed, W.F., Hassan, N.H.: Antibacterial activity of silver nanoparticles synthesized from latex and leaf extract of *Ficus sycomorus*. Ind. Crops Prod. **62**, 228–234 (2014)

50. Shameli, K., Ahmad, M.B., Zargar, M., Wan Yunus, W.M.Z., Ibrahim, N.A., Sha-banzadeh, P., Ghaffari-Moghadam, M.: Synthesis and characterization of silver/montmorillonite/chitosan bionanocomposites by chemical reduction method and their antibacterial activity. Int. J. Nanomed. **6**, 271–284 (2011)

51. Adewoye, S.O., Lateef, A.: Assessment of the microbiological quality of *Clarias gariepinus* exposed to an industrial effluent in Nigeria. Environmentalist **24**, 249–254 (2004)

52. Lateef, A.: The microbiology of a pharmaceutical effluent and its public health implications. World J. Microbiol. Biotechnol. **20**, 167–171 (2004)

53. Lateef, A., Oloke, J.K., Gueguim-Kana, E.B.: Antimicrobial resistance of bacterial strains isolated from orange juice products. Afr. J. Biotechnol. **3**, 334–338 (2004)

54. Lateef, A., Oloke, J.K., Gueguim-Kana, E.B.: The prevalence of bacterial resistance in clinical, food, water and some environmental samples in Southwest Nigeria. Environ. Monit. Assess. **100**, 59–69 (2005)

55. Lateef, A., Yekeen, T.A.: Microbial attributes of a pharmaceutical effluent and its genotoxicity on *Allium cepa*. Int. J. Environ. Stud. **63**, 535–536 (2006)

56. Lateef, A., Oloke, J.K., Gueguim-Kana, E.B., Pacheco, E.: The microbiological quality of ice used to cool drinks and foods in Ogbomoso metropolis, Southwest. Nigeria. Internet J. Food Safety **8**, 39–43 (2006)

57. Lateef, A., Yekeen, T.A., Ufuoma, P.E.: Bacteriology and genotoxicity of some pharmaceutical wastewaters in Nigeria. Int. J. Environ. Health **1**, 551–562 (2007)

58. Kaiser, J.P., Zuin, S., Wick, P.: Is nanotechnology revolutionizing the paint and lacquer industry? A critical opinion. Sci. Total Environ. **442**, 282–289 (2013)

59. Shanmugam, C., Sivasubramanian, G., Parthasarathi, B., Baskaran, K., Balachander, R., Parameswaran, V.R.: Antimicrobial, free radical scavenging activities and catalytic oxidation of benzyl alcohol by nano-silver synthesized from the leaf extract of *Aristolochia indica* L.: a promenade towards sustainability. Appl. Nanosci. (2015). doi:10.1007/s13204-015-0477-8

60. Bhakya, S., Muthukrishnan, S., Sukumaran, M., Muthukumar, M.: Biogenic synthesis of silver nanoparticles and their antioxidant and antibacterial activity. Appl. Nanosci. (2015). 10.1007/s13204-015-0473-z

61. Priyadarshini, K.A., Murugan, K., Panneerselvam, C., Ponarulselvam, S., Hwang, J.S., Nicoletti, M.: Biolarvicidal and pupicidal potential of silver nanoparticles synthesized using *Euphorbia hirta* against *Anopheles stephensi* Liston (Diptera: culicidae). Parasitol. Res. **111**, 997–1006 (2012)

62. Patil, C.D., Borase, H.P., Patil, S.V., Salunkhe, R.B., Salunke, B.K.: Larvicidal activity of silver nanoparticles synthesized using *Pergularia daemia* plant latex against *Aedes aegypti* and *Anopheles stephensi* and non-target fish *Poecillia reticulata*. Parasitol. Res. **111**, 555–562 (2012)

3

Optimized reduction conditions for the microfluidic synthesis of 1.3 ± 0.3 nm Pt clusters

M. Jakir Hossain[1,2] · Md. Saidur Rahman[1] · M. S. Rahman[2] · M. A. Ali[3] ·
N. C. Nandi[1] · P. Noor[1] · K. N. Ahmed[1] · S. Akhter[2]

Abstract Recently, small (<2 nm) and monodispersed Pt clusters has gained much attention due to their high catalytic activity in the aerobic oxidations. However, the chemical synthesis of small Pt clusters is not trivial; high temperature is often required to completely reduce the $Pt^{4+/2+}$ ions to Pt^0, which accelerates the growth of the Pt clusters. Here, we discussed a very simple microfluidic reduction of Pt^{4+} to Pt^0 by $NaBH_4$ in the presence of PVP that produces <2 nm Pt clusters in any variable reduction conditions. The microfluidic reduction conditions were optimized for the synthesis of possible smallest Pt clusters in terms of five reaction parameters: (1) temperature, (2) concentration of H_2PtCl_6, (3) molar ratio of $NaBH_4$ to Pt^{4+} ions, (4) molar ratio of PVP-monomer to Pt^{4+} ions, and (5) molecular weight/chain length of PVP. We found that possible smallest particles with average diameter 1.3 ± 0.3 nm were produced when aqueous solutions of H_2PtCl_6 (4 mM) and $NaBH_4$ (40 mM) containing PVP (160 mM) were injected into the micromixer placed in an icebath at a flow rate of 200 mL/h. The produced particles were characterized by UV–visible absorption spectrophotometry, powder X-ray diffractometry and transmission electron microscopy.

Keywords Platinum · Clusters · PVP · $NaBH_4$ · Reduction

Introduction

Colloidal clusters of noble metal in the 1–10 nm size regime have recently attracted considerable attention in many areas of research due to their nobility in physical and chemical properties [1, 2]. Those noble properties are of fundamental interest to both homogeneous and heterogeneous catalysis, which are significantly different from their bulk counterparts [3, 4]. Both catalytic activity [5] and selectivity [6] are known to be influenced by the size and as well as the shape of the clusters [7–11], and therefore the synthesis of well-controlled sizes and shapes of particles could be critical for this purpose.

Many procedures have been reported for the synthesis of colloidal Pt clusters. Commonly used procedures include Pt^{4+} or Pt^{2+} salts reduction by borohydride [12–16], hydrogen [17, 18], alcohol [19–24], glycol [25, 26], or ethylene glycol [27, 28] in the presence of a stabilizer or on a solid support. Despite a large choice of synthetic protocols, an accurate control of the particle size, simple and reproducible synthetic strategies are important to investigate their physical and chemical properties.

Typically, metal clusters provide highly active centers, and due to their higher surface energies and larger surfaces, they are not in a thermodynamically stable state [29, 30]. To produce homogeneous and stable metal particles usually soluble polymers are widely employed as supports because of their availability, enhanced stabilization properties and resistance to particle sintering or agglomeration [31]. Poly(vinylpyrrolidone) (PVP) is the most widely used capping agent to control not only the size but also the shape

✉ M. Jakir Hossain
smjakir080@gmail.com;

[1] Industrial Botany Research Division, BCSIR Laboratories Chittagong, Sholashahar, Chittagong 4220, Bangladesh

[2] Forest Chemistry Division, Bangladesh Forest Research Institute, Chittagong 4211, Bangladesh

[3] Department of Agricultural Chemistry, Bangladesh Agricultural University, Mymensingh 2202, Bangladesh

of the metal clusters where PVP molecules interact strongly through their carbonyl group by oxygen atom with the metal surface for their enhanced stabilization [32, 33]. Chemical syntheses offer a versatile route by assembling atoms and particles from the atomic or molecular state to the macroscopic scale [34]. The characteristics of the crystals can be controlled by the thermodynamics and kinetics of the synthesis [35]. Even though great progress has been made, there is still a necessity to develop chemical synthetic methods that can tailor the morphology of Pt crystals at different scales.

In this article, a simple microfluidic reduction of Pt^{4+} to Pt^0 by $NaBH_4$ in the presence of PVP were discussed that can provide a large-scale synthesis of Pt clusters hydrosol. The sizes of the produced clusters were <2 nm in any variable condition of the reaction parameters. The reduction conditions were optimized for the synthesis of possible smallest Pt clusters. The as produced particles were characterized by UV–visible absorption spectrophotometry, powder X-ray diffractometry (XRD) and transmission electron microscopy (TEM).

Experimental

Chemicals and solvents

Hexachloroplatinicacid hexahydrate ($H_2PtCl_6 \cdot 6H_2O$, Sigma-Aldrich), PVP (average molecular weight 3500–3,60,000, E-Merck) and sodium borohydride ($NaBH_4$, Wako Pure Chemical Industries Ltd.) of an analytical grade were purchased and used as received. Deionized water was used to prepare aqueous solutions and final washings of glassware's.

Preparation of Pt:PVP clusters

The microfluidic preparation of Pt clusters was described in detail previously [22]. In brief, freshly prepared ice-chilled aqueous solution of H_2PtCl_6 (4 mM, 35 mL) containing PVP (80 mM with respect to monomer) and ice-chilled aqueous solution of reducing agent $NaBH_4$ (20 mM) also containing PVP (80 mM with respect to monomer) were injected into the micromixer with the help of automatically actuated two syringe pumps with a constant flow rate of 200 mL/h by placing the micromixer in an ice bath. The Pt^{4+} ion to PVP-monomer final molar ratio was 1:40. The eluted hydrosol of produced Pt:PVP through the outlet were collected in a conical flask equipped with a magnetic stirrer that was also placed in an ice bath. The schematic view of the microfluidic preparation of Pt:PVP clusters is shown in Fig. 1. The as produced particles were purified by the diafiltration technique where hemodialyzer was used as

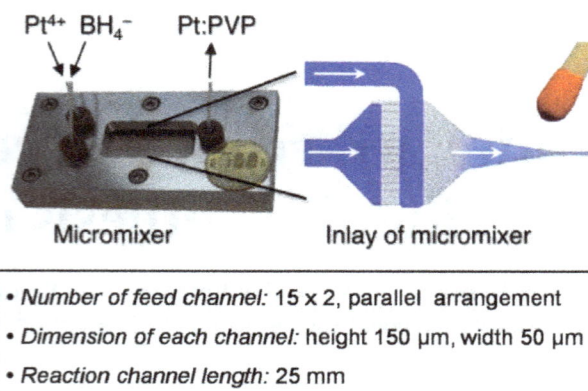

- **Number of feed channel**: 15 x 2, parallel arrangement
- **Dimension of each channel**: height 150 µm, width 50 µm
- **Reaction channel length**: 25 mm

Fig. 1 Schematic view of the preparation of Pt:PVP clusters by micromixer

an ultrafiltration membrane for the removal of unwanted ions, molecules or solvents.

Characterization

Using quartz cuvette, UV–visible absorption spectra of colloidal solutions were carried out on a Shimadzu UV-1800 spectrophotometer (from 200 to 1100 nm). Transmission electron micrographs (TEM) were performed using a JEM-2100F instrument operated at an accelerating voltage of 200 kV. The samples for the TEM analysis were prepared by casting a drop of ethanolic dispersion (~0.3 mM, on Pt-atom basis) on a carbon-coated TEM grid and left to air dry. X-ray diffraction (XRD) measurements of the clusters were performed using a diffractometer (D8 ADVANCE, Bruker) with Cu K_α radiation (1.5418 Å) operated at 40 kV and 40 mA. Fine powder samples were used for these measurements. The diffraction patterns were simulated using a TOPAS-4 program.

Results and discussion

Optimization of reduction conditions

For the efficient production of small and monodisperse PVP-stabilized Pt clusters in microfluidic method, reduction conditions were optimized for five reaction parameters, namely (1) temperature, (2) concentration of precursor metal ions, H_2PtCl_6, (3) molar ratio of $NaBH_4$ to Pt^{4+} ions, (4) molar ratio of PVP-monomer to Pt^{4+} ions, and (5) molecular weight of PVP.

A temperature effect on microfluidic synthesis was observed at 0, 20, 40, 60 and 80 °C where both of the solutions were met together at this temperature inside micromixer. It was found that the average diameters of the

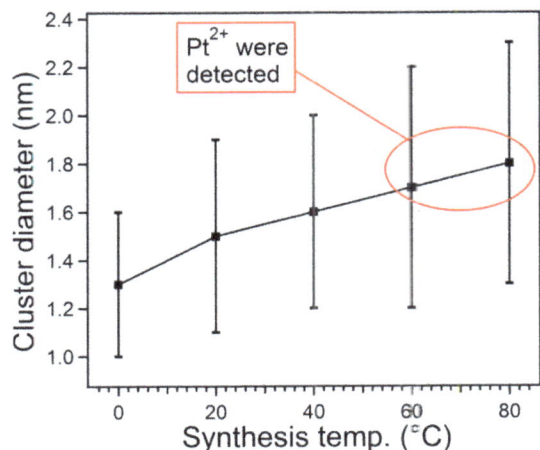

Fig. 2 Effect of temperature on synthesis of Pt:PVP clusters by microfluidic method

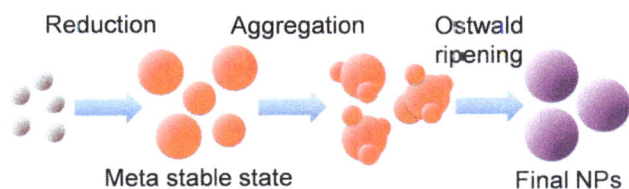

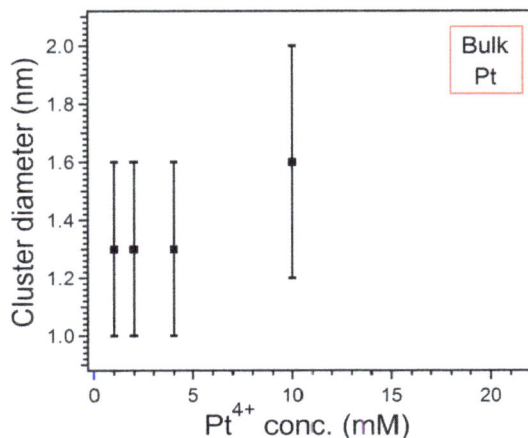

Fig. 3 Effect of Pt^{4+} concentration on synthesis of Pt:PVP clusters by microfluidic method

Scheme 1 Enlargement of Pt clusters due to temperature effect in microfluidic preparation

metal clusters were increased gradually with increasing the synthetic temperature from 0 to 80 °C as shown in Fig. 2. With increasing temperate, nucleation rate usually increases and results in increased number of cluster seeds and also increases the normal degradation rate of borohydride. At elevated temperature, formation of larger clusters indicated that higher temperature enhanced the relative population of stable particles by degradation of other less-stable particles and reunited them via thermal degradation as shown in Scheme 1 [37]. This can be explained with Ostwald ripening process, which is a thermodynamically driven spontaneous process because larger particles are more energetically stable than the smaller one [38, 39]. As the system tries to lower its overall energy, molecules on the surface of small (energetically unfavorable) particles will tend to detach and diffuse through the solution and then attach to the surface of larger particles. Therefore, the number of smaller particles continues to shrink, while larger particles continue to grow. On the other hand at higher temperature, the rate of degradation of borohydride in aqueous solution was increased which resulted incomplete reduction of Pt^{4+} ions where particles growth step was prolonged by slowed down the reduction step.

An effect of H_2PtCl_6 concentration on the size of final Pt clusters was studied ranging from 1.0 to 10.0 mM. It is relevant to know an optimum concentration, which will provide high and quality yield with the view of an efficient reduction and large amount preparation. No significant difference was observed in Pt clusters size up to 4.0 mM H_2PtCl_6 solution concentration as shown in Fig. 3. Similar size $(1.3 \pm 0.3 \text{ nm})$ and monodispersed clusters were produced. However, as the concentration was increased (10 mM), comparatively larger Pt clusters $(1.6 \pm 0.5 \text{ nm})$ were produced. In addition, unreacted Pt^{2+} ions were also detected in the ultrafiltrate of cluster hydrosol. The higher concentration of H_2PtCl_6 may not provide homogeneous solutions, which led into an incomplete reduction. We determined the amount of unreduced Pt ions in the ultrafiltrate by ICP measurement. We observed that nearly 90 % of precursor Pt^{4+} ions were reduced to Pt^0 whereas about 10 % was unreduced which remained in the ultrafiltrate as Pt^{2+} ions. At much higher concentration of Pt^{4+}, bulk Pt was formed along with unreduced Pt^{2+} ions. That was probably due to the higher concentration of $NaBH_4$ that destabilized the protective nature of PVP, which resulted bulk platinum formation.

$NaBH_4$ is a strong reducing agent that readily reduces H_2PtCl_6 to colloidal platinum. The proposed overall chemical reactions are:

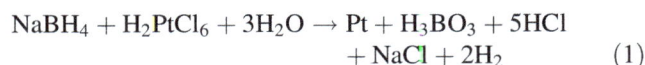

$$NaBH_4 + H_2PtCl_6 + 3H_2O \rightarrow Pt + H_3BO_3 + 5HCl + NaCl + 2H_2 \qquad (1)$$

$$NaBH_4 + H_2O \rightarrow H_2 + \text{boron hydrolysis products} \qquad (2)$$

According to the above-mentioned balance chemical equation [40, 41], for complete reduction of all Pt^{4+} ions present in solution to Pt^0, a 1:1 molar ratio is required between borohydride and Pt^{4+} ions. Unreduced Pt^{2+} ions were detected even at 1:3 molar ratio of Pt^{4+} to $NaBH_4$, although according to the balance chemical equation this ratio was much higher for the complete reduction of all Pt ions present in the solution. The formed Pt^0 acts as a highly

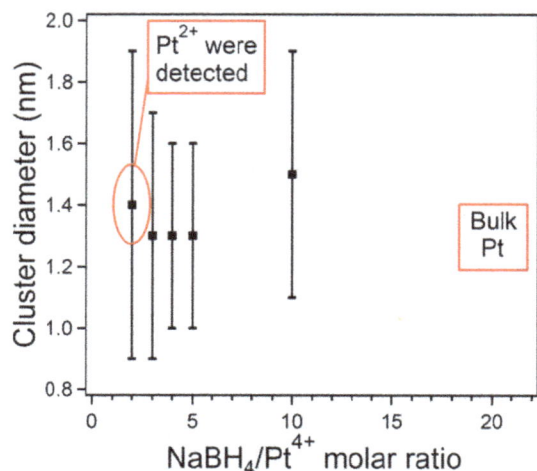

Fig. 4 Effect of NaBH$_4$ to Pt^{4+} molar ratio on the synthesis of Pt:PVP clusters by microfluidic method

Fig. 5 Effect of PVP-monomer to Pt^{4+} molar ratio on synthesis of Pt:PVP clusters by microfluidic method

active catalysts for the hydrolysis of NaBH$_4$ [22], so that a large excess of NaBH$_4$ could not ensure the complete reduction of all precursor ions as shown in Fig. 4. We observed that minimum molar ratio 4 is required for complete reduction of all Pt^{4+} ions present in the solution by microfluidic method. This higher amount of NaBH$_4$ is required because of the degradation during reduction although produced Pt^0 were separated and fresh borohydride and Pt^{4+} ions were always fed. But at much higher ratio of NaBH$_4$ >20 resulted bulk platinum formation. Probably higher ratio of NaBH$_4$ destabilized the protective nature of PVP, which resulted bulk platinum formation.

Reduction of H$_2$PtCl$_6$ to Pt^0 by MeOH in the presence of PVP is suggested to be proceeded in two steps where first step is the conversion of [Pt^{+IV}Cl$_6$]$^{2-}$ to [Pt^{+II}Cl$_4$]$^{2-}$ [42]. Although we did not studied the reaction mechanism, the presence of Pt^{2+} ions in the ultrafiltrate even after addition of 100 times NaBH$_4$ compared to microfluidic system is one of the proof to follow the reaction accordance with the two-step reduction mechanism, steps (3) and (4) as was suggested by Ingelsten et al. [43]:

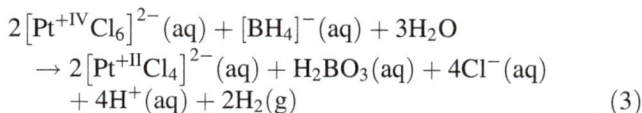

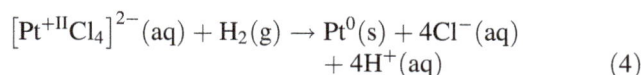

$$2\left[Pt^{+IV}Cl_6\right]^{2-}(aq) + [BH_4]^-(aq) + 3H_2O$$
$$\rightarrow 2\left[Pt^{+II}Cl_4\right]^{2-}(aq) + H_2BO_3(aq) + 4Cl^-(aq)$$
$$+ 4H^+(aq) + 2H_2(g) \tag{3}$$

$$\left[Pt^{+II}Cl_4\right]^{2-}(aq) + H_2(g) \rightarrow Pt^0(s) + 4Cl^-(aq)$$
$$+ 4H^+(aq) \tag{4}$$

The molar ration of PVP-monomer to Pt^{4+} from 5 to 200 was studied in microfluidic reduction, which is shown in Fig. 5 where the average gram molecular weight of PVP was 40,000. It was found that average size of the Pt clusters was increased with decreasing the PVP-monomer to Pt^{4+} ratio from 20 to 5. The degree of monodispersity of the clusters was also decreased. It was probably, this amount of

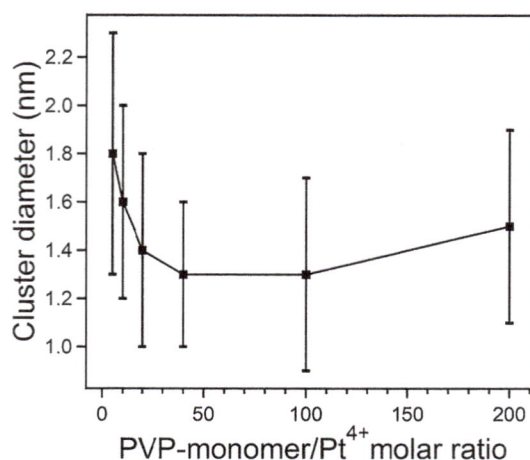

PVP was not sufficient to well stabilize the smaller Pt clusters during preparation. On the other hand, there were no significant differences between the size and size distribution of clusters were observed when the molar ratios of PVP-monomer to Pt^{4+} ions were tuned from 40 to 100. But at much higher ratio of PVP-monomer, larger particles were observed. It was probably because of the higher viscosity of the solution, which hindered the well mixing of both solutions. With increasing the PVP content, the probability of bare surface for accessing the substrates to the clusters surface decreases, so the minimum ratio 40 was chosen for smaller particles formation and as well as maximizing the bare surfaces of the clusters. Interestingly, it was observed that at much lower PVP-monomer to Pt^{4+} ratio (10–5), clusters were less stable for long time storage, usually for more than a year. After purification of the produced particles, one part was freeze dried, pulverized and stored in an airtight vial at room temperature. Another part was colloidal hydrosol, which was also stored in an airtight vial at room temperature. The stability of the clusters was tested by UV–visible absorption spectrophotometry and TEM micrographs observation. Basically, there was no difference between before and after storage.

An effect of the average gram molecular weight (MW) of the stabilizing polymer that is chain length of PVP was investigated on the size and size distribution of Pt clusters from 3500 to 360,000 for a constant PVP-monomer to Pt-atom ratio of 40, which is shown in Fig. 6. It was found that a molecular weight of 40,000 provided the smallest particles of 1.3 ± 0.3 nm. With increasing or decreasing the average gram MW of PVP, produced particles were larger than 1.3 ± 0.3 nm. It was assumed that in case of small-chained PVP, the relative number of polymer chain per seed of clusters was much higher than required to well stabilize that initially produced small but less

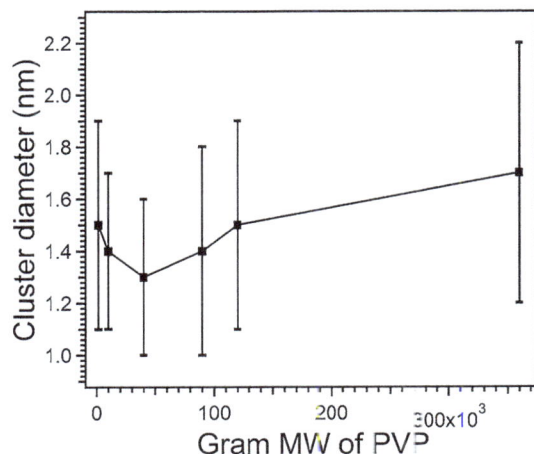

Fig. 6 Effect of PVP chain length on synthesis of Pt:PVP clusters by microfluidic method

stable particles by steric hindrance those were finally reunited to produce larger clusters. On the other hand, in case of longer chain polymer, insufficient number of polymer chain per cluster that was actually required destabilized the smaller particles formation, which is required for sufficient stabilization.

Size distributions

Figure 7 shows the representative TEM image of Pt:PVP clusters where particles were found to be comprised of spherical in shape. More than 300 clusters diameter was measured and plotted as histogram. An average diameter (d_{TEM}) and distribution of the clusters were determined from the best fitting of the histograms using Gaussian function. To the best of our knowledge, this is the unique method for the preparation of the smallest Pt:PVP clusters

in aqueous phase reduction of $Pt^{4+/2+}$ ions by $NaBH_4$ with an average diameter 1.3 ± 0.3 nm reported so far [2–8, 16–18, 22, 36].

Optical spectra for cluster formation

Figure 8 shows the normalized UV–visible absorption spectra of the Pt clusters hydrosol measured at 0.2 mM concentration of Pt^0 and precursor solution spectra (Pt^{4+} ions) was measured at 1.0 mM concentration. Before reduction, the H_2PtCl_6 solution with or without PVP were pale yellow. After reduction with $NaBH_4$, it was instantly turned into dark brown regardless of the particles size and shape. The formed Pt:PVP cluster hydrosol exhibited an exponential-like profile in the UV–visible absorption spectrum with no obvious absorption peak, which is in consistence with earlier studies, whereas the original precursor solution optical spectra were completely different. This feature is attributed for intra- and inter-band optical transitions, suggesting the formation of Pt clusters [20, 22, 29, 30]. These measurements further confirmed that the Pt^{4+} ions were completely reduced to Pt^0 forming the nano-clusters.

Aqueous phase reduction of $Pt^{4+/2+}$ ions to Pt^0 by $NaBH_4$ is the most suitable method for synthesizing small and monodisperse Pt clusters but the produced Pt^0 readily reacts with borohydride (BH_4^-) causing incomplete reduction of the precursor ions, which results a very low yield along with larger particles formation [19–22]. Hence, fresh BH_4^- should be supplied continuously to reduce metal ions and separate the produced Pt^0 from the reaction mixture. Using a fluidic system, the hydrolysis of BH_4^- can be prevented by separating the produced Pt^0, which will facilitate the reduction of the metal ions. In addition, microfluidic system ensures a homogeneous mixing of

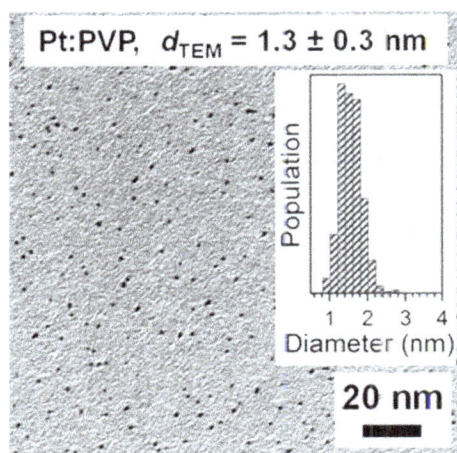

Fig. 7 Representative TEM micrograph of Pt:PVP clusters and inset shows the size distribution of the clusters

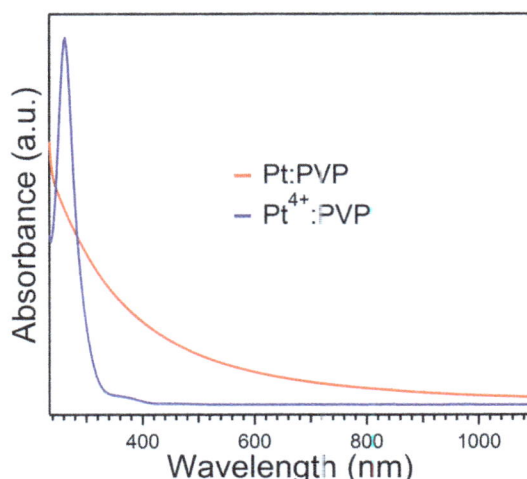

Fig. 8 UV–visible absorption spectra of Pt:PVP clusters and precursor solution

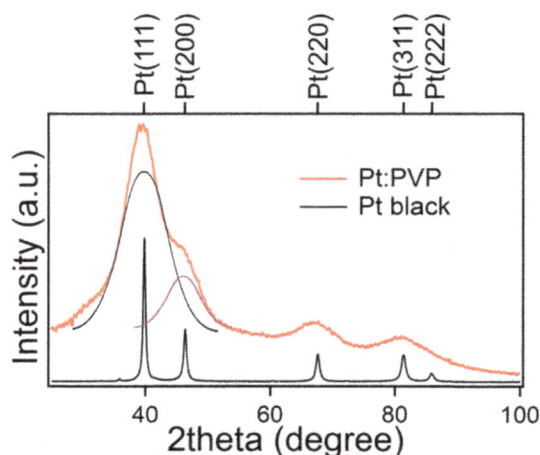

Fig. 9 XRD patterns of microfluidic Pt:PVP and bulk Pt

metal ions and BH_4^-, which is expected to produce monodispersed Pt clusters.

Structural characterization

The XRD pattern of produced Pt:PVP clusters and a bulk Pt metal is shown in Fig. 9. The observed four peaks in the spectrum can be designated to (111), (200), (220) and (311) planes of platinum clusters with an fcc phase structure, which suggests the purity of the produced platinum crystals and no platinum oxide or reaction residue were detectable. The average diameters of the Pt crystallites (d_{XRD}) were calculated using the Scherrer equation to the Pt (111) diffraction peak, which was found to be 1.2 nm. The XRD diffraction patterns were simulated using TOPAS-4 program.

$$d_{XRD} = \frac{k\lambda}{\beta \cos\theta} \quad \text{Scherrer equation}$$

where k is a dimensionless constant called shape factor (for spherical particles its value is equal to 0.89), λ is the X-ray wavelength in nanometer (nm), θ is the Bragg diffraction angle of the Pt (111) plane, and β is the full width at half-maximum (FWHM) of the Pt (111) diffraction peak. The value of the crystallite size determined from XRD was consistent with the value determined from TEM analysis. To the best of our knowledge, this is the smallest Pt:PVP crystallite synthesized in aqueous phase reduction of Pt ions reported so far [19–21].

The broadened and low diffraction peak at Pt (111) as well as little shift to the smaller angle (from 40°) suggests the formation of a large number of vacancies associated with higher population of low-coordination sites in smaller Pt clusters [32]. The shifting of diffraction peak to the smaller angle also suggests the expansion of the lattice parameter (interatomic distances) from the perfect single crystal lattice (for bulk Pt $a_0 = 0.39231$ nm) and also

increase in the unit cell volume [44]. Lattice expansion with crystal size reduction has also been observed for Pd nanoparticles [45] and Ag nanoparticles [46]. A probable explanation for this effect might be the use of PVP as stabilizer, which has the ability to donate electron to the vacant d-orbital of metal atoms, hence leading to Pt–Pt bond expansion of the surface-near atoms by increasing electron density [2, 22, 47].

A change of interatomic distances and unit cell parameters considerably influences the catalytic activity of Pt-based nanoparticles. An expansion of the Pt–Pt interatomic distances demonstrated an enhanced catalytic activity on oxide reduction, as well as a significantly improved resistance to CO poisoning [48]. Similar beneficial effects were also reported for Pd particles [49]. We also expect the similar enhanced catalytic activity for our microfluidic Pt:PVP particles.

Conclusion

Mixing is a crucial factor for producing small and monodispersed metal clusters. Hence, micromixer is a simple and powerful tool that helps to mix solutions homogeneously nearly at the molecular level. It also prevents the degradation of $NaBH_4$ by preventing the mixing of produced clusters from reactants by separating them through fluidic system that enables to produce small and monodispersed Pt clusters. From the reduction behavior and nucleation/growth mechanism based on an empirical model, the microfluidic-borohydride reduction for Pt^{4+} ions occurs in two steps. In the first step, Pt^{4+} converts to Pt^{2+}, which is subsequently reduced to Pt^0 by the produced H_2 gas in the second step. Optimizing the reduction conditions, Pt nano-clusters of possible smallest diameter was synthesized and that was 1.3 ± 0.3 nm of TEM diameter. The crystal size determined by XRD was consistent with TEM diameter and that was 1.2 nm having fcc crystal geometry same as bulk platinum. Utilizing this technique, other nano-clusters of transition metals could be synthesized.

Acknowledgements This research work was done under the Post-doctoral Fellowship Programme of Bangladesh Council of Science and Industrial Research (BCSIR), Bangladesh and financial support thereby. Thanks to BFRI authority for their kind cooperation. Finally, special thanks to Sophisticated Analytical Instrument Facility (SAIF), Shillong, India for the TEM micrographs.

References

1. Hirai, H., Toshima, N.: In: Tailored Metal Catalysts. Iwasawa Y., (Ed) Reidel Publishing, Dordrecht (1986)

2. Tsunoyama, H., Ichikuni, N., Sakurai, H., Tsukuda, T.: Effect of electronic structures of Au clusters stabilized by poly (N-vinyl-2-pyrrolidone) on aerobic oxidation catalysis. J. Am. Chem. Soc. 131, 7086–7093 (2009)

3. Xiao, C.X., Cai, Z.P., Wang, T., Kou, Y., Yan, N.: Aqueous-phase Fischer–Tropsch synthesis with a ruthenium nanocluster catalyst. Angew. Chem. 120, 758–761 (2008)

4. Xiao, C.X., Wang, H.Z., Mu, X.D., Kou, Y.: Ionic-liquid-like copolymer stabilized nanocatalysts in ionic liquids I. Platinum catalyzed selective hydrogenation of o-chloronitrobenzene. J. Catal. 250, 25–32 (2007)

5. Song, S., Liu, R., Zhang, Y., Feng, J., Liu, D., Xing, Y., Zhao, F., Zhang, H.: Colloidal noble-metal and bimetallic alloy nanocrystals: a general synthetic method and their catalytic hydrogenation properties. Chem. A Eur. J. 16, 6251–6256 (2010)

6. Yan, N., Zhao, C., Luo, C., Dyson, P.J., Liu, H., Kou, Y.: One-step conversion of cellobiose to C$_6$-alcohols using a ruthenium nanocluster catalyst. J. Am. Chem. Soc. 128, 8714–8715 (2006)

7. Mu, X.D., Meng, J.Q., Li, Z.C., Kou, Y.: Rhodium nanoparticles stabilized by ionic copolymers in ionic liquids: long lifetime nanocluster catalysts for benzene hydrogenation. J. Am. Chem. Soc. 127, 9694–9695 (2005)

8. Zhao, C., Wang, H.Z., Yan, N., Xiao, C.X., Mu, X.D., Dyson, P.J., Kou, Y.: Ionic-liquid-like copolymer stabilized nanocatalysts in ionic liquids: II. Rhodium Catal. Hydrog. Arenes J. Catal. 250, 33–40 (2007)

9. Zhao, C., Gan, W., Fan, X., Cai, Z., Dyson, P.J., Kou, Y.: Aqueous-phase biphasic dehydroaromatization of bio-derived limonene into p-cymene by soluble Pd nanocluster catalysts. J. Catal. 254, 244–250 (2008)

10. He, L., Liu, H., Xiao, C.X., Kou, Y.: Liquid-phase synthesis of methyl formate via heterogeneous carbonylation of methanol over a soluble copper nanocluster catalyst Green Chem. 10, 619–622 (2008)

11. Somorjai, G.A.: Introduction to Surface Chemistry and Catalysis. Wiley, New York (1994)

12. White, R.J., Luque, R., Budarin, V.L., Clark, J.H., Macquarrie, D.J.: Supported metal nanoparticles on porous materials. Methods and applications. Chem. Soc. Rev. 38, 481–494 (2009)

13. Astruc, D., Lu, F., Aranzaes, J.R.: Nanoparticles as recyclable catalysts: the frontier between homogeneous and heterogeneous catalysis. Angew. Chem. Int. Ed. 44, 7852–7872 (2005)

14. Rioux, R.M., Song, H., Grass, M., Habas, S., Niesz, K., Hoefelmeyer, J.D., Yang, P., Somorjai, G.A.: Monodisperse platinum nanoparticles of well-defined shape: synthesis, characterization, catalytic properties and future prospects. Top. Catal. 39, 167–174 (2006)

15. Chen, A., Holt-Hindle, P.: Platinum-based nanostructured materials: synthesis, properties, and applications. Chem. Rev. 110, 3767–3804 (2010)

16. Xu, S., Yang, Q.: Well-dispersed water-soluble Pd nanocrystals: Facile reducing synthesis and application in catalyzing organic reactions in aqueous media. J. Phys. Chem. C 112, 13419–13425 (2008)

17. Zhang, Q., Xie, J., Yang, J., Lee, J.Y.: Monodisperse icosahedral Ag, Au, and Pd nanoparticles: size control strategy and superlattice formation. ACS Nano 3, 139–148 (2008)

18. Wiley, B., Sun, Y., Xia, Y.: Synthesis of silver nanostructures with controlled shapes and properties. Acc Chem. Res. 40, 1067–1076 (2007)

19. Van Rheenen, P.R., McKelvy, M.J., Glaunsinger, W.S.: Synthesis and characterization of small platinum particles formed by the chemical reduction of chloroplatinic acid. J. Solid State Chem. 67, 151–169 (1987)

20. Knecht, M.R., Weir, M.G., Myers, V.S., Pyrz, W.D., Ye, H., Petkov, V., Buttrey, D.J., Frenkel, A.I., Crooks, R.M.: Synthesis and characterization of Pt dendrimer-encapsulated nanoparticles: effect of the template on nanoparticle formation. Chem. Mater. 20, 5218–5228 (2008)

21. Borodko, Y., Thompson, C.M., Huang, W., Yildiz, H.B., Frei, H., Somorjai, G.A.: Spectroscopic study of platinum and rhodium dendrimer (PAMAM G4OH) compounds: structure and stability. J. Phys. Chem. C 115, 4757–4767 (2011)

22. Hossain, M.J., Tsunoyama, H., Yamauchi, M., Ichikuni, N., Tsukuda, T.: High-yield synthesis of PVP-stabilized small Pt clusters by microfluidic method. Catal. Today 183, 101–107 (2012)

23. Gobby, D., Angeli, P., Gavriilidis, A.: Mixing characteristics of T-type microfluidic mixers. J. Micromech. Microeng. 11, 126 (2001)

24. Soleymani, A., Kolehmainen, E., Turunen, I.: Numerical and experimental investigations of liquid mixing in T-type micromixers. Chem. Eng. J. 135, S219–S228 (2008)

25. Schwesinger, N., Frank, T., Wurmus, H.: A modular microfluid system with an integrated micromixer. J. Micromech. Microeng. 6, 99 (1996)

26. Munson, M.S., Yager, P.: Simple quantitative optical method for monitoring the extent of mixing applied to a novel microfluidic mixer. Anal. Chim. Acta 507, 63–71 (2004)

27. Hessel, V., Hardt, S., Löwe, H., Schönfeld, F.: Laminar mixing in different interdigital micromixers: I. Experimental characterization. AIChE J. 49, 566–577 (2003)

28. Hardt, S., Schönfeld, F.: Laminar mixing in different interdigital micromixers: II. Numerical simulations. AIChE J. 49, 578–584 (2003)

29. Rioux, R.M., Song, H., Hoefelmeyer, J.D., Yang, P., Somorjai, G.A.: High-surface-area catalyst design: synthesis, characterization, and reaction studies of platinum nanoparticles in mesoporous SBA-15 silica. J. Phys. Chem. B 109, 2192–2202 (2005)

30. Teranishi, T., Hosoe, M., Tanaka, T., Miyake, M.: Size control of monodispersed Pt nanoparticles and their 2D organization by electrophoretic deposition. J. Phys. Chem. B 103, 3818–3827 (1999)

31. Campelo, J.M., Luna, D., Luque, R., Marinas, J.M., Romero, A.A.: Sustainable preparation of supported metal nanoparticles and their applications in catalysis. Chem. Sus. Chem. 2, 18–45 (2009)

32. Toshima, N., Harada, M., Yonezawa, T., Kushihashi, K., Asakura, K.: Structural analysis of polymer-protected palladium/platinum bimetallic clusters as dispersed catalysts by using extended X-ray absorption fine structure spectroscopy. J. Phys. Chem. 95, 7448–7453 (1991)

33. Borodko, Y., Humphrey, S.M., Tilley, T.D., Frei, H., Somorjai, G.A.: Charge-transfer interaction of poly(vinylpyrrolidone) with platinum and rhodium nanoparticles. J. Phys. Chem. C 111, 6288–6295 (2007)

34. Elechiguerra, J.L., Reyes-Gasga, J., Yacaman, M.J.: The role of twinning in shape evolution of anisotropic noble metal nanostructures. J. Mater. Chem. 16, 3906–3919 (2006)

35. Teranishi, T., Miyake, M.: Size control of palladium nanoparticles and their crystal structures. Chem. Mater. 10, 594–600 (1998)

36. Lim, B., Jiang, M., Tao, J., Camargo, P.H., Zhu, Y., Xia, Y.: Shape-controlled synthesis of Pd nanocrystals in aqueous solutions. Adv. Funct. Mater. 19, 189–200 (2009)

37. Leszczyńska, A., Njuguna, J., Pielichowski, K., Banerjee, J.R.: Polymer/montmorillonitenanocomposites with improved thermal properties: part I. Factors influencing thermal stability and mechanisms of thermal stability improvement. Thermochim. Acta **453**, 75–96 (2007)

38. Shao-Horn, Y., Sheng, W.C., Chen, S., Ferreira, P.J., Holby, E.F., Morgan, D.: Instability of supported platinum nanoparticles in low-temperature fuel cells. Top. Catal. **46**, 285–305 (2007)

39. Chen, Z., Waje, M., Li, W., Yan, Y.: Supportless Pt and PtPd nanotubes as electrocatalysts for oxygen-reduction reactions. Angew. Chem. Int. Ed. **46**, 4060–4063 (2007)

40. Garron, A., Świerczyński, D., Bennici, S., Auroux, A.: New insights into the mechanism of H_2 generation through $NaBH_4$ hydrolysis on Co-based nanocatalysts studied by differential reaction calorimetry. Int. J. Hydrog. Energy **34**, 1185–1199 (2009)

41. Gonçalves, A., Castro, P., Novais, A., Fernandes, V.R., Rangel, C.M., Matos, H.: Dynamic modeling of hydrogen generation via hydrolysis of sodium borohydride. Reactions **1**, 2 (2007)

42. Duff, D.G., Edwards, P.P., Johnson, B.F.: Formation of a polymer-protected platinum sol: a new understanding of the parameters controlling morphology. J. Phys. Chem. **99**, 15934–15944 (1995)

43. Ingelsten, H.H., Bagwe, R., Palmqvist, A., Skoglundh, M., Svanberg, C., Holmberg, K., Shah, D.O.: Kinetics of the formation of nano-sized platinum particles in water-in-oil microemulsions. J. Colloid Interface Sci. **241**, 104–111 (2001)

44. Leontyev, I.N., Kuriganova, A.B., Leontyev, N.G., Hennet, L., Rakhmatullin, A., Smirnova, N.V., Dmitriev, V.: Size dependence of the lattice parameters of carbon supported platinum nanoparticles: X-ray diffraction analysis and theoretical considerations. RSC Adv. **4**, 35959–35965 (2014)

45. Kuhrt, C., Anton, R.: On the origin of a lattice expansion in palladium and Pd@Au vapour deposits on various substrates. Thin Solid Films **198**, 301–315 (1991)

46. Onodera, S.: Lattice parameters of fine copper and silver particles. J. Phys. Soc. Jpn. **61**, 2190–2193 (1992)

47. Qiu, L., Liu, F., Zhao, L., Yang, W., Yao, J.: Evidence of a unique electron donor-acceptor property for platinum nanoparticles as studied by XPS. Langmuir **22**, 4480–4482 (2006)

48. Alayoglu, S., Nilekar, A.U., Mavrikakis, M., Eichhorn, B.: Ru–Pt core–shell nanoparticles for preferential oxidation of carbon monoxide in hydrogen. Nat. Mater. **7**, 333–338 (2008)

49. Suo, Y., Zhuang, L., Lu, J.: First-principles considerations in the design of Pd-alloy catalysts for oxygen reduction. Angew. Chem. Int. Ed. **46**, 2862–2864 (2007)

4

Synthesis optimization of carbon-supported ZrO$_2$ nanoparticles from different organometallic precursors

Pankaj Madkikar[1] · Xiaodong Wang[2] · Thomas Mittermeier[1] ·
Alessandro H. A. Monteverde Videla[3] · Christoph Denk[1] · Stefania Specchia[3] ·
Hubert A. Gasteiger[1] · Michele Piana[1]

Abstract We report here the synthesis of carbon-supported ZrO$_2$ nanoparticles from zirconium oxyphthalocyanine (ZrOPc) and acetylacetonate [Zr(acac)$_4$]. Using thermogravimetric analysis (TGA) coupled with mass spectrometry (MS), we could investigate the thermal decomposition behavior of the chosen precursors. According to those results, we chose the heat treatment temperatures (T_{HT}) using partial oxidizing (PO) and reducing (RED) atmosphere. By X-ray diffraction we detected structure and size of the nanoparticles; the size was further confirmed by transmission electron microscopy. ZrO$_2$ formation happens at lower temperature with Zr(acac)$_4$ than with ZrOPc, due to the lower thermal stability and a higher oxygen amount in Zr(acac)$_4$. Using ZrOPc at $T_{HT} \geq 900$ °C, PO conditions facilitate the crystallite growth and formation of distinct tetragonal ZrO$_2$, while with Zr(acac)$_4$ a distinct tetragonal ZrO$_2$ phase is observed already at $T_{HT} \geq 750$ °C in both RED and PO conditions. Tuning of ZrO$_2$ nanocrystallite size from 5 to 9 nm by varying the precursor loading is also demonstrated. The chemical state of zirconium was

analyzed by X-ray photoelectron spectroscopy, which confirms ZrO$_2$ formation from different synthesis routes.

Keywords Carbon-supported zirconia nanoparticles · Thermogravimetric analysis · X-ray diffraction · Transmission electron microscopy · X-ray photoelectron spectroscopy

Introduction

Bulk zirconia has been explored over decades. ZrO$_2$ exists in three phases, viz., monoclinic (a room temperature stable phase), tetragonal (stable above 1100 °C), and cubic (stable above 2300 °C); the latter two are thus called high-temperature phases [1]. Various mechanisms are reported which explain the stabilization of high-temperature ZrO$_2$ phases at room temperature [2, 3]. ZrO$_2$ stabilized with dopants like yttria, magnesia, and alumina has a wide range of applications in solid-oxide fuel cells, thermal barrier coatings, and biomedical implants [4–5]. In comparison to the work done on bulk ZrO$_2$, carbon-supported ZrO$_2$ nanoparticles are not much explored. Recently, works on the applications of carbon-supported ZrO$_2$ nanoparticles were reported. Sulfated ZrO$_2$ supported on multiwalled carbon nanotubes (MWCNTs) as a support for platinum is an example; the resulting catalyst is claimed to exhibit a higher methanol oxidation reaction (MOR) activity than unsulfated Pt-ZrO$_2$/MWCNT and commercial Pt/C [7]. Sulfated-ZrO$_2$ acts as a co-catalyst of Pt, resulting in could be a relatively cheap anode catalyst in comparison to PtRu for MOR. Another example of application are ZrO$_2$-C hybrid supports for Pt electrocatalysts to increase the stability of noble metal during the course of potential cycling [8]. Here it is reported, that the increase in durability is due to nanometric ZrO$_2$

✉ Pankaj Madkikar
pankaj.madkikar@tum.de

[1] Chair of Technical Electrochemistry, Department of Chemistry and Catalysis Research Center, Technische Universität München, 85748 Garching, Germany

[2] Johnson Matthey Catalysts (Germany) GmbH, Bahnhofstr. 43, 96257 Redwitz, Germany

[3] Department of Applied Science and Technology, Politecnico di Torino, Corso Duca degli Abruzzi 24, 10129 Turin, Italy

which inhibits the migration and aggregation of Pt during cycling. Similar ZrO_2-C hybrid supports for Pd catalysts are reported to show higher activity and durability than Pd/C in formic acid electro-oxidation [9]. In the latter study, the authors concluded that the physical characteristics of ZrO_2 could promote dispersion of Pd nanoparticles and the presence of ZrO_2 could change the interaction of Pd with the support material, resulting in increased activity and reduced CO poisoning effect on Pd. There were also several other articles published which clearly indicated that ZrO_2 serves as co-catalyst in energy conversion [10–13]. In 2013, Seo et al. reported the synthesis of valve-metal-oxide nanoparticles by an electrodeposition technique as oxygen-reduction-reaction (ORR) electrocatalysts [14]. Sebastián et al. have also reported facile synthesis of Zr- and Ta-based ORR-active methanol-tolerant catalysts for direct alcohol fuel cells (DAFCs) [15]. In addition, Ota research group has also reported ORR activity for ZrO_2 [16, 17]. This material has been considered as a promising non-noble metal catalyst for PEMFCs (proton exchange membrane fuel cells) because of its availability, its observed ORR activity, and stability in the strong acidic environment of the PEMFC [18]. In a few conference proceedings articles, zirconium oxy-phthalocyanine (ZrOPc) has been used as a starting precursor for ZrO_2 nanoparticles synthesis [19–21], however, lacking detailed information and study of that synthesis. Carbon-supported ZrO_2 is not restricted to only DAFCs and PEMFCs applications; it has also used as a cathode in microbial fuel cell (MFC) as reported by Mecheri et al. [22].

Looking into this growing interest in carbon-supported valve-metal oxide nanoparticles like zirconia and lacking the scientific detailed information on its synthesis using ZrOPc as a precursor, this work provides a study on the preparation of ZrO_2 nanoparticles under different conditions. Two precursors, zirconium oxyphthalocyanine (ZrOPc) and zirconium acetylacetonate [$Zr(acac)_4$], were chosen based on previous studies [19, 23] especially for their difference in atomic constitutions and solubility in organic solvents. ZrOPc contains less molecular oxygen but is nitrogen-rich, while, $Zr(acac)_4$ is an oxygen-rich but nitrogen-free precursor. Furthermore, ZrOPc is barely soluble in common organic solvents like chloroform, acetone, ethanol, etc., in contrast to $Zr(acac)_4$, which is soluble [24, 25]. Starting from these highly different precursors, we aimed at comparing the structure, size, and ORR activity of the resulting carbon supported ZrO_2 nanoparticles.

The organometallic precursors were first supported on carbon. The obtained precursor was heat-treated at different temperatures ranging from 350 to 1000 °C, using two different gas conditions, i.e., reducing (RED) (5% H_2 in Ar) or partially oxidizing (PO) (0.5% O_2 in 2.5% H_2 in a mixture of N_2 and Ar). Additionally, ZrO_2 loading variation was also carried out by varying the starting amount of ZrOPc and its

effect on the oxide nanoparticles was checked. Our final aim was to check the ORR activity of the supported nanoparticles and find a possible correlation between their activity and size–structure. We have reported the electrochemical results in an electrochemistry-oriented journal [26].

Experimental procedure

Synthesis of carbon supported ZrO_2 nanoparticles

Zirconium oxy-phthalocyanine (ZrOPc) was synthesized as reported by Tomachynski et al. (Refer to the supplementary information for the chemical analyses of the produced ZrOPc) [27]. Zirconium acetylacetonate [$Zr(acac)_4$, 98%] and chloroform ($CHCl_3$, ≥99.9%) were purchased from Sigma-Aldrich. Ketjenblack E-type (KB) carbon support was bought from Tanaka Kikinzoku Kogyo K.K. Argon 5.0 (Ar, 99.999%), hydrogen W5 (5% H_2 in Ar), and synthetic air (a mixture of 20.5% O_2 and 79.5% N_2) were supplied by Westfalen AG. All the commercial chemicals and gases were used as received without further purification.

For the synthesis of carbon-supported ZrO_2 nanoparticles, initially the precursors were deposited on KB by a method developed on the similar guidelines described in US 2011/0034325 A1 patent [28]. The general scheme is depicted in Fig. 1. Typically, 504 mg of ZrOPc (Zr = 0.8 mmol) or 220 mg of $Zr(acac)_4$ (Zr = 0.45 mmol) were added to 200 ml $CHCl_3$ in a 500 ml round-bottom flask. Considering different precursor losses in two different $CHCl_3$ separation methods for both precursors, the initial Zr mmoles are different to obtain similar Zr-loadings after supporting on carbon. The mixture was then sonicated for 5 min in an ice-cold ultrasonic bath. After this a uniform dispersion/solution was formed, which was dark blue colored in the case of ZrOPc and colorless in $Zr(acac)_4$. Thereafter, 403 mg of KB and an additional 200 ml of $CHCl_3$ were added to the flask. Ice-cold bath sonication was continued for another 2.5 h and the dispersion was observed to be uniform. To ensure the maximum deposition of ZrOPc or $Zr(acac)_4$ on the carbon support, the sonicated dispersion was further kept under continuous stirring at 20 °C for 48 h. Afterwards, the carbon-supported precursor was isolated from $CHCl_3$ by centrifugation (*Eppendorf, Centrifuge 5810 R*) in the case of ZrOPc/KB and by rotovaporation (*Heidolph, Hei-VAP Value*) in the case $Zr(acac)_4$/KB. The collected residue was dried at RT, then ground to fine powder in a mortar with pestle, and further dried in a temperature controlled vacuum oven at 70 °C overnight, to ensure complete removal of $CHCl_3$.

Thermogravimetric analysis (TGA) of both ZrOPc/KB and $Zr(acac)_4$/KB was performed on a *Mettler Toledo TGA/ DSC 1* instrument to check the thermal stability and

Fig. 1 Scheme of the steps and conditions involved in the synthesis of supported ZrO$_2$ nanoparticles

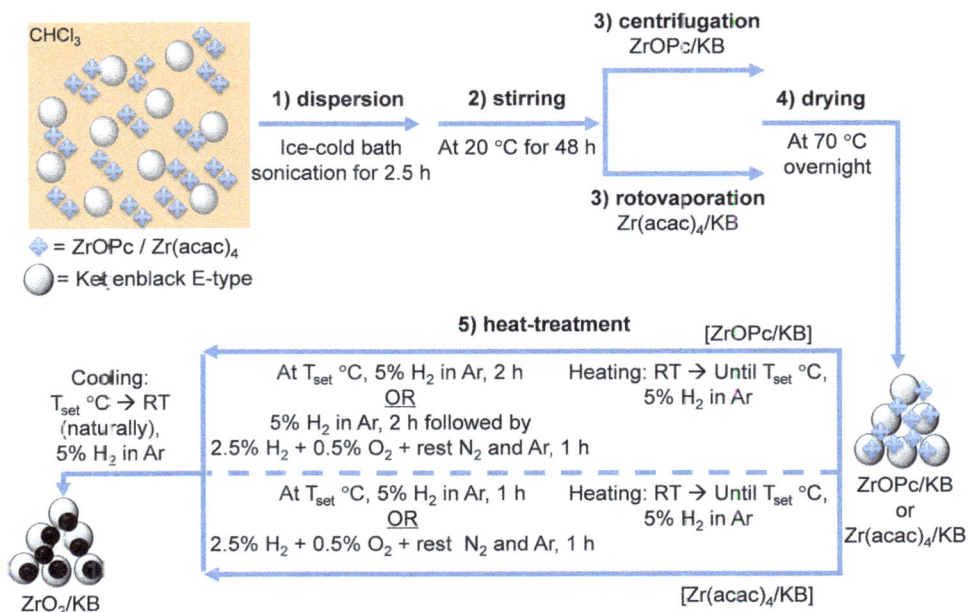

Fig. 1 Scheme of the steps and conditions involved in the synthesis of supported ZrO$_2$ nanoparticles

degradation pattern of the precursors in both Ar and H$_2$/Ar atmospheres, adopting a heating rate of 20 °C/min in the analysis. In addition, TGA-MS analysis of unsupported ZrOPc and Zr(acac)$_4$ was carried out with *Pfeiffer Vacuum Thermostar* mass spectrometer (MS) in Ar atmosphere with a heating rate of 10 °C/min to obtain the fragmentation pattern of the precursors. Unsupported organometallic precursors were used in MS analysis to avoid signals from carbon, which might overlap with the signals from molecular fragments and cause possible complications in the data interpretation.

From the TGA of the carbon supported precursors (see discussion of Figs. 2, 5) we chose the desired gas conditions (PO or RED) and temperatures (350–1000 °C) for the final heat treatment in a quartz tube furnace (*HTM Reetz*). PO conditions were selected to be similar to those reported by Yin et al. [20].

In the above depositions of ZrOPc and Zr(acac)$_4$ on KB, the targeted ZrO$_2$ loading on carbon was 13 wt.% ZrO$_2$/KB for both. Additionally, a loading variation study was conducted with carbon supported ZrOPc, to investigate its effect on the resulting particle size and structure. Samples with three loadings of ZrO$_2$ on KB, i.e., 5, 10, and 15 wt.%, were synthesized by varying the relative amount of ZrOPc with respect to KB during the supporting process and heat-treating in PO conditions at 950 °C.

Size–structural characterization of supported ZrO$_2$ nanoparticles

All heat-treated samples were analyzed by a *STOE* X-ray powder diffractometer (XRD), equipped with molybdenum

(Mo) Kα1 ($\lambda = 0.7093$ Å, 50 kV, 40 mA) X-ray source and a one-dimensional silicon strip detector Mythen 1 K (*Dectris*). The measurements were conducted in a Debye–Scherrer geometry with a 2θ range of 2°–50° and a step size of 0.015° 2θ. Crystallite sizes were determined using the Scherrer equation on the (111) reflection (13.6°–13.8° 2θ), correcting the values for instrumental broadening. Transmission electron microscopy (TEM) measurements were conducted on the samples to evaluate their particle size and distribution. The measurements were performed with a *JEOL JEM 2010* transmission electronic microscope equipped with a tungsten cathode, operated at an acceleration voltage of 120 kV. Holey carbon-coated TEM grids were used for sample mounting. Several images were collected at magnifications from 100,000 to 500,000 with a CCD camera. The software ImageJ® was used for particle size analysis in which diameter of at least 100 individual particles was measured. Further the number average (D_{average}) and the standard deviation (SD) were computed. Sauter's diameter (D_{Sauter}) (surface−volume diameter) was calculated using Eq. 1, where l_i is the number of particles having a diameter (d_i).

$$D_{\text{Sauter}} = \frac{\sum_{i=1}^{n} l_i d_i^3}{\sum_{i=1}^{n} l_i d_i^2} \tag{1}$$

Standard deviation for Sauter's diameter (SD$_{\text{Sauter}}$) was calculated as per Eq. 2, where SD is the standard deviation from TEM particle size analysis.

$$SD_{\text{Sauter}} = \left(\frac{\partial D_{\text{Sauter}}}{\partial d} \right) \cdot SD \tag{2}$$

Which is further expressed in Eq. 3

$$SD_{\text{Sauter}} = \left[3 - \frac{\left(2 \sum_{i=1}^{n} d_i^3 \right) \left(D_{\text{average}} \right)}{\left(\sum_{i=1}^{n} d_i^2 \right)^2} \right] \cdot SD \qquad (3)$$

X-ray photoelectron spectroscopy (XPS) was performed on selected samples to determine the oxidation state of Zr in the synthesized catalysts. The analysis was carried out using a *Physical Electronics PHI 5000 Versa Probe* electron spectrometer system with monochromated aluminum (Al) Kα X-ray source at 1486.60 eV operated at 25 W, 15 kV, and 1 mA anode current. To reduce any possible charging effects of X-rays, a dual-beam charge neutralization method was applied, combining both low energy ions and electrons. The samples were previously outgassed in an ultrahigh vacuum chamber at 2.5×10^{-6} Pa for 12 h. Survey scans, as well as narrow scans (high-resolution spectra) were recorded with a 100 μm X-ray diameter spot size. The X-ray was used with a take-off angle of 45° for all samples. The survey spectra were collected from 0 to 1200 eV. The narrow Zr 3d spectra were collected from 174 to 194 eV. All of the spectra were obtained under identical conditions and calibrated against a value of the C 1s binding energy of 284.5 eV [29]. Measures on

selected samples were repeated at least three times on different spots; furthermore, on one sample the measurement was repeated from the beginning on a different portion of it, to estimate the precision of the values obtained. All measurements were affected by a standard deviation of about 0.4 eV. A commercial pure monoclinic nanometric (5–25 nm) ZrO_2 (*PlasmaChem GmbH*) was used as reference for the XPS data. Multipak 9.0 software was used to obtain semi-quantitative atomic percentage compositions. The peak position and areas were evaluated using symmetrical Gaussian–Lorentzian equations (in the fraction of 70 and 30%, respectively) with a Shirley-type background.

Results and discussion

Thermogravimetric analysis

ZrOPc/KB

The TGA profiles of ZrOPc/KB in Ar and H_2/Ar are shown in Fig. 2. No difference in ZrOPc decomposition was observed upon heat treatment under inert and reductive gas conditions. From both the weight loss curves a two-step decomposition process can be identified, with the first one between 180 and 350 °C, showing an inflection point (maximum weight loss rate) at 280 °C, and the second one between 450 and 650 °C and a maximum weight loss rate at 550 °C. It can be seen that after ~ 750 °C nearly all ZrOPc has degraded, with weight loss being practically independent of the temperature.

TGA-MS of unsupported ZrOPc is shown in Fig. 3. In the first step at ~ 300 °C benzonitrile (C_6H_5CN) (m/z = 103) fragments from the skeleton start breaking. The observed typical fractionation pattern of benzonitrile (m/z = 104, 76, 63, 50 not shown) [30] allows for a definite identification of m/z = 103 with benzonitrile. When the temperature reaches ~ 500 °C, ZrOPc decomposition is still incomplete. This is confirmed by both TGA and MS signals which show rapid sample weight loss and C_6H_5CN signals, respectively. However, when the sample

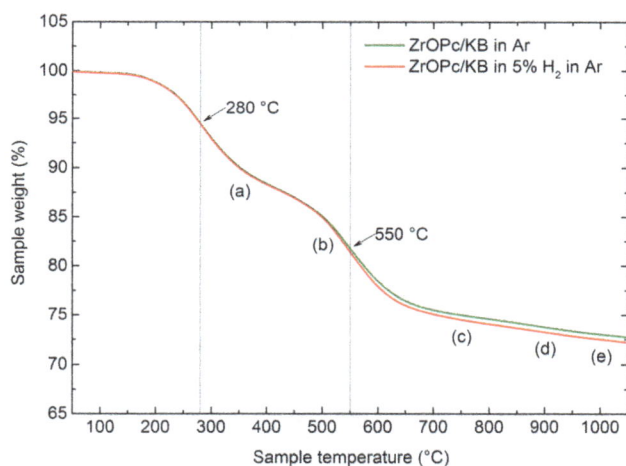

Fig. 2 Thermogravimetric analysis of ZrOPc/KB in pure Ar and 5% H_2/Ar atmospheres

Fig. 3 a Thermogravimetric and **b** mass spectrometry analysis of evolution products from unsupported ZrOPc in pure Ar atmosphere

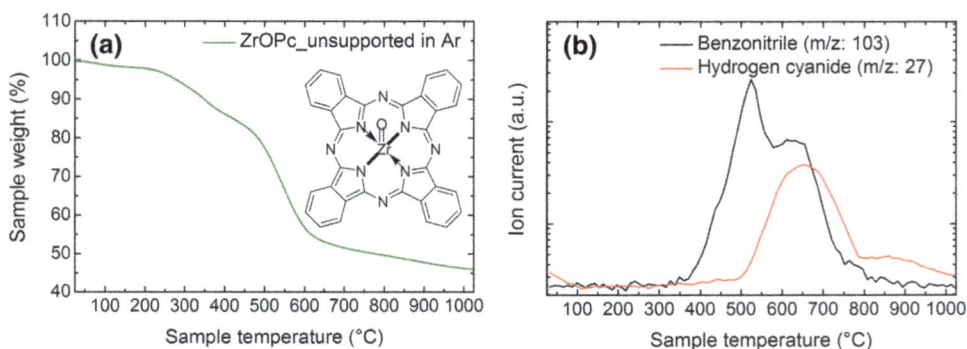

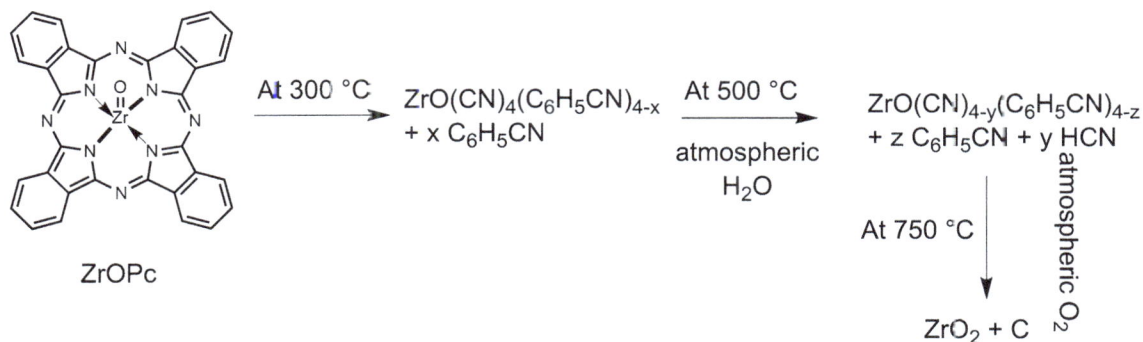

Fig. 4 Proposed mechanism for ZrOPc thermal degradation in pure Ar atmosphere

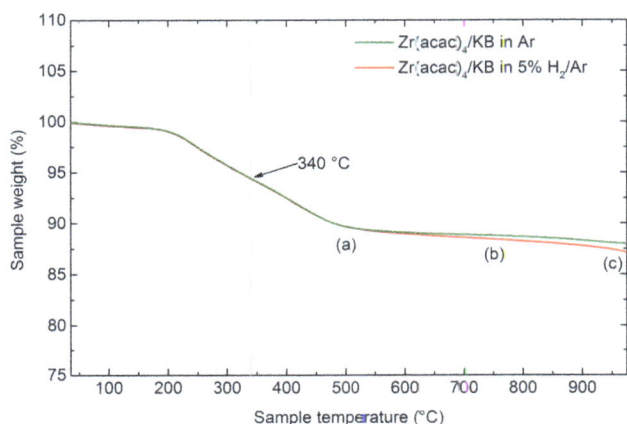

Fig. 5 Thermogravimetric analysis of $Zr(acac)_4$/KB in pure Ar and 5% H_2/Ar atmospheres

temperature reaches ~750 °C, nearly all the C_6H_5CN groups are detached from zirconium.

In addition to C_6H_5CN, hydrogen cyanide (HCN) (m/z = 27) and its fragments (m/z = 26, 12, 13 not shown) are detected between 500 and 800 °C [31], coming from the nitrile groups which chelate the zirconium atom. It can be concluded that molecular fragments in ZrOPc are stable until ~750 °C. ZrOPc thermal degradation proceeds through the proposed mechanism (Fig. 4). After 750 °C, ZrO_2 (confirmed by XRD, refer to the supplementary

information), and carbon from degraded molecular fragments of the intermediates remains. Ideally, no ZrO_2 should be formed as the TGA was done in pure Ar atmosphere, but atmospheric H_2O and O_2 through minor leaks in the instrument is the source of HCN and oxide formation.

From Fig. 2, it is clear that at 500 °C, decomposition of ZrOPc is not complete but after around 700 °C ZrOPc/KB has relatively stable weight. Five different temperatures were selected (a) 350 °C (b) 500 °C (c) 750 °C (d) 900 °C, and (e) 1000 °C to study the effect of heat-treatment temperatures on ZrOPc/KB (as marked in Fig. 2).

$Zr(acac)_4$/KB

Figure 5 shows the weight loss curve of $Zr(acac)_4$/KB upon heating in Ar and 5% H_2/Ar. One can easily see that the molecule degradation seems to follow the same path in both inert and reductive atmospheres. Only one-step degradation starting at ~200 °C was observed here; with the maximum weight loss rate at ~340 °C and a continuous weight loss until ~500 °C.

Thus, it shows less thermal stability compared to ZrOPc (Fig. 2). Unsupported $Zr(acac)_4$ was also analyzed with TGA-MS (Fig. 6). Between ~180 and ~250 °C we observe the highest and rapid weight loss ($\sim45\%$ of the initial weight, Fig. 6a, region I), thus most of the

Fig. 6 a Thermogravimetric and **b** mass spectrometry analysis of evolution products from unsupported $Zr(acac)_4$ in pure Ar atmosphere

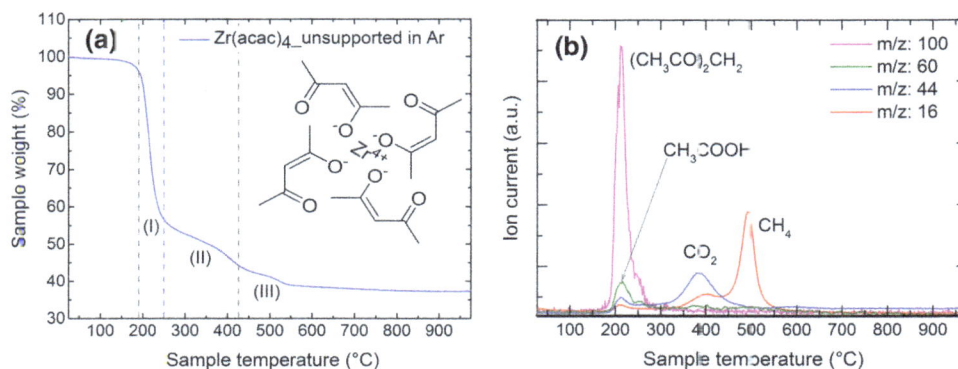

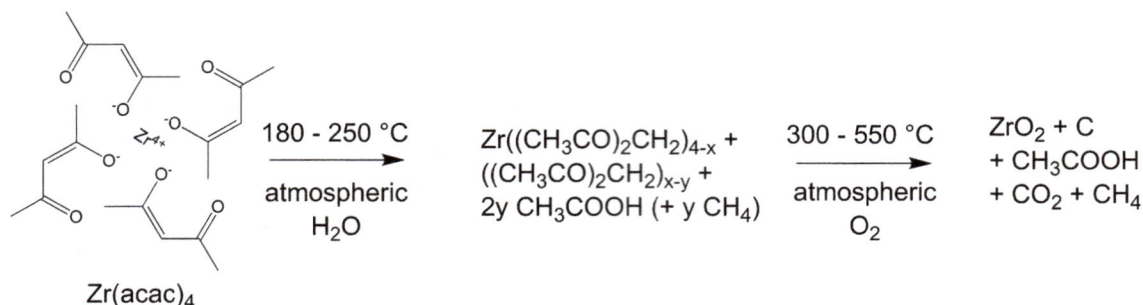

Fig. 7 Proposed mechanism for $Zr(acac)_4$ thermal degradation in pure Ar atmosphere

$Zr(acac)_4$ has degraded. From the MS-analysis we attribute this mainly to the detachment of acetylacetone $[(CH_3CO)_2CH_2]$ (m/z = 100, Fig. 6b and m/z = 85, 72, 58, 43 not shown) [32]. At the same temperature range acetic acid (CH_3COOH) is unambiguously identified by the characteristic mass signal at m/z = 60 (Fig. 6b), that in acetylacetone MS-pattern has a comparably low intensity [33].

Mass signals at m/z = 44 and 16 (Fig. 6b) are detected; they are attributed to acetylacetone and/or acetic acid (part of their standard MS-pattern). We cannot exclude the evolution of acetone (CH_3COCH_3) (m/z = 58) since it has

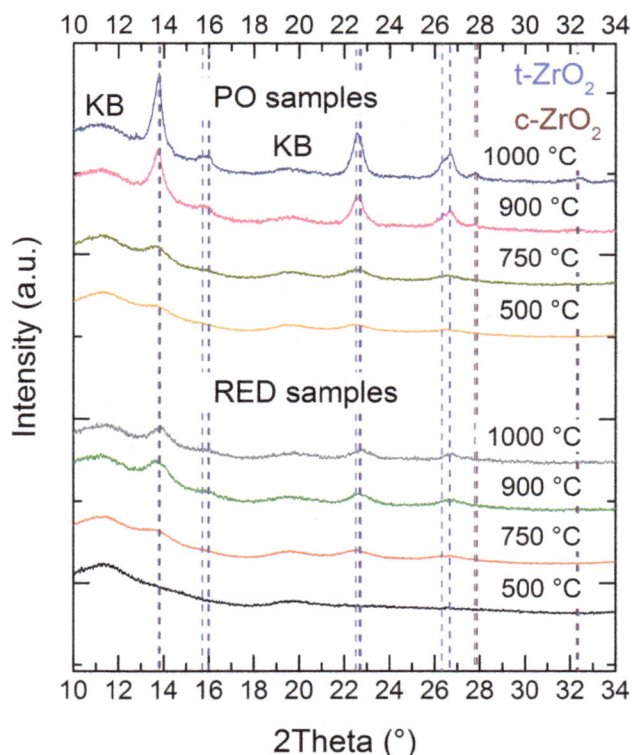

Fig. 8 X-ray diffractograms of ZrOPc/KB heat-treated at different temperatures in PO and RED gas conditions. KB is Ketjenblack carbon support with reflexes at positions similar to graphite (PDF no. 00-056-0159)

a MS-pattern that almost completely superimposes on the lower m/z mass-pattern of acetylacetone [34]. At a temperature slightly higher than ~ 250 °C the acetylacetone mass signal approaches zero (see Fig. 6b). Between 300 and 450 °C, carbon dioxide (CO_2) (m/z = 44) and acetic acid are observed as degradation products of $Zr(acac)_4$ [35] (Fig. 6a, region II). In the same temperature range, a mass signal at m/z = 16 (Fig. 6b) is detected; this is attributed to carbon dioxide and/or acetic acid as it is a part of their standard lower m/z MS-pattern. When the sample temperature reaches 450–550 °C there is a final weight loss of $\sim 4\%$ which is attributed to methane (CH_4) (m/z = 16), confirmed by its fragments (m/z = 15, 14, 13 not shown) [36] (Fig. 6a, region III). The observed molecular breakage pathway is in agreement with the reported literature [37]. $Zr(acac)_4$ degradation also yields ZrO_2 (confirmed by XRD, refer to the supplementary information) and carbon (Fig. 7).

As with ZrOPc, residual carbon in $Zr(acac)_4$ degradation is also attributed to the intermediate fragments.

At a temperature higher than 500 °C, $Zr(acac)_4$/KB attains a stable weight. Three different temperatures were selected (a) 500 °C (b) 750 °C, and (c) 950 °C to study the effect of heat-treatment conditions on the final product (Fig. 5).

Heat treatment of carbon-supported precursors

ZrOPc/KB

Due to the high thermal stability of ZrOPc (discussion from Figs. 2, 3, 4, 5, 6), we maintained 2 h of RED conditions before PO conditions. This was to ensure complete removal of the organic fragments from ZrOPc.

XRD patterns of the heat treated samples are shown in Fig. 8. For the sample heat treated at 350 °C under PO conditions no ZrO_2 formation was observed (diffractogram not shown), which was further confirmed by TEM measurements, where no particles were found. For samples heat-treated in PO gas conditions, the formation of ZrO_2 phase was observed for treatment at 500 °C and above.

With the increase in treatment temperature, full width half maximum (FWHM) of the reflections decreases, clearly indicating an increase in nanoparticle crystallinity.

For the samples heat-treated in RED conditions (Fig. 8), no diffraction pattern related to ZrO_2 was detected for samples treated at 350 (diffractogram not shown) and 500 °C. This result fits well with the TGA data of ZrOPc/KB which clearly show that at 350 and 500 °C the phthalocyanine macrocycle degradation was incomplete with some organic residues remaining which might hinder oxidation of the metal center (see mass spectrometer data in Fig. 3). For higher synthesis temperatures ($\geq$750 °C), broad ZrO_2 reflections were observed, which clearly indicated the formation of ZrO_2. In particular, for heat-treatment temperatures $\geq$900 °C, ZrO_2 reflections from RED samples show higher FWHM compared to that of PO samples prepared at the same temperature. The results suggest that, at synthesis temperatures $\geq$900 °C PO gas conditions facilitates the formation of larger crystallites.

For the heat-treated samples correct phase assignment for the synthesized ZrO_2 was not trivial.

This was because the reflections were broad and the standard tetragonal (t) and cubic (c) phase reflections were nearly overlapping. The main difference is only a shoulder reflection at 26.32° for t-ZrO_2 (marked by a red asterisk, Fig. 9). The reflections of the high temperature

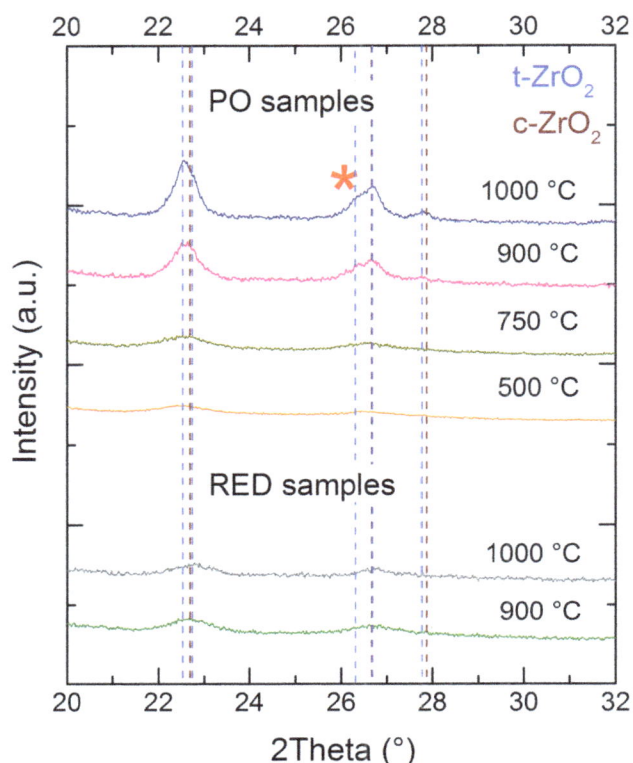

Fig. 9 ZrO_2 phase identification in heat-treated ZrOPc/KB samples

PO samples (900 and 1000 °C) fit very well with the reference data of a t-ZrO_2 phase (PDF no. 01-072-7115). The particle size is estimated to be dominantly below 20 nm based on the broadening of the reflections. No specific ZrO_2 phase was assignable for the samples synthesized at lower temperatures (500 and 750 °C) in PO conditions due to broad reflections. Also for all the samples synthesized in RED conditions no definite ZrO_2 phase is assignable. Thus, at synthesis temperatures $\geq$900 °C, PO gas conditions are important in formation of nanocrystals in which t-ZrO_2 phase can be clearly identified.

Figure 10 shows TEM micrographs of the supported ZrO_2 nanoparticles prepared from ZrOPc/KB. For the samples prepared under PO gas conditions (Fig. 10a–d, f), a gradual increase in particle size with heat treatment temperature was observed. This is further confirmed in Table 1, showing that the average size of synthesized ZrO_2 nanoparticles enlarged from 4 to 9 nm with an increase in synthesis temperature from 500 to 1000 °C. Comparing the heat-treated ZrOPc/KB sample under RED conditions at 1000 °C with samples heat-treated in PO conditions, it is observed that PO conditions clearly facilitate particle growth (Fig. 10d–f). Since Zr in phthalocyanine is chelated by N_4 of the complex, a possible reason for the faster ZrO_2 growth in PO heat treatment is the assistance of dilute oxygen in degrading the N_4 chelate. Evaluation of TEM micrographs of PO samples confirms the particle size trend from XRD analysis (Figs. 8, 10).

Zr(acac)₄/KB

From the diffractograms, ZrO_2 formation is confirmed for samples synthesized in both PO and RED conditions (Fig. 11). These results fit well with the data obtained from TGA for Zr(acac)₄/KB, which clearly show that acetylacetonate degradation is nearly complete at 500 °C (Fig. 5). This means that at this temperature the metal atom can easily get oxidized. Thus, lower thermal stability and/or higher molecular oxygen from Zr(acac)₄ makes ZrO_2 formation easier.

Further, as the heat treatment temperature is increased, an increase in the ZrO_2 crystallite size is observed. In addition, FWHM of ZrO_2 reflections from samples heat-treated at 950 °C under PO and RED conditions seem very similar.

For the samples heat-treated at 750 and 950 °C in PO and RED conditions, a t-phase is assigned to the synthesized ZrO_2 due to the presence of the shoulder reflection at 26.32° (PDF no. 01-072-7115) (marked by a violet asterisk, Fig. 12). No specific ZrO_2 phase is assignable to samples synthesized at 500 °C in PO and RED conditions due to very broad reflections. Thus, from the observations on XRD patterns, we can conclude that ZrO_2 synthesized from ZrOPc at $\geq$900 °C under

Fig. 10 TEM images of supported ZrO$_2$ nanoparticles from ZrOPc/KB heat-treated at different temperatures under PO conditions (**a–d**): **a** 500 °C; **b** 750 °C; **c** 900 °C; **d** 1000 °C; **e** under RED conditions 1000 °C and **f** particle size distribution plot of the samples shown here

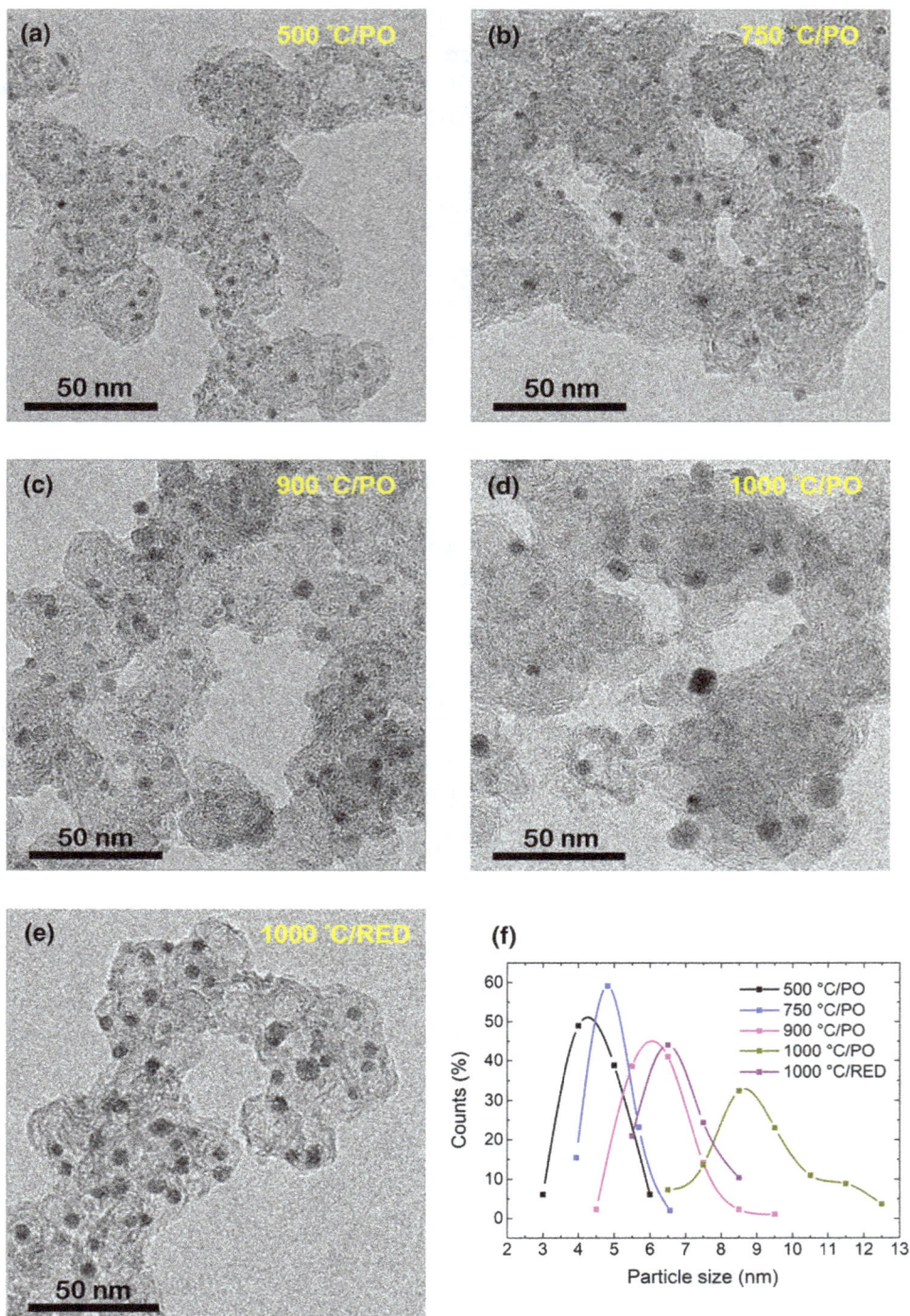

PO conditions and from Zr(acac)$_4$ at $\geq$750 °C under both PO and RED conditions is isostructural.

In Fig. 13a, b, d, an increase in the average particle size from 6.5 to $\sim$10 nm can be observed in the samples as the heat-treatment temperature is increased from 750 to 950 °C. Comparing b and c, no major difference in the particle size is seen. The similarity in particle size is further confirmed in Fig. 13d and Table 1. This clearly indicates

that heat-treatment gas conditions are not influencing particle growth, which is in contrast with ZrOPc/KB samples Fig. 10f. This difference is attributed to lower Zr(acac)$_4$ thermal stability and its higher oxygen content. Observations from TEM analysis are in congruence with the trends observed in diffractograms of Zr(acac)$_4$ samples (Fig. 11). Dispersion of ZrO$_2$ nanoparticles from Zr(acac)$_4$ and ZrOPc precursor is similar (Figs. 10, 13).

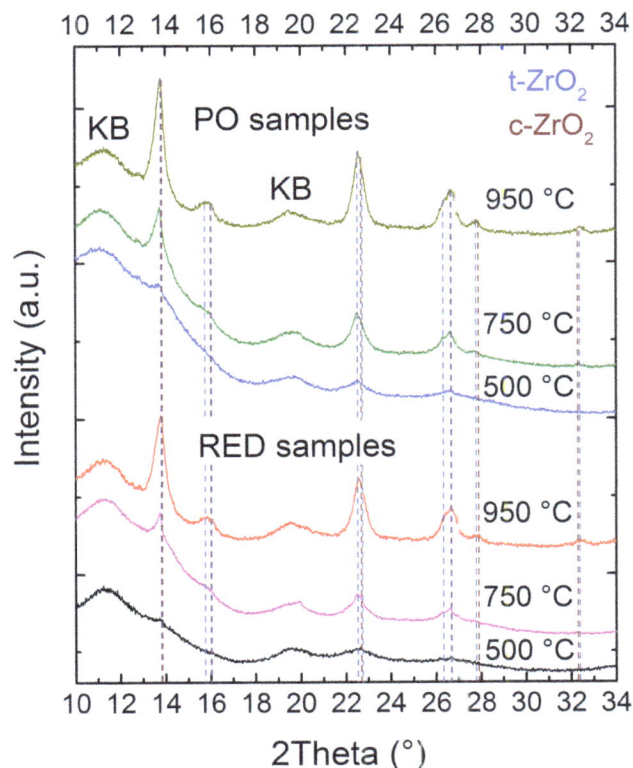

Fig. 11 X-ray diffractograms of Zr(acac)$_4$/KB heat-treated at different temperatures in PO and RED gas conditions. KB is Ketjenblack carbon support with reflexes at positions similar to graphite (PDF no. 00-056-0159)

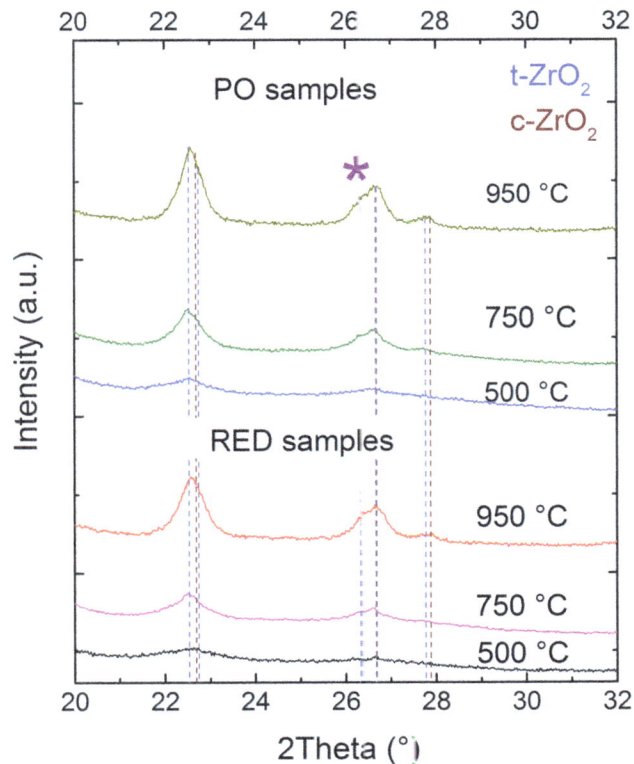

Fig. 12 ZrO$_2$ phase identification in heat-treated Zr(acac)$_4$/KB samples

ZrO$_2$ crystallite size calculations, particle size analysis from TEM and its Sauter's diameter (surface-volume diameter) have been summarized in Table 1.

From Table 1, it is evident that crystallite size calculations from XRD and particle size analysis from TEM are significantly in agreement. This infers that ZrO$_2$ nanoparticles are nanocrystals, and the amorphous phase should be negligible in the samples which are synthesized at temperatures $\geq$750 °C. There is also a good agreement between average particle size and Sauter's diameter calculations, which further confirms that all the samples have a narrow size distribution.

Variation of ZrO$_2$ loading

t-ZrO$_2$ phase formation is apparent (PDF no. 01-072-7115) in 15 and 10 wt.% ZrO$_2$/KB samples, while it is not attributable in 5 wt.% sample due to very broad reflections (Fig. 14). As the loading increases, FWHM of ZrO$_2$ reflections decreases. This infers an increase in the oxide particle size.

This increase in ZrO$_2$ particle size is confirmed in the TEM analysis (see Fig. 15), with a particle size ranging from $\sim$5 to 8.5 nm in the samples as the loading increases from 5 to 15 wt.%. These observations are complementary with the FWHM trends from XRD (Fig. 14). Based on TEM micrographs the nanoparticles are well-dispersed on the support without obvious agglomeration and the dispersion looks similar for all the three samples. An attempt to calculate the average inter-particle distance (AID) for the above samples has been performed. In the literature about carbon supported platinum catalysts, several approaches have been followed to estimate the inter-particle distance on the support surface. Meier et al. proposed one of these methods, which as a rule-of-thumb is an estimate for AID (adapted here for carbon supported zirconia nanoparticles) (Eq. 4) [38]:

$$AID = \sqrt{\frac{\pi}{3\sqrt{3}} \cdot 10^{-3} \cdot \rho_{ZrO_2} \cdot \frac{100 - L_{ZrO_2}}{L_{ZrO_2}} \cdot A_{carbon} \cdot D_{average}^3} - D_{average}$$

(4)

where $\rho_{ZrO_2} = 6.1$ g/cm^3 is the density of tetragonal zirconia, L_{ZrO_2} is the loading of ZrO$_2$ on the catalyst powder in percent, $A_{carbon} = 800$ m^2/g for Ketjenblack E-type [39], and $D_{average}$ has the same meaning as defined before.

Fig. 13 TEM images of
supported ZrO_2 nanoparticles
from $Zr(acac)_4$/KB heat-treated
at different temperatures under
PO conditions (**a**, **b**): **a** 750 °C;
b 950 °C; **c** under RED
conditions 950 °C; **d** and
particle size distribution plot of
the samples shown here

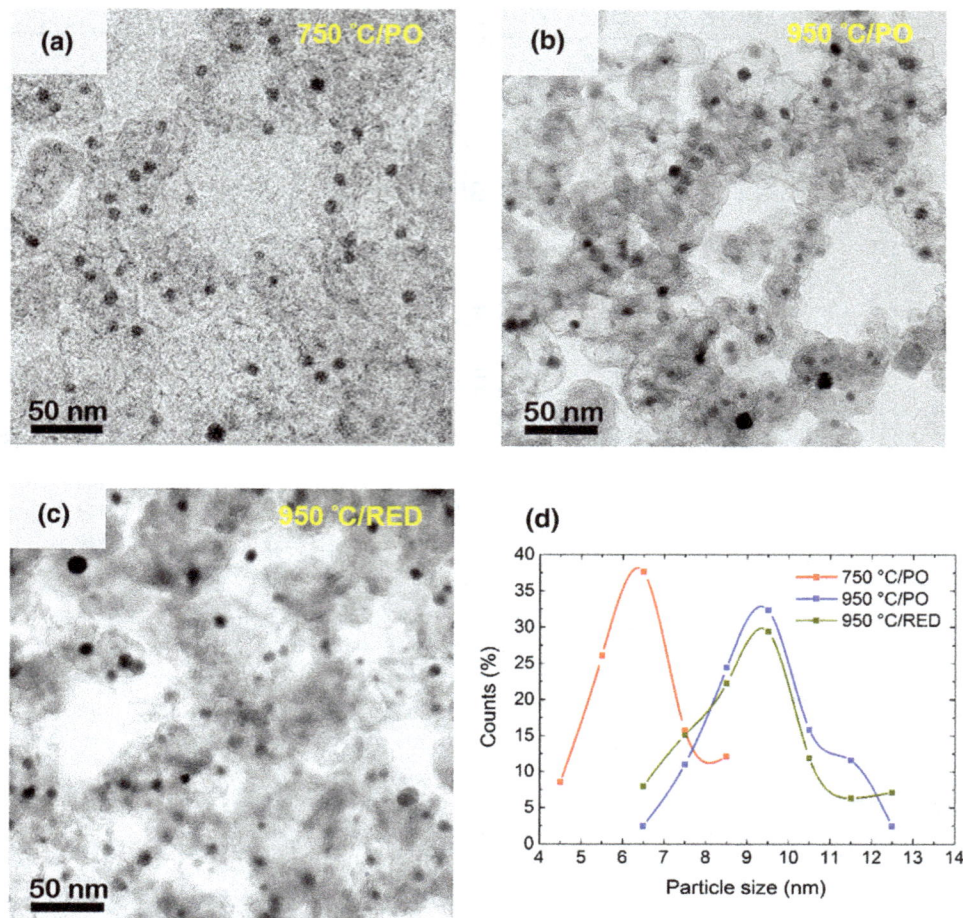

Table 1 Crystallite size based
on Scherrer equation ($k = 0.94$)
($D_{Scherrer}$), average particle size
($D_{average}$) analysis with standard
deviation (SD), and Sauter's
diameter (D_{Sauter}) with standard
deviation for Sauter's diameter
(SD_{Sauter}) from particle size
analysis of TEM images of
ZrOPc/KB and $Zr(acac)_4$/KB
samples heat-treated at different
temperatures and gas conditions

	Sample (°C)	$D_{Scherrer}$ (nm)	$D_{average} \pm$ SD (nm)	$D_{Sauter} \pm SD_{Sauter}$ (nm)
Heat-treated ZrOPc/KB				
PO	500	4	4 ± 1	5 ± 1
	750	4.5	5 ± 1	5 ± 1
	900	8	6.5 ± 1	7 ± 1
	1000	9	9 ± 2	10 ± 2
RED	500	–	–	–
	750	4	4 ± 1	4 ± 1
	900	4.5	5.5 ± 1	6 ± 1
	1000	5	7 ± 2	7 ± 2
Heat-treated $Zr(acac)_4$/KB				
PO	500	5	3.5 ± 1	3.5 ± 1
	750	7	6.5 ± 2	7 ± 2
	950	8.5	10 ± 2	10 ± 2
RED	500	–	3.5 ± 1	3.5 ± 1
	750	5.5	5 ± 1	5 ± 1
	950	8	9 ± 2	10 ± 2

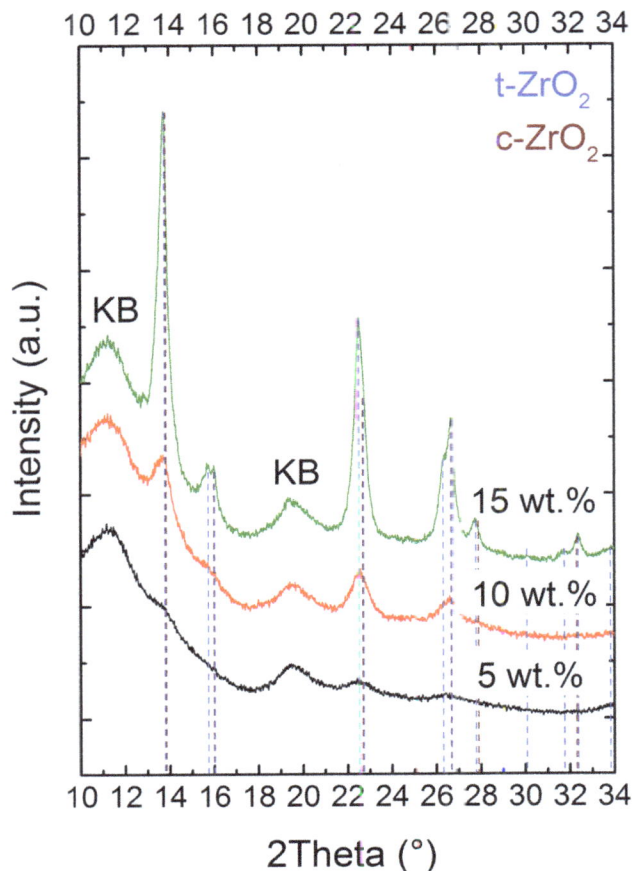

Fig. 14 X-ray diffractograms of ZrOPc/KB heat-treated at 950 °C in PO gas conditions with varied ZrO$_2$ loadings. KB is Ketjenblack carbon support with reflexes at positions similar to graphite (PDF no. 00-056-0159)

If one calculates the AID for the samples shown if Fig. 15, the values range from ≈80–95 nm (Table 2). Thus, although the loadings differ by a maximum factor of three, the apparent particle density on the support surface remains only weakly influenced.

From Table 2 we can also conclude in this case that the synthesized nanoparticles are nanocrystals which grow in size as ZrO$_2$ loading increases. Thus, metal loading could be used to tune the size of the supported nanocrystals.

X-ray photoelectron spectroscopy

X-ray photoelectron spectroscopy (XPS) is a sensitive tool for analyzing the chemical state of Zr cations in ZrO$_2$. Figure 16 shows the high-resolution Zr 3d spectra of samples prepared in various conditions. All of the spectra show the typical doublet structure in 3d$_{5/2}$ and 3d$_{3/2}$

components, due to the spin–orbit splitting of the 3d level. The peaks resulted ranging from 182.3 to 183.4 eV for Zr 3d$_{5/2}$, and from 184.7 to 185.7 eV for Zr 3d$_{3/2}$, as shown in shaded bands in Fig. 16. In particular, sample ZrOPc/KB 900 °C/PO (which is one of the most interesting catalyst on the electrochemical point of view [26]) shows the measured spectra with Gaussian–Lorentzian fits and a Shirley type background. The deconvolution of the Zr 3d spin–orbit doublet is in agreement with the existence of Zr^{4+} [40, 41]. Similar analyses has been performed for other samples. Zr 3d$_{5/2}$ values are extracted by taking the peak positions after data-fitting and reported in Fig. 17, together with the standard deviation (±0.4 eV).

After comparing the binding energy of Zr^{4+} of pure monoclinic ZrO$_2$ (internal reference: 181.9 eV) with those reported in literature 181.9–182.1 eV [41–44] (green shaded band in Fig. 17), we could confirm that all the measurements were in agreement with reference values. The measured binding energies of Zr 3d$_{5/2}$ in all the samples lie inside or close to the range of Zr^{4+} species in pure tetragonal ZrO$_2$ 182.1–182.8 eV [40, 41, 44–47] (yellow shaded band in Fig. 17).

Thus, the formation of ZrO$_2$ is confirmed by XPS for all the examined synthesis routes. Considering the trends of the binding energy of the Zr 3d$_{5/2}$ peak with the synthesis conditions (Fig. 17) and the standard deviation linked with the XPS measurements, it is not possible to obtain a clear trend between Zr 3d$_{5/2}$ with temperature, heat treatment gas conditions, and oxide particle size. In addition, it is impossible to link the shift of the binding energy with the presence of oxygen vacancies. The presence of suboxides must be excluded since in XRD we do not detect them and similar average ZrO$_2$ particle size from XRD and TEM demonstrate the only presence of pure ZrO$_2$ (Figs. 8, 11; Table 1).

Conclusions

Successful synthesis of pure carbon supported ZrO$_2$ nanoparticles has been reported in this paper. A thorough comparative study on the synthesis of metal-oxide nanoparticles from two different precursors namely, ZrOPc and Zr(acac)$_4$ has been done. Our aim is to optimize the nanoparticle size and crystallinity of the samples for the possible application as electrocatalysts for the oxygen reduction reaction in PEMFCs. Using thermogravimetric analysis coupled with mass spectrometry, we could show and confirm the thermal-decomposition behavior of the

Fig. 15 TEM images of carbon supported ZrO$_2$ nanoparticles from ZrOPc/KB with different loadings: **a** 5 wt.%; **b** 10 wt.%; **c** 15 wt.% and **d** particle size distribution plot of the samples shown here

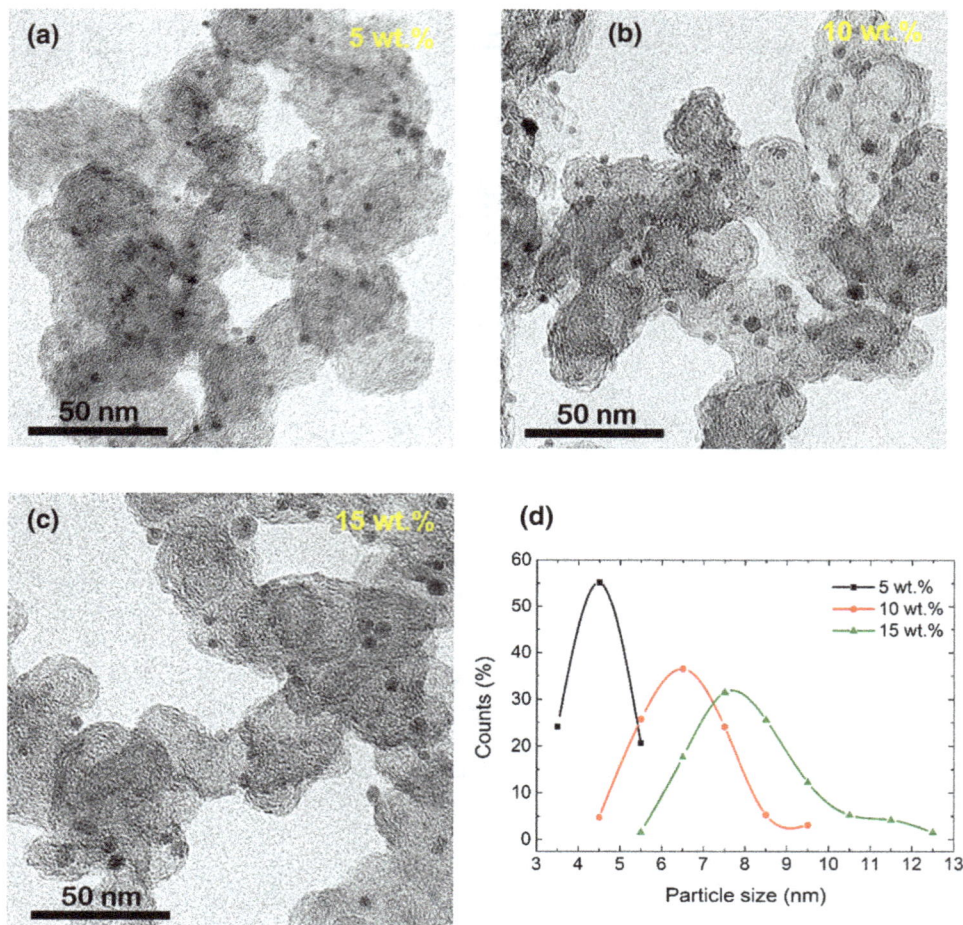

Table 2 Crystallite size based on Scherrer equation ($k = 0.94$) ($D_{Scherrer}$), average particle size ($D_{average}$) analysis with standard deviation (SD), and Sauter's diameter (D_{Sauter}) with standard deviation for Sauter's diameter (SD$_{Sauter}$) from particle size analysis of TEM images of ZrOPc/KB samples with varied loadings, but heat-treated at same conditions. Average inter-particle distance (AID) with standard deviation (SD$_{AID}$) from Eq. 4

Sample (wt.%)	$D_{Scherrer}$ (nm)	$D_{average} \pm$ SD (nm)	$D_{Sauter} \pm$ SD$_{Sauter}$ (nm)	AID $\pm$ SD$_{AID}$ (nm)
5	4.5	5 ± 1	5 ± 1	79 ± 25
10	5.5	7 ± 1	7 ± 1	88 ± 20
15	8.5	8.5 ± 2	9 ± 2	93 ± 36

chosen precursors. A clear correlation between the results from thermal analysis of precursors and the size- and structure of the nanoparticles obtained after heat-treatment at different temperatures was clearly seen. We showed that ZrO$_2$ formation happens at a lower temperature with Zr(acac)$_4$ than with ZrOPc, due to the lower thermal stability of acetylacetonate precursor and a higher content of oxygen in comparison to phthalocyanine. With ZrOPc at heat-treatment temperatures ≥900 °C, PO conditions facilitate crystallite growth and formation of distinct t-ZrO$_2$, but with Zr(acac)$_4$ a distinct t-ZrO$_2$ phase

formation is observed already at temperatures ≥750 °C in both PO and RED conditions, due to the presence of a stoichiometrical excess of oxygen already in the precursor. The oxide nanoparticles in all the samples are well-distributed on the carbon support without evident agglomeration. After the size- and structural- analysis of the oxide nanoparticles, it is concluded that the oxide nanoparticles are nanocrystals and the amorphous phase is negligible in samples heat-treated at temperatures ≥750 °C. From the loading variation of zirconium, we show that metal loading can also be used to tune the size of oxide nanocrystals.

Fig. 16 High resolution XPS spectra of Zr 3d core level of carbon supported ZrOPc and $Zr(acac)_4$ heat-treated at different temperatures in PO and RED gas conditions. The shaded bands highlight the binding energies variability of the samples

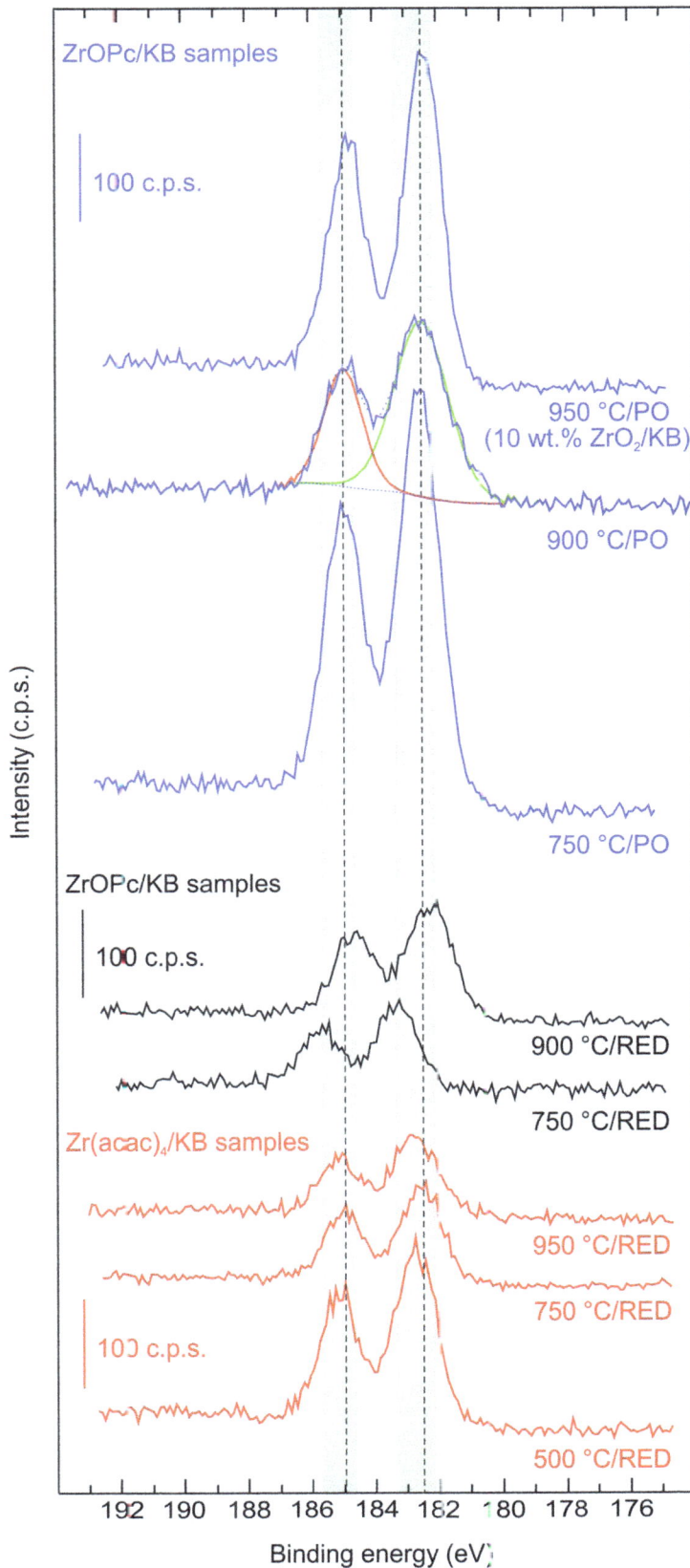

Fig. 17 Binding-energy shift of the Zr 3d$_{5/2}$ peak of heat-treated samples as a function of the synthesis temperature and gas conditions, together with the measured value of the commercial m-ZrO$_2$. The ranges of binding energies from the literature for m-ZrO$_2$ [41–44] and t-ZrO$_2$ [40, 41, 44–47] are depicted by the shaded green and yellow bands, respectively

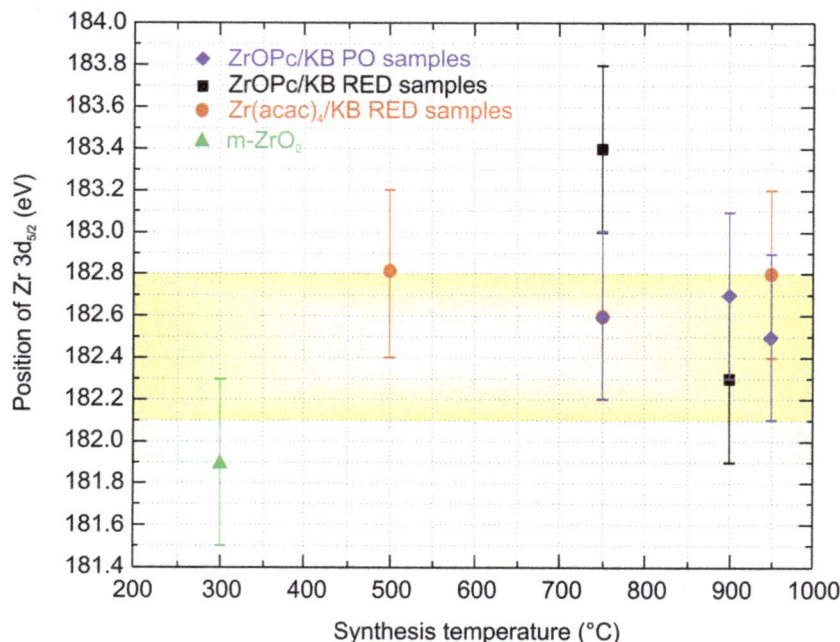

From XPS analysis, it is clear that Zr species in samples from different synthesis routes are in the pure ZrO$_2$ state. No clear trend between Zr 3d$_{5/2}$ binding energy and synthesis temperature or gas conditions was found.

Acknowledgements For these results we acknowledge for funding the Fuel Cells and Hydrogen Joint Undertaking under Grant Agreement Duramet no 278054 as part of the Seventh Framework Programme of the European Community for research, technological development and demonstration activities (FP7/2007–2013). The authors are thankful to Dr. Viktor Ya. Chernii from Institute of General and Inorganic Chemistry (Ukraine) for his scholarly support on ZrOPc synthesis and purification. In addition, we thank Dr. Marianne Hanzlik for the superb TEM images. Sincere thanks to Dr. Hans Beyer and Michael Metzger for their assistance in TGA and MS, respectively. Dr. Salvatore Guastella is greatly acknowledged for XPS measurements.

References

1. Earnshaw, A., Greenwood, N.: Chemistry of the elements, 2nd edn. Butterworth-Heinemann, Oxford (1998)
2. Shukla, S., Seal, S.: Mechanisms of room temperature metastable tetragonal phase stabilisation in zirconia. Int. Mater. Rev. **50**, 45–64 (2005)
3. French, R., Glass, S., Ohuchi, F., Xu, Y., Ching, W.: Experimental and theoretical determination of the electronic structure and optical properties of three phases of ZrO2. Phys. Rev. B **49**, 5133–5142 (1994)
4. Minh, N.: Ceramic fuel cells. J. Am. Ceram. Soc. **76**, 563–588 (1993)
5. Schulz, U., Leyens, C., Fritscher, K., Peters, M., Saruhan-Brings, B., Lavigne, O., Dorvaux, J.-M., Poulain, M., Mévrel, R., Caliez, M.: Some recent trends in research and technology of advanced thermal barrier coatings. Aerosp. Sci. Technol. **7**, 73–80 (2003)
6. Kosmač T, Oblak Č, Jevnikar P, Funduk N, Marion L.: Strength and reliability of surface treated Y-TZP dental ceramics. J. Biomed. Mater. Res. **53**, 304–313 (2000)
7. Guo, D., Qiu, X., Zhu, W., Chen, L.: Synthesis of sulfated ZrO2/MWCNT composites as new supports of Pt catalysts for direct methanol fuel cell application. Appl. Catal. B **89**, 597–601 (2009)
8. Lv, H., Cheng, N., Peng, T., Pan, M., Mu, S.: High stability platinum electrocatalysts with zirconia–carbon hybrid supports. J. Mater. Chem. **22**, 1135–1141 (2012)
9. Qu, W., Wang, Z., Sui, X., Gu, D., Yin, G.: ZrC-C and ZrO2-C as novel supports of Pd catalysts for formic acid electrooxidation. Fuel Cells **13**, 149–157 (2013)
10. Malolepszy, A., Mazurkiewicz, M., Stobinski, L., Lesiak, B., Kövér, L., Tóth, J., Mierzwa, B., Borodzinski, A., Nitze, F., Wågberg, T.: Deactivation resistant Pd–ZrO2 supported on multiwall carbon nanotubes catalyst for direct formic acid fuel cells. Int. J. Hydrog. Energy **40**, 16724–16733 (2015)
11. Rutkowska, I., Kulesza, P.: Electrocatalytic oxidation of ethanol in acid medium: enhancement of activity of vulcan-supported platinum-based nanoparticles upon immobilization within nanostructured zirconia matrices. Funct. Mater. Lett. **7**, 1440005 (2014)
12. Cheng-Lan, L., Yu-Chi, Y.: Platinum nanoparticles supported on zirconia–carbon black nanocomposites for methanol oxidation reaction. Res. Chem. Intermed. **40**, 2207–2215 (2014)
13. Wang, R., Wang, K., Wang, H., Wang, Q., Key, J., Linkov, V., Ji, S.: Nitrogen-doped carbon coated ZrO2 as a support for Pt nanoparticles in the oxygen reduction reaction. Int. J. Hydrog. Energy **38**, 5783–5788 (2013)
14. Seo, J., Cha, D., Takanabe, K., Kubota, J., Domen, K.: Electrodeposited ultrafine NbOx ZrOx and TaOx nanoparticles on carbon black supports for oxygen reduction electrocatalysts in acidic media. ACS Catal. **3**, 2181–2189 (2013)
15. Sebastián, D., Baglio, V., Sun, S., Tavares, A., Aricò, A.: Facile synthesis of Zr- and Ta-based catalysts for the oxygen reduction reaction. Chin. J. Catal. **36**, 484–489 (2015)
16. Liu, Y., Ishihara, A., Mitsushima, S., Kamiya, N., Ota, K.: Zirconium oxide for PEFC cathodes. Electrochem. Solid-State Lett. **8**, A400–A402 (2005)

17. Ohgi, Y., Ishihara, A., Matsuzawa, K., Mitsushima, S., Ota, K.: Zirconium oxide-based compound as new cathode without platinum group metals for PEFC. J. Electrochem. Soc. **157**, B885–B891 (2010)
18. Shao, M.: Electrocatalysis in fuel cells: A non- and low- platinum approach. Springer, London (2013)
19. Yin, S., Ishihara, A., Kohno, Y., Matsuzawa, K., Mitsushima, S., Ota, K.: Preparation of highly active Zr Oxide-based oxygen reduction electrocatalysts as PEFC cathode. ECS Trans. **50**, 1785–1790 (2013)
20. Yin, S., Ishihara, A., Kohno, Y., Matsuzawa, K., Mitsushima, S., Ota, K.: Enhancement of oxygen reduction activity of zirconium oxide-based cathode for PEFC. ECS Trans. **58**, 1489–1494 (2013)
21. Okada, Y., Ishihara, A., Arao, M., Matsumoto, M., Imai, H., Kohno, Y., Matsuzawa, K., Mitsushima, S., Ota, K.: Improvement of the electrocatalytic activity of zirconium oxide-based catalyst for ORR. ECS Trans. **64**, 231–238 (2014)
22. Mecheri, B., Iannaci, A., D'Epifanio, A., Mauri, A., Licoccia, S.: Carbon-supported zirconium oxide as a cathode for microbial fuel cell applications. ChemPlusChem **81**, 80–85 (2016)
23. Lee, S., Zhang, Z., Wang, X., Pfefferle, L., Haller, G.: Characterization of multi-walled carbon nanotubes catalyst supports by point of zero charge. Catal. Today **164**, 68–73 (2011)
24. Leznoff, C., Lever, A.: Phthalocyanines: properties and applications. Wiley, New York (1989)
25. Stary, J., Liljenzin, J.: Critical evaluation of equilibrium constants involving acetylacetone and its metal chelates. Pure Appl. Chem. **54**, 2557–2592 (1982)
26. Mittermeier, T., Madkikar, P., Wang, X., Gasteiger, H., Piana, M.: ZrO2 based oxygen reduction catalysts for PEMFCs: towards a better understanding. J. Electrochem. Soc. **163**, F1543–F1552 (2016)
27. Tomachynski, L., Chernii, V., Volkov, S.: Synthesis of dichloro phthalocyaninato complexes of titanium, zirconium, and hafnium. Russ. J. Inorg. Chem. **47**, 208–211 (2002)
28. Catanorchi S., Piana M.: High performance ORR (oxygen reduction reaction) pgm (pt group metal) free catalyst. Google Patents. http://www.google.com.ar/patents/US20110034325 (2011). Accessed 17 April 2015
29. Liu, G., Li, X., Ganesan, P., Popov, B.: Studies of oxygen reduction reaction active sites and stability of nitrogen-modified carbon composite catalysts for PEM fuel cells. Electrochim. Acta **55**, 2853–2858 (2010)
30. National Institute of Standards and Technology. Benzonitrile. http://webbook.nist.gov/cgi/cbook.cgi?ID=C100470&Units=SI&Mask=200#Mass-Spec. Accessed 03 Mar 2015
31. Cornu, A., Massot, R.: Compilation of mass spectral data/Index de Spectres de Masse. Heyden & Sons, London (1966)
32. National Institute of Standards and Technology. Acetylacetone. http://webbook.nist.gov/cgi/cbook.cgi?ID=C123546&Mask=200#Mass-Spec. Accessed 17 April 2015
33. National Institute of Standards and Technology. Acetic acid. http://webbook.nist.gov/cgi/cbook.cgi?ID=C64197&Mask=200#Mass-Spec. Accessed 17 April 2015
34. National Institute of Standards and Technology. Acetone. http://webbook.nist.gov/cgi/cbook.cgi?ID=67-64-1&Units=SI&cMS=on. Accessed 17 April 2015
35. National Institute of Standards and Technology. Carbon dioxide. http://webbook.nist.gov/cgi/cbook.cgi?ID=C124389&Units=SI&Mask=200#Mass-Spec. Accessed 17 April 2015
36. National Institute of Standards and Technology. Methane. http://webbook.nist.gov/cgi/cbook.cgi?Formula=CH4&NoIon=on&Units=SI&cMS=on. Accessed 17 April 2015
37. Jasim, F.: Simultaneous thermal analysis of zirconium(IV) acetylacetonate in a helium atmosphere. J. Therm. Anal. **37**, 149–153 (1991)
38. Meier, J., Galeano, C., Katsounaros, I., Witte, J., Bongard, H., Topalov, A., Baldizzone, C., Mezzavilla, S., Schüth, F., Mayrhofer, K.: Design criteria for stable Pt/C fuel cell catalysts. Beilstein J. Nanotechnol. **5**, 44–67 (2014)
39. Meini, S., Piana, M., Beyer, H., Schwämmlein, J., Gasteiger, H.: Effect of carbon surface area on first discharge capacity of Li-O2 cathodes and cycle-life behavior in ether-based electrolytes. J. Electrochem. Soc. **159**, A2135–A2142 (2012)
40. Ram, S., Mondal, A.: X-ray photoelectron spectroscopic studies of Al3 + stabilized t-ZrO2 of nanoparticles. Appl. Surf. Sci. **221**, 237–247 (2004)
41. Basahel, S., Ali, T., Mokhtar, M., Narasimharao, K.: Influence of crystal structure of nanosized ZrO2 on photocatalytic degradation of methyl orange. Nanoscale Res. Lett. **10**, 1–13 (2015)
42. Majumdar, D., Chatterjee, D.: X-ray photoelectron spectroscopic studies on yttria, zirconia, and yttria-stabilized zirconia. J. Appl. Phys. **70**, 988–992 (1991)
43. Guittet, M., Crocombette, J., Gautier-Soyer, M.: Bonding and XPS chemical shifts in ZrSiO4 versus SiO2 and ZrO2: charge transfer and electrostatic effects. Phys. Rev. B **63**, 125117 (2001)
44. Kuratani, K., Uemura, M., Mizuhata, M., Kajinami, A., Deki, S.: Novel fabrication of high-quality ZrO2 ceramic thin films from aqueous solution. J. Am. Ceram. Soc. **88**, 2923–2927 (2005)
45. Ardizzone, S., Bianchi, C.: XPS characterization of sulphated zirconia catalysts: the role of iron. Surf. Interface Anal. **30**, 77–80 (2000)
46. Alvarez, M., López, T., Odriozola, J., Centeno, M., Domínguez, M., Montes, M., Quintana, P., Aguilar, D., González, R.: 2,4-Dichlorophenoxyacetic acid (2,4-D) photodegradation using an Mn +/ZrO2 photocatalyst: XPS, UV–vis, XRD characterization. Appl. Catal. B **73**, 34–41 (2007)
47. Brenier, R., Mugnier, J., Mirica, E.: XPS study of amorphous zirconium oxide films prepared by sol–gel. Appl. Surf. Sci. **143**, 85–91 (1999)

5

Antibiofilm activity of biogenic copper and zinc oxide nanoparticles-antimicrobials collegiate against multiple drug resistant bacteria: a nanoscale approach

C. Ashajyothi[1] · K. Handral Harish[2] · Nileshkumar Dubey[2] · R. Kelmani Chandrakanth[1]

Abstract The synthesis of biogenic nanoparticles from non-chemical resources has increased the drive toward understanding infection biology. Accordingly, we aimed to address the symbiotic antibiofilm effect of biogenic copper and zinc oxide nanoparticles with antimicrobials against multidrug resistant (MDR) pathogens. The minimum inhibitory concentration (MIC) of copper nanoparticles (CuNPs) and zinc oxide nanoparticles (ZnONPs) at the range from 2 to 128 µg/ml was calculated against Gram-positive and Gram-negative pathogenic bacteria using a broth dilution method. Both nanoparticles have prime antibacterial activity compared with standard antibiotics (excluding against *P.aeruginosa* MTCC 741). A qualitative assessment of biofilm formation and collegial effect was performed using a modified test tube and the microtiter plate-based method by measuring the optical density and time kill of nanoparticles. The results demonstrated efficient antibiofilm activity of CuNPs in its lowest concentration than ZnONPs and antibiotics itself. In addition, significant enhancing antibiofilm effect was also shown by CuNPs in the presence of third generation antibiotics against Gram-negative and Gram-positive bacteria. A scanning electron microscopy (SEM) analysis was used to investigate the effect of the nanoparticles on morphological changes of *Staphylococcus aureus*. Current data highlights,

biogenic CuNPs and ZnONPs could be used as an adjuvant for antibiotics in the treatment of bacterial infections.

Keywords Biogenic nanoparticles · Test tube method · Microtiter plate method · Scanning electron microscopy · Antibiofilm activity

Introduction

Biofilms are complex communities of microorganisms that show resistance to the action of antibiotics and the human immune system, due to their resistant nature and stability [1, 2]. Biofilm infections are difficult to eradicate, especially in the case of multidrug resistant pathogens [3]. Recently, the number of infections associated with budding antibiotic resistant bacteria has increased [4]. Remarkably, bacteria–host interactions could raise the rate of infections in which pathogens rapidly kill the host. Both Gram-positive and Gram-negative bacteria can form a biofilm on medical devices, such as catheters, mechanical heart valves, and prosthetic joints [5]. Biofilm-related diseases are typically persistent infections characterized by slow development, and these diseases have an ability to resist both a host's immune system and a transient response to antimicrobial therapy [6].

Staphylococcus aureus, *Staphylococcus epidermidis*, *Escherichia coli*, *Klebsiella pneumonia*, and *Pseudomonas aeruginosa* are the most common biofilm forming bacteria causing human disease, such as infection lesion in endocarditis, cystic fibrosis, and otitis media with effusion [7]. Biofilms have also been identified in most indwelling medical device infections and in biliary tract infections, periodontitis, and ophthalmic infections [8].

✉ R. Kelmani Chandrakanth
ckelmani@gmail.com

[1] Department of Biotechnology, Gulbarga University, Gulbarga 585106, Karnataka, India

[2] Oral Sciences Disciplines, Faculty of Dentistry, National University of Singapore, Singapore 117510, Singapore

Nowadays, treatment of biofilms with antibiotics has been shown to be ineffective, since many agents fail to reach the target cells embedded deep inside the biofilm matrix. An alternative approach is needed to control the diseases involving biofilms [9].

Various methods, such as use of bacteriophages and designing of semi-synthetic analogs of natural products to prevent bacterial biofilms, were considered to address the ineffectiveness of antibiotics. In contrast to above-revealed conventional methods, the nanotechnology-based approach is one such efficient approach to combat biofilm formation. The use of nanoparticles has been considered as a feasible solution to stop infectious diseases due to their antimicrobial properties. There are numerous reports explaining the multifaceted potential of nanoparticles as antimicrobial agents. The ability of metals to target multiple sites in an organism makes them superior to conventional antibiotics [10].

Currently, nanoparticles are considered as active and safe drugs to boost the antibacterial activity of conventional antibiotics, through exploit these new antimicrobials effectively with important antibiotics in synergistic combination therapy against pathogenic microorganism [11]. The combination of nanoparticles with existing antibiotics seems to be very enthralling option by combining the two treatment modalities. Recent studies have revealed that the combining nanoparticle with antibiotics not only reduces the toxicity of both agents toward human cells by decreasing the requirement for high dosages but also enhances and restores their bactericidal properties [12].

Among the metal nanoparticles, CuNPs are superconductive, easily available, and cost-effective metal well known for its variety of applications. CuNPs are also considered as an effective nanoparticle against plant and animal pathogens [13]. Most pathogens, including strains of *Clostridium difficile*, *Salmonella enterica*, *Campylobacter jejuni*, *Escherichia coli* 0157:H7, *Pseudomonas aeruginosa*, *Enterobacter aerogenes*, *Staphylococcus aureus*, methicillin-resistant *S. aureus* (MRSA), and vancomycin-resistant *Enterococcus* (VRE), are killed when exposed to the surfaces of copper and copper alloys have been reported by Wilks et al. 2005 and Casey et al. 2010 in their study [14, 15].

According to Borkow et al. (2010) to prevent the bacterial contamination on medical devices, CuNPs have also been used as antimicrobial coating agents [16]. Inspite of their bioactivity, the antibiofilm potential of CuNPs is rarely explored [17]. Only a few reports are published in the last year on, the use of copper-containing nanoparticles in the treatment of biofilms [9, 18].

According to Begum et al. (2009) and Guy Applerot et al. (2012), using the inorganic metal oxide nanoparticles, such as TiO_2, ZnO, MgO, and CaO as an antibacterial agent, has a major advantages due to its stability under harsh process conditions but also generally regarded as safe materials to human beings and animals compared to organic materials such as conventional antibiotics [19, 20]. Studies on antibacterial activity among the various metal oxides nanoparticles, zinc oxide nanoparticles, have been found to be highly toxic. Many studies have shown that selective toxic nature of ZnONPs toward bacteria shows the minimal effect on human cells, which is suggested their potential uses in agricultural, food industries, diagnostics, surgical devices, and nanomedicine-based antimicrobial agents [21–24]. Among the several metal oxide nanoparticles, ZnONPs are emerged as booming nanoparticle due to their attractive characteristics and ideal properties in various biomedical applications.

The synergistic activity of ZnONPs with more than 25 different antibiotics against *S. aureus and E. coli.* concludes that ZnONPs can enhance antibacterial activities of penicillins, cephalosporins, aminoglycosides, glycopeptides, macrolides, lincosamides, gentamicin, clarithromycin, ofloxacin, and ceftriaxone and tetracycline [25, 26].

Our objective in this investigation is mainly focused on use of biologically synthesized CuNPs and ZnONPs to probe the antibiofilm activity of antimicrobials against the panel of Gram-negative and Gram-positive human pathogens.

Materials and method

Biogenic metal nanoparticles

Biogenic nanoparticles: copper (CuNPs) and zinc oxide nanoparticles (ZnONPs) were biologically synthesized from non-pathogenic *Enterococcus faecalis* by extracellular enzymatic method. Organism was obtained from Medical Biotechnology and Phage Therapy Laboratory (MBPT), Department of Biotechnology, Gulbarga University, Gulbarga.

Enterococcus faecalis culture was inoculated in sterile Luria–Bertani broth (HiMedia, Mumbai, India) and incubated at 37 °C for 72 h. Culture was centrifuged at 10,000 rpm for 10–15 min to separate supernatant from pellet. The bacterial supernatant was added separately to the reaction vessels' containing 100 mM (v/v) copper sulfate ($CuSO_4$, HiMedia, Mumbai, India) and zinc sulfate

(ZnSO$_4$. 7H$_2$O, HiMedia, Mumbai, India), and controls (only with bacterial supernatant) are maintained separately to each reaction. The reaction was carried out for 24 h, at 37^0 C, pH: 7 in rotary shaker at 120 rpm in dark condition [27, 28]. Furthermore, morphology and crystalline nature of nanoparticles were confirmed and characterized through TEM (Tecnai 20 G2, CSIR-CECRI, Karaikudi, India) and XRD (PW3040/60 X'pert PRO, CSIR-CECRI, Karaikudi, India) analysis.

Collection of bacterial strains

Clinical isolates, *E. coli* 03, *K. Pneumonia* 125, methicillin-resistant *S. aureus* 20 (MRSA) were taken from stock cultures of the Medical Biotechnology and Phage Therapy Laboratory (MBPT), Department of Biotechnology, Gulbarga University, Gulbarga. Standard MTCC cultures, *E. coli* MTCC 9537, *K. pneumonia* MTCC 109, *S. aureus* MTCC 96, *P. Aeruginosa* MTCC 741, *S. flexneri* MTCC 1457, and *E. faecalis* NCIM 5025 (Microbial Type Culture Collection and Gene Bank, Chandigarh, India) were stored in Luria–Bertani broth cultures with sterile glycerol at -20 °C (20 %, v/v) for further studies.

Minimum inhibitory concentration (MIC) of CuNPs, ZnONPs, and antibiotics

Minimal inhibitory concentration (MIC) was defined as the lowest concentration of an antimicrobial agent that is needed to inhibit the growth of a microorganism after 24 h of incubation. The CLSI 2012 M100-S22 practice was implemented to determine the MIC and to assess the

Table 1 Concentrations of CuNPs, ZnONPs, and antibiotics used against different pathogenic bacteria

Pathogenic bacteria	CuNPs (μg/ml)	ZnONPs (μg/ml)	Antibiotics (μg/ml)	
E. coli 03	10	10	Ceftriaxone	12
E.coli MTCC 9537	12	16	Ceftriaxone	12
K.pneumonia 125	18	08	Ceftazidime	14
K.pneumonia MTCC 109	10	12	Ceftazidime	14
S. aureus 20	16	16	Gentamicin	10
S.aureus MTCC 96	20	18	Gentamicin	10
E.faecalis NCIM 5025	20	18	Gentamicin	10
P.aeruginosa MTCC 741	64	64	Ceftazidime	14
S.flexneri MTCC 1457	12	10	Ceftazidime	14

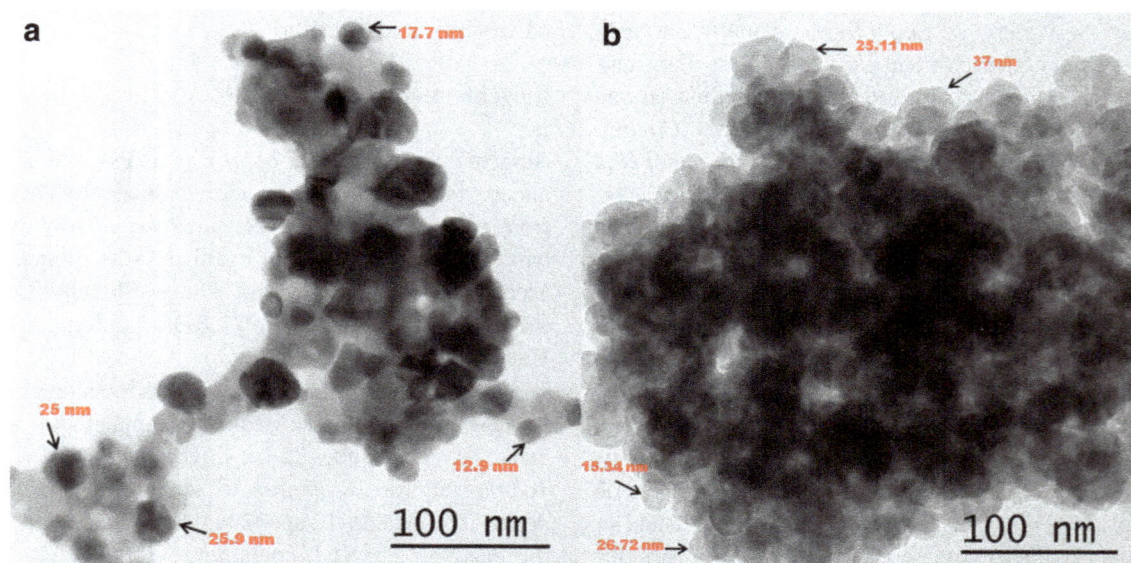

Fig. 1 TEM images of biogenic CuNPs (**a**) and ZnONPs (**b**) synthesized from Gram-positive non-pathogenic bacterium *Enterococcus faecalis*

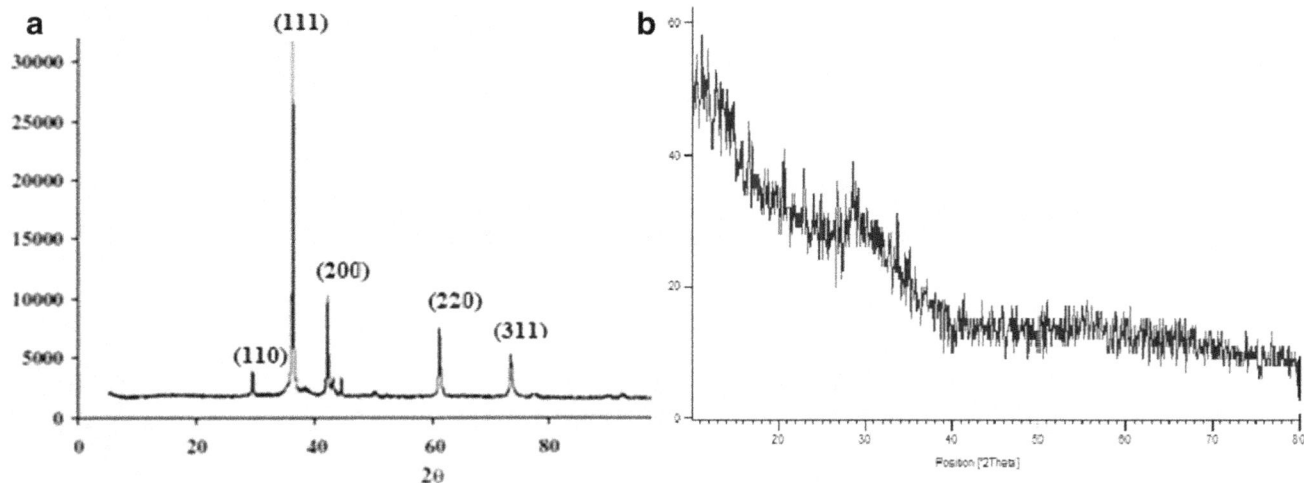

Fig. 2 XRD pattern of biogenic CuNFs (**a**) and ZnONPs (**b**)

Table 2 MIC levels of CuNPs and ZnONPs nanoparticles and antibiotics in different pathogenic bacteria

Type of pathogen	Pathogenic bacteria	MIC level of CuNPs (µg/ml)	MIC level of ZnONPs (µg/ml)	MIC level of antibiotics in (µg/ml)
Gram-negative bacteria	E. coli C3	08	08	10
	E.coliMTCC 9537	10	16	11
	K.pneumonia 125	16	04	10
	K.pneumonia MTCC 109	08	08	13
	P.aeruginosa MTCC 741	≥68	≥64	12
	S.flexneri MTCC 1457	10	09	13
Gram-positive bacteria	S. aureus 20	≥16	08	09
	S.aureus MTCC 96	18	16	09
	E.faecalis NCIM 5025	18	≥16	09

Fig. 3 MIC levels of biogenic nanoparticles and antibiotics in different pathogenic bacteria

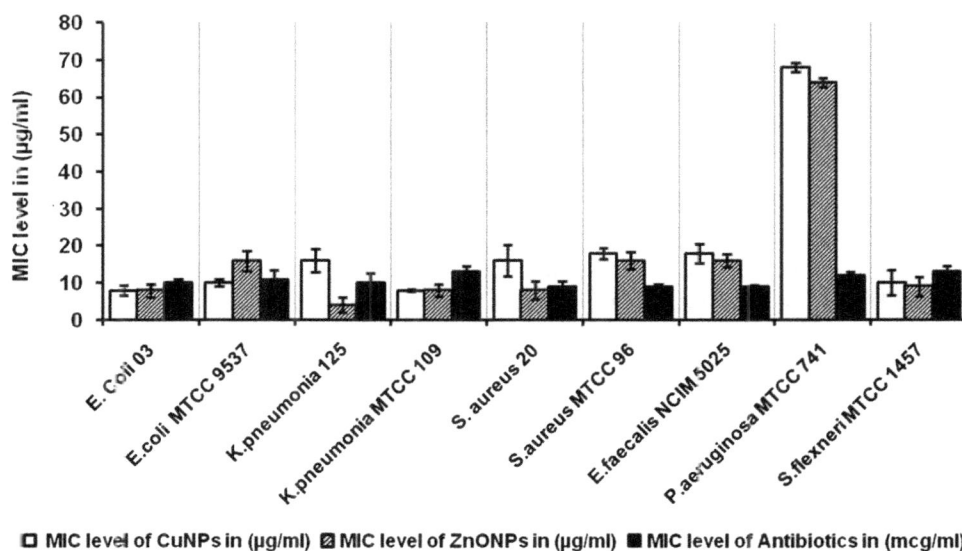

□ MIC level of CuNPs in (µg/ml) ▨ MIC level of ZnONPs in (µg/ml) ■ MIC level of Antibiotics in (mcg/ml)

Fig. 4 Scanning electron microscopy of *S.aureus* 20 (MRSA) untreated control cells (**a**), cells treated with gentamicin antibiotic (**b**), cells treated with CuNPs (**c**), cells treated with CuNPs + Gentamicin (**d**), cells treated with Zinc ZnONPs (**e**), cells treated with ZnONPs + Gentamicin (**f**)

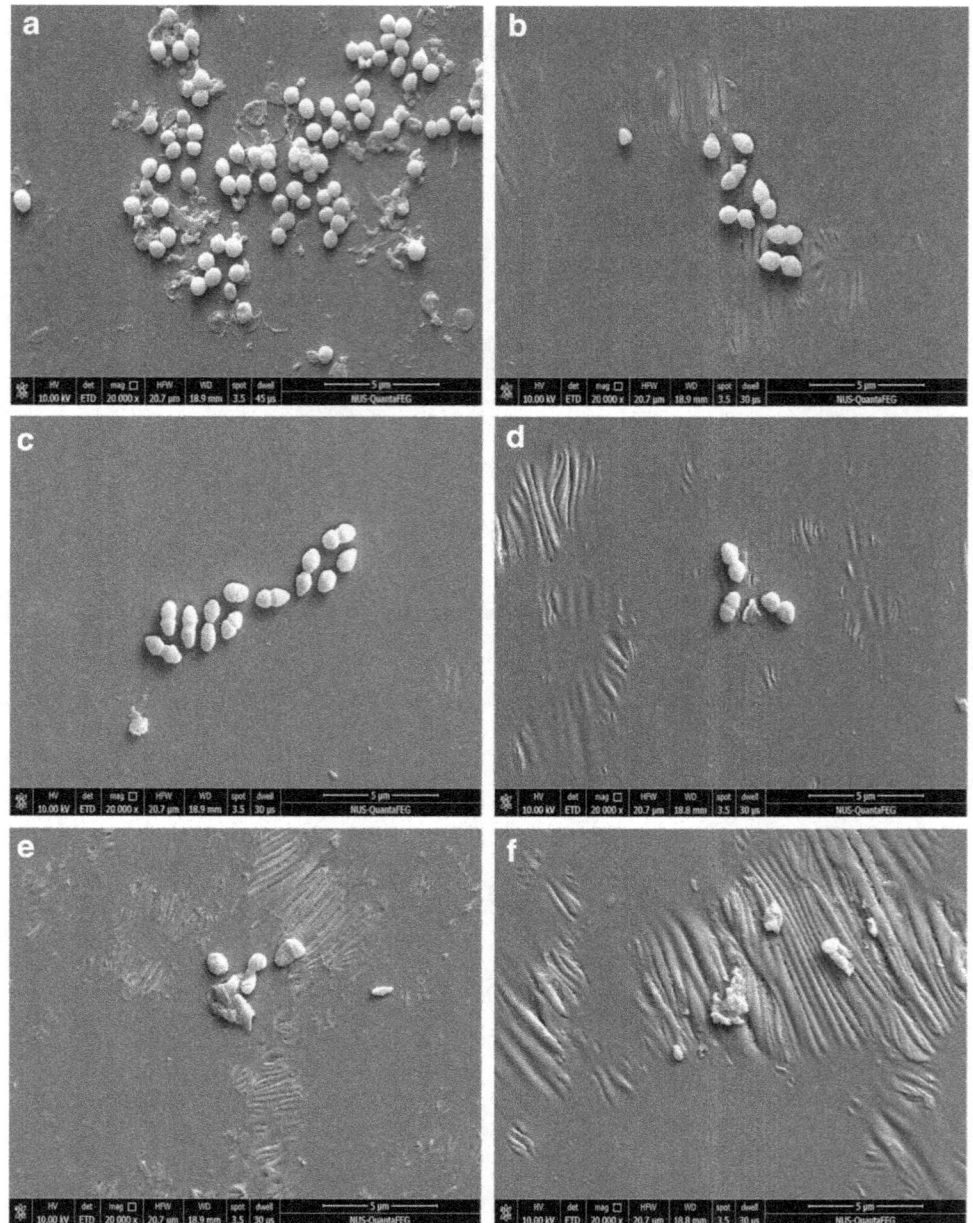

competence of CuNPs and ZnONPs in controlling pathogenic bacteria (Gram-positive and Gram-negative) by broth dilution method [29]. Each 6 h bacterial test strain was cultured in Luria–Bertani broth that was supplemented with 2, 4, 8, 16 up to 128 μg/ml of nanoparticles and antibiotics separately, and each culture was incubated at 37 °C for 18 and 24 h. Absorbance was measured at 600 nm using BioPhotometer Plus (Eppendorf AG, Hamburg, Germany).

Scanning electron microscopy

The lethal effect of nanoparticles on the surface of bacteria was imaged by performing scanning electron microscopy of the

CuNPs, ZnONPs, and antibiotic to treated test strain. They were examined using an FEI QUANTA 650 FEG Scanning Electron Microscope (National University of Singapore, Singapore).

Antibiofilm effect of CuNPs and ZnONPs in combination with antibiotic against clinical MDR's and standard MTCC cultures

Antibiofilm assay by test tube method

A biofilm formation assay [30] and antibiofilm activity of the biogenic nanoparticles used in combination with standard antimicrobials were detected by the simple and modified test tube method and estimated by a

Table 3 Percentage inhibition of CuNPs in combination with antibiotics in the test tube method

Pathogenic bacteria	Copper nanoparticles			Copper sulfate			Antibiotic			Copper nanoparticles + antibiotic		
	18 h (%)	24 h (%)	48 h (%)	18 h (%)	24 h (%)	48 h (%)	18 h (%)	24 h (%)	48 h (%)	18 h (%)	24 h (%)	48 h (%)
E. coli 03	65	72.8	88.8	0.8	1.2	2.5	36.2	20.3	22.5	78	90	92
E.coli MTCC 9537	80.2	82	85.5	0.07	1.52	3.33	–	–	–	76.5	86	88
K.pneumonia 125	80.5	79.6	89.4	1.98	2.48	6.31	39.7	30.2	33.6	83.2	90.6	91.5
K.pneumonia MTCC 109	72.8	78	87.1	–	–	–	18.9	15.8	10	86.5	88.8	91.4
S. aureus 20	76.2	85	90.3	–	–	–	19.8	15.8	14.5	82.1	87.5	91.6
S. aureus MTCC 96	79.2	89	91.6	–	–	0.2	17.9	14.9	15.8	90.5	92.6	94.7
E.faecalis NCIM5 025	84	82.6	85.5	17.2	14.6	16.3	29.7	24.8	0.5	90.6	90.8	95.9
P.aeruginosa MTCC 741	19.8	–	11.1	–	–	–	–	–	–	–	3.8	10.9
S.flexneri MTCC 1457	50.6	45.8	55.7	9.6	4.9	6.3	29.9	28.9	24.2	75.6	67.9	71.5

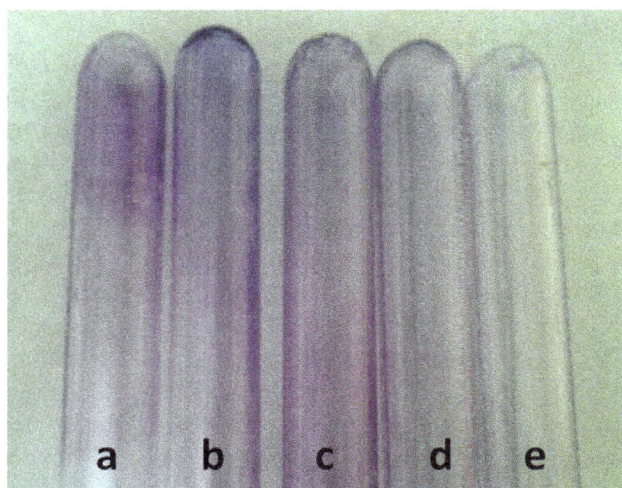

Fig. 5 Antibiofilm assay by the test tube method for *S. aureus* MTCC 96 at 48 h. (*a* untreated, *b* copper sulfate treated, *c* antibiotic treated, *d* CuNPs treated, *e* CuNPs + Antibiotic)

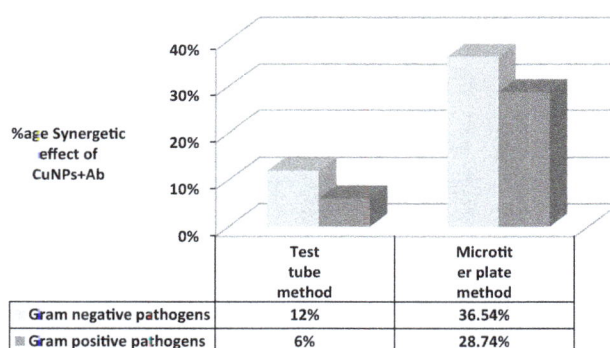

Fig. 6 Graphical representation of the percentage of synergetic effect of CuNPs + Ab shown in the test tube and microtiter plate method

spectrophotometer assay [31]. MDR and standard MTCC bacterial cultures were mixed with 2 ml of tryptic soy broth (TSB), and for each culture, the following five separate plastic tubes were used; Tube 1: bacterial culture + TSB + Cu/ZnO nanoparticles from 50 mg/ml stock; Tube 2: bacterial culture + TSB + Antibiotics from 30 mg/ml stock; Tube 3: bacterial culture + TSB + 35 μg/ml of copper sulfate/zinc sulfate solution from 100 mM stock; Tube 4: bacterial culture + TSB + Cu/ZnO nanoparticles + antibiotics; and Tube 5: bacterial culture + TSB (control; as shown in

Table 1). Experiments were designed separately for CuNPs and ZnONPs.

Antibiofilm assay by microtitre plate method

Clinical isolates and MTCC cultures were grown overnight at 37 °C in TSB supplemented with 0.2 % glucose [32]. The cultures were diluted 1:100 in medium, and 200 μl of cell suspensions were used to inoculate separate wells of sterile flat-bottomed 96-well polystyrene microtiter plates (Corning Inc., Corning, NY, USA). For each organism, the five wells were maintained separately and repeated in triplicate. Additions of 0.3 μl of copper/zinc oxide nanoparticles and antibiotics (concentrations as shown in Table 1) and copper sulfate/zinc sulfate solution from 100 mM stock were incubated for 24 h at 37 °C without

Table 4 Percentage inhibition of CuNPs in combination with antibiotics in the microtiter plate method

Pathogenic bacteria	CuNP (%)	CuNP + 2 %glucose (%)	CuSO$_4$ (%)	CuSO$_4$ + 2 %glucose (%)	Ab (%)	Ab + 2 %glucose (%)	CuNP + Ab (%)	CuNP + Ab +2 %glucose (%)
E. coli 03	54	52.3	20	18.9	22	20.5	66.9	59.8
E.coli MTCC 9537	31.7	26.33	12.8	10.6	7.4	5.3	45.9	38.94
K.pneumonia 125	57.6	34.7	14.8	10.7	39.13	20.2	69.5	62.5
K.pneumonia MTCC 109	46.04	41.7	18.7	11.9	28	22.1	75	71.2
S. aureus 20	44.72	38.97	2.78	0.8	13.2	10	64.8	63.8
S.aureus MTCC 96	46	27	12.8	9.76	10.2	5	76.8	65.9
E.faecalis NCIM 5025	45	49.6	3.87	1.89	38	29	73.2	69.5
P.aeruginosa MTCC 741	35.7	30.55	12.8	7.9	0	26.7	56.4	63.8
S.flexneri MTCC 1457	70.8	56.9	6.88	2.45	21	19	89.5	75.6

Where, *CuNPs* copper nanoparticles, *CuSO$_4$* copper sulfate, *Ab* antibiotic

shaking. One hundred microliters of destaining solution was measured at 490 nm using a microtiter plate reader (iMark Microplate Reader S/N 12883 Biorad Pvt Ltd. India). Media without inoculums were used as control. The percentage of biofilm inhibition (1) and the percentage of synergetic effect (2) were calculated using the following equations [31]:

%age of biofilm inhibition

$$= \frac{(OD490 \text{ in control} - OD490 \text{ in treatment})}{OD490 \text{ in control}} \times 100 \quad (1)$$

%age of synergetic effect

$$= \frac{[\% \text{biofilm inhibition for (NPs + Ab)} - \% \text{biofilm inhibition for NPs}]}{\% \text{ biofilm inhibition for (NPs + Ab)}} \times 100 \quad (2)$$

(where: NPs; biogenic CuNPs and ZnONPs, Ab; antibiotic).

Results and discussion

In the present investigation, the Gram-positive non-pathogenic bacterium *Enterococcus faecalis* was found notable in producing CuNPs and ZnONPs of different sizes ranging from 1 to 100 nm in distribution. TEM analysis reports the presence of biosynthesized CuNPs and ZnONPs from *E. faecalis* with core shell morphology of size 12–90 nm and spherical in shape for CuNPs and ZnONPs ranging from 16 to 96 nm with marginal variation and aggregate form (shown in Fig. 1).

Crystalline nature of the biogenic CuNPs and ZnONPs was confirmed by X-ray diffraction analysis. The XRD pattern clearly shows that the extracellular synthesis of CuNPs and ZnONPs formed by the reduction of sulfate ions from 100 mM copper sulfate and zinc sulfate using culture supernatant of *E. faecalis*. CuNPs exhibited four prominent Bragg reflections around 38.19°, 44.22°, 64.65°, and 77.7° (*Fig.* 2a). The fraction between the intensity of the (111) plane higher than the (200), (220), and (311) diffraction peaks. Intensity of the (111) facets for the very sharp diffraction peak at 38.19° is considered for the face centered cubic structure [33]. The (111) facet is extremely reactive and stable due to high rate of electron transfer. The XRD facets of the CuNPs compared and indexed with standard copper which was published by JCPDS file (JCPDS card No: 41-0254). The mean size of CuNPs was calculated using the Debye–Scherer equation by determining the width of the (111) and the similar Bragg reflection was found to be around 32.54 nm. The absence of diffraction peak in ZnONPs sample (shown in Fig. 2b) confirms the amorphous character of the sample.

Table 5 Percentage of synergetic effect of CuNPs shown in the test tube and microtiter plate method

Pathogenic bacteria	Antibiotics	%Age of synergetic effect in Test tube method			%Age of synergetic effect in Microtitre plate method for 72 h	
		18 h (%)	24 h (%)	48 h (%)	CuNP + Ab (%)	CuNP + Ab +2 %glucose (%)
E. coli 03	Ceftriaxone	16 6	19.1	3.4	19.2	12.5
E.coli MTCC 9537	Ceftriaxone	–	4.6	2.8	30.9	32.3
K.pneumonia 125	Ceftazidime	3.24	12.14	2.2	17.12	44.4
K.pneumonia MTCC 109	Ceftazidime	15 8	12.16	4.7	38.6	41.4
S. aureus 20	Gentamicin	7	2.8	1.4	30.9	38.9
S.aureus MTCC 96	Gentamicin	12 4	3.8	3.2	40.10	24.7
E.faecalis NCIM 5025	Gentamicin	7.2	9	10.8	38.5	28.6
P.aeruginosa MTCC 741	Ceftazidime	–	–	–	36.7	52.11
S.flexneri MTCC 1457	Ceftazidime	33	32.5	22	20.8	24.7

Table 6 Percentage inhibition of ZnONPs in combination with antibiotics in the test tube method

Pathogenic bacteria	ZnONPs			Zinc sulfate			Antibiotic			ZnONPs + antibiotic		
	18 h (%)	24 h (%)	48 h (%)	18 h (%)	24 h (%)	48 h (%)	18 h (%)	24 h (%)	48 h (%)	18 h (%)	24 h (%)	48 h (%)
E. coli 03	80.3	85.4	89.7	4.89	3.6	7.5	36.2	20.3	22.5	82.7	88.8	89.7
E.coli MTCC 9537	80.7	86.8	88	–	–	–	–	–	–	92.3	89.9	91.1
K.pneumonia 125	93.8	89	93.6	9.6	9.8	9.4	39.7	30.2	33.6	99.1	98.5	95.7
K.pneumonia MTCC 109	96.8	92.7	91.4	–	–	–	18.9	15.8	10	98.5	96.4	94.2
S. aureus 20	97.5	96.4	94.8	10	–	–	19.8	15.8	14.5	98.7	95.4	97.9
S.aureus MTCC 96	95.9	94.8	93.7	14.8	15.2	15.8	17.9	14.9	15.8	97.9	97.5	95.8
E.faecalis NCIM 5025	94.7	89.6	90.8	9.6	9.6	9.1	29.7	24.8	0.5	96.8	94.8	93.8
P.aeruginosa MTCC 741	10.7	–	9	–	–	–	–	–	–	7.8	3.8	3.0
S.flexneri MTCC 1457	97.5	96.7	93.6	3.8	2.8	2.3	29.9	28.9	24.2	96.9	95.8	93.7

Where, *ZnONPs* zinc oxide nanoparticles, *ZnSO₄* zinc sulfate

The antimicrobial activities of the biogenic CuNPs and ZnONPs against Gram-negative and Gram-positive bacteria were estimated through MIC by the broth dilution method. The tested concentrations for biogenic CuNPs and ZnONPs were from 2 to 128 µg/ml, as shown in Table 2 and Fig. 3. The results demonstrated that effective doses of biogenic CuNPs, ZnONPs, and antibiotics for both Gram-positive and Gram-negative bacteria are different. Biogenic nanoparticles are efficient inhibitors against both Gram-positive and Gram-negative bacteria in contrast to antibiotics. MIC values of CuNPs and ZnONPs against Gram-negative bacteria include *E. coli* 03, *E. coli* MTCC 9537, *K. pneumonia* 125, *K. pneumonia* MTCC 109, and *S. flexneri* MTCC 1457 ranging from 8 to 16 µg/ml. In addition, 18 to ≥68 µg/ml of CuNPs and ZnONPs showed inhibition kinetics against Gram-positive pathogens, including methicillin-resistant *S. aureus* 20, *S. aureus* MTCC 96, *E. faecalis* NCIM 5025, and the Gram-negative bacteria *P. Aeruginosa* MTCC 741. This disparity could be due to differences in the membrane structure and the composition of the cell wall, thereby affecting access of the CuNPs and ZnONPs. Cell walls of both Gram-positive and Gram-negative bacteria have negative charge because of the presence of teichoic acids and lipopolysaccharides, respectively [30]. Many researchers found that the antibacterial effect of nanoparticles was more prominent against Gram-negative bacteria than Gram-positive bacteria. This could be due to the excess of negative charges

Table 7 Percentage inhibition of ZnONPs in combination with antibiotics in the microtiter plate method

Pathogenic bacteria	ZnONP (%)	ZnONP + 2 %glucose (%)	$ZnSO_4$ (%)	$ZnSO_4$ + 2 %glucose (%)	Ab (%)	Ab + 2 %glucose (%)	ZnONP + Ab (%)	ZnONP + Ab + 2 %glucose (%)
E. coli 03	65.8	60.6	28.6	18.6	22	20.5	78.55	70.5
E.coli MTCC 9537	72.8	68.99	12.7	5.88	7.4	5.3	86.9	82.77
K.pneumonia 125	68.4	59.7	8.96	3.85	39.13	20.2	88.67	79.67
K.pneumonia MTCC 109	78.5	65.6	6.78	4.7	28	22.1	89.5	72.8
S. aureus 20	72.7	65.5	14.7	5.89	13.2	10	90.5	78.99
S.aureus MTCC 96	77.4	64.8	15.9	13	10.2	5	85.3	80.4
E.faecalis NCIM 5025	89.4	76.7	17.9	12.7	38	29	92.8	83.7
P.aeruginosa MTCC 741	45.8	40.7	9.67	8.2	0	26.7	76.5	70.77
S.flexneri MTCC 1457	87.6	74.8	13.8	8.56	21	19	92.3	87.8

ZnONPs zinc oxide nanoparticles, *$ZnSO_4$* zinc sulfate, *Ab* antibiotic

on the Gram-negative bacteria which assists the burly interaction between nanoparticles and cell wall components of the bacteria [34]. Absolutely, the exact mechanism of inhibition by the nanoparticles on the microorganisms depends on their small size and high surface area to volume ratio (S/V), which permit them to interact closely with the membranes of the microbe [34]. Furthermore, the biosynthesized CuNPs and ZnONPs in this study displayed the promising antibacterial activity against Gram-negative subsequent to Gram-positive bacteria, which could be attributed to their size less than 100 nm and greatest surface area to volume ratio; therefore, the contact with bacteria is the greatest. This could be the reason why they exhibit the best antibacterial activity.

These results also specify that there were no significant antibacterial activities observed at concentrations less than 8 µg/ml by any biogenic nanoparticles. In contrast, MIC levels of antibiotics were more similar to the nanoparticles against several Gram-negative bacteria. The dissimilarity in the MIC results against both Gram-positive and Gram-negative bacteria might be due to differences in cell wall structure. Thus, it was concluded that CuNPs and ZnONPs inhibited the growth of all the tested microorganisms.

Morphological analysis by scanning electron microscopy

Scanning electron microscopy was used to determine the morphological changes of the *S. aureus* 20, after treatment with antibiotic, nanoparticles (NPs) alone, and antibiotics with nanoparticle (Antibiotic + NPs). Cells without any treatment (control) showed normal morphology, with a multilayered surface consisting of the outer membrane (Fig. 4a). In contrast, the cells exposed to >9 µg/ml concentration of gentamicin and >16 µg/ml of CuNPs for 24 h showed increased cell size and change in cell shape (Fig. 4b). The cells treated with in combination of CuNPs + Antibiotic (Fig. 4c, d) showed deformed morphology lacking a cytoplasmic membrane. For the cells treated with ZnONPs + Antibiotic (>8 + >9 µg/ml), the outer membrane was progressively lost, and the cytoplasm tended to spill out of the cell leading to cell death, which corresponded to the final stage of cell disruption, plasmolysis, and partial disappearance of the cytoplasmic membrane (in Fig. 4e, f). Finally, SEM studies proved that CuNPs and ZnONPs used in combination with gentamicin had the highest antibacterial activity when compared with CuNPs, ZnONPs, and antibiotics alone treatment.

Antibiofilm assay

Various applications of nanoparticle-based therapies have gained attraction across several biomedical fields. Due to

Table 8 Percentage of synergetic effect of ZnONPs showed in the test tube and microtiter plate method

Pathogenic bacteria	Antibiotics	%Age of synergetic effect in Test tube method			%Age of synergetic effect in Microtiter plate method for 72 h	
		18 h (%)	24 h (%)	48 h (%)	ZnONP + Ab (%)	ZnONP + Ab + 2 % glucose (%)
E. coli 03	Ceftriaxone	2.9	3.7	0	16.23	14
E.coli MTCC 9537	Ceftriaxone	13	3.4	3.4	16.22	15.7
K.pneumonia 125	Ceftazidime	5.3	9.6	2.1	22.8	0.06
K.pneumonia MTCC 109	Ceftazidime	1.72	3.8	2.9	12.2	9.8
S. aureus 20	Gentamicin	1.2	–	3.1	19.6	17
S.aureus MTCC 96	Gentamicin	2	2.7	2.17	9.2	19.4
E.faecalis NCIM 5025	Gentamicin	2.1	5.4	3.1	3.6	8.3
P.aeruginosa MTCC 741	Ceftazidime	–	–	–	40.13	42.48
S.flexneri MTCC 1457	Ceftazidime	–	–	–	5.09	14.8

the resistant nature of biofilms, eradication of biofilm-related diseases/infection is challenging [35]. Efforts are being made to use penetrating capacity of nanoparticles in biofilm studies [36–38]. Further application of nanotechnology could be a way to combat biofilm infections.

This study investigated inhibition of biofilm activity by biosynthesized nanoparticles. Activity of Gram-positive and Gram-negative bacteria was ceased under in vitro conditions, subsequently leading to the inhibition of biofilm formation. Both CuNPs and ZnONPs have been used to inhibit the initial stage of biofilm formation.

The results for both test tube and microtiter plate wells showed that for all the bacterial strains tested (except for *P.aeruginosa* MTCC 741), biologically synthesized CuNPs and ZnONPs inhibited the activity of biofilm formation at its irreversible adhesion stage (also known as Initial stage). Interestingly, an inhibition of initial stage biofilm activity was observed at the MIC values of CuNPs and ZnONPs.

Furthermore, this study also revealed the synergistic effect of CuNPs and ZnONPs of antibiofilm activity against different pathogenic bacteria in the presence of antibiotics. Biofilm production has been reported in all strains. The results from the test tube method indicated that CuNPs alone reduce the biofilm activity by approximately ≥9 % in Gram-negative and ≥2 % in Gram-positive bacteria. Combination of CuNPs and antibiotics showed more effective biofilm inhibition activity in Gram-negative and Gram-positive bacteria by 12 % and 6 %, respectively (shown in Tables 3 and 5, Figs. 5 and 6). Samples treated with antibiotics and positive control alone showed negligible activity on biofilm prevention/inhibition. In the microtiter method, 36.54 % and 28.74 % of antibiofilm activity was recorded for Gram-negative and Gram-positive bacteria, respectively (Tables 4, 5; Fig. 6).

Antibiofilm activity of ZnONPs in combination with antibiotic showed comparatively less activity against

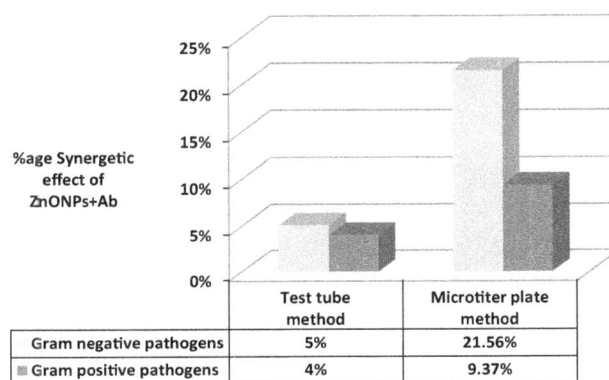

Fig. 7 Graphical representation of the percentage of synergetic effect of ZnONPs + Ab shown in the test tube and microtiter plate method

Gram-negative and Gram-positive bacteria in both the test tube and the microtiter plate method compared to CuNPs. In the test tube method, ZnONPs showed 5 and 4 % of antibiofilm activity against Gram-negative and Gram-positive bacteria, respectively (shown in Table 6). Enhanced antibiofilm activity was reported from the microtiter plate method (21.56 % for Gram-negative and 9.37 % for Gram-positive bacteria, as shown in Table 7). Using the microtiter plate assay method, we found that the synergetic effects of CuNPs and ZnONPs (Table 8; Fig. 7) with specific antibiotic in the presence of 2 % glucose are relatively high compared with the treatment without glucose. Due to bonding interactions, increased synergistic effects among antibiotics and nanoparticles were observed. In addition, the large surface area and presence of functional groups, such as hydroxyl, amino, etc., lead nanoparticles to interact with antibiotics by chelating reactions [39]. However, the mechanistic action of NPs with antibiotics in biofilm-related studies has yet to be demonstrated. For the NP therapies, the results indicates, microtiter plate assay method is an accurate and reproducible method for antibiofilm screening, and assay serves as a reliable quantitative tool for determining the antibiofilm potential of nanoparticles in combination with antibiotic agents against several clinical isolates. Collectively, these findings conclude that the enhanced synergistic effect of biosynthesized nanoparticles in combination with antibiotics against pathogenic bacteria could be used as potent adjuvant therapy against several bacterial infections.

The difference in the inhibitory activity may also be explained by several factors, including efficacy in antimicrobial activity, biosorption-dependent manner, physical properties, such as the size of the nanoparticle, penetration abilities, and other chemical properties effecting the affinity between the materials and the biofilms [40]. The results suggest that CuNPs were better antibiofilm agents against the Gram-negative and Gram-positive bacteria than ZnONPs.

Conclusion

This study was designed to elucidate the enhanced antibiofilm effects of the third generation antibiotics with biogenic CuNPs and ZnONPs (as shown in Fig. 8). The need for higher dosage of NPs and antibiotics could be reduced by the synergistic action of antimicrobial agents, and this phenomenon also minimizes side effects. This study demonstrated improvement of the bactericidal property of nanoparticles by understanding their synergistic

effect with other antimicrobial agents to improve their efficacy against various pathogenic microbes. The increased antibiofilm activity of CuNPs was more promising than that of ZnONPs for targeting Gram-negative and Gram-positive bacteria. The increased inhibition activity of CuNPs on bacteria is associated with release of free ions from nanoparticles. In addition, the potentiality is, furthermore, enhanced by its small size (12–90 nm) and high surface area to volume ratio which permits them to interact intimately with microbial membranes. Antimicrobial activity is due to its affinity to instability between its oxidation states. Differentiating copper ions from other trace metals results in the production of hydroxyl radicals that subsequently bind with DNA molecules and lead to disorder of the helical structure by crosslinking within and between the nucleic acid strands and damage essential proteins by binding to the sulfhydryl amino and carboxyl groups of amino acids and denatures the protein. The exact mechanism behind is still not known and needs to be further investigated. Based on all of these studies, antimicrobial characteristics of CuNPs is by denaturing affect of Cu ion on proteins and enzymes in microbes [41]. In addition, NPs could be used as an adjuvant therapy for the treatment of various infectious diseases caused by Gram-negative and Gram-positive bacteria. Thus, our findings support the notion that NPs have effective antibiofilm activity that could be used to enhance the action of existing antibiotics against Gram-negative and Gram-positive bacteria.

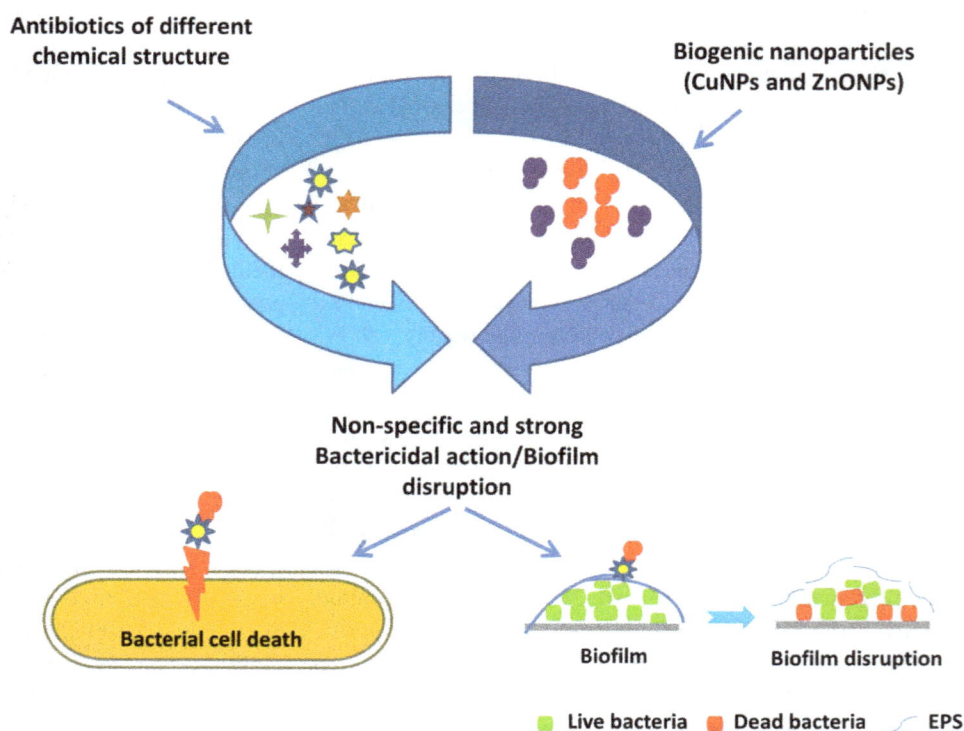

Fig. 8 Graphical abstract on "synergistic antibacterial and antibiofilm activity of biogenic CuNPs and ZnONPs-antimicrobials against pathogenic bacteria"

Author's contribution KCR and AC conceived and preformed the experiments. HHK and NKD assisted with experiments. NKD contributed in SEM imaging and analysis of results. All authors contributed in interpreting results, arranging tables, illustrations and preparing the manuscript. KCR improved experimental design and manuscript. All authors read and approved the final manuscript.

Acknowledgments This work was funded by University Grants Commission (Grant Number: MRP-MAJOR-BIOT-2013-15248), New Delhi, India and the authors gratefully acknowledge the Department of Biotechnology, Gulbarga University, Gulbarga for providing the facilities to pursue this research.

Compliance with ethical standards

Disclosure statement No competing financial interests exist.

References

1. Høiby, N., Bjarnsholt, T., Givskov, M., Molin, S., Ciofu, O.: Antibiotic resistance of bacterial biofilms. Int. J. Antimicrob. Agents **35**(4), 322–332 (2010)
2. Hall-Stoodley, L., Costerton, J.W., Stoodley, P.: Bacterial biofilms: from the Natural environment to infectious diseases. Nat. Rev. Micro **2**(2), 95–108 (2004)
3. Markowska, K., Grudniak, A.M., Wolska, K.I.: Silver nanoparticles as an alternative strategy against bacterial biofilms. Acta Biochim. Pol. **60**(4), 523–530 (2013)
4. Ventola, C.L.: The antibiotic resistance crisis: part 1: causes and threats. P t **40**(4), 277–283 (2015)
5. Chen, M., Yu, Q., Sun, H.: Novel strategies for the prevention and treatment of biofilm related infections. Int. J. Mol. Sci. **14**(9), 18488–18501 (2013)
6. Parsek, M.R., Singh, P.K.: Bacterial biofilms: an emerging link to disease pathogenesis. Ann. Rev. Microbiol. **57**, 677–701 (2003)
7. Donlan, R.M., Costerton, J.W.: Biofilms: survival mechanisms of clinically relevant microorganisms. Clin. Microbiol. Rev. **15**(2), 167–193 (2002)
8. Yokoi, N., Okada, K., Sugita, J., Kinoshita, S.: Acute conjunctivitis associated with biofilm formation on a punctal plug. Jpn. J. Ophthalmol. **44**(5), 559–560 (2000)
9. LewisOscar, F., MubarakAli, Davoodbasha, Nithya, C., Priyanka, R., Gopinath, V., Alharbi, N.S., Thajuddin, N.: One pot synthesis and antibiofilm potential of copper nanoparticles (CuNPs) against clinical strains of *Pseudomonas aeruginosa*. Biofouling **31**(4), 379–391 (2015)
10. Sondi, I., Salopek-Sondi, B.: Silver nanoparticles as antimicrobial agent: a case study on *E. coli* as a model for Gramnegative bacteria. J. Colloid Interface Sci. **275**, 177–182 (2004)
11. Kathiresan, K., Manivannan, S., Nabeel, M.A., Dhivya, B.: Studies on silver nanoparticles synthesized by a marine fungus, *Penicillium fellutanum* isolated from coastal mangrove sediment. Colloids Surf. B **71**, 133–137 (2009)
12. Allahverdiyev, A.M., Kon, K.V., Abamor, E.S., Bagirova, M., Rafailovich, M.: Coping with antibiotic resistance: combining nanoparticles with antibiotics and other antimicrobial agents. Expert Rev. Anti Infect. Ther. **9**(11), 1035–1052 (2011)
13. Cho, K.H., Park, J.E., Osaka, T., Park, S.G.: The study of antimicrobial activity and preservative effects of nanosilver ingredient. Electrochim. Acta **51**, 956–960 (2005)
14. Wilks, S.A., Michels, H., Keevil, C.W.: The survival of *Escherichia coli* O157 on a range of metal surfaces. Int. J. Food Microbiol. **105**, 445–454 (2005)
15. Casey, A.L., Adams, D., Karpanen. T.J., Lambert, P.A., Cookson, B.D., Nightingale, P.: Role of copper in reducing hospital environment contamination. J. Hosp. Infect. **74**, 72–77 (2010)
16. Borkow, G., Gabbay, J., Dardik, R., Eidelman, A.I., Lavie, Y., Grunfeld, Y., Ikher, S., Huszar, M., Zatcoff, C., Marikovsky, M.: Molecular mechanisms of enhanced wound healing by copper oxide-impregnated dressings. Wound Repair Regen. **18**, 266–275 (2010)
17. Eshed, M., Lellouche, J., Gedanken, A., Banin, E.: A Zn-doped CuO nanocomposite shows enhanced antibiofilm and antibacterial activities against Streptococcus mutans compared to nanosized CuO. Adv Funct. Mat. **24**, 1382–1390 (2014)
18. Christena, L.R., Mangalagowri, V., Pradheeba, P., Ahmed, K.B.A., Shalini, B.I.S., Vidyalakshmi, M.: Copper nanoparticles as an efflux pump inhibitor to tackle drug resistant bacteria. RSC Adv. **5**, 12899–12909 (2015)
19. Begum, A.N., Mondal, S., Basu, S., Laskar, A.R., Mandal, D.: Colloids Surf. B **71**, 113–118 (2009)
20. Applerot, G., Lellouche, J., Perkas, N., Nitzan, Y., Gedanken, A., Banin, E.: ZnO nanoparticle-coated surfaces inhibit bacterial biofilm formation and increase antibiotic susceptibility. RSC Advances. **2**, 2314–2321 (2012)
21. Brayner, R., Ferrari-Iliou, R., Brivois, N., Djediat, S., Benedetti, M.F., Fievet, F.: Toxicological impact studies based on *Escherichia coli* bacteria in ultrafine ZnO nanoparticles colloidal medium. Nano Lett. **6**, 866–870 (2006)
22. Thill, A., Zeyons, O., Spalla, O., Chauvat, F., Rose, J., Auffan, M., Flank, A.M.: Cytotoxicity of CeO_2 nanoparticles for *Escherichia coli* physico-chemical insight of the cytotoxicitymechanism. Environ. Sci. Technol. **40**, 6151–6156 (2006)
23. Reddy, K.M., Feris, K., Bell, J., Wingett, D.G., Hanley, C., Punnoose, A.: Selective toxicity of zinc oxide nanoparticles to prokaryotic and eukaryotic systems. Appl. Phys. Lett. **90**, 2139021–2139023 (2007)
24. Zhang, L.L., Jiang, Y.H., Ding, Y.L., Povey, M., York, D.: Investigation into the antibacterial behaviour of suspensions of ZnO nanoparticles (zno nanofluids). J. Nanopart. Res. **9**, 479–489 (2007)
25. Thati, V., Roy, A.S., Ambika Prasad, M.V.N., Shivannavar, C.T., Gaddad, S.M.: Nanostructured Zinc oxide enhances the activity of antibiotics against *Staphylococcus aureus*. J. Biosci. Tech. **1**, 64 (2010)
26. Luo, Z., Wu, Q., Xue, J., Ding, Y.: Selectively enhanced antibacterial effects and ultraviolet activation of antibiotics with ZnO nanorods against *Escherichia coli*. J. Biomed. Nanotechnol. **9**, 69 (2013)
27. Ashajyothi, C., Jahanara, K., Chandrakanth, K.R.: Biosynthesis and characterization of copper nanoparticles from *Enterococcus faecalis*. Int. J. Pharma Biosci. **5**(4), 204–211 (2014)
28. Ashajyothi, C., Manjunath, R., Narasanna, K., Chandrakanth, R.: Antibacterial activity of Biogenic Zinc oxide nanopaticals synthesized from *Enterococcus faecalis*. Int. J. Chemtech Res. **69**(5), 3131–3136 (2014)
29. Clinical and Laboratory Standards Institute, *Performance standards for antimicrobial susceptibility testing*. Performance standards for antimicrobial susceptibility testing. Twenty-Second informational supplement. Document M100- S22, CLSI. 2012, Wayne, PA
30. Mathur, T., Khan, S.S., Upadhyay, D.J., Fatma, T., Rattan, A.: Detection of biofilm formation among the clinical isolates of *staphylococci*: an evaluation of three different screening methods. Indian J. Med. Microbiol. **24**(25), 9 (2006)
31. Ashajyothi C, Manjunath, K., Chandrakanth, R.: Prevention of

multiple drug resistant bacterial biofilm by synergistic action of biogenic silver nanoparticle and antimicrobials. J. Microbiol. Biotech. Res. **5**(1), 7 (2015)

32. Mohamed, J.A., Huang, W., Nallapareddy, S.R., Teng, F., Murray, B.E.: Influence of origin of isolates, especially endocarditis isolates, and various genes on biofilm formation by *Enterococcus faecalis*. Infect. Immun. **72**(6), 3658–3663 (2004)

33. Ramyadevi, J., Jeyasubramanian, K., Marikani, A., Rajakumar, G., AbdulRahuman, A.: Synthesis and antimicrobial activity of copper nanoparticles. Mater. Lett. **71**, 114–116 (2012)

34. Eman, A., Rasha A., Ahmed: Synthesis of copper nanoparticles with various sizes and shapes: application as a superior non-enzymatic sensor and antibacterial agent. Int. J. Electrochem. Sci. **11**, 4712–4723 (2016)

35. Lewis, K.: Riddle of biofilm resistance. Antimicrob. Agents Chemother. **45**, 8 (2001)

36. Li, Xiaoning, Yeh, Y.C., Giri, K., Mount, R., Landis, R.F., Prakash, Y.S., Rotello, V.M.: Control of nanoparticle penetration into biofilms through surface design. Chem. Commun. **51**(2), 282–285 (2015)

37. Ikuma, K., Decho, A.W., Lau, B.L.T.: When nanoparticles meet biofilms—interactions guiding the environmental fate and accumulation of nanoparticles. Frontiers in Microbiology **6**, 591 (2015)

38. Wang, L.-S., Gupta, A., Rotello, V.M.: Nanomaterials for the Treatment of Bacterial Biofilms. ACS Infect. Dis. **2**(1), 3–4 (2016)

39. Dhas, S.P., Mukherjee, A., Chandrasekaran, N.: Synergistic effect of biogenic silver nanocolloid in combination with antibiotics: a potent therapeutic agent. Int. J. Pharm. Pharm. Sci. **5**(1), 292–295 (2013)

40. Park, H.J., Kim, H.Y., Cha, S, Ahn C.H., Roh, J, Park, S, Kim S, Choi K, Yi J, Kim Y, Yoon J.: Removal characteristics of engineered nanoparticles by activated sludge. Chemosphere. **92**(5), 524–528 (2013)

41. Yoon, K., Byeon, J.H., Park, J., Hwang, J.: Susceptibility constants of *E. coli* and Bacillus subtilis to Ag and Cu nanoparticles. Sci. Total Environ. **37**(3), 572–575 (2007)

6

A novel nitrogen dioxide gas sensor based on TiO$_2$-supported Au nanoparticles: a van der Waals corrected DFT study

Amirali Abbasi[1,2,3] · Jaber Jahanbin Sardroodi[1,2,3]

Abstract The interactions of nitrogen dioxide molecule with TiO$_2$-supported Au nanoparticles were investigated using density functional theory. Surface Au atoms on the TiO$_2$-supported Au overlayer were found to be the most favorable binding sites, thus making the adsorption process very strong. Both oxygen and nitrogen atoms of the NO$_2$ molecule can bind to the Au surface by forming strong chemical bonds. The adsorption of NO$_2$ molecule on the considered structures gives rise to significant changes in the bond lengths, bond angles, and adsorption energies of the complex systems. The results indicate that NO$_2$ adsorption on the TiO$_2$-supported Au nanoparticle by its oxygen atoms is energetically more favorable than the NO$_2$ adsorption by its nitrogen atom, indicating the strong binding of NO$_2$ to the TiO$_2$-supported Au through its oxygen atoms. Thus, the bridge configuration of TiO$_2$/Au + NO$_2$ is found to be the most stable configuration. Both oxygen and nitrogen atoms of NO$_2$ move favorably towards the Au surface, as confirmed by significant overlaps in the PDOSs of the atoms that forming chemical bonds. This study not only suggests a theoretical basis for gas-sensing properties of the TiO$_2$-supported Au nanoparticles, but also offers a rational approach to develop nanostructure-based chemical sensors with improved performance.

Keywords Density functional theory · NO$_2$ · TiO$_2$-supported Au nanoparticle · PDOS

Introduction

TiO$_2$ is one of the most broadly studied transition metal semiconductors with outstanding properties, such as non-toxicity, high catalytic efficiency, and extensive bandgap [1]. Until now, various kinds of well-known applications have been proposed for TiO$_2$, such as photo-catalysis, gas sensor devices, organic dye-sensitized solar cells, water splitting, and air pollution control [2–5]. Anatase, rutile, and brookite are the most important polymorphs of TiO$_2$ [6]. Of the three polymorphs of TiO$_2$, the rutile form is found to be the most stable phase. There is not any detailed theoretical investigation on the physical and chemical properties of brookite because of its metastable property. This meta-stability results in some troubles during the synthesis of brookite [7]. The improved reactivity of anatase is comparable with that of rutile and brookite phases [8–14]. Anatase has been extensively studied due to its enhanced activity in some photo-catalysis reactions, such as TiO$_2$-supported metal particle reactions, compared to the rutile and brookite phases [15–17]. Unfortunately, as a most promising material, the widely application of TiO$_2$-based gas sensors is influenced by its wide bandgap (3–3.2 eV). This results in the absorption of a small percentage of the incoming solar light (3–5%). An enormous amount of effort has been invested in enhancing the optical response of TiO$_2$ by nitrogen doping [8].

✉ Amirali Abbasi
a_abbasi@azaruniv.edu

1 Molecular Simulation Laboratory (MSL), Azarbaijan Shahid Madani University, Tabriz, Iran

2 Department of Chemistry, Faculty of Basic Sciences, Azarbaijan Shahid Madani University, Tabriz, Iran

3 Computational Nanomaterials Research Group (CNRG), Azarbaijan Shahid Madani University, Tabriz, Iran

Recently, gold was considered as an inactive metal, which possesses less activity than the other metals in many reactions. Haruta and co-workers showed that gold particles can increase the combustion of CO molecule and promote different catalytic reactions [18]. The gold particles supported by metal oxides (oxide-supported gold particles) have gained more attention due to their higher activities in the surface processes [19–22]. This leads to the structures with enhanced catalytic activity and higher stability [23, 24]. There are a large number of important reactions, in which the oxide-supported Au overlayers play a key role, including the epoxidation of C_3H_6 [25], reduction of NO_x molecules [26], and dissociation of SO_2 molecule [27]. TiO_2 has been considered as one of the most appropriate support materials for gold particles [28, 29]. The interactions of gold nanoparticles with TiO_2 (rutile and anatase) have been widely studied in the last few years. Vittadini et al. considered the adsorption behaviors of gold clusters on the TiO_2 anatase (101) surfaces [30]. Metiu and co-workers investigated the adsorption site and electronic structures of the TiO_2 rutile-supported Au nanoparticles [31].

The adsorption of the O_2 and CO_2 on gold nanoparticles supported by TiO_2 has been investigated by DFT calculations [32]. The main source of nitrogen dioxide emission is internal combustion engines, burning fossil fuels. It also results from cigarette smoke, kerosene heaters, and vehicle engines and stoves. Thus, finding an efficient sensor for the removal of this toxic molecule is an important issue to public health and environmental protection [33]. An ideal semiconductor oxide-based gas sensor should have properties, such as high sensitivity to the expected toxic material, low price fabrication, and compatibility with modern electronic devises. Among different gas sensors, the oxide-supported gold nanoparticles have been characterized as efficient sensor materials because of their higher activities. Therefore, establishing multi-component structures in sensor materials has long been regarded as the best strategy for improving the sensing performance of TiO_2 particles. The mechanism of gas sensing for the removal of toxic NO_2 molecules by metal oxide-based sensors is represented in Fig. 1. We have decided to perform a DFT study of the interaction of NO_2 molecule with the TiO_2-supported Au overlayers to fully exploit the gas-sensing capabilities of these nanocomposites.

The consecutive adsorption of NO_2 molecules on the TiO_2-supported Au overlayers probably produces N_2 molecule formed from the central nitrogen atoms of the two adsorbed NO_2 molecules. This is a consequence of the formation of weak chemical bonds between oxygen atoms of NO_2 molecule and Ti sites of the adsorbent. This leads to weakening of the bond between central nitrogen and the side oxygen atoms of the adsorbed NO_2 molecules. Based

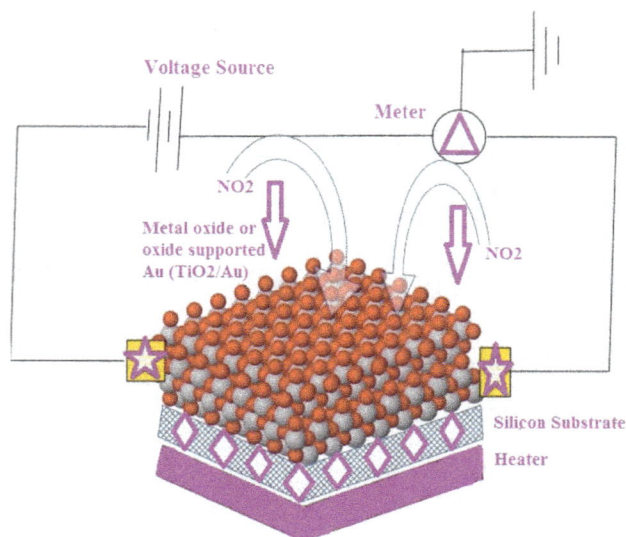

Fig. 1 Schematic drawing of a typical metal oxide-based gas sensor

on this fact, we can conclude desorption of NO_2 molecule from the surface of the TiO_2-supported Au overlayer. The next NO_2 molecule then can be adsorbed on the considered nanocomposite, and this consecutive process was repeated over and over again to obtain the enhanced sensing performance of the adsorbent material. Figure 1 shows a schematic structure of a metal oxide-based gas sensor. We have also commented on the charge analysis of the complex system according to the Mulliken population analysis. In this study, the main objective is to perform a systematic investigation on the adsorption behaviors of the TiO_2-supported Au nanoparticles as potentially efficient gas sensors for NO_2 detection.

Computational methods

Details of computation

All of DFT calculations [34, 35] were carried out using the Open source Package for Material eXplorer (OPENMX3.8) [36]. OPENMX is an efficient software package for nano-scale materials simulations based on norm-conserving pseudo-potentials and pseudo-atomic localized basis functions [37, 38]. To optimize the structures, the pseudoatomic orbitals (PAOs) centered on atomic sites were used as basis sets. The calculations were done with a considered energy cutoff of 150 Ry. Pseudo-atomic orbitals were constructed by minimal basis sets (three s-state, three p-state, and one d-state radial functions) for the Ti, (three s-state, three p-state, two d-state, and one f-state radial functions) for the Au, and (two s-state, two p-state radial functions) for O and N atoms, within cut-off radii of basis functions set to the values of seven for Ti, nine for Au, and five for O and N (all in Bohrs). The total energy of the system was computed within the Perdew–Burke–Ernzerhof (PBE) form of the generalized gradient approximation (GGA) exchange–correlation potential [39]. Mulliken population analysis was also conducted to fully analyze the charge transfer between NO_2 and nanocomposite. To optimize the adsorption configurations of the TiO_2-supported Au overlayers with adsorbed NO_2 molecules, all atoms of the system are entirely relaxed until the force on each atom is less than 0.01 eV/Å. The size of the simulation box containing pristine TiO_2-supported Au nanoparticles is 20 Å × 20 Å × 30 Å, being larger than the realistic size of the composite system.

Three possible orientations of NO_2 towards the TiO_2-supported Au nanoparticles are studied in this work. XCrysDen program was used for visualization of the figures presented in this study [40]. The total number of atoms of the nanocomposite in the considered box is 88 atoms (16 Au, 48 O, and 24 Ti atoms) of undoped TiO_2-supported Au overlayer. The effects of vdW interactions were also taken into account in this study. Both LDA and GGA methods cannot describe the vdW interactions in the systems, such as NO_2 adsorption, on the TiO_2-supported Au nanoparticles. Thus, an inclusion of additional functional into standard DFT methods would be required, which correctly describes the effects of vdW interactions. Grimme's DFT-D2 [41] and DFT-D3 methods [42, 43] were used in this study to correct the adsorption energies for dispersion energy. The adsorption energy, E_{ad}, is estimated as the following equation:

$$E_{ad} = E_{(composite+adsorbate)} - E_{composite} - E_{adsorbate} \quad (1)$$

where $E_{(composite\ +\ adsorbate)}$ is the total energy of the TiO_2-supported Au overlayers with adsorbed NO_2, $E_{composite}$ is the energy of bare TiO_2-supported Au overlayer, and $E_{adsorbate}$ represents the energy of a free gas-phase NO_2 molecule. As distinct from this equation, the adsorption energies of energy favorable configurations are negative.

Modeling TiO_2-supported Au nanoparticles

We have constructed TiO_2 anatase nanoparticle using a 3 × 2 × 1 supercell of TiO_2 anatase. The considered unit cell is available at "American Mineralogists Database" webpage [44] and reported by Wyckoff [45]. The size of the studied nanoparticles was chosen following Lei et al. [46] and Liu et al. [47]. The results published by Lei et al. [46] show that the smaller the particle is, the higher the

Table 1 Calculated surface energies (in J/m^2) for the anatase (0 0 1) and (1 0 1) surfaces, calculated based on GGA pseudo-potential

Surface	(0 0 1)	(1 0 1)
Calculated	0.96	0.49
Literature	0.98	0.49

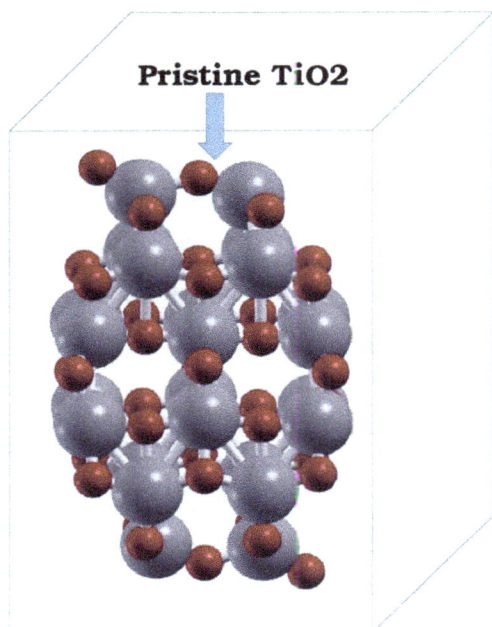

Fig. 2 Representation of a pristine TiO_2 anatase nanoparticle, two dangling oxygen atoms were used to set a 1:2 atomic charge ratio between the oxygen and titanium atoms

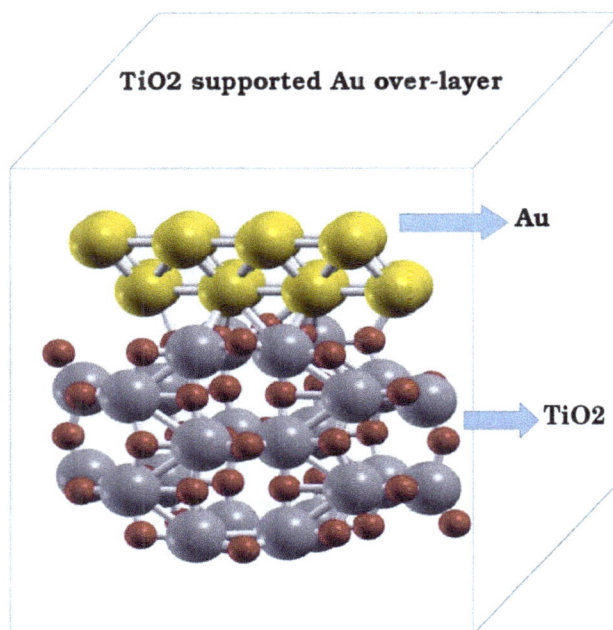

Fig. 3 Optimized geometry of undoped TiO_2-supported Au overlayer. The *yellow, gray,* and *red balls* denote gold, titanium, and oxygen atoms, respectively. TiO_2 was demonstrated to an appropriate support material for gold particles

Fig. 4 Optimized geometry configurations of the NO_2 molecule adsorbed on the undoped TiO_2-supported Au overlayers in different orientations, **a** NO_2 adsorption on the *top*-Au sites by its oxygen atoms (configuration A), **b** NO_2 adsorption on the *top*-Au site by its nitrogen atom (configuration B), and **c** NO_2 adsorption on the *side*-Au sites by its oxygen atoms (configuration C)

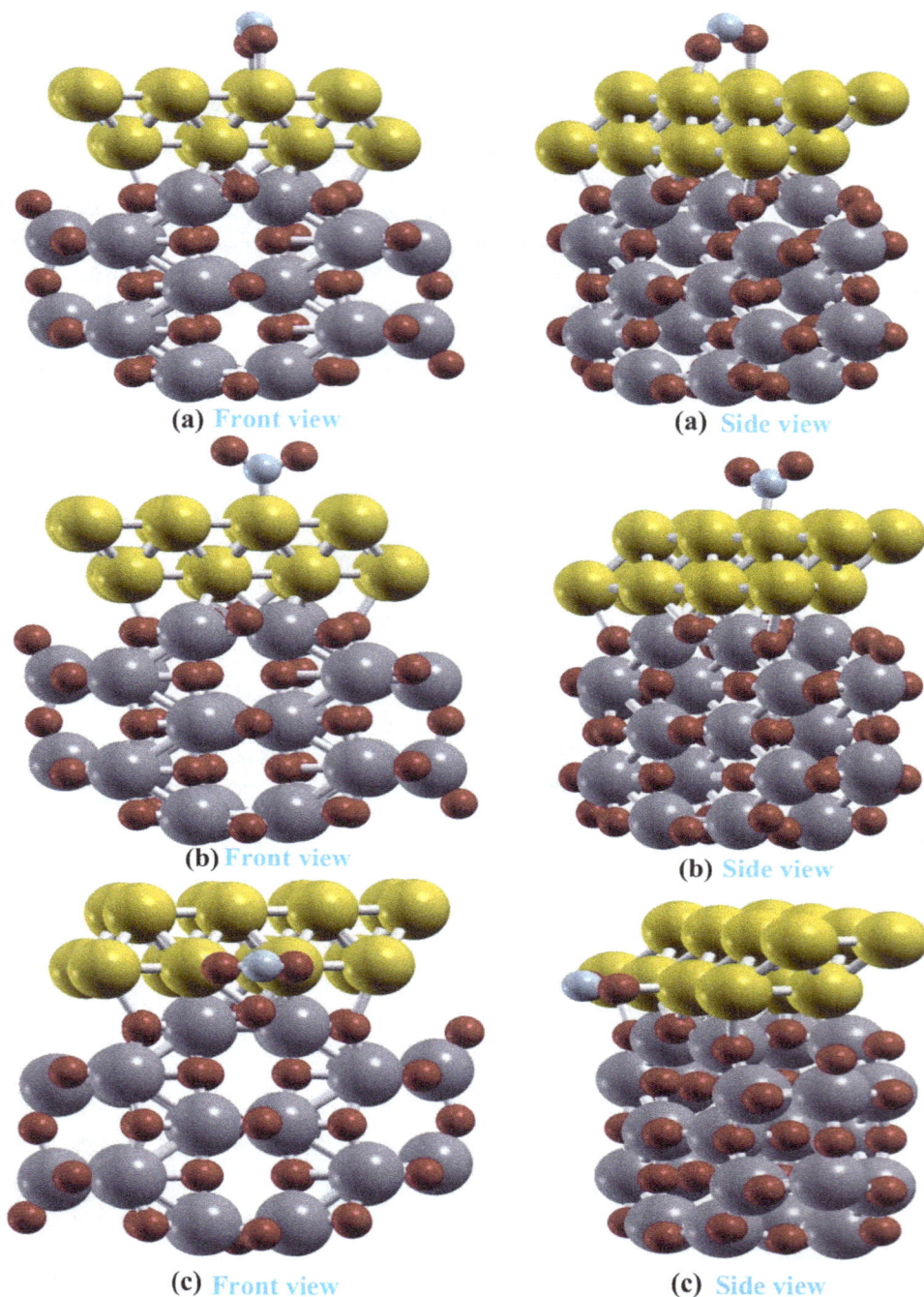

(a) Front view   (a) Side view

(b) Front view   (b) Side view

(c) Front view   (c) Side view

average energy is. They have explained that the nanoparticles containing 72 atoms have the lowest energy (the highest stability among the different types of nanoparticles. The optimized structure of the pristine TiO_2 nanoparticle is shown in Fig. 2. The constructed structure of pristine TiO_2 nanoparticle was geometrically optimized and then coupled with Au nanoparticle to model a metal oxide-supported Au overlayer. The calculated structural parameters for the chosen Au nanoparticle are listed in Table S1. The atomic number ratio between titanium and oxygen atoms should

Table 2 Bond lengths (in Å) and angles (in degrees) of NO_2 molecule adsorbed on the TiO_2-supported Au nanoparticles

Complex type	Au–O_1	Au–O_2	Au–N	N–O_1	N–O_2	O–N–O
Undoped						
A	2.23	2.36	–	1.32	1.34	120.4
B	–	–	2.19	1.30	1.31	122.8
C	2.34	2.36	–	1.32	1.32	120.7
Non-adsorbed	–	–	–	1.20	1.20	134.3

Fig. 5 Optimized geometry configurations of the NO_2 molecule adsorbed on the bare TiO_2 and Au nanoparticles

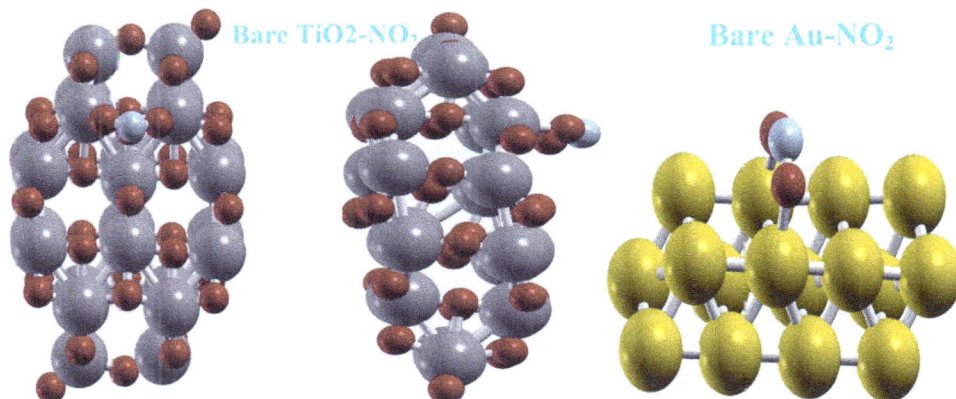

Table 3 Adsorption energies (in eV) and Mulliken charge values (in e) for NO_2 molecule adsorbed on bare TiO_2, bare Au, and TiO_2-supported Au overlayers

Type of complex	Adsorption energy (eV)			Mulliken charge (normal basis sets)	Mulliken charge (large basis sets)
	PBE	DFT-D2	DFT-D3		
Undoped					
A	−2.18	−4.12	−8.04	0.15	0.37
B	−1.41	−2.70	−2.08	0.25	0.47
C	−1.87	−3.60	−7.06	0.16	0.38
Bare TiO_2	−0.72	−1.44	−2.86	+0.05	0.27
Bare Au	−1.23	−2.24	−4.06	+0.04	0.26

be set as 1:2, which was obtained by setting two dangling oxygen atoms in TiO_2. Spin polarization is not used for the optimization of pristine TiO_2 particles due to the even electron number of pure particles. During the optimization process, "Cluster" method was used as efficient eigenvalue solver. For electronic structure calculations, the convergence criterion of 1.0×10^{-6} Hartree was used, whereas the criterion for geometry optimization was set at 1.0×10^{-4} Hartree/bohr. In addition, "Opt" was used as geometry optimizer, which presents a robust scheme for optimization of solid-state structures based on cluster method.

The surface energies for TiO_2 anatase were computed and summarized in Table 1. The calculated data from GGA are in reasonable agreement with the experimentally reported data or other computational works [48]. It indicates that GGA pseudo-potential possesses a reasonable accuracy for calculating the properties of anatase particles. The energy calculations were carried out at the Γ point. The considered simulation box has the dimension of 20 Å × 15 Å × 30 Å. To reduce the additional interactions between neighbor particles, a 11.5 Å distance in three directions was considered.

TiO_2 includes two types of titanium atoms, namely, five-fold coordinated (5f-Ti) and six-fold coordinated (6f-Ti), as well as two types of oxygen atoms, indicated by three-fold coordinated (3f-O) and two-fold coordinated (2f-O) atoms [49, 50]. For the bent geometry of the NO_2 molecule, the calculated N–O bond length and O–N–O bond angle are about 1.20 Å and 134.3°, respectively, which agree reasonably with the previous gas-phase data [51]. Au nanoparticle was supported by TiO_2 anatase to model a TiO_2-supported Au overlayer. Figure 3 displays the equilibrium structure of the undoped TiO_2-supported Au nanoparticle.

Results and discussion

Structural parameters and adsorption energies

Three possible orientations of the NO_2 molecule towards the TiO_2-supported Au overlayers were considered, in which the NO_2 molecule can bind to the surface of Au atoms either by its nitrogen or by oxygen atoms. The relevant configurations of NO_2 adsorption on the TiO_2-supported Au nanoparticles are shown in Fig. 4, as indicated by adsorption types A–C. We found that the NO_2 interaction with Au atoms is stronger than the interaction with TiO_2 nanoparticle. Thus, the surfaces of Au atoms are strongly favored during the adsorption process. Over the TiO_2-supported Au nanoparticle, the NO_2 molecule preferentially interacts with the Au nanoparticle. The interaction by oxygen atoms leads to a bridge configuration of NO_2 on the nanocomposite. As a closer comparison, it is of eminent importance to describe the adsorption configurations and relative orientations in detail. Configuration A shows the adsorption of NO_2 on the top-site Au atoms of the TiO_2-supported Au, while configuration C represents the interaction of NO_2 with the lateral-site Au atoms. In configuration B, we can see that NO_2 molecule interacts with the top-site Au atoms by its nitrogen atom, providing a single contacting point between NO_2 and nanocomposite. Configurations A and C indicate a double contacting point between NO_2 and TiO_2-supported Au. Figure S1 also displays the top views of NO_2 molecule adsorbed on the TiO_2-supported Au overlayers.

Table 2 summarizes the lengths and distances for the newly formed Au–O bonds, N–O bonds of the adsorbed NO_2 molecule, and O–N–O bond angles of NO_2 after the adsorption process. Based on the obtained results, we found that the N–O bonds of the adsorbed NO_2 molecule were elongated due to the considerable electronic density shifts

from the Au–Au bonds of Au nanoparticle and N–O bonds of the NO_2 molecule to the newly formed Au–O and Au–N bonds between the nanocomposite and NO_2. Thus, the adsorption process results in weakening the N–O bonds of the NO_2 molecule. The O–N–O bond angles of NO_2 were increased compared to those of non-adsorbed NO_2 molecule. This increase in the bond angles could be mostly attributed to the elongation of the N–O bonds of the adsorbed NO_2 molecule. In configuration B, the bond angle increase and geometry changing could be ascribed to the formation of new Au–N bond. The formation of new bond is a key reason, which is responsible for changing the sp hybridization of the nitrogen atom in the NO_2 molecule to hybridization with higher p contribution (near-sp^2). Consequently, the p characteristics of bonding molecular orbitals of the nitrogen atom in the adsorbed NO_2 molecule become higher.

For clear comparison, the adsorption configurations of NO_2 molecule on pristine Au and TiO_2 nanoparticles are also represented in Fig. 5, indicating less stable adsorption of NO_2 on the considered bare nanoparticles. Table 3

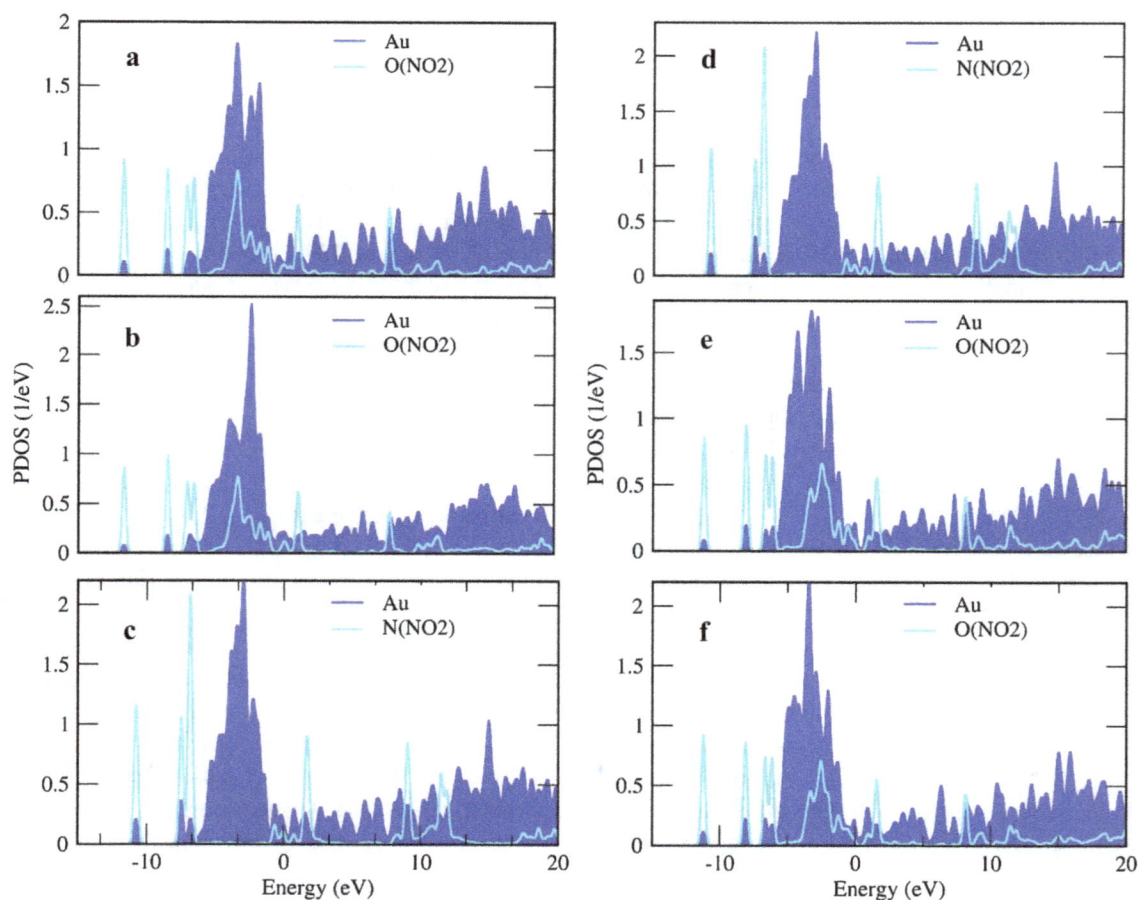

Fig. 6 PDOSs for the adsorption complexes of the undoped TiO_2-supported Au nanoparticles with adsorbed NO_2 molecules. (**a, b**) complex A; (**c, d**) complex B; (**e, f**) complex C

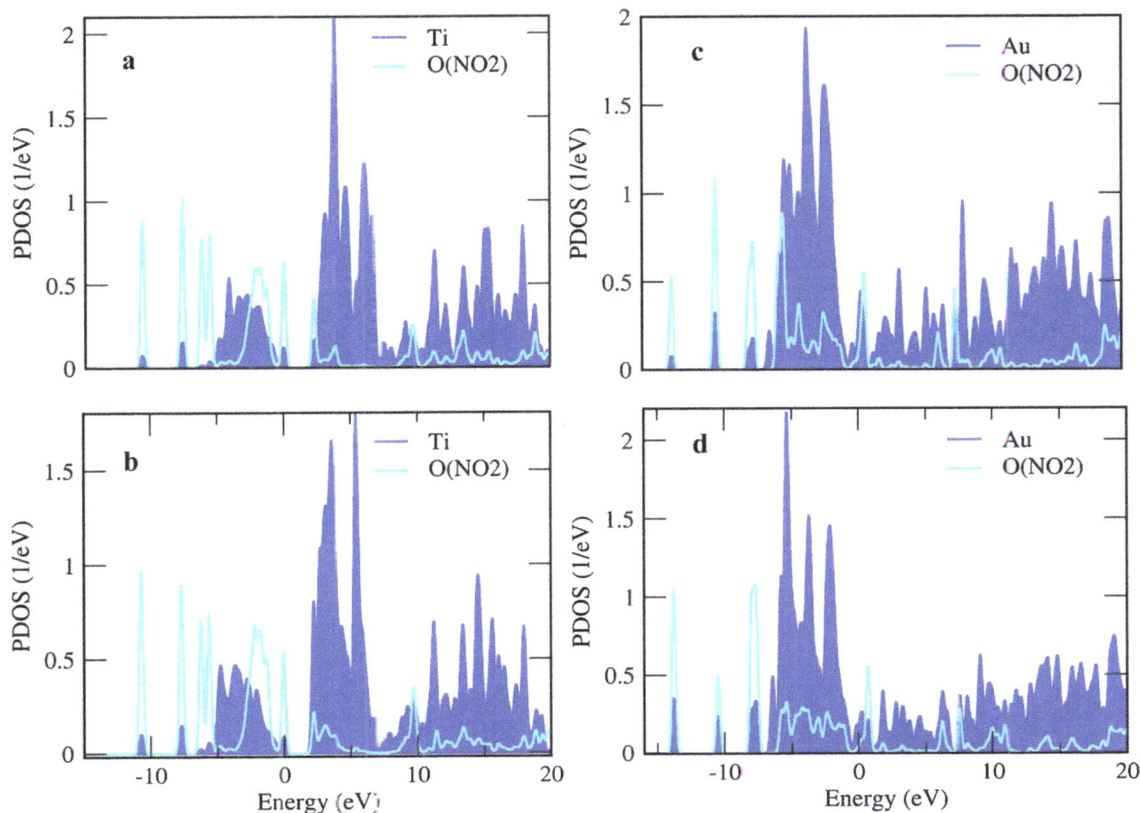

Fig. 7 PDOSs for the adsorption complexes of bare TiO_2 and Au nanoparticles with adsorbed NO_2 molecules. (**a, b**) Bare TiO_2–NO_2; (**c, d**) Bare Au–NO_2

summarizes the adsorption energies of NO_2 molecules adsorbed on the pristine TiO_2-supported Au nanoparticles. Of the three configurations, configuration A has the highest value of adsorption energy, thus making it the most favorable adsorption configuration and, consequently, the most likely binding site to be located on the TiO_2-supported Au.

Therefore, the adsorption of NO_2 on the TiO_2-supported Au nanoparticle (configuration A) is more favorable in energy than the adsorptions in configurations B and C. For both adsorption types A and C, the adsorption energy is higher (more negative) than the adsorption energy of adsorption type B. The reason is that the configurations A and C provide a double contacting point between the NO_2 and TiO_2-supported Au, whereas configuration B shows a single contacting point. NO_2 molecule was strongly coordinated to the TiO_2-supported Au by its two oxygen atoms. In other words, two oxygen atoms of the NO_2 molecule can interact with the TiO_2-supported Au overlayer more efficiently. The adsorption energies calculated from DFT-D2 and DFT-D3 methods are significantly larger than those obtained from the standard DFT calculations. Tamijani et al. [52] reported the results of the adsorption of noble-gas atoms on the TiO_2 (110) surface-based van der Waals

corrected DFT approach and clearly demonstrated the increase in the adsorption energies caused by vdW interactions.

The adsorption energies are considerably increased when the adsorption energies are corrected for dispersion energy, as shown in Table 3. We have calculated the adsorption energies for NO_2 molecule on the bare Au and TiO_2 nanoparticles. As can be seen from Table 3, the adsorption energy of NO_2 molecule on the Au nanoparticle is about -1.23 eV and that of pristine TiO_2 is -0.72 eV, and NO_2 adsorption on the TiO_2-supported Au nanocomposite is found to be -1.48 eV. The higher adsorption energy gives rise to a strong interaction between the adsorbent and adsorbed molecules, and its more negative sign also represents an energy favorable process. Thus, NO_2 adsorption on the TiO_2-supported Au nanocomposite is more energetically favorable than the adsorption on the bare Au and TiO_2 nanoparticles. In the TiO_2-supported Au overlayers, the interactions of NO_2 and TiO_2 are stronger than those between NO_2 and bare TiO_2 nanoparticles, indicating that Au nanoparticle is conducive to the interaction of NO_2 molecule with TiO_2 nanoparticles. In other words, Au nanoparticle enhances NO_2 detection by means of the TiO_2-supported Au nanocomposite-based sensors.

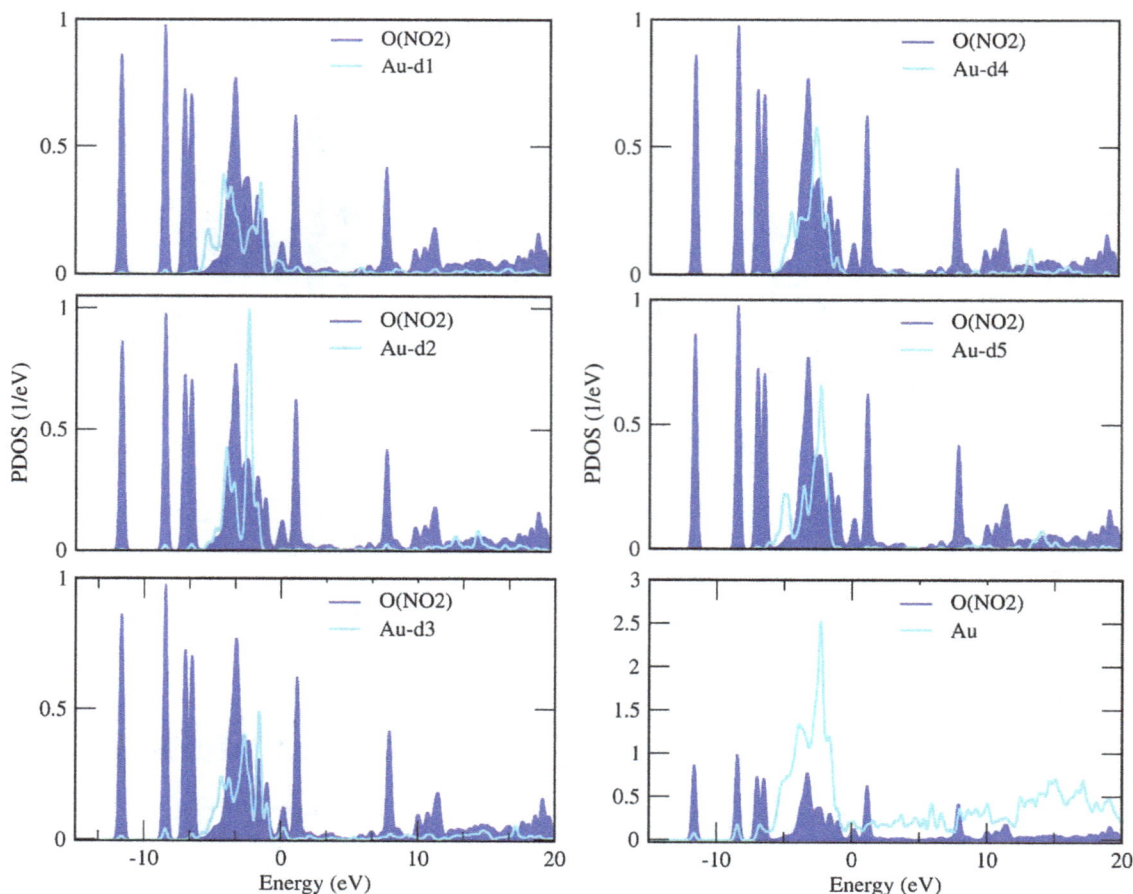

Fig. 8 PDOSs of the oxygen atom of NO_2 molecule and five d orbitals of the Au atom for the TiO_2-supported Au overlayers with adsorbed NO_2 molecules (configuration A)

Therefore, the results of adsorption energies suggest that the TiO_2-supported Au nanoparticle is a good candidate to be utilized in sensing of toxic NO_2 molecules in the environment.

Electronic structures

Figure 6 displays the projected density of states (PDOSs) for NO_2 adsorbed on the pristine TiO_2-supported Au overlayers. Panels (a, b) in this figure show the PDOSs of the Au atom of gold nanoparticle and oxygen atoms of the NO_2 molecule (configuration A). The great overlaps between the PDOSs of these two atoms indicate that the Au atoms form chemical bonds with the oxygen atoms of NO_2. In panels (c, d), we can see also the PDOSs of the Au atom of gold nanoparticle and the nitrogen atoms (configuration B), indicating substantial overlaps and thus formation of chemical Au–N bond. For configuration C, the pertinent PDOSs of the oxygen atoms of NO_2 molecule and the Au atoms are displayed in panels (e, f). As distinct from these PDOSs, the large overlaps show that both oxygen atoms of NO_2 molecule

form chemical bonds with the Au atoms of the gold nanoparticle. This delivers a double interaction point between the NO_2 and TiO_2-supported Au. The interaction of NO_2 molecule with bare Au and TiO_2 nanoparticles was also examined, and the relevant PDOSs are shown in Fig. 7, representing noticeable overlaps between the PDOSs of the interacting atoms. This implies that the gold and titanium atoms form chemical bonds with the oxygen atoms of NO_2.

We also presented the PDOSs of five d orbitals of the Au atom and the oxygen atoms of the NO_2 molecule (configuration A). Figures 8, 9 show the PDOSs of the oxygen atom of NO_2 and different d orbitals of the Au atom, representing higher overlaps between the PDOSs of the oxygen with d^3 orbital of the Au atom. This indicates that the oxygen atom has a substantial mutual interaction with the d^3 orbital of the Au atom. Similarly, Fig. 10 displays the PDOSs of the nitrogen atom of the NO_2 and d orbitals of the Au atom, demonstrating noticeable overlaps between the nitrogen atom and d^2 orbital. For configuration C, the calculated PDOSs of the atom and different orbitals of Au atom are represented in Fig. S2.

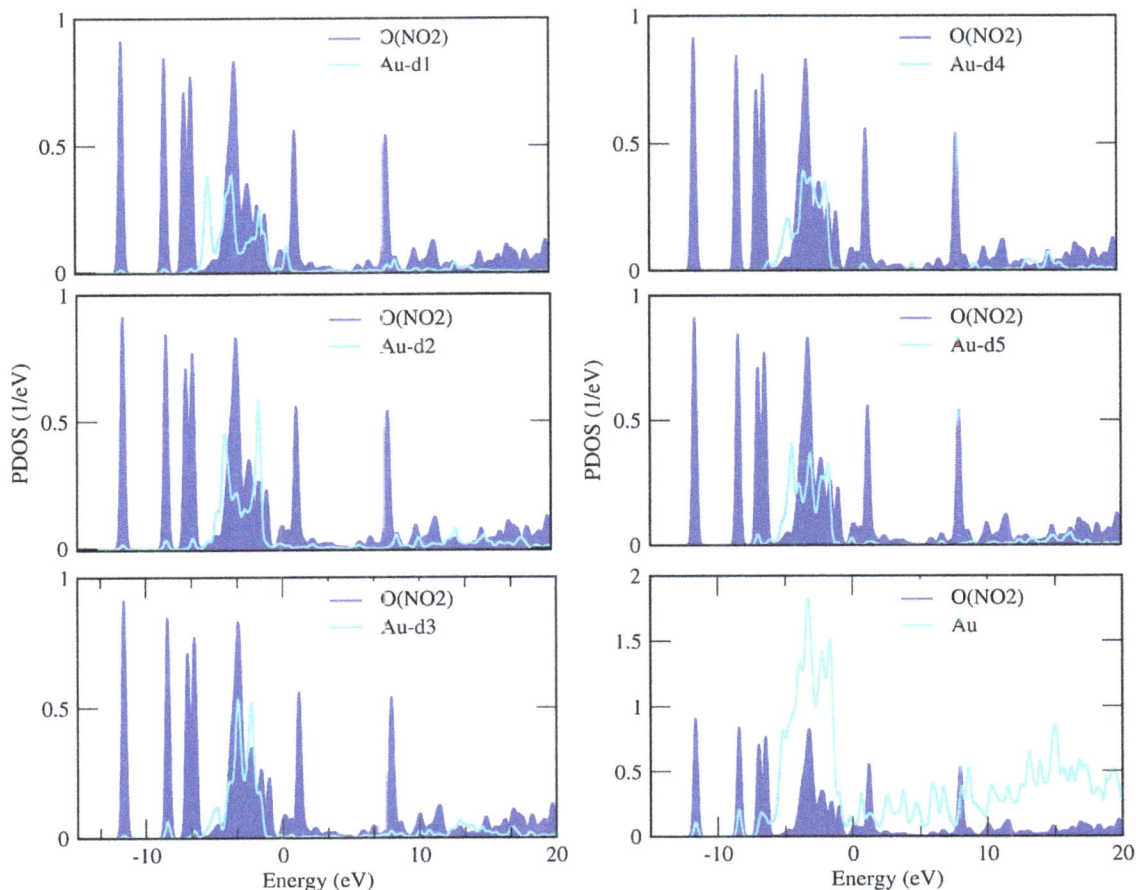

Fig. 9 PDOSs of the oxygen atom of NO_2 molecule and five d orbitals of the Au atom for the TiO$_2$-supported Au overlayers with adsorbed NO_2 molecules (configuration A)

We geometrically optimized the structure of TiO$_2$ anatase, and then calculated its band structure. The energy band structure of pristine anatase system is shown in Fig. 11. This figure represents that the calculated band structure of the valence band maximum (VBM) and the conduction band minimum (CBM) was both positioned at the G point. It also represents that pristine TiO$_2$ was a direct-gap semiconductor material. The calculated bandgap (BG) energy was 2.16 eV for TiO$_2$ anatase, which was slightly lower than the experimental result of 3.2 eV. Important to note is that, the electronic structure calculation using GGA pseudo-potential usually underestimates energy bandgaps [53, 54]. This band-gap underestimation is mostly ascribed to the well-known limitation of the exchange–correlation functional in describing excited states. Here, valance band corresponds to the O 2p orbitals and the conduction band arises from Ti 3d orbitals.

Mulliken charge analysis

The Mulliken population analysis was also conducted in this work to fully describe the charge exchange between the NO_2 molecule and TiO$_2$-supported Au overlayer. This method of charge analysis provides a means of estimating partial atomic charges from calculations implemented by computational chemistry packages. The calculated Mulliken charge values are listed in Table 3. This method assigns an electronic charge to a given atom in the considered system, that is, the gross atom population (GAP). The charge difference, ΔQ_A, is a measure of the difference between the number of electrons on the isolated free atom (Z_A) and the gross atom population:

$$\Delta Q_A = Z_A - GAP_A. \tag{2}$$

For instance, configuration A represents a sizeable charge transfer of about 0.15 |e| (e, the electron charge) from the TiO$_2$-supported Au nanoparticle to the NO_2 molecule, implying that NO_2 plays an important role as a charge acceptor. It is worth mentioning that the charge exchange between adsorbent and adsorbed molecule affects the conductivity of the system, being a great strategy to design more efficient and more appropriate sensor devices for the detection of NO_2 in the environment.

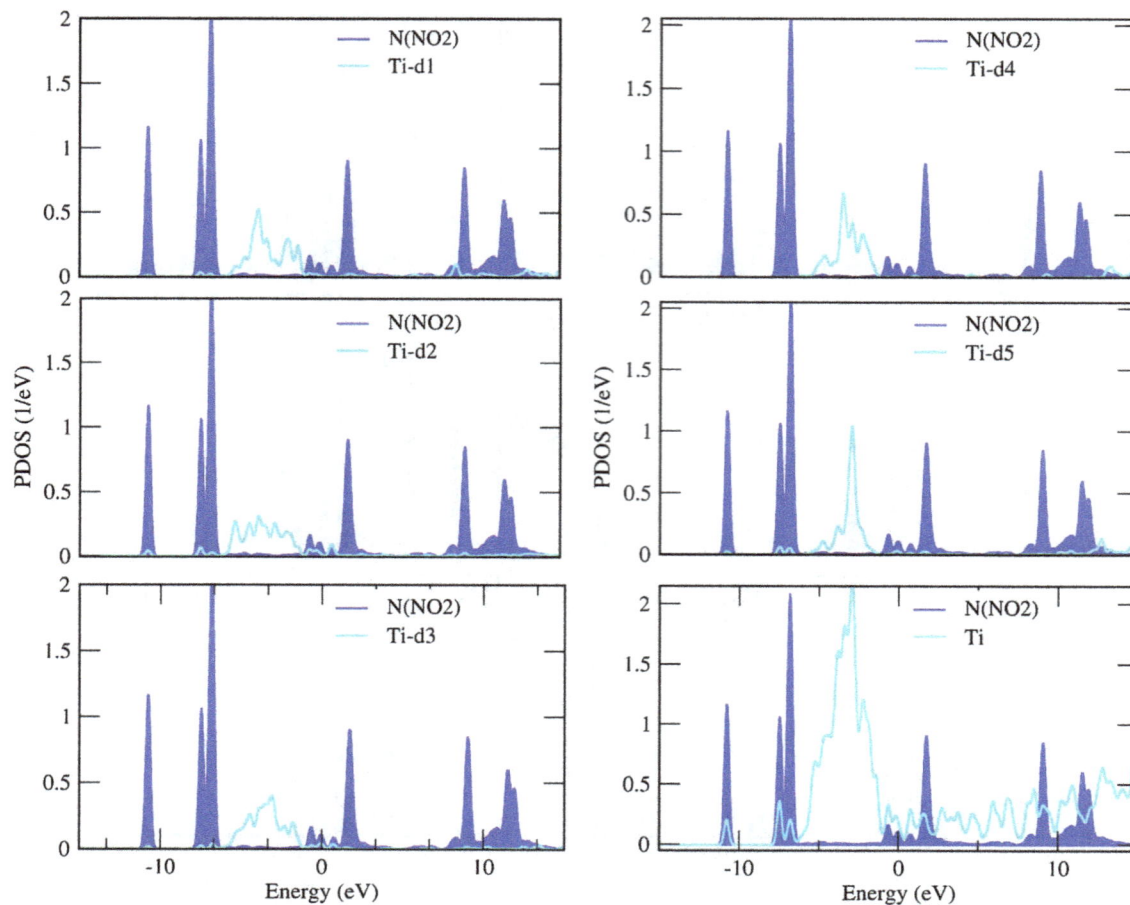

Fig. 10 PDOS of the nitrogen atom of NO_2 molecule and five d orbitals of the Au atom for the TiO_2-supported Au overlayers with adsorbed NO_2 molecules (configuration B)

Fig. 11 Electronic band structure of pristine (undoped) TiO_2 anatase

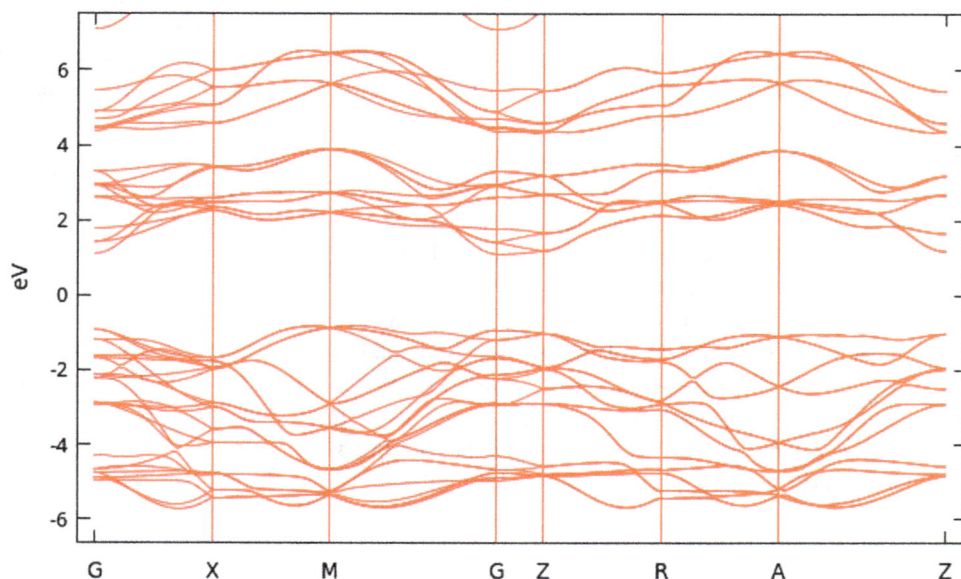

One of the most important problems in Mulliken charge analysis is the intense sensitivity of Mulliken charges to the basis set choice. Fundamentally, a comprehensive basis set for a molecule can be covered by placing a large set of functions on a single atom. In the Mulliken scheme, all the electrons would then be assigned to the single atom. Therefore, it is well known that the Mulliken charge approach has no complete basis set limit, as the precise value which strongly depends on the way the limit is approached. Consequently, an efficient convergence for charges does not exist, and different basis set families may produce extremely different results. To overcome this problem, many modern approaches can be tried for estimating net atomic charges, such as electrostatic potential and natural population methods [55]. We have also calculated the Mulliken charges with the large basis sets of higher accuracy and then found that increasing basis set can simply modify Mulliken charges by approximately 0.22 e. The obtained results are summarized in Table 3. This table shows that the strong basis sets give rise to an increase in the Mulliken charge values.

Conclusions

DFT calculations were carried out to investigate the sensing performance of undoped TiO_2-supported Au nanoparticles for the detection of NO_2 molecules. The adsorption behaviors of NO_2 on the TiO_2-supported Au nanoparticles were investigated in detail. The results show that the N–O bonds of the adsorbed NO_2 molecule were elongated after the adsorption process, which indicates the weakening N–O bonds of the NO_2 molecule. The results also suggest that the interaction of the NO_2 molecule with the TiO_2-supported Au overlayer through its oxygen atoms is energetically more favorable than the interaction of NO_2 through its nitrogen atom. This interaction provides a double interacting point between the NO_2 and TiO_2-supported Au, suggesting the strong adsorption of NO_2 over the substrate. The current results suggest that Au nanoparticle in the TiO_2-supported Au nanocomposites affects the final configuration of TiO_2 nanoparticles with adsorbed NO_2 molecules and, therefore, strengthens the interaction between NO_2 and TiO_2. The substantial overlaps between the PDOSs of the Au and oxygen atoms, as well as, Au and nitrogen atoms indicate the formation of chemical bonds between them. Mulliken population analysis reveals a noticeable charge transfer from the TiO_2-supported Au to the NO_2 molecule, indicating the acceptor characteristic of the NO_2 molecule. Based on the inclusion of vdW interactions, we found that the adsorption energies become larger. However, our findings thus suggest that the TiO_2-supported Au nanoparticles can be utilized as potentially efficient gas sensors for NO_2 recognition.

Acknowledgements This work was supported by Azarbaijan Shahid Madani University.

References

1. Fujishima, A., Honda, K.: Electrochemical photolysis of water at a semiconductor electrode. Nature **238**, 37–38 (1972)
2. Fujishima, A., Hashimoto, K., Watanabe, T.: TiO_2 photocatalysis: fundamentals and applications. Bkc, Tokyo (1999)
3. Abbasi, A., Sardroodi, J.J.: N-doped TiO_2 anatase nanoparticles as a highly sensitive gas sensor for NO_2 detection: insights from DFT computations. J. Environ. Sci. Nano. **3**, 1153–1164 (2016)
4. Abbasi, A., Sardroodi, J.J.: Modified N-doped TiO_2 anatase nanoparticle as an ideal O_3 gas sensor: insights from density functional theory calculations. J. Comp. Theor. Chem. **600**, 2457–2469 (2016)
5. Fujishima, A., Zhang, X., Tryk, D.A.: TiO_2 photocatalysis and related surface phenomena. J. Surf. Sci. Rep. **63**, 515–582 (1992)
6. Batzilla, M., Morales, E.H., Diebold, U.: Surface studies of nitrogen implanted TiO_2. J. Chem. Phys. **339**, 36–43 (2007)
7. Isley, S.L., Penn, R.L.: Relative brookite and anatase content in sol–gel synthesized titanium dioxide nanoparticles. J. Phys. Chem. B **110**, 15134 (2006)
8. Rumaiz, A.K., Woicik, J.C., Cockayne, E., Lin, H.Y., Jaffari, G.H., Shah, S.I.: Oxygen vacancies in N doped anatase TiO_2: experiment and first-principles calculations. J. Appl. Phys. Lett. **95**, 262111 (2009)
9. Buonsanti, R., Grillo, V., Carlino, E., Giannini, C., Kipp, T., Cingolani, R., Cozzoli, P.D.: Nonhydrolytic synthesis of high-quality anisotropically shaped brookite TiO_2 nanocrystals. J. Am. Chem. Soc. **130**, 11223–11233 (2008)
10. Cassaignon, S., Koelsch, M., Jolivet, J.P.: Selective synthesis of brookite, anatase and rutile nanoparticles: thermolysis of $TiCl_4$ in aqueous nitric acid. J. Mater. Sci. **42**, 6689–6695 (2007)
11. Di Paola, A., Addamo, M., Bellardita, M., Cazzanelli, E., Palmisano, L.: Preparation of photocatalytic brookite thin films. Thin Solid Films **515**(7), 3527–3529 (2007)
12. Djaoued, Y., Bruning, R., Bersani, D., Lottici, P.P., Badilescu, S.: Sol–gel nanocrystalline brookite-rich titania films. Mater. Lett. **58**(21), 2618–2622 (2004)
13. Iskandar, F., Nandiyanto, A.B.D., Yun, K.M., Hogan, C.J., Okuyama, K., Biswas, P.: Enhanced photocatalytic performance of brookite TiO_2 macroporous particles prepared by spray drying with colloidal templatings. Adv. Mater. **19**, 1408–1412 (2007)
14. Kobayashi, M., Petrykin, V.V., Kakihana, M.: One-step synthesis of TiO2 (B) nanoparticles from a water-soluble titanium complex. Chem. Mater. **19**, 5373–5376 (2007)
15. Li, J.G., Ishigaki, T., Sun, X.D.: Anatase, brookite, and rutile nanocrystals via redox reactions under mild hydrothermal conditions: phase-selective synthesis and physicochemical properties. J. Phys. Chem. C **111**, 4969–4976 (2007)
16. Reddy, M.A., Kishore, M.S., Pralong, V., Caignaert, V., Varadaraju, U.V.: Room temperature synthesis and Li insertion into nanocrystalline rutile TiO_2. Electrochem. Commun. **8**(8), 1299–1303 (2006)
17. Shibata, Y., Irie, H., Ohmori, M., Nakajima, A., Watanabe, T., Hashimoto, K.: Comparison of photochemical properties of brookite and anatase TiO_2 films. Phys. Chem. Chem. Phys. **6**, 1359–1362 (2007)

18. Haruta, M., Kobayashi, T., Sano, H., Yamada, N.: Novel gold catalysts for the oxidation of carbon monoxide at a temperature far below 0 & #xB0;C. J. Chem. Lett. **16**(2), 405–408 (1987)

19. Landon, P., Collier, P.J., Papworth, A.J., Kiely, C.J., Hutchings, G.J.: Direct formation of hydrogen peroxide from H_2/O_2 using gold catalysts. Chem. Commun. **18**, 2058 (2002)

20. Molina, L.M., Hammer, B.: Some recent theoretical advances in the understanding of the catalytic activity of Au. Appl. Catal. A Gen. **291**, 21–31 (2005)

21. Okumura, M., Tsubota, S., Haruta, M.: Preparation of supported gold catalysts by gas-phase grafting of gold acethylacetonate for low-temperature oxidation of CO and of H_2. J. Mol. Catal. A: Chem. **199**, 73–84 (2003)

22. Lopez, N., Norskov, J.K.: Catalytic CO oxidation by a gold nanoparticle: a density functional study. J. Am. Chem. Soc. **124**, 11262–11263 (2002)

23. Chen, M.S., Goodman, D.W.: Structure–activity relationships in supported Au catalysts. Catal. Today. **111**, 22–33 (2006)

24. Kung, H.H., Kung, M.C., Costello, C.K.: Supported Au catalysts for low temperature CO oxidation. J. Catal. **216**, 425–432 (2003)

25. Hayashi, T.M., Tanaka, K., Haruta, M.: Selective vapor-phase epoxidation of propylene over Au/TiO_2 catalysts in the presence of oxygen and hydrogen. J. Catal. **178**, 566–575 (1998)

26. Salama, T., Ohnishi, R., Shido, T., Ichikawa, M.: Highly selective catalytic reduction of NO by H_2 over Au 0 and Au(I) impregnated in NaY zeolite catalysts. J. Catal. **162**(2), 169–178 (1996)

27. Rodriguez, J.A., Liu, G., Jirsak, T., Hrbek, J., Chang, Z.P., Dvorak, J., Maiti, A.: Activation of gold on titania: adsorption and reaction of SO_2 on $Au/TiO2$ (110). J. Am. Chem. Soc. **124**, 5242–5250 (2002)

28. Chen, M.S., Goodman, D.W.: The structure of catalytically active gold on titania. Science **306**(5694), 252–255 (2004)

29. Cosandey, F., Madey, T.E.: Growth, morphology, interfacial effects and catalytic properties of au on TiO_2. Surf. Rev. Lett. **8**, 73 (2001)

30. Vittadini, A., Selloni, A.: Small gold clusters on stoichiometric and defected TiO_2 anatase (101) and their interaction with CO: a density functional study. J. Chem. Phys. **117**, 353–361 (2002)

31. Chrétien, S., Metiu, H.: O_2 evolution on a clean partially reduced rutile TiO_2 (110) surface and on the same surface precovered with Au_1 and Au_2: the importance of spin conservation. J. Chem. Phys. **127**, 084704 (2007)

32. Molina, L.M., Rasmussen, M.D., Hammer, B.: Adsorption of O_2 and oxidation of CO at Au nanoparticles supported by TiO_2 (110). J. Chem. Phys. **120**(16), 7673 (2004)

33. Lin, C., Wen, G., Liang, A., Jiang, Z.: A new resonance Rayleigh scattering method for the determination of trace O_3 in air using rhodamine 6G as probe. J. RSC. Adv. **3**, 6627–6630 (2013)

34. Hohenberg, P.M., Kohn, W.: Inhomogeneous electron gas. J. Phys. Rev. **16**, B864–B868 (1964)

35. Kohn, W., Sham, L.: Self-consistent equations including exchange and correlation effects. J. Phys. Rev. **140**, A1133–A1138 (1965)

36. Ozaki, T., Kino, H., Yu, J., Han, M.J., Kobayashi, N., Ohfuti, M., Ishii, F., et al.: The code OpenMX, pseudoatomic basis functions, and pseudopotentials are available on a web site 'http://www.openmxsquare.org' (2017). Accessed 2 Mar 2017

37. Ozaki, T.: Variationally optimized atomic orbitals for large-scale electronic structures. J. Phys. Rev. B **67**, 155108 (2003)

38. Ozaki, T., Kino, H.: Numerical atomic basis orbitals from H to Kr. J. Phys. Rev. B **69**, 195113 (2004)

39. Perdew, J.P., Burke, K., Ernzerhof, M.: Generalized gradient approximation made simple. J. Phys. Rev. Lett. **78**, 1396 (1981)

40. Kok1j, A.: Computer graphics and graphical user interfaces as tools in simulations of matter at the atomic scale. J. Comput. Mater. Sci. **28**, 155–168 (2003)

41. Grimme, S.: Semiempirical GGA-type density functional constructed with a long-range dispersion correction. J. Comput. Chem. **27**, 1787–1799 (2006)

42. Grimme, S., Antony, J., Ehrlich, S., Krieg, H.: A consistent and accurate ab initio parametrization of density functional dispersion correction (DFT-D) for the 94 elements H–Pu. J. Chem. Phys. **132**, 154104 (2010)

43. Grimme, S., Ehrlich, S., Goerigk, L.: Effect of the damping function in dispersion corrected density functional theory. J. Comput. Chem. **32**, 1456–1465 (2011)

44. Downs, R.T.: Web page at: http://rruff.geo.arizona.edu/AMS/amcsd.php (2014). Accessed 9 May 2014

45. Wyckoff, R.W.G.: Crystal structures, 2nd edn. Interscience Publishers, New York (1963)

46. Lei, Y., Liu, H., Xiao, W.: First principles study of the size effect of TiO_2 anatase nanoparticles in dye-sensitized solar cell. Modelling Simul. Mater. Sci. Eng. **18**, 025004 (2010)

47. Liu, J., Dong, L., Guo, W., Liang, T., Lai, W.: CO adsorption and oxidation on N-doped TiO_2 nanoparticles. J. Phys. Chem. C **117**, 13037–13044 (2013)

48. Lazzeri, M., Vittadini, A., Selloni, A.: Structure and energetics of stoichiometric TiO_2 anatase surfaces. Phys. Rev. B. **63**, 155409 (2001)

49. Wu, C., Chen, M., Skelton, A.A., Cummings, P.T., Zheng, T.: Adsorption of arginine–glycine–aspartate tripeptide onto negatively charged rutile (110) mediated by cations: the effect of surface hydroxylation. ACS Appl. Mat. Interfaces **5**, 2567–2579 (2013)

50. Liu, J., Liu, Q., Fang, P., Pan, C., Xiao, W.: First principles study of the adsorption of a NO molecule on N-doped anatase nanoparticles. J. Appl. Surf. Sci. **258**, 8312–8318 (2012)

51. Breedon, M., Spence, R.M., Yarovsky, I.: Adsorption of NO_2 on oxygen deficient ZnO (2110) for gas Sensing applications: a DFT study. J. Phys. Chem. C **14**(39), 16603–16610 (2010)

52. Tamijani, A.A., Salam, A., de-Lara-Castells, P.: Adsorption of noble-gas atoms on the TiO2 (110) surface: an ab initio assisted study with van der Waals corrected DFT. J. Phys. Chem. C **120**(32), 18126–18139 (2016)

53. Longa, M., Cai, W., Wang, Z., Liu, G.: Correlation of electronic structures and crystal structures with photocatalytic properties of undoped, N-doped and I-doped, TiO_2. Chem. Phys. Lett. **420**, 71–76 (2006)

54. Gao, H., Zhou, J., Dai, D., Qu, Y.: Photocatalytic activity and electronic structure analysis of N-doped anatase TiO_2: a combined experimental and theoretical study. Chem. Eng. Technol. **32**(9), 867–872 (2009)

55. Reed, A.E., Weinstock, R.B., Weinhold, F.: Natural population analysis. J. Chem. Phys. **83**(2), 735–746 (1985)

Biologically synthesized silver nanoparticles by aqueous extract of *Satureja intermedia* C.A. Mey and the evaluation of total phenolic and flavonoid contents and antioxidant activity

Somayeh Firoozi[1] · Mina Jamzad[1] · Mohammad Yari[2]

Abstract Developing low cost and environmentally friendly methods for metallic nanoparticles is an increasing need. Using plants towards synthesis of nanoparticles are beneficial with the presence of bio-molecules in plants, which can act as capping/stabilizing and reducing agents. In the present attempt, we describe rapid biosynthesis of silver nanoparticles by *Satureja intermedia* C. A. Mey (Lamiaceae) aqueous extract. Synthesized nanoparticles were characterized by UV–Visible spectroscopy, X-ray diffraction (XRD), transmission electron microscopy (TEM) and the chemical groups in plant extract were detected by Fourier Transform Infra-Red (FT-IR) spectroscopy. The XRD study showed crystalline nature and face cubic center shape for nanoparticles. TEM study showed that the mean diameter and standard deviation for the silver nanoparticles were 29.29 ± 28.18 nm. Total phenolic and flavonoid contents and radical scavenging activity of the aqueous extract and SNPs/extract mixture, were also evaluated in this study. It can be concluded that the aerial parts of *S. intermedia* is a good source of phenolic compounds, a potent antioxidant and a valuable choice for bio-reduction and biosynthesis of silver nanoparticles.

Keywords *Satureja intermedia* · Silver nanoparticles · Phenolic compounds · Flavonoids · Radical scavenging effect

---

✉ Mina Jamzad
minajamzadiau@gmail.com; m.jamzad@qodsiau.ac.ir

[1] Department of Chemistry, Shahr-e-Qods Branch, Islamic Azad University, Tehran, Iran

[2] Department of Chemistry, Islamshahr Branch, Islamic Azad University, P.O. Box: 33135-369, Islamshahr, Iran

Introduction

Nanotechnology is one of the most fascinating research areas in modern material science. In general, particles with a size less than 100 nm are referred to as nanoparticles. Entirely novel and enhanced characteristics such as size, distribution and morphology have been revealed by these particles in comparison to the larger particles of bulk material [1]. Nanoparticles are gaining importance in the fields of biology, medicine and electronics owing to their unique physical and biological properties [2]. Among the metallic nanoparticles, Silver has been enormously utilized for its diverse applications in the fields of bio-labeling, opt biosensors, polarizing filters, electrical batteries, cancer cell imaging, drug delivery systems etc. [3]. Silver has long been recognized as having an inhibitory effect toward many microorganisms [4] while it is not toxic to human cells in low concentrations [5]. The most widely used and known applications of silver and silver nanoparticles are in the medical industry; for example, in topical ointments to prevent infection of burns or open wounds and also in medical devices and implants [6].

Many techniques of synthesizing silver nanoparticles (SNPs) have been reported in the literature; and chemical reduction is the most commonly used method for the preparation of SNPs [7]. Reducing agents commonly used in chemical reduction are borohydride, citrate, ascorbate, and elemental hydrogen [8]. Most of these methods are extremely expensive and also involve the use of toxic and hazardous chemicals, which may pose potential environmental and biological risks. Since noble metal nanoparticles are widely applied to areas of human contact, there is a growing need to develop environmental friendly processes for nanoparticle synthesis that do not use toxic chemicals [9].

Nowadays, green synthesis of nanoparticles from plants is an utmost emerging field in nanotechnology. Previously, noble nanoparticles were synthesized by using various plant materials like: *Mentha piperita* [10]; *Ipomoea pescaprae* [11]; *Ocimum sanctum* [12]; *Amaranthus dubius* [13] etc. The exact mechanism of SNPs synthesis mediated by plant extracts is not yet fully understood. It is expected that plants with higher reducing capacity are more potent in reducing metallic ions to metallic nanoparticles [14]. Phenolic compounds are the major constituents of antioxidants of most plant species and their antioxidant activity is mainly due to their redox properties. As a result, they can act as reducing agents in neutralizing free radicals [15, 16] and the reduction of metallic ions to metallic nanoparticles [17].

In this study, we evaluated total phenolic and flavonoid contents in aqueous extract of *Saturaja intermedia* C. A. Mey (Lamiaceae). Results showed that the extract is rich of these biomolecules and the plant can be a good choice for bio-reduction processes. So we investigated biosynthesis of SNPs mediated by the aqueous extract of *S. intermedia* and found an easy and rapid procedure for this purpose. Anti-oxidant activity of the extract and the SNPs/extract mixture, were also evaluated in this study.

Experimental

Chemicals

1, 1-Diphenyl-2-picrylhedrazyl (DPPH) and Gallic acid were prepared from Sigma-Aldrich (US). Quercetin, Folin-Ciocalteu reagent, Aluminum chloride, Sodium bicarbonate, Sodium acetate, Butylated hydroxyl toluene (BHT), Silver nitrate and all the solvents were purchased from Merck (Germany).

Instrumental

UV–Visible spectrophotometer (CECIL, CE 7800, UK); X-Ray diffractometer (Intel, EQUINX -3000, France); FT-IR spectrophotometer (Perkin-Elmer, Spectrum100, Germany); Transmission electron microscope (TEM) Σ1GMAVP, Zeiss, Germany); Ultrasonic (Elma, S15H, Germany); Centrifuge (EBA20, Hettich, Germany).

Plant material

The aerial parts of *Satureja intermedia* C. A. Mey (including leaves, stems and flowers), was collected during flowering stage from Gardaneh Almas (2350–2400 m), between Astara and Ardabil, (Iran), on June 2014. Voucher specimen (No: 83139) has been deposited at the Herbarium of Research Institute of Forests and Rangelands (TARI), Tehran, Iran.

Preparation of the extracts

Dried and powdered aerial parts of *Satureja intermedia* (20 g) were soaked in de-ionized water (200 mL) and boiled for 10 min. After filtration, the extract was concentrated to 1/3 initial volume and then centrifuged at 4000 rpm for 15 min and kept in a dark bottle at 22 °C for further uses.

Total phenolic contents

Determination of total phenolic contents was carried out following the Folin-Ciocalteu method by Singleton and Rossi [18]. Briefly, 100 μL of the extract (20 mg/mL) was mixed up with 0.75 mL of Folin-Ciocalteu reagent (previously diluted 10-fold with distilled water) and allowed to stand at room temperature (22 °C) for 5 min. Then 0.75 mL of sodium bicarbonate (60 mg/mL) solution was added and mixed thoroughly. Finally, the sample was measured spectrophotometrically at 765 nm after 90 min at room temperature. The experiment was also repeated for the mixture including synthesized silver nanoparticles and the plant extract. A calibration curve was plotted for the standard solutions of Gallic acid (0–100 ppm) with the standard curve equation ($Y = 0.0105X + 0.0138, R^2 = 0.9955$). Total phenolic contents of the samples were expressed in terms of Gallic acid equivalent (mg/L and or mg/g). Experiments were performed in triplicate and expressed as mean ± standard deviation (SD).

Total flavonoid contents

Total flavonoid contents were determined by aluminum chloride colorimetric method which is based on the formation of a complex flavonoid-aluminum having the maximum absorption at 415 nm [19]. Briefly, 0.5 mL of the extract and the mixture including extract and silver nanoparticles (20 mg/mL) were dissolved in methanol (1.5 mL) separately, and then 10 % aluminum chloride (0.1 mL) and 1.0 M sodium acetate (0.1 mL) were added to the samples. Finally, distilled water (2.8 mL) was added, and the solutions were incubated at room temperature. After half an hour the absorbance of the reaction mixtures was measured at 415 nm by a UV–Visible spectrophotometer. The calibration curve was plotted by employing the same procedure for the standard solutions of quercetin (0–100 ppm) with the standard curve equation

$(Y = 0.0673X + 0.0051, R^2 = 0.9961)$. Flavonoid contents of the extract and SNPs/extract mixture were expressed in terms of quercetin equivalent (mg/L and or mg/g). Experiments were performed in triplicate and expressed as mean $\pm$ standard deviation (SD).

Synthesis of silver nanoparticles

A 100 mL aliquot of a 0.01 M solution of $AgNO_3$ was gradually added to 20 μL of the aqueous extract of Satureja intermedia. The mixture was kept in an ultrasonic during the addition, and then was stirred in a magnetic stirrer (500 rpm) at room temperature for 48 h. Silver nanoparticles were gradually obtained during the reaction. The solution turned light yellow after 2 h and then a dark brown color appeared, indicating the formation of SNPs. The synthesized nanoparticles were filtered with a membrane filter paper (0.2 μm) and washed by de-ionized water. Finally, SNPs were dried for one hour in an oven at 100 °C and kept at room temperature for further evaluations.

DPPH radical scavenging assay

Electron donation ability of the extract, SNPs/extract mix and BHT as a standard were measured from the bleaching of the purple-colored methanol solution of DPPH. To determine the radical scavenging ability, the method reported by Bondet et al. was used [20]. Briefly, 2.5 mL of DPPH solution in methanol (40 μg/mL) (freshly prepared), was added to 10 μL of the samples. After 30 min incubation in dark, absorbance of the test tubes, were taken by a spectrophotometer at 517 nm. The percentage of scavenged DPPH was calculated using the following equation:$\%I = 100 \left[\frac{A_c - A_s}{A_c} \right]$, where A_c is the absorbance of control (containing all reagents except the test samples), and A_s is the absorbance of sample. Experiments were done in duplicate and the average was calculated for the absorbance of each test tube.

Characterization of silver nanoparticles

UV–Visible absorbance spectroscopy of synthesized SNPs was performed. The sample was prepared by diluting 50 μL of the mixture (collected at the end of reaction) in 3 mL of de-ionized water. The wave length was ranged from 200 to 800 nm and de-ionized water was used as blank. Fourier-Transform Infra-Red (FT-IR) spectra was carried out by KBr pellet method to identify the possible chemical functional groups and bonds in the synthesizing medium containing the plant extract responsible for the reduction and capping SNPs. Crystalline nature of metallic silver nanoparticles was examined using an X-ray diffractometer, equipped with Cu Kα radiation source using Ni as filter at a setting of 30 kV/30 mA. Transmission electron microscopy (TEM) was performed for the determination of morphology, size and crystalline nature of the synthesized SNPs.

Results and discussion

Total phenolic and flavonoid contents of the aqueous extract of S. intermedia and SNPs/extract mix were evaluated (Table 1). As it is seen the plant is rich in flavonoids (21.123 ± 0.0698 mg/L; 2.006 ± 0.0087 mg/g) and phenolic compounds (25.289 ± 0.0698 mg/L; 2.398 ± 0.0028 mg/g). Total flavonoids and phenolic compounds in silver suspension were found (3.758 ± 0 mg/L; 0.357 ± 0 mg/g) and (5.352 ± 0.078 mg/L; 0.507 ± 0.0031 mg/g), respectively. Each sample had three replicates and data were shown as mean $\pm$ standard deviation (SD).

In the next part of our study, green synthesis of silver nanoparticles through S. intermedia aqueous extract was carried out. The appearance of pale yellow to dark brown coloration of the reaction mixture indicated the biosynthesis of silver nanoparticles (Fig. 1). It is well known that silver nanoparticles exhibit striking colors (light yellow to brown) due to the excitation of surface Plasmon vibrations in the particles [21].

Table 1 Total phenolic and flavonoid contents of Satureja intermedia aqueous extract and silver nanoparticles/extract mixture

	Phenolic contents (mg/L)[a]	Phenolic contents (mg/g)[b]	Flavonoid contents (mg/L)[c]	Flavonoid contents (mg/g)[d]
Extract	25.289 ± 0.0698	2.398 ± 0.0028	21.123 ± 0.0698	2.006 ± 0.0087
SNPs/extract	5.352 ± 0.078	0.507 ± 0.0031	3.758 ± 0	0.357 ± 0

Values are expressed as mean $\pm$ SD, n = 3

[a] Total phenolic content in terms of Gallic acid equivalent (mg of Gallic acid/L of the extract and or SNPs/extract mixture)

[b] Total phenolic content in terms of Gallic acid equivalent (mg of Gallic acid/g of dried plant material)

[c] Total flavonoid content in terms of quercetin equivalent (mg of Quercetin/L of the extract and or SNPs/extract mixture)

[d] Total flavonoid content in terms of quercetin equivalent (mg of Quercetin/g of dried plant material)

Fig. 1 Color changes before
(**a**) and after (**b**) biosynthesis of
SNPs by *Satureja intermedia*
C.A. Mey (**c**) aqueous extract

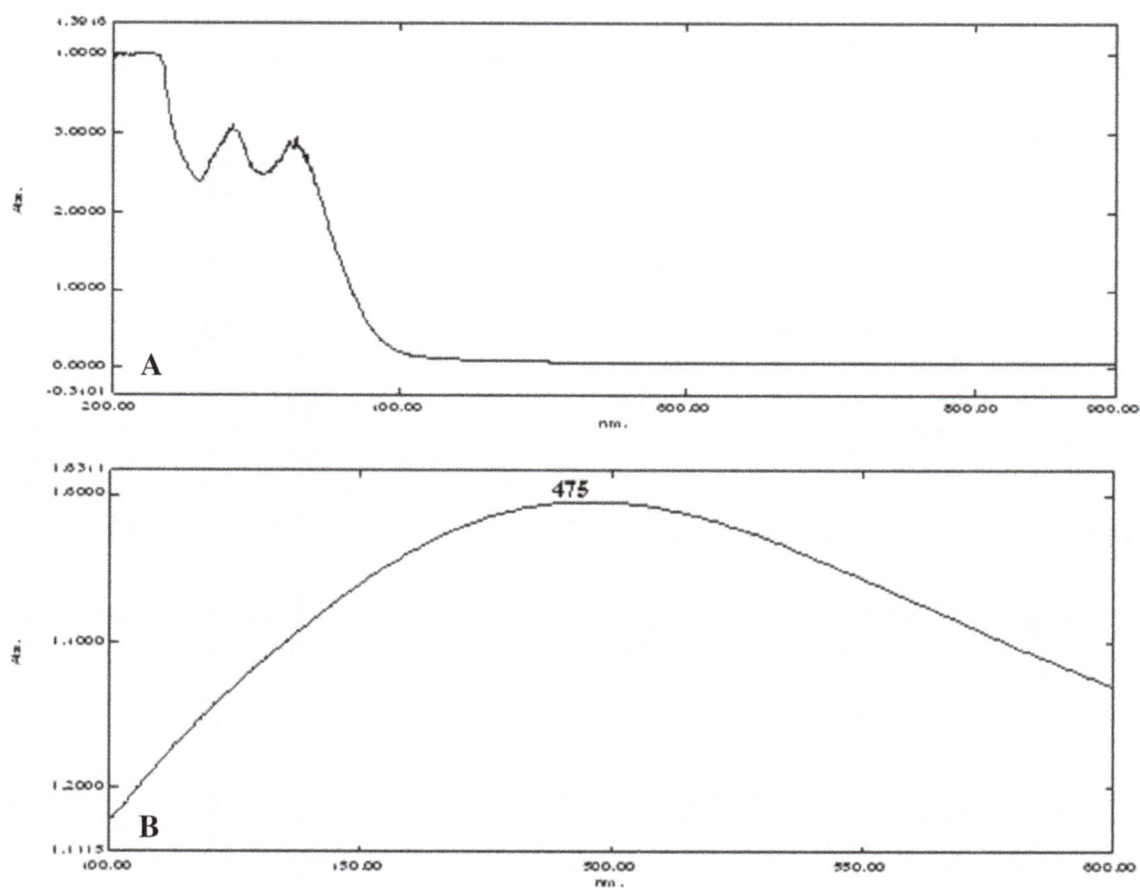

Fig. 2 UV-Vis absorption spectra of **a** *Satureja intermedia* C. A. Mey and **b** SNPs/*Satureja intermedia* extract mixture

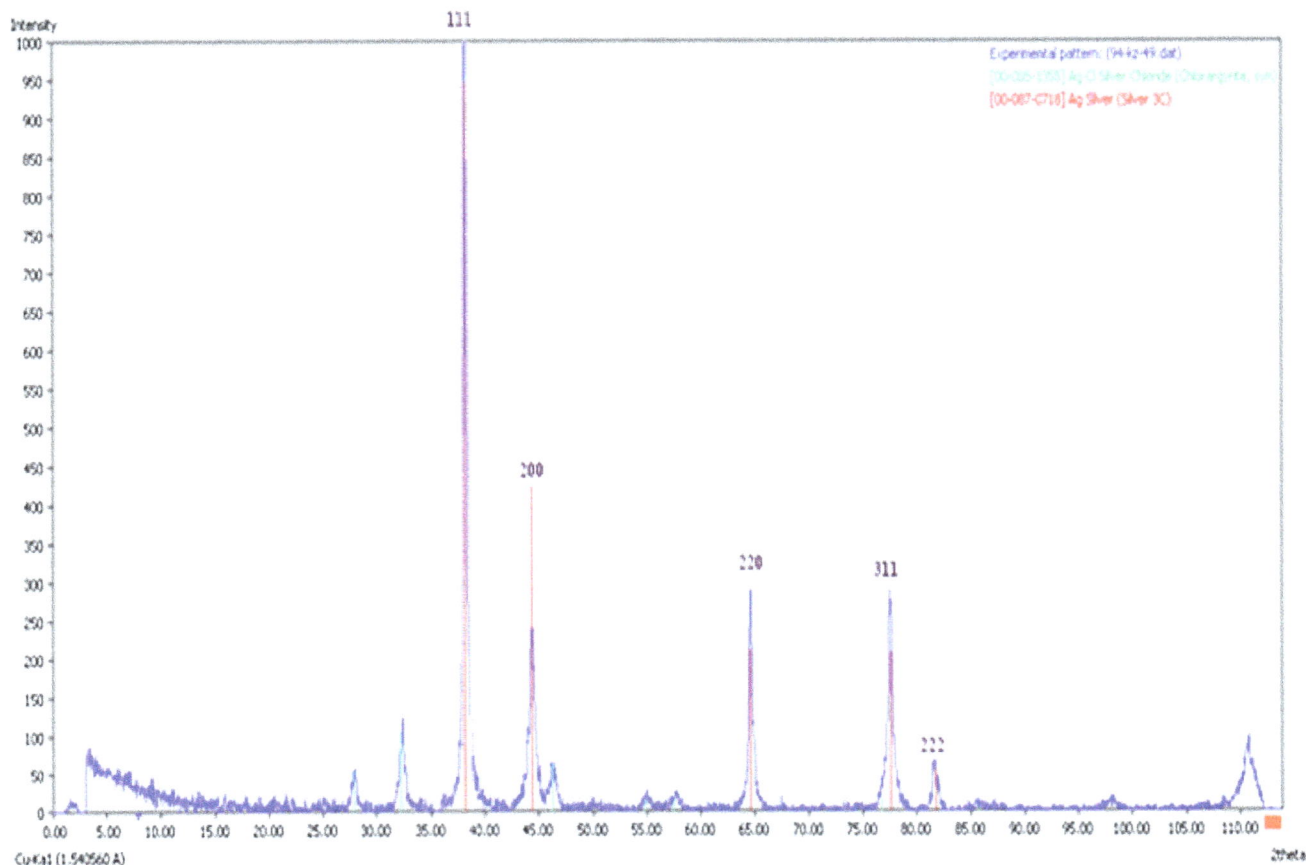

Fig. 3 X-ray diffraction spectrum of silver nanoparticles synthesized by *Satureja intermedia* C. A. Mey aqueous extract

UV–Visible spectra analysis

UV–Vis spectrum of SNPs showed a strong surface Plasmon resonance centered at 475 nm (Fig. 2) confirmed the nanocrystalline character of the particles [22]. It is generally recognized that UV–Vis spectroscopy could be used to examine size and shape-controlled nanoparticles in aqueous suspensions [23]. Silver nanoparticles have free electrons, which give rise to an SPR absorption band due to the combined vibration of electrons of metal nanoparticles in resonance with the light wave [24, 25].

XRD analysis

X-ray diffraction analysis was carried out to confirm the nature of nanoparticles (Fig. 3). The high intense peaks at 2θ degrees of 38.18, 44.54, 64.86, 77.55 and 81.54 can be attributed to the (111), (200), (220), (311) and (222) Bragg reflections, respectively, which confirm the face centered cubic (FCC) structure of SNPs. The intensity of peaks reflected the high degree of crystallinity of the silver nanoparticles. However, the diffraction peaks are broad which indicating that the crystallite size is very small.

Apart from these, there were also few other sharp peaks, which might be due to the existence of the organic phytochemicals in the mix [26]. The average size of SNPs was calculated 25.05 nm, according to Debye–Scherrer equation: $(D = K\omega/\beta \cos\theta)$. The equation uses the reference peak width at angle θ, where ω is the X-ray wavelength (1.540560 Å), β is the width of XRD peak at half height and K is the shape factor with value 0.9.

TEM study

A typical transmission electron microscope (TEM) image of the nanoparticles formed is presented in Fig. 4. The result showed narrow particle size distributions with diameters in range of (1.11–57.47) nm. Result was established base on 429 individual measurements. The mean diameter and standard deviation of silver nanoparticles were found 29.29 ± 28.18 (nm).

FT-IR analysis

Fourier transform infrared spectroscopy (FT-IR), is a technique which is used to analyze the chemical

Fig. 4 TEM images and corresponding size of SNPs synthesized by *Satureja intermedia* C. A. Mey extract

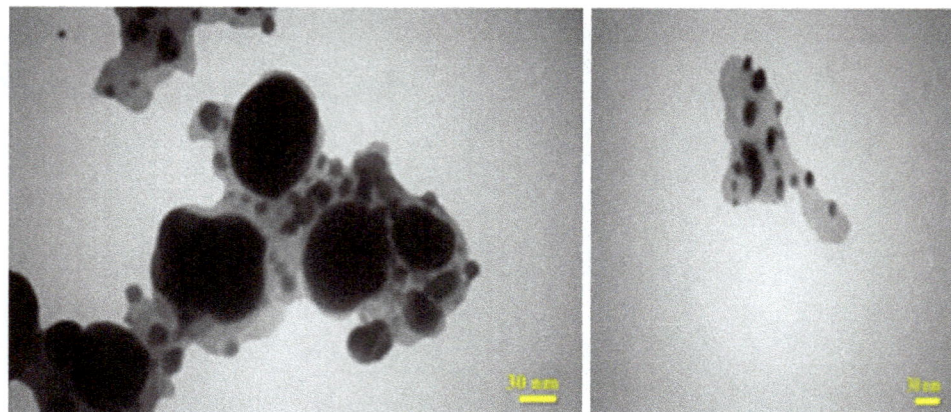

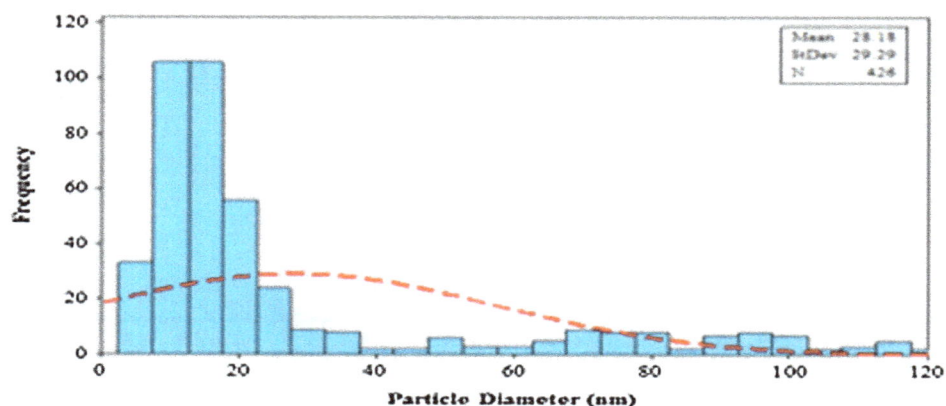

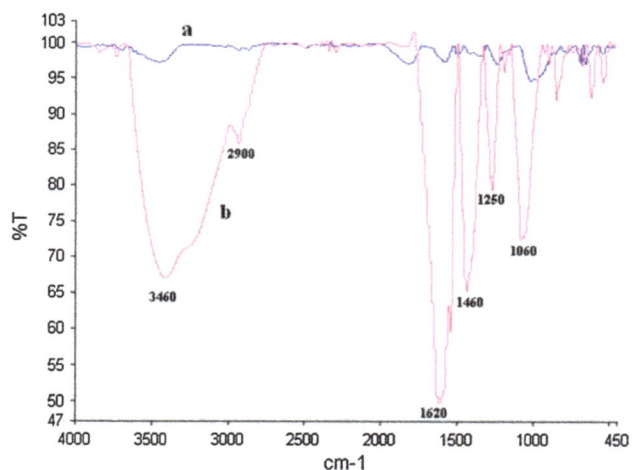

Fig. 5 FT-IR analysis of **a** SNPs/*Satureja intermedia* C.A. Mey extract mixture and **b** *Satureja intermedia* aqueous extract

Fig. 6 Radical scavenging effect of *Satureja intermedia* C.A. Mey aqueous extract and synthesized Silver nanoparticles in comparison with BHT

composition of many organic chemicals. FT-IR spectrum of plant extract before and after synthesis of SNPs was carried out to identify the possible bio molecules responsible for the capping and stabilization of nanoparticles (Fig. 5). The broad peak at about 3400 cm^{-1} corresponds to stretching vibrations of O–H bonds in alcohols, phenols

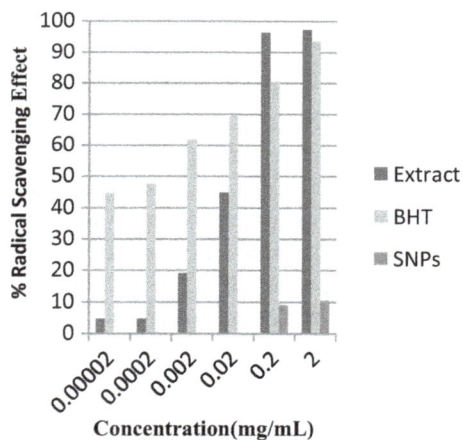

and N–H bond of amides. The strong band at 1620 cm^{-1} is attributed to the C=C stretch in aromatic rings attributed in polyphenols and also may correspond to C=O stretch in amides. The bonds at 1000–1300 corresponds to C–C, C–O and C–N stretching vibrations in alcohols, phenols, esters, carboxylic acids and amides. This suggests that flavonoids,

Phenolic compounds and proteins present in aqueous extract of the plant species could be responsible for the reduction of silver ions and for the stabilization of the phythosynthesized SNPs.

DPPH scavenging assay

Radical scavenging effect of the aqueous extract and SNPs/extract mixture, were evaluated by DPPH scavenging assay. As shown in Fig. 6 the aqueous extract of *S. intermedia* exhibited higher scavenging activity in concentrations 2 and 0.2 (µg/mL) compared to SNPs/extract mixture and DPPH (as a standard). By reducing the concentration to 0.02 µg/mL and lower, BHT was more effective than the aqueous extract and the mixture of SNPs/extract. We can observe a correlation between the concentration of the extract and radical scavenging effect. It means that by reducing the concentration, hence reducing the level of biomolecules in the extract, the antioxidant activity will drop. Total antioxidant capacity of aqueous plant extract defines the electron supplying capacity of the extract. It could be related to SNPs formation rate, since SNPs formation relies on the reduction of Ag+ in which the electron is supplied by the molecules in the extract.

There are many reports in the literature which shows the relation between phenolic content, antioxidant activity and potent in green synthesis of SNPs. Goodarzi et al. evaluated the antioxidant potential, total reducing capacity and SNPs synthetic potential of methanolic leaf extracts of seven plant species. They revealed that plants with high antioxidant potentials also showed higher total phenolic contents and total reducing capacity. In fact, the order of the plants reducing capacity was similar to that of their antioxidant potential [14]. Subramanian et al. demonstrated that the stem bark extract of *Shorea roxburghii* contain high level of total phenolic compounds and radical scavenging activity. They also revealed that the plant extract could be used as an efficient green reducing agent for the production of SNPs [27]. Ahmad et al. reported that the phenolic compounds in pineapple, exhibit excellent antioxidant activity and these phenols can react with a free radical to form the phenoxy radicals. Therefore, the use of natural antioxidants for the synthesis of SNPs seems to be a good alternative which can be due to its benign composition [28].

Although it is still under dispute, various biomolecules existing in aqueous plant extracts such as polyphenols, polysaccharides, proteins, etc. have been proposed to take role in SNPs formation [29, 30]. A majority of such biomolecules known as antioxidants [31, 32] are, in fact, successfully employed in the chemical synthesis of SNPs.

Conclusion

High amounts of flavonoids and phenolic compounds and a high potent of radical scavenging activity were evaluated for *S. intermedia* aqueous extract. A simple green synthesis of stable silver nanoparticles using *Satureja intermedia* aqueous extract was also reported in this study. These nanoparticles were synthesized with an average size of 29.29 ± 28.18 and spherical in shape and were characterized by XRD, TEM, UV–Visible and FT-IR spectroscopy. This eco-friendly method could be a competitive alternative to the conventional physical/chemical methods used for synthesis of silver nanoparticles. Plants with high antioxidant and reducing capacities are not only useful for the green synthesis of metallic NPs, but also for the prevention or reduction of the harmful effects of reactive oxygen species (ROS), generated during normal cellular metabolism of plants and animals.

Acknowledgments Authors are grateful to Dr. Ziba Jamzad for her help in plant collection and identification.

References

1. Van Den Wildenberg, W.: Roadmap report on nanoparticles, Spain. W&W Espana SL, Barcelona (2005)
2. Morones, J.R., Elechiguerra, J.L., Camacho, A., Holt, K., Kouri, J.B., Ramirez, J.T., Yacaman, M.J.: The bactericidal effect of silver nanoparticles. Nanotechnology. 16, 2346–2353 (2005)
3. Hang, X., Ngo, Y.W., Yu, X.: YiL.: DNA aptamer functionalized nanomaterials for intracellular analysis, cancer cell imaging and drug delivery. Ann. Mat. Pura Appl. 16, 429–435 (2012)
4. Allafchian, A.R., Mirahmadi-Zare, S.Z., Jalali, S.A.H., Hashemi, S.S., Vahabi, M.R.: Green synthesis of silver nanoparticles using *phlomis* leaf extract and investigation of their antibacterial activity. J. Nanostruc. Chem. 6, 129–135 (2016)
5. Pal, S., Tak, Y.K., Song, J.M.: Dose the antibacterial activity of silver nanoparticles depend on the shape of the nanoparticles? A study of the gram-negative bacterium *Escherichia coli*. Appl. Environ. Microbiol. 73(6), 1712–1720 (2007)
6. Becker, R.O.: Silver ions in the treatment of local infections. Met. Based Drugs 6, 297–300 (1999)
7. Ahmad, A., Mukherjee, P.S., Senapati, S., Mandal, D., Khan, M.I., Kumar, R., Sastry, M.: Extracellular biosynthesis of silver nanoparticles using the fungus *Fuscriumoxysporum*. Colloid Surf. B. 28(4), 313–318 (2003)
8. Kaler, A., Jain, S., Banerjee, C.B.: Green and Rapid synthesis of anticancerous silver nanoparticles by *Saccharomyces boulardii* and insight into mechanism of nanoparticle synthesis. Biomed. Res. Int. 2013, 1–8 (2013)
9. Jae, Y.S., Beom, S.K.: Rapid biological synthesis of Silver nanoparticles using plant Leaf extracts. Bioproc. Biosyst. Eng. 32, 79–84 (2009)
10. Mubarakali, D., Thajuddin, N., Jeqanathan, K., Gunasekaran, M.: Plant extract mediated synthesis of silver and gold nanoparticles and its antibacterial activity against clinically isolated pathogens. Colloid Surf. B. 85(2), 360–365 (2011)
11. Satyavani, K., Gurudeeban, S., Ramanathan, T., Balasubrama-

nian, T.: Biomedical potential of silver nanoparticles synthesized from callicells of *Citrulluscolocynthis*. J. Nanobiotechnol. **9**, 43–50 (2011)

12. Bramachari, G., Sarkar, S., Ghosh, R., Barman, S., Mandal, N.C., Jash, S.K.: Sunlight-induced rapid and efficient biogenic synthesis of silver nanoparticles using aqueous leaf extract of *Ocimum sanctum* Linn. with enhanced antibacterial activity. Org. Med. Chem. Lett. **4**(1), 18–34 (2014)

13. Jannathul Firdhouse, M., Lalitha, P.: Biocidal potential of biosynthesized silver nanoparticles against fungal threats. J. Nanostruct. Chem. **5**(1), 25–33 (2015)

14. Goodarzi, V., Zamani, H., Bajuli, L., Moradshah, A.: Evaluation of antioxidant potential and reduction capacity of some plant extracts in silver nanoparticle synthesis. Mol. Biol. Res. Commun. **3**(3), 165–174 (2014)

15. Kohkonene, M.P., Hopia, A.I., Vourela, H.J., Pihlaja, K., Kujala, T.S., Heinonen, M.: Antioxidant activity of plant extract containing phenolic compounds. J. Agric. Food Chem. **47**, 3954–3962 (1999)

16. Rice-evans, C.A., Miller, N.J., Bowell, P.G., Bramley, P.M., Pridham, J.B.: The relevant antioxidant activities of plant-derived polyphenolic flavonoids. Free Radic. Res. **22**, 375–383 (1995)

17. Schwarz, K., Bertelsen, G., Nissen, L.R., Gardner, P.T., Heinonen, M.I., Hopia, A., Huynh-Ba, T., Lambelet, P., Mc Phail, D., Skibsted, L.H., Tijburg, L.: Investigation of plant extracts for the protection of processed foods against lipid oxidation. Comparison of antioxidant assay based on radical scavenging, lipid oxidation and analysis of the principal antioxidant compounds. Eur. Food Res. Technol. **212**, 319–328 (2001)

18. Singleton, V.L., Rossi, J.A.: Colorimetry of total phenolics with phosphomolybdic-phosphotungstic acid reagents. Am. J. Enol. Viticult. **16**, 144–158 (1965)

19. Chang, C.C., Yang, M.H., Wen, H.M., Chern, J.C.: Estimation of total flavonoid content in Propolis by two complementary colorimetric methods. J. Food Drug Anal. **10**(3), 178–182 (2002)

20. Bondet, V., Brand-Williams, W., Berset, C.: Kinetics and mechanisms of antioxidant activity using the DPPH free radical method. LWT-Food Sci. Technol. **30**, 609–615 (1997)

21. Prabha, S., Supraja, N., Garud, M., Prasad, T.N.V.K.V.: Synthesis, characterization and antimicrobial activity of Alstonia scholaris bark-extract-mediated silver nanoparticles. J. Nanostruc. Chem. **4**(4), 161–170 (2014)

22. Schneider, S., Halbig, P., Grau, H., Nickel, U.: Reproducible preparation of silver sols with uniform particle size for application in surface enhanced Raman Spectroscopy. Photochem. Photobiol. **60**(6), 605–610 (1994)

23. Wiley, B.J., Im, S.H., Li, Z.Y., McLellan, J., Siekkinen, A., Xia, Y.: Maneuvering the surface plasmon resonance of silver nanostructures through shape-controlled synthesis. J. Phys. Chem. **110**, 15666–15675 (2006)

24. Noginov, M.A., Zhu, G., Bahoura, M., Adegoka, J., Small, C., Ritzo, B.A., Drachev, V.P., Shalaev, V.M.: The effect of gain and absorption on surface Plasmon in metal nanoparticles. Appl. Phys. B-Lasers O. **86**, 455–460 (2006)

25. Dubey, M., Bhadauria, S., Kushwah, B.S.: Green synthesis of nanosilver particles from extract of Eucalyptus hybrid(Safeda) leaf. Dig. J. Nanomater. Bios. **4**, 537–543 (2009)

26. Shankar, S.S., Ahmad, A., Parsricha, R., Sastry, M.: Bioreduction of chloroaurate ions by geranium leaves and its endophytic fungus yields gold nanoparticles of different shapes. J. Mater. Chem. **13**, 1822–1826 (2003)

27. Subramanian, R., Subbramaniyan, P., Raj, V.: Antioxidant activity of the stem bark of *Shorea roxburghii* and its silver reducing power. Springer Plus **2**, 28 (2013)

28. Ahmad, N., Sharma, S.: Green synthesis of silver nanoparticles using extracts of *Anana comosus*. Green Sustain. Chem. **2**, 141–147 (2012)

29. Shan, B., Yizhong, Z., Sun, M., Corke, H.: Antioxidant capacity of 26 spice extracts and characterization of their phenolic constituents. J. Agric. Food Chem. **53**, 7749–7759 (2015)

30. Szydlowska- Czerniak, A., Amarowicz, R., Szlyk, E.: Antioxidant capacity of rapeseed meal and rapeseed oils enriched with meal extract. Eur. J. Lipid Sci. Technol. **112**, 750–760 (2010)

31. Martinez-Tome, M., Jimenez, A.M., Ruggieri, S., Frega, N., Strabbioli, R., Murcia, M.A.: Antioxidant properties of mediterranean spices compared with common food additives. J. Food Protect. **64**, 1412–1419 (2001)

32. Li, H.B., Cheng, K.W., Wong, C.C., Fan, K.W., Chen, F., Jiang, Y.: Evaluation of antioxidant capacity and total phenolic content of different fractions of selected microalgae. Food Chem. **102**, 771–776 (2001)

8

Seagrass-mediated silver nanoparticles synthesis by *Enhalus acoroides* and its α-glucosidase inhibitory activity from the Gulf of Mannar

P. Senthilkumar[1] · D. S. Ranjith Santhosh Kumar[1] · B. Sudhagar[1] ·
M. Vanthana[1] · M. Hajistha Parveen[1] · S. Sarathkumar[1] · Jeslin Cheriyan Thomas[1] ·
A. Sandhiya Mary[1] · Chandramouleeswaran Kannan[1]

Abstract *Enhalus acoroides* (Linnaeus f.) Royle belongs to the Family: Hydrocharitaceae is an abundantly growing seagrass in coastal areas of Gulf of Mannar. Sea grasses contain very good potent therapeutic properties for challenging diseases. The current investigation is to green synthesis Ag-NPs from *Enhalus acoroides* and characterize silver nanoparticles (Ag-NPs) and evaluated α-glucosidase inhibition activity. Furthermore, the characterization of the same using ultraviolet–visible spectroscopy (UV–Vis), Fourier transform infrared (FT-IR) spectroscopy, X-ray diffraction (XRD), energy dispersive X-ray (EDX), and followed by Transmission Electron Microscopy (TEM). Surface Plasmon resonance exhibited the development of Ag-NPs in UV–Visible spectra at 419 nm. The FTIR examination was done to find and read the moiety accountable aimed at the bioconversion of silver ions. The crystalline form was observed in the XRD examination. The synthesized Ag-NPs were polydispersed spherical Ag-NPs as confirmed by EDAX and stabilized in the solution with the spherical shapes further confirmed by TEM analysis designate in the reading of 2–100 nm. The findings of the study shown that Ag-NPs of promisingly proved have strong α-glucosidase inhibitory activity by which Ag-NPs with an IC$_{50}$ 47 µg/ml. This current research report is first to account on the green synthesis of silver nanoparticles from seagrass *E. acoroides*, because there is no literature survey and investigations on *Enhalus acoroides*. Hence, we suggest that Ag-NPs synthesized from *E.*

acoroides might be a significant resource of α-glucosidase inhibitors preparation that may benefit diabetes treatment.

Keywords Seagrass · *Enhalus acoroides* · Gulf of Mannar · Ag-NPs · α-glucosidase activity

Introduction

Ocean offers a unique environment for the sustenance of aquatic organisms. This complex environment changes constantly with the concentration and dilution of chemicals, deposition, and erosion of sediments. This place of the ocean is probably the most dynamic place on the earth [1]. Marine has devised innumerable progressions for the synthesis of nano-level inorganic resources which take donated towards the growth of reasonably novel and essentially unknown part of study according to the green synthesis of nanoproducts [2]. Green-synthesized Ag-NPs possess a variety of uses since of their significant physical and chemical properties [3]. Incidentally, the green synthesis of metal nanoparticles takes to initiate to be economical and environmentally responsible [4]. Coastal plants are a significant, nontoxic and definitely obtainable materials used for the production of nanoparticles by the extensive diversities of compounds possibly support in reduction [5]. The bioconversion of Ag-NPs using coastal plants is much smaller number. Exactly, though there is comparatively very slight otherwise not any works proceeding of green synthesis of Ag-NPs using seagrasses [6].

The current investigation describes the bioconversion of Ag-NPs by using extracts of seagrass at straight daylight state. The characterization of green-synthesized Ag-NPs using the standard devices.

Enhalus acoroides (Linnaeus f.) Royle (Class: Monocots Order: Alismatales, Family: Hydrocharitaceae, Genus:

✉ P. Senthilkumar
senthilkumar1185@gmail.com

[1] PG and Research Department of Biotechnology, Kongunadu Arts and Science College, Coimbatore 640 025, Tamilnadu, India

Enhalus) is a lavishly growing seagrass in south Indian coast, available in intertidal part of Gulf of Mannar. *E. acoroides* are the marine vascular and flowering plants inhabiting the shallow coastal waters in tropical and temperate zones. *E. acoroides* exhibited valuable antioxidants [7]. It showed significant major sterol and fatty acid component. Qi et al. 2008 [8] examined the chemical constituents and antifeedant, antibacterial, and the antilarval activities of ethanol extracts of *E. acoroides* from South China and recorded eleven pure compounds including four flavonoids and five sterols.

Hence, the present investigation on the synthesis and characterization of Ag-NPs using seagrass *E. acoroides* and its effects on α-glucosidase inhibitory is the first research study.

Materials and methods

Collection and processing of Seagrass materials

E. acoroides (Linnaeus f.) Royle, a seagrass was collected Thondi (Lat. 9°45′N and Lang 79°3′E) is situated Gulf of Mannar, Tamilnadu, South India coastal region during the month of January 2016. *E. acoroides* leaves were collected at 3–5 m depth using the SCUBA diving equipment model (SCUBA EA 2/07), and the young leave reaches their maximum size. All leaves were present covered with a dense growth of diatoms. Epiphytic diatoms were removed by scarping the individual leaves in 1 cm areas with the tip region in spatula.

Preparation of the Extract

The samples of healthy and fresh *E. acoroides* were transferred to the laboratory and cleaned carefully by fresh water to eliminate adhering materials and epiphytes. Seagrass samples were then dissected into leaves, stems, and roots. The dissected samples were shade dried for 1 week at room temperature. The *E. acoroides* samples were again cleaned by sterilized distilled water to eliminate the salted materials remaining on the external parts of the samples. The fine pieces of dried samples (25 g) were taken and boiled with sterilized distilled water (100 ml) for not more than 5 min. The green-colored aqueous extract was delivered to Whatman No. 1 filter paper, and the remains were kept at 4 °C for future study.

Green synthesis of silver nanoparticles

The silver nitrate (AgNO$_3$) as analytical grade (AR) was obtained from E. Merck (India). The characteristic bioconversion of Ag-NPs, 10 ml of *E. acoroides* extract, was mixed with 90 ml of aqueous silver nitrate solution (1 mM) in conical flask (250 ml) and it lights at room temperature for 48 h at 120 rpm. Appropriate controls were kept through the carry out the study. The solution obtained was transferred to an amber colored bottle to prevent autoxidation of silver. The reaction mixture with silver ion solution and seagrass extract was noted for color change through visual observation.

Characterization of silver nanoparticles

The Ag-NPs were described by means of UV–Vis spectrophotometer (UV-100 Cyberlab USA) in the frequencies between 300 and 800 nm. The functional group analysis of the silver nanoparticles was carried out by FTIR spectra that were noted for the Ag-NPs by FTIR Nicolet Avatar 660 (Nicolet, USA). For the determination of the elemental materials, the Ag-NPs solutions were examined using energy dispersive X-ray spectroscopy (EDX): XRD and transmission electron microscopy (TEM) were done to predict the structure of the Ag-NPs by placing of 1 µl of the sample carbon films supported by copper grids, air dried, and viewed at 100 kV (JEOL 1010 TEM).

In vitro α-glucosidase enzyme inhibition assay

The preliminary anti-diabetic activity was done by α-glucosidase inhibition assay of the modified method of [9]. The α-glucosidase enzyme solution having P-nitrophenyl-αglucopyranoside (2.9 mM), the silver nanoparticles (0.25 ml) with different combinations (10–100 µg/ml). The 0.6 U/ml in yeast α-glucosidase were added in sodium phosphate buffer (pH 6.9). The controls having with dimethyl sulfoxide, α-glucosidase, and pNPG, whereas in positive controls, acarbose substituted with silver nanoparticles. Combinations without α-glucosidase, silver nanoparticles, and positive controls functioned as blanks. All the solutions were kept to water bath (25 °C for 5 min). The absorbance of the resultant p-nitrophenol was reading at 405 nm by UV–Vis spectrophotometer (UV-100Cyberlab USA). The percentage of α-glucosidase inhibition was measured as follows:

$$(1 - B/A) \times 100,$$

where A is the absorbance of control, and B is the absorbance of samples containing silver nanoparticles. The inhibitory concentration of the silver nanoparticles necessary to inhibit the action of the α-glucosidase by 50 % (IC$_{50}$) was calculated graphically.

Results and discussion

The current research on bioconversion of silver nanoparticles using the seagrass *E. acoroides*. The aqueous extract was used for Ag-NPs biosynthesis from silver nitrate solution. The first indication of positive Ag-NPs formation was the appearance of light yellow to brown gradually. It was interesting to note that the Ag-NPs formation occurred

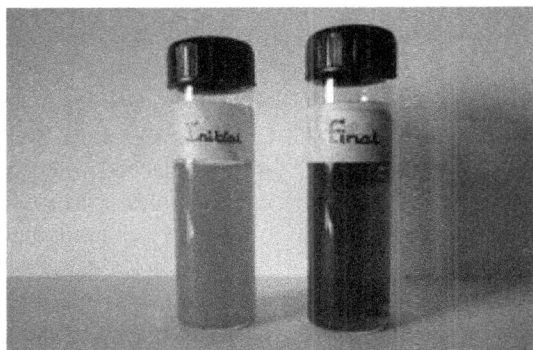

Fig. 1 Green biosynthesis of silver nanoparticles from *E. acoroides*

only in the presence of light [10–13]. The controls were remaining same after the reactions take place, so it is reasonable to conclude that the development of Ag-NPs in the present of light dependent (Fig. 1).

The Ag-NPs obtained were described by UV–Vis spectroscopy to monitor the reactions of Ag-NPs developed in the reaction time (for 48 h of incubation) has the characteristic absorbance showed λmax at 419 nm (Fig. 2); the appearance of value exhibited that the Ag-NPs are relatively small, polydispersed, and spherical Ag-NPs. In addition, no precipitation or agglomeration was found during the reaction and up to 3 months after the experiment suggesting that the biosynthesized Ag-NPs are stable.

The FTIR analysis spectrum exhibited sharp absorbance between 500 and 4000 cm^{-1} (Fig. 3). The two sharp peak absorption at 1645 and 3435 cm^{-1} indicating the probable relations between proteins and silver nanoparticles. The

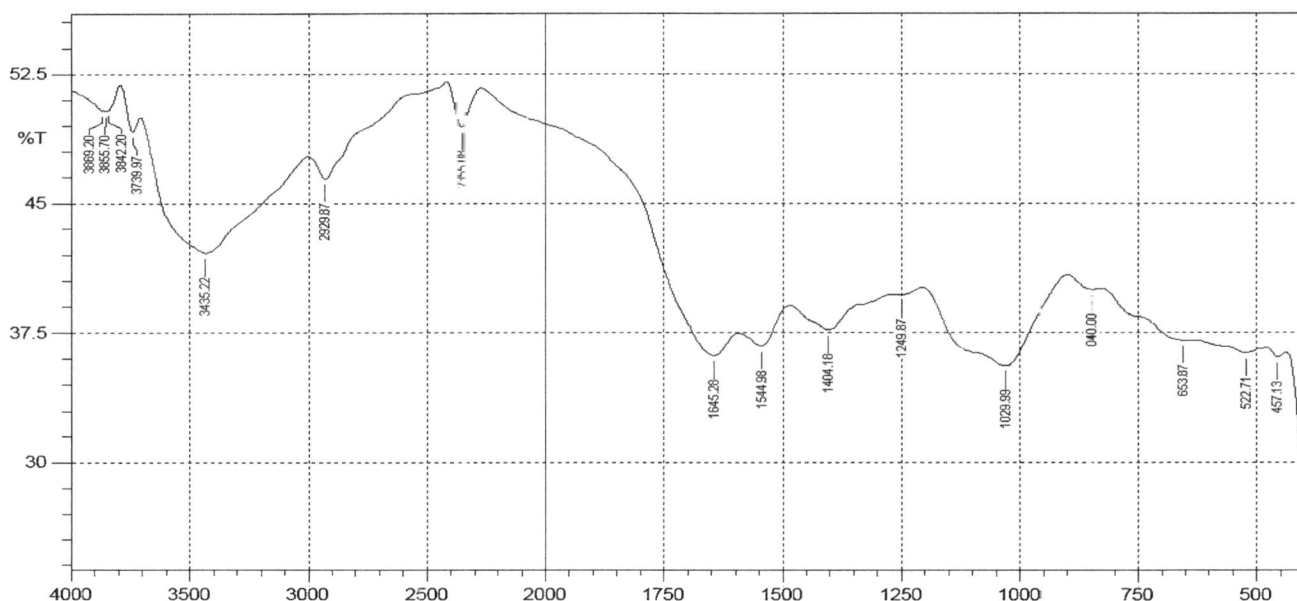

Fig. 2 UV–Vis absorption spectra of silver nanoparticles synthesized by *E. acoroides*

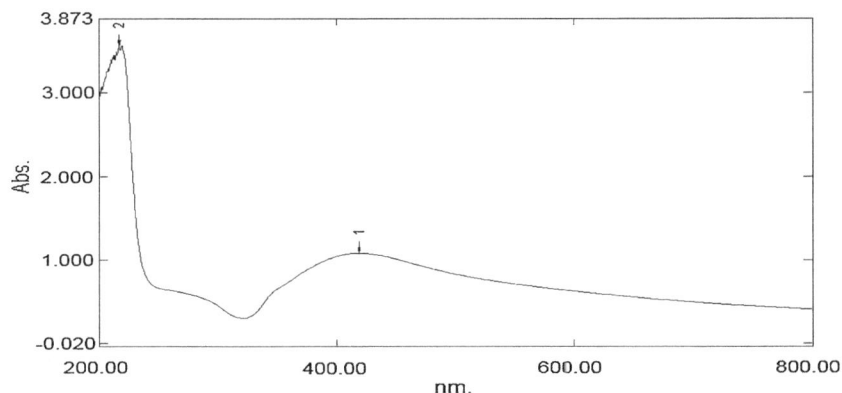

Fig. 3 FTIR spectrum of silver nanoparticles synthesized by *E. acoroides*

absorption peak at 1645 cm^{-1} possibly will on account of the amide bond coming from the carbonyl group of the protein [14]. There is another peak of the spectrum at 3435, 2929, 2355, and 1544 which could be the Intramolecular H bonds, alkanes, amines, and C=C stretching vibrations on—conjugated [15].

The green-synthesized Ag-NPs obtained using *E. acoroides* extract was again proved by specific peaks detected in the XRD (Fig. 4) image at 2θ = 28.09°, 34.2, 39.5, 44.4, 57.7, and 59.05. A small number of strong extra and, however, unassigned peaks were likewise detected by the surrounding area of the distinctive peaks of Ag. These sharp Bragg peak influences have brought about bioorganic

compounds/protein(s) contains in the *E. acoroides*. The XRD results clearly demonstrations that the Ag-NPs are crystalline form [16].

The elemental composition of the bio-functionalized *E. acoroides* Ag-NPs was used by EDXA and revealed in Fig. 5. The area-profile study of the Ag-NPs exhibited strong peaks of Ag at 2.5 keV that are specific of Ag-NPs, combined by Cl and O signals by underneath 1 keV. EDXA study exhibited that CI and O signatures were the results of X-ray release by proteins attached with the Ag-NPs outer region, and then, the boundless particles were separated by centrifugation followed by constant cleaning [17, 18].

A TEM analysis was carried out in the organization Ag-NPs from *E. acoroides* that were formed and represented in Fig. 6. The results of TEM showed a perfect indication about the shape and dimension of the Ag-NPs. Bio-conversed Ag-NPs formed were predominantly polydispersed and the particles shaped with spherical, hexagonal, and triangular. The nanotriangle shapes revealed to have a big surface area with the range 2–100 nm. The Ag-NPs that became reduced by the assistance of aqueous extract *E. acoroides* had sizes small sufficient to be electron apparent and imaged as polydispersed small- and large-sized nanoparticles with a different diameter. The existent study approves the occurrences of Ag-NPs in *E. acoroides*, and the results were similar when compared with the earlier reviews of nanoparticles [19, 20].

The *invitro* α-glucosidase inhibitory studies demonstrated that Ag-NPs from *E. acoroides* had α-glucosidase

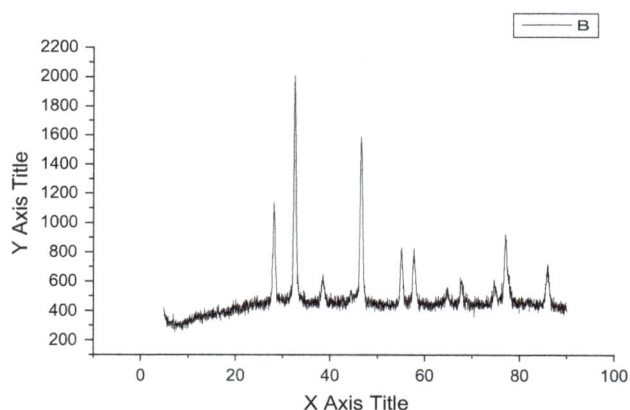

Fig. 4 XRD patterns of silver nanoparticles synthesized by treating *E. acoroides*

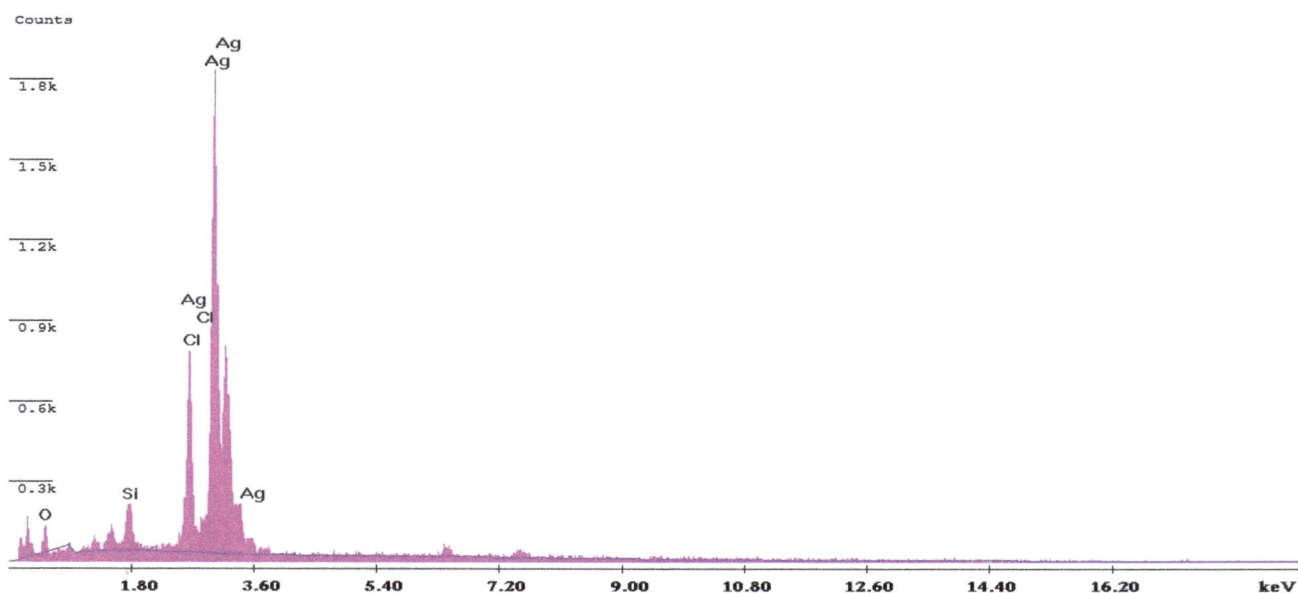

Fig. 5 Energy dispersive X-ray spectrum of silver nanoparticles synthesized by *E. acoroides*

Fig. 6 TEM micrograph of silver nanoparticles synthesized by *E. acoroides*

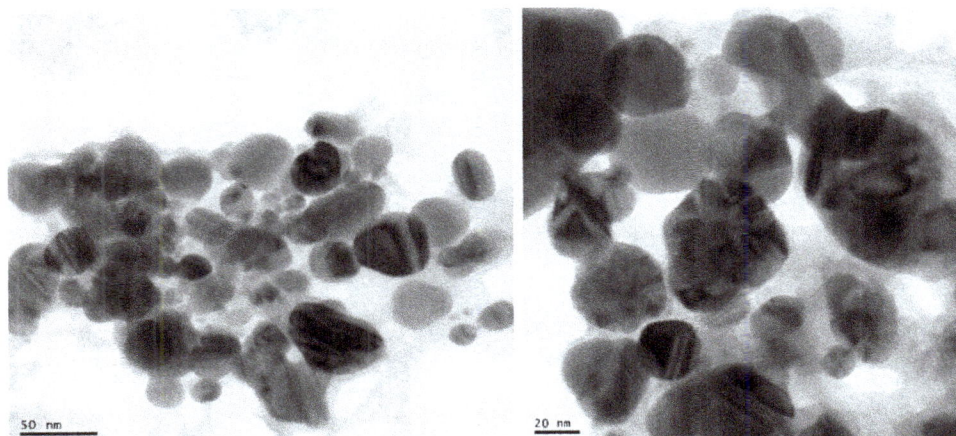

Fig. 7 Invitro α-glucosidase enzyme inhibition activity of silver nanoparticles synthesized from *E. acoroides*

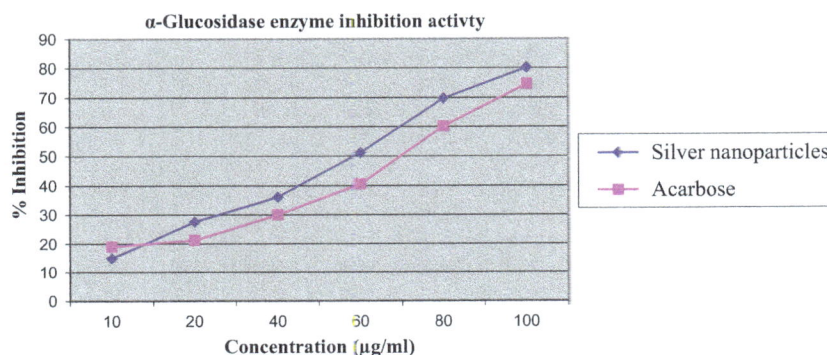

α-Glucosidase enzyme inhibition activty

inhibitory activity. Ag-NPs exhibited a strong inhibitory potential with an IC_{50} value of 47 μg/ml, while acarbose exhibited the with an IC_{50} value estimated at 72 μg/ml inhibitory activity of the α-glucosidase enzyme (Fig. 7). The α-glucosidase inhibitors performed an anti-nutrient that controls the ingestion and absorption of carbohydrates [21]. The concentration-dependent reduction is been shown in the percentage inhibition of alpha-glucosidase activity at 100, 80, 60, 40, 20, and 10 μg/ml concentration of Ag-NPs. The significant inhibition of alpha-glucosidase activity obstructs the function of alpha-glucosidase in the small intestine, which is economical in the reduction in carbohydrate consumption. The major attribution to alpha-glucosidase inhibitors is the decrease in postprandial hyperglycemic levels of the diabetic patients [22]. Metal nanoparticles are already proven as alpha-glucosidase inhibitors [23, 24].

Conclusion

Our study is the first to report on the green synthesis of silver nanoparticles from seagrass *E. acoroides* and suggests their strong inhibition of digestive enzyme α-glucosidase. The efficacy of the silver nanoparticles to inhibit glucosidase and serve as an anti-diabetic agent has been ascertained, and thus, these nanoparticles can be studied as anti-diabetic agents in vivo. Overall, even though further research in toxicity and in vivo study is required, we propose that Ag-NPs synthesized from *E. acoroides* might be a potential resource of α-glucosidase inhibitors formulation that may benefit diabetes treatment.

Acknowledgments The authors are grateful to the authorities of Kongunadu Arts and Science College, Coimbatore, Tamilnadu, India for providing facilities and for their encouragement. The authors would like to acknowledge Department of Nano Science and Technology, Karunya University for the XRD and EDX analysis. We extend our thanks to Dr. Anuradha Ashok, Nanotech Research Facility, PSG Institute of Advanced Studies, Tamil Nadu, India for the HRTEM analysis.

References

1. Chapman, V.J., Chapman D.J.: Seaweeds and their uses. (3rd Eds.) Chapman and Hall, London (1980)
2. Mohanpuria, P., Rana, K.N., Yadav, S.K.: Biosynthesis of nanoparticles: technological concepts and future applications. J. Nanopart. Res. **10**, 507–517 (2008)

3. Kasthuri, J.S., Veerapandian, S., Rajendiran, N.: Biological synthesis of silver and gold nanoparticles using apiin as reducing agent. Colloids Surf. B **68**, 55–60 (2009)

4. Casida, J.E., Quistad, G.B.: Insecticide targets: learning to keep up with resistance and changing concepts of safety. Agric. Chem. Biotechnol. **43**, 185–191 (2005)

5. Prathna, T.C., Mathew, L., Chandrasekaran, N., Raichur, AM., Mukherjee, A.: Biomimetic Synthesis of Nanoparticles: Science, Technology and Applicability, Biomimetics, Learning from Nature, Amitava Mukherjee (Ed.). ISBN: 978-953-307-025-**4**, (2010)

6. Kumar, Senthil: P., Sudha, S.: biosynthesis of silver nanoparticles from *Dictyota bartayresiana* extract and their antifungal activity, Nano. Biomed. Eng. **5**, 72–75 (2013)

7. Ragupathi Raja Kannan, R., Arumugam, R., Meenakshi, S., Anantharaman, P.: Thin layer chromatography analysis of antioxidant constituents of seagrasses of Gulf of Mannar Biosphere Reserve, South India. Int. J. Chem. Tech. Res. **2**(3): 1526–1530 (2010)

8. Qi S.H, Zhang, S., Qian, P.Y., Wang, B.G.: Antifeedant, antibacterial and antilarval compounds from the South China seagrass *Enhalus acoroides*. Bot. Mar **51**, 441–447 (2008)

9. Matsui, T., Tanaka, T., Tamura, S., Toshima, A., Tamaya, K., Miyata, Y., Tanaka, K.: Alpha-glucosidase inhibitory profile of catechins and a flavins. J. Agric. Food Chem. **55**, 99–105 (2007)

10. Mukherjee, P., Roy, M., Mandal, B.P., Dey, G.K., Mukerjee, P.K., Ghatak, J.: Green synthesis of highly stabilized nanoparticles by a non—pathogenic and agriculturally important fungus *T. Asperellum*. Nanotechnology. **19**, 7 (2008)

11. Kalishwaralal, K., Deepak, V., Ramkumarpandian, S., Nellaiah, H., Sangiliyandi, G.: Extracellular biosynthesis of silver nanoparticles by the culture supernatant of *Bacillus licheniformis*. Mater. Lett. **62**, 4411–4413 (2008)

12. Senthilkumar, P., Bhuvaneshwari, J., Prakash, Lakshmi P., Ranjith SK.: Green synthesis and characterization of silver nanoparticles from aqueous extract brown seaweed of *Padina boergesenii* and its antifungal activity. World J Pharm Sci. **4** (10), 1858–1870 (2015)

13. Lateef, A., Adelere, I.A., Gueguim-Kana, E.B., Asafa, T.B., Beukes, L.S.: Green synthesis of silver nanoparticles using keratinase obtained from a strain of *Bacillus safensis* LAU 13. Int. Nano Lett. **5**, 29–35 (2015)

14. Macdonald, I.D.G., Smith, W.E.: Orientation of cytochrome C absorbed on a citrate—reduced silver colloid surface. Langmuir **12**, 706–713 (1996)

15. Jilie, K., Shaoning, Y.U.: Fourier transform infrared spectroscopic analysis of protein analysis of protein secondary structures. Acta Biochim. Biophys. Sin. **39**(8), 549–559 (2007)

16. Shankar, S.S., Rai, A., Ahmad, A., Sastry, M.: Rapid synthesis of Au, Ag, and bimetallic Au core-Ag shell nanoparticles using Neem (*Azadirachta indica*) leaf broth. J. Colloid Interface Sci. **275**, 496–502 (2004)

17. Akhtar, M.S., Panwar, J., Yun, Y.S.: Biogenic Synthesis of Metallic Nanoparticles by Plant Extracts. ACS Sustain Chem. Eng. **3**(1), 591–602 (2013)

18. Bhat, R., Sharanabasava, V.G., Deshpande, R., Shetti, U., Sanjeev, G., Venkataraman, A.: Photo-bio-synthesis of irregular shaped functionalized gold nanoparticles using edible mushroom *Pleurotus florida* and its anticancer evaluation. J. Photochem. Photobiol. B **5**(125), 63–69 (2013)

19. Elumalai, E.K., Prasad, T.N.V.K.V., Hemachandran, J., Viviyan Therasa, S., Thirumalai, T., David, E.: Extracellur synthesis of silver nanoparticles using leaves of *Euphorbia hirta* and their antibacterial activities, J. Pharm. Sci. Res. **2**(9), 549–554 (2010)

20. Ranjith Santhosh Kumar, D.S., Lakshman Kumar,B., SenthilKumar, P., Chandirasekar,R., UthyaKumar.: Biomimetics of silver nanoparticles from *Ganoderma lucidum* (Curtis) P.Karst and its anticancer potential on breast cancer cells. Int. J. Adv. Multidiscip. Res. **2**(11), 0903–0909 (2015)

21. Narkhede, M.B., Ajimire, P.V., Wagh, A. E., Manoj Mohan., Shivashanmugam, A.T.: *In vitro* antidiabetic activity of *Caesalpina digyna* (R.) methanol root extract. Asian J. Plant Sci. Res, **1**(2), 101–106 (2011)

22. Truscheit, E., Frommer, W., Junge, B., Muller, L., Schmidt, D.D., Wingender, W.: Chemistry and biochemistry of microbial α-glucosidase inhibitors. Angew. Chem. Int. Ed. Engl. **20**, 744–761 (1981)

23. Senthilkumar, P., Bhuvaneshwari., Janani., Prakash., Lakshmi Priya., Ranjith Santhosh Kumar: Potent α-glucosidase inhibitory activity of green synthesized gold nanoparticles from the brown seaweed *Padina boergesenii*. Int. J. Adv. Multidiscip. res. **2**(11) 0917–0923 (2015)

24. Ganesh Kumar, V.K., Gokavarapu, S.D., Rajeswari, A., Stalin Dhas, T., Karthick, V., Kapadia, Z., Shrestha, T., Barathy, I.A., Roy, A., Sinha, S.: Facile green synthesis of gold nanoparticles using leaf extract of antidiabetic potent *Cassia auriculata*.Colloids. Surf. B. **87**, 159 (2011)

9

Green-nanochemistry for safe environment: bio-friendly synthesis of fluorescent monometallic (Ag and Au) and bimetallic (Ag/Au alloy) nanoparticles having pesticide sensing activity

Md Niharul Alam[1] · Sreeparna Das[1] · Shaikh Batuta[1] · Debabrata Mandal[2] · Naznin Ara Begum[1]

Abstract Aqueous methanol (water:methanol 20:80) extract of leaves (AMEL) of Indian curry leaf plant was found to be highly efficient in the rapid and controlled synthesis of stable and fluorescent monometallic (Ag and Au) and also bimetallic (Ag/Au alloy) nanoparticles with wide spectrum of task specific morphologies under sonochemical condition. The nanoparticles synthesized by the present economically viable and environment-friendly protocol showed characteristic fluorescence activity. This was exploited in the fluorometric sensing of the dithiocarbamate pesticide, Mancozeb in aqueous medium. The surface chemistry of these nanoparticles was extensively studied to understand their sensing activity. The naturally occurring flurophoric/chromophoric compounds (carbazole alkaloids and polyhydroxy flavcnoid) present in AMEL instilled (in situ) strong and characteristic fluorescent behavior to the synthesized nanoparticles which opened up their utility as the fluorometric sensors and detectors for pesticides in aqueous medium.

Keywords Monometallic (Ag and Au) nanoparticles · Bimetallic (Ag/Au alloy) nanoparticles · Indian curry leaf plant · Fluorometric sensing · Mancozeb · Dithiocarbamate pesticides

Introduction

Energy, environment, and human health have emerged as the main concerns not only in the research arena, but also in all aspects of our lives. In this connection, nanomaterials (e.g., metal nanoparticles) with their unique structure-dependent properties are emerging as a good promise in offering solutions in each of these priority areas. Metal nanoparticles (NPs) are being explored enormously in recent time because of their distinctive catalytic, electronic, optical, and structural properties. Subsequently, these NPs are being explored extensively to develop novel catalysts, sensors/biosensors, nanoelectronic devices, and medical diagnostic tools. The usefulness of these NPs depends critically on their morphology, composition (alloy or core–shell), and surface structure [1–3]. Thus, the design and development of simple, but energy-efficient, economic, and eco-friendly synthetic protocols for metal NPs with tailor-made structures, capable of serving specific task and biocompatibility, are the highly cherished goals for the researchers working in the field of nanoscience and nanotechnology.

Though a vast number of chemical and physical methods of synthesis are available for metal NPs, these methods are not free from drawbacks [4–6]. Sometimes, the reactants, precursors, and solvents used in the chemical synthetic methods are found to be toxic and potentially hazardous [4, 5]. Formation of toxic by-product is another problem associated with these methods [4–6]. On the other hand, in the case of physical method of synthesis of metal NPs, high temperature and pressure are required leading to the vast consumption of energy [7].

✉ Naznin Ara Begum
naznin.begum@visva-bharati.ac.in

[1] Department of Chemistry, Visva-Bharati (Central University), Santiniketan 731 235, India

[2] Department of Chemistry, University College of Science and Technology, University of Calcutta, 92, Acharya Prafulla Chandra Road, Kolkata 700 009, India

Therefore, researchers working in the field of nanochemistry are in an incessant quest for the new and alternative synthetic routes which are more dexterous, economic, hazard free, and environmentally viable and can yield metal NPs with desirable structural activities. In this context, an alternative synthetic strategy is being developed in the recent time which is based upon the principles of Green Chemistry [8–10]. Very often, these green chemical synthetic protocols (applicable at room temperature, pressure, and in very simple laboratory setup) take advantage of non-toxic, green multifunctional agents (GMAs) derived from the biological sources ranging from unicellular organisms to higher plants [6, 8–15].

In our previous studies, we have observed that the plant extract is very unique as GMA, because it is the source of wide spectrum of bioactive natural products which not only actively take part in the NP synthesis process, control the morphologies and surface structure/chemistry of the Ag and Au NPs synthesized by the GMA, but also impose their chromophoric/fluorophoric behavior to the synthesized NPs to make them either photoactive or fluorescent [6, 16, 17]. Recently, there is a remarkable rise of the use of the fluorescent metal NPs for sensing/biosensing process [18, 19]. These are very much significant in the development of disease (cancer) diagnostic tools and tumor biomarkers [18, 19]. On the other hand, metal NPs having characteristic fluorophoric/chromophoric activities also have immense applications in the colorimetric and fluorometric sensing of pesticides [19–22]. This opened up a new direction towards the synthesis of metal NPs having surfaces functionalized with fluorescent molecules or appropriate sensing molecular moieties [19–22]. However, in most of the cases, this type of metal NPs is synthesized by the conventional chemical method, and mostly, synthetic fluorophores have been used as adsorbates [19–22].

India is a country of diverse range of medicinal and aromatic plants which are the integral part of Indian traditional medicine. One of such plants is Indian curry leaf plant (*Murraya koenigii* Spreng.; Family: Rutaceae) and it is non-toxic and less expensive, therefore, easily available. Leaves of this traditionally used medicinal plant are extensively used in Indian cuisine, and as a whole, this plant is the rich sources of wide spectrum of multifunctional and biologically active natural products [23, 24].

In the present paper, we have demonstrated the excellent efficacy of the aqueous methanol (hydro-alcohol solvent) (water:methanol 20:80) extract of leaves (AMEL) of Indian curry leaf plant along with its active components, e.g., koenigine (**A**), koenidine (**B**), girinimbine (**C**), mahanimbine (**D**), and quercetin (Fig. 1) in the rapid and controlled synthesis of stable monometallic Ag and Au NPs and also Ag/Au bimetallic alloy NPs under sonochemical condition. Our method is based on a non-toxic, very cheap, and widely abundant GMA, i.e., leaves of Indian curry leaf plant are edible. Moreover, collection of leaves did not destruct the tree. At the same time, this protocol can be performed at room temperature and pressure using a very easy lab setup. There is also not any possibility of the formation of toxic by-products. Thus, it is simple, economic, and environment friendly.

These synthesized NPs showed characteristic fluorescence activity. Exploring the applicability of the NPs synthesized by plant-based GMA and having tailored structural properties is essential to assess the usefulness of an NP synthesis protocol. In the present case, we have found that AMEL itself along with its active chemical constituents (which have the characteristic fluorescent activity) controlled the surface chemistry of the synthesized NPs and also imposed specific fluorescent behavior (in situ) to them which opened up their possible utility as the eco-friendly and easily synthesizable fluorometric sensors and detectors for hazardous dithiocarbamate pesticides/fungicide, such as Mancozeb (Fig. 1), which are extensively used in agriculture industries in aqueous medium [21, 22].

Experimental section

Materials

Chloroauric acid ($HAuCl_4$) and silver nitrate ($AgNO_3$) (Sigma Aldrich) were used as the sources of Au^{3+} and Ag^+ ions, respectively. Rest of the chemicals used for the present work was of analytical grade. All analyses were done in Milli-Q (Milli-Q Academic with 0.22 mm Millipak R-40) water. Mancozeb was obtained from Indofil Industries Ltd., India.

Instrumentation

The formation and growth of the NPs were examined with the help of UV–Vis spectroscopy. Absorption spectra of the sample solutions were recorded on Perkin Elmer Lambda 35 spectrophotometer, where as fluorescence spectra of all the experimental solutions were recorded on a Perkin Elmer LS55 fluorimeter. All spectroscopic measurements were done at 25 °C with an excitation wavelength (λ_{ex}) of 308 nm to obtain maximum fluorescence intensity. The spectrum of each of the experimental solutions remained unchanged for sufficiently long periods of time during which the spectroscopic experiments were finished. Therefore, the possibility of the decomposition of any of the experimental solutions which could change the spectroscopic results can be safely ruled out.

Fig. 1 Chemical structures of koenigine (**A**), koenidine (**B**), girinimbine (**C**), mahanimbine (**D**), quercetin and Mancozeb

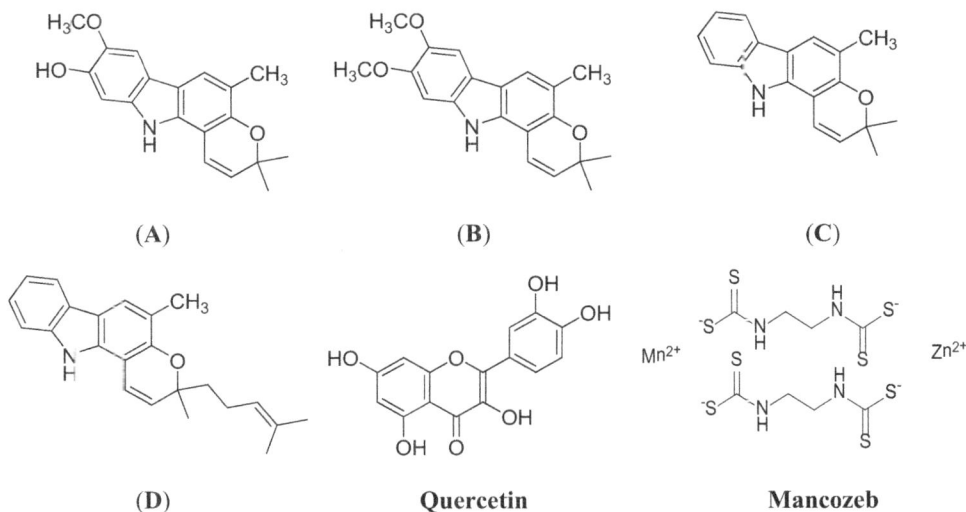

(A)

(B)

(C)

(D)

Quercetin

Mancozeb

Thermal responses of the NPs synthesized by the present protocol were estimated by Thermo gravimetric analysis (TGA). TGA of NP samples was done using a Pyris Diamond TG/DTA (Perkin Elmer, STA-6000) thermal analyzer. The experiment was set in the temperature range of 40–900 °C and at a heating rate of 15 °C min^{-1} under nitrogen atmosphere.

Fourier-transformation-infrared (FT-IR) spectra of the experimental samples were recorded on a Shimadzu FTIR-8400S PC instrument. Prior to FT-IR measurements, NP solutions were centrifuged at 14,000 rpm for 30 min followed by their drying in vacuum.

Melting points of the isolated compounds (**A, B, C,** and **D**) were determined by an electro-thermal apparatus and were uncorrected. All the compounds were purified to A. R. grade before NP synthesis. Purity of the compounds was routinely tested with the help of thin layer chromatography (TLC). TLC studies of these compounds were carried out in silica gel GF 254 pre-coated plates. The structures of the isolated compounds were elucidated and confirmed by comparing their melting point, IR, and ^{1}H NMR data. ^{1}H NMR spectroscopy was recorded at 400 MHz in a Brucker Avance-400 spectrometer in DMSO-d_6 or CDCl$_3$ solution.

Shape and size of the synthesized particles were studied by transmission electron microscopy (TEM). For the preparation of samples for TEM, the NP solution was drop-coated onto the carbon-coated copper grids of size 400 mesh. The films on the grids were dried prior to the TEM measurement by a JEOL JEM-2100 instrument.

For further characterization of the synthesized NPs, powder X-ray diffraction (XRD) analysis was done. For the preparation of XRD samples, NP solutions were centrifuged at 14,500 rpm for 30 min and the supernatant was discarded. Then, NPs were dispersed in water and vortexed. After repeating these steps for three times, the residue part was dried in vacuum. The dry powder obtained

was spreaded evenly on a quartz slide to perform the XRD studies. The XRD patterns were recorded using the Rigaku Ultima IV diffractometer attached with D/tex ultra detector and CuKα source operating at 50 mA and 40 kV. The scan range was fixed at $2\theta = 25°–85°$ with a stepwise size of 0.01°.

Zeta-potential measurement of the experimental solutions was done by Malvern Zetasizer Nano ZS-ZEN 3600 (Malvern Instruments Ltd, UK) instrument. Disposable cuvettes (1 mL volume) specific for this instrument were used for this purpose. Prior to the experiment, all NP samples were diluted appropriately with Milli-Q water to observe optimum signal intensity. Five replications were done for each sample.

Collection of leaves of Indian curry leaf plant, the extraction procedure to get aqueous methanol extract (AMEL) of dried and pulverized leaves, and isolation and identification of the active chemical constituents of AMEL are discussed in detail in the Electronic Supplementary Material (page S2).

This gummy mass obtained from AMEL (GAMEL) obtained after evaporation of the solvent under reduced pressure was stored at 4 °C and diluted appropriately in Milli-Q water by sonication before its use as green multi-functional agent (GMA) in the synthesis of three types of NPs: monometallic Ag, Au, and bimetallic Ag/Au NPs (discussed in later section). We have quantified the total flavonoid and polyphenol contents of AMEL using the standard colorimetric methods [25, 26] which are found to be 102.5 mg quercetin equivalent g^{-1} and 214.3 mg gallic acid equivalent g^{-1}. Detailed procedures are discussed in the Electronic Supplementary Material (page S3). In our previous work, we have observed that leaves of Indian curry leaf plants are rich in polyphenols, flavonoids, such as quercetin, and quercetin-3-glucoside [16]. Other reported flavonoids present in aqueous methanol extract of leaves of this plant are myricetin-3-galactoside, quercetin-O-pentohexoside, quercetin-3-diglucoside, quercetin-3-O-rutinoside, quercetin-3-acetylhexoside, kaempferol-O-glucoside, and kaempferol-aglucoside [24].

We have also isolated and identified four carbazole alkaloids by column chromatography of the gummy mass obtained from AMEL. Detailed isolation procedure has been given in the Electronic Supplementary Material (page S3). These isolated compounds are koenigine (**A**), koenidine (**B**), girinimbine (**C**), and mahanimbine (**D**) (Fig. 1).

Method of synthesis of monometallic (Ag and Au) and bimetallic (Ag/Au) NPs by AMEL and its chemical constituents (A, B, C, D, and quercetin) as GMA

In general, the synthesis of Ag NP was initiated by adding 200 μL of 0.1 (M) aqueous solution of AgNO$_3$ to 10 mL of aqueous solution of GAMEL (400 μg mL^{-1}) in a 100 mL conical flask. The final concentration of Ag$^+$ ions in the reaction mixture was maintained at 2×10^{-3} (M). The pH of the mixture was adjusted at 9 (optimum pH for the reaction) by adding dilute aqueous solution of NaOH. A series of trial experiments were run at several lower and higher pH ranges to get an idea about the optimum pH for this reaction. The conical flask containing the reaction mixture was placed in an ultrasonic bath (Branson 1510), and the reaction mixture was sonicated at 40 kHz at room temperature. The reaction mixture turned to golden yellow color within a minute indicating the onset of formation of Ag NPs. The progress of the reaction was followed by monitoring the absorbance of the reaction mixture at regular interval of times. The absorption peak is assigned to the surface plasmon resonance (SPR) band of Ag NP formed by the reduction of Ag$^+$ ions by GMA.

We have followed similar method for the synthesis of Au NPs except in this time, and we have added 200 μL of 0.1 (M) aqueous HAuCl$_4$ solution instead of AgNO$_3$ aqueous solution. In this case, the final concentration of Au^{3+} ions was 2×10^{-3} (M). Within 2 min, a pink coloration was observed indicating the onset formation of Au NPs.

For the synthesis of bimetallic Ag/Au NPs, different amounts of 0.1 (M) aqueous AgNO$_3$ and 0.1 (M) aqueous HAuCl$_4$ solutions were added to 10 mL of aqueous solution of GAMEL (400 μg mL^{-1}), while other reaction parameters were kept unaltered (Table 1). In each case, characteristic change of the color of reaction mixture was observed within 2 min which was the primary indication of the formation of bimetallic alloy NPs.

In the following section, we have discussed the method of synthesis of the NPs by individual chemical constituents of AMEL isolated by us.

In each case, the stock solution of each of the compounds (**A**)/(**B**)/(**C**)/(**D**) was prepared by dissolving 2 mg of the compound in 3 mL of ethanol. For Ag NP synthesis by (**A**), 50 μL of its stock solution was added to 5 mL of 10 mM SDS solution (to avoid the precipitation of the organic compounds in aqueous medium) and the addition of aqueous NaOH solution was done to adjust the pH at 9 (optimum pH). 50 μL of 0.05 (M) aqueous solution of AgNO$_3$ was added to this reaction mixture, so that the final Ag$^+$ ion concentration became 0.5×10^{-3} (M). This reaction mixture was sonicated at 40 kHz at room temperature, and golden yellow coloration was developed within 5 min indicating the onset of formation of Ag NP. Similar method was found to be useful for the synthesis of Ag NPs either by (**B**)/(**C**) or (**D**).

The compounds (**A**), (**B**), (**C**), and (**D**) were also found to be efficient in the synthesis of Au NPs. Similar method has been followed for this synthesis except that 50 μL of 0.05 (M) aqueous HAuCl$_4$ solution was used instead of AgNO$_3$ aqueous solution. In this case, the final concentration of Au^{3+} ions became 0.5×10^{-3} (M). Within 5 min of sonication, a blue coloration was observed which indicated the onset formation of Au NPs.

Similarly, for bimetallic Ag/Au NPs (1:1) synthesis by the compound (**A**)/(**B**)/(**C**)/(**D**), 50 μL of the stock solution of the respective compound was added to 5 mL of 10 mM SDS solution followed by simultaneous addition of 25 μL of 0.05 M aqueous AgNO$_3$ and 25 μL of 0.05 M aqueous HAuCl$_4$ solution keeping other parameters unaltered (Table 2). Color generation was observed within 5 min which indicated the bimetallic alloy formation.

We have also shown the efficacy of quercetin, abundantly found polyhydroxy flavonoid in leaves of Indian Curry leaf plant towards the synthesis of Au, Ag, and Ag/Au NPs. For Ag NP synthesis by quercetin, 100 μL of 0.05 (M) aqueous solution of AgNO$_3$ was added to the 10 mL aqueous solution of quercetin (0.025 mg mL^{-1}). Aqueous NaOH solution was added to the reaction mixture to adjust the pH at 9 (optimum pH). The final Ag$^+$ ion concentration became 0.5×10^{-3} (M). This mixture was sonicated at 40 kHz at room temperature, and golden yellow coloration was developed within 5 min indicating the onset of formation of Ag NP (Table 2). Similar method using 100 μL of 0.05 (M) aqueous HAuCl$_4$ solution was followed for the synthesis of monometallic Au NPs. For bimetallic Ag/Au NP (1:1) synthesis, 50 μL of 0.05 (M) aqueous AgNO$_3$ and 50 μL of 0.05 (M) aqueous HAuCl$_4$ solution were added simultaneously (Table 2) and sonicated as stated earlier. In these two cases, we kept all other reaction parameters unchanged. Color was developed within 5 min which indicated the formation of NPs. In all the cases, the progress of the reaction was monitored by measuring the absorbance of the reaction mixture with time.

Table 1 Synthesis of monometallic Ag, Au, and bimetallic Ag/Au alloy NPs using AMEL as GMA

Synthesized NP	Set	Conc of Ag$^+$ (M)	Conc. of Au^{3+} (M)	Position of SPR band (nm)	Inset of formation (min)
Monometallic Ag	a	2×10^{-3}	0	408	2
Bimetallic Ag/Au alloy	b	1.5×10^{-3}	0.5×10^{-3}	432	2
Bimetallic Ag/Au alloy	c	1×10^{-3}	1×10^{-3}	459	2
Bimetallic Ag/Au alloy	d	0.5×10^{-3}	1.5×10^{-3}	487	2
Monometallic Au	e	0	2×10^{-3}	522	2

Table 2 Synthesis of monometallic Ag, Au and bimetallic Ag/Au alloy NPs by individual compounds

Synthesized NP	Set	Compound	Conc. of Ag$^+$ (M)	Conc. of Au^{3+} (M)	Position of SPR band (nm)	Inset of formation (min)
Monometallic Ag	f	Koenigine (A)	0.5×10^{-3}	0	421	5
Bimetallic Ag/Au alloy	g		0.25×10^{-3}	0.25×10^{-3}	488	5
Monometallic Au	h		0	0.5×10^{-3}	564	5
Monometallic Ag	i	Koenidine (B)	0.5×10^{-3}	0	424	5
Bimetallic Ag/Au alloy	j		0.25×10^{-3}	0.25×10^{-3}	482	5
Monometallic Au	k		0	0.5×10^{-3}	565	5
Monometallic Ag	l	Girinimbine (C)	0.5×10^{-3}	0	426	5
Bimetallic Ag/Au alloy	m		0.25×10^{-3}	0.25×10^{-3}	506	5
Monometallic Au	n		0	0.5×10^{-3}	572	5
Monometallic Ag	o	Mahanimbine (D)	0.5×10^{-3}	0	422	5
Bimetallic Ag/Au alloy	p		0.25×10^{-3}	0.25×10^{-3}	511	5
Monometallic Au	q		0	0.5×10^{-3}	537	5
Monometallic Ag	r	Quercetin	0.5×10^{-3}	0	406	5
Bimetallic Ag/Au alloy	s		0.25×10^{-3}	0.25×10^{-3}	466	5
Monometallic Au	t		0	0.5×10^{-3}	525	5

Method of fluorometric sensing of Mancozeb, a dithiocarbamate pesticide by the synthesized NPs

We have explored the intense fluorescence activity of the synthesized NPs solutions as a measuring tool for sensing study of the dithiocarbamate pesticide, Mancozeb in aqueous medium. We have prepared 296.11 µM (100 ppm) aqueous stock solution of Mancozeb. 2.5 mL of the NP solution [sets a–c (Table 1); f–h, r–t (Table 2)] was taken in a fluorescence cuvette followed by successive addition of different volumes (10–300 µL) of stock solution of Mancozeb to it, and fluorescence spectrum was recorded after each addition. The final concentration of Mancozeb in the reaction mixture was varied from 1.18 to 31.73 µM (0.39–10.71 ppm). The enhancement of fluorescence emission intensity or 'fluorescence turn on' of the NP solution on the addition of Mancozeb confirmed the sensing activity of the respective NP solution.

Results and discussion

Formation and growth of monometallic (Ag and Au) and bimetallic (Ag/Au) NPs synthesized by AMEL and its chemical constituents as GMA

UV–Vis spectroscopy was used to confirm the formation of the NPs by the reduction of the corresponding metal ion in aqueous solutions when exposed to the present GMA. Figure 2i(set a) shows the result of the reaction between Ag$^+$ ions and the aqueous solution of GAMEL at pH 9. The curve denoted by broken line in Fig. 2i represents the absorption spectrum of aqueous solution of GMA in the absence of Ag$^+$ ions. Upon sonication of this reaction mixture at room temperature, SPR band for Ag NP appeared at 408 nm within 2 min and the intensity of this band was found to be increased with time. Finally, a saturation was observed after 7 min [Fig. 2ii(set a)]. Figure 2i(set e) shows the result of formation of the Au NPs when Au^{3+} ions were reduced by the GMA. The absorbance maxima of Au NPs formed by GMA was recorded and plotted against time, as shown in Fig. 2ii(set e). Within 2 min of the addition of Au^{3+} ion with continuous sonication at room temperature, the reaction mixture turned pink with the appearance of a main peak at 522 nm.

The efficacy of the AMEL was further tested in the synthesis of three types of alloy NPs: Ag/Au (3:1), (1:1), and (1:3) NPs. Ag^{+} and Au^{3+} ions present in the same solution were simultaneously reduced by the GMA to form bimetallic Ag/Au alloy NPs which were also stabilized by the same GMA [Fig. 2i(sets b–d)]. The formation of bimetallic Ag/Au alloy NP formation was established from the fact that the absorption spectrum showed only one plasmon band in place of two individual bands for Ag and Au NPs [2, 3, 5, 27, 28]. Figure 2i(sets b–d) shows the normalized absorption spectra of the reaction mixture for the simultaneous reduction of the various concentration of Ag^{+} and Au^{3+} ions by AMEL (detailed concentration range is given in Table 1). For set b, after completion of the reduction the absorption maximum appeared at 432 nm and for sets c and d, the corresponding maxima were observed at 459 and 487 nm, respectively. It was observed that, only one absorbance peak was obtained for each of the bimetallic NP solutions. Moreover, in all the cases, the absorbance maxima were found to be located at the positions in between the SPR bands associated with the monometallic Ag NPs (408 nm) and Au NPs (522 nm). This is in agreement with the previously reported data [3, 27]. Such absorption spectra cannot be observed if it was a case of simple physical mixture of monometallic Ag and Au NP solutions [29]. Moreover, the spectra in Fig. 1i(sets b–d) did not have any resemblance to those exhibited by the bimetallic Ag/Au core–shell NPs. In general, two characteristic absorption peaks are observed for the bimetallic Ag/Au core–shell NPs [3, 27]. Moreover, as shown in Fig. 2iii, the SPR peak position, i.e., absorbance maxima of the bimetallic system, was found to be gradually red shifted linearly with the increase of concentration of Au^{3+} ions ($y = 406.2 + 56.2x$); which further supports the formation of Ag/Au bimetallic alloy NPs [3, 27]. The UV–Vis absorption data thus satisfactorily confirmed the simultaneous reduction of Ag^{+} and Au^{3+} ions by GMA in aqueous medium to produce the homogeneous bimetallic alloy NPs. This was further confirmed by the TEM images (discussed in later section).

Formation and growth of monometallic (Ag and Au) and bimetallic (Ag/Au) NPs by isolated compounds (A, B, C, and D) and quercetin

Plant extract is the concoction of several types of chemical compounds (natural products) which are the characteristics of a specific plant genus. These chemical constituents impart characteristic redox and stabilizing/capping activities to the plant extract making it a green multifunctional agent (GMA) in the field of synthesis and stabilization of NPs [5, 14]. Use of plant-based GMA is being explored extensively now-a-days. However, one of the major problems associated with these types of methods is lack of fine tuning of the NP morphology and reproducibility which are essential for practical implementation of any synthetic protocol. From our previous work [3, 4, 6, 16, 27], we have observed that nature and content of these chemical constituents are major controlling parameters for these types of protocols. Therefore, identification of these compounds is very much necessary for properly scaling up of these synthetic protocols.

In our previous work, we have observed that aqueous extract of Indian curry leaf plant is rich in polyphenols and flavonoids [16]. However, in the present case, we have observed that in addition to polyphenols and flavonoids, AMEL is also rich in carbazole alkaloids, such as koenigine (**A**), koenidine (**B**), girinimbine (**C**), and mahanimbine (**D**).

Therefore, for the first time, we have explored the carbazole-based secondary metabolites of plant in the rapid synthesis of stable monometallic Ag, Au, and bimetallic Ag/Au (1:1) NPs.

Formation and growth of the NPs were monitored with the help of UV–Vis spectroscopy, and the results are shown in Fig. 3i–iv. The curves with broken line represent the absorption spectra of koenigine (**A**), koenidine (**B**), girinimbine (**C**), and mahanimbine (**D**), respectively, in the absence of metal ions. In all the cases, the appearance of characteristic absorption spectra clearly confirmed the formation of Ag NP, Au NP, and alloy Ag/Au (1:1) NP (Fig. 3i–iv). Position of SPR bands and the time for onset of formation of respective NPs were shown in Table 2(sets f–q).

In each case of the Ag NP synthesis by the isolated compounds (**A**, **B**, **C**, and **D**), sharp SPR bands were obtained in the range of 421–426 nm [Fig. 3i–iv; Table 2(sets f–q)]. However, in the case of Au NPs synthesized by the same systems, much broader SPR bands (indicating the formation of diverse type NPs) were observed (Fig. 3i–iv). These were further confirmed on the basis of TEM analysis. Only one absorbance peak was observed for the formation of bimetallic alloy Ag/Au (1:1) NPs by each of these individual compounds, and absorbance maxima were found to be located at the position intermediate to those usually observed for monometallic Ag and Au NPs. Such absorption spectra could not be obtained if simply a physical mixture of Ag and Au was formed (Fig. 3i–iv; Table 2).

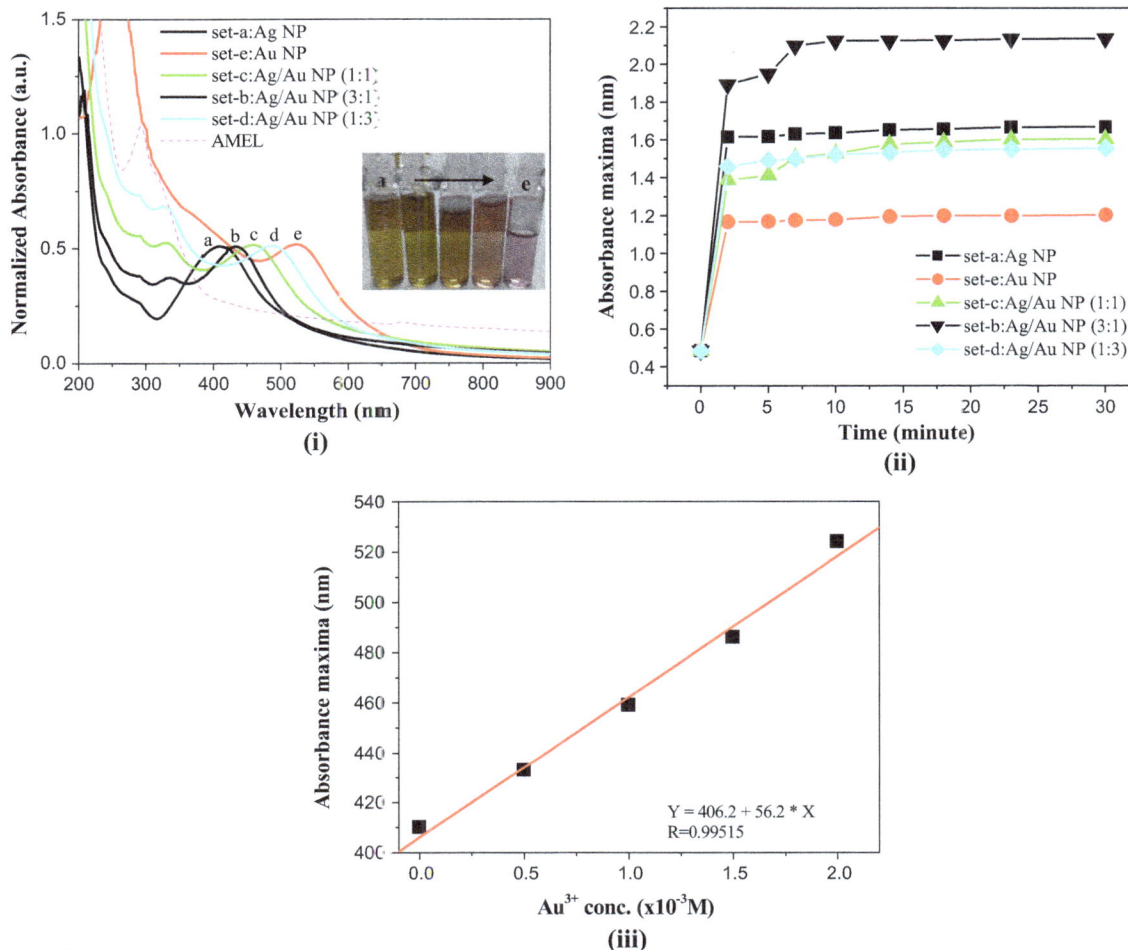

(set-a)

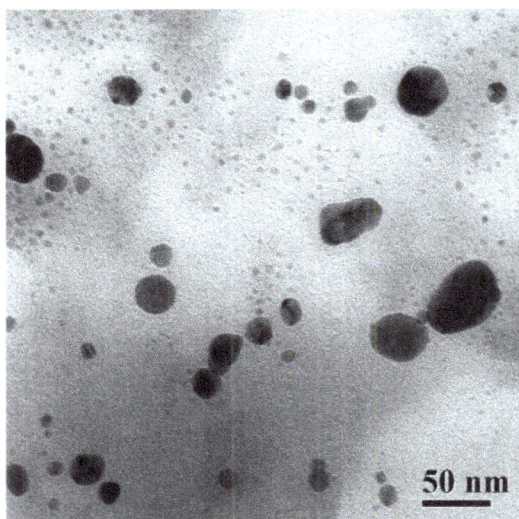

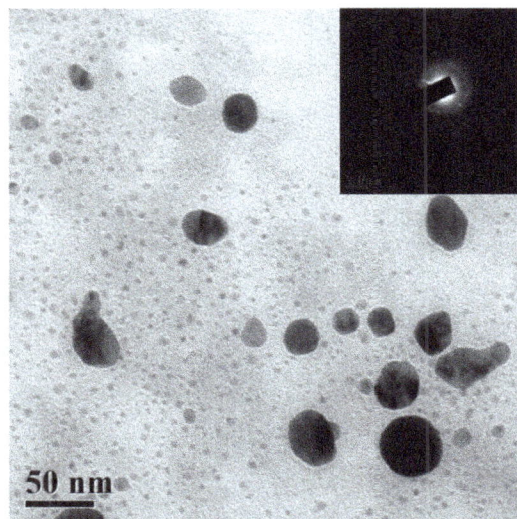

Fig. 2 i UV–visible spectra and **ii** change in peak absorbance with time for different sets of synthesized NPs: Ag NP (*set a*), Au NP (*set e*), and Ag/Au alloy NP prepared at different Ag/Au molar ratios 3:1 (*set b*), 1:1 (*set c*), and 1:3 (*set d*). *Broken line* in **i** represents absorbance of AMEL, and used as GMA. Inset shows the color of the corresponding NP solutions. **iii** Positions of surface plasmon resonance band maxima was plotted as a function of the molar fraction of Au. **iv** TEM images of Ag NPs (*set a*), Ag/Au alloy NPs prepared at different Ag/Au molar ratios 3:1 (*set b*), 1:1 (*set c*), 1:3 (*set d*), and Au NPs (*set e*). *Inset* in **iv** shows the SAED pattern of the corresponding NPs. **v** XRD patterns and **vi** EDX profiles for Ag NP (*set a*), Au NP (*set e*), and Ag/Au alloy NP (1:1) (*set c*)

(set-b)

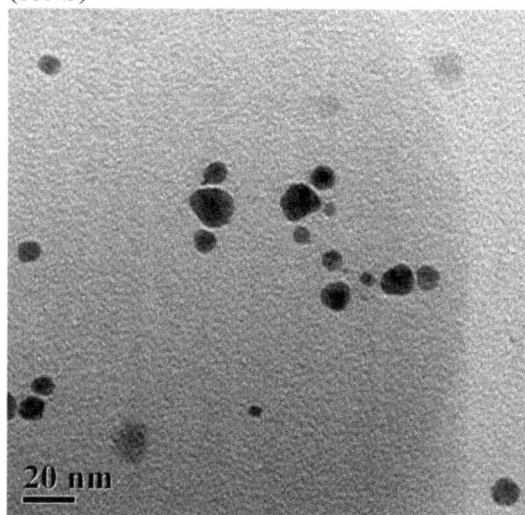
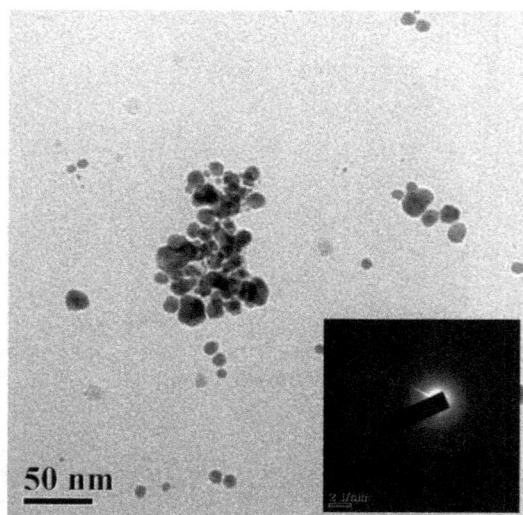

(set-c)

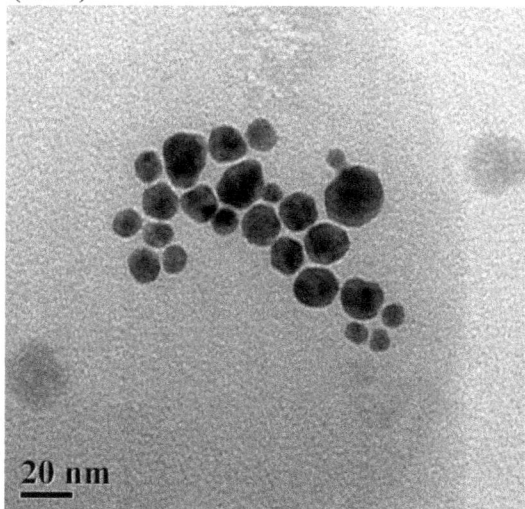
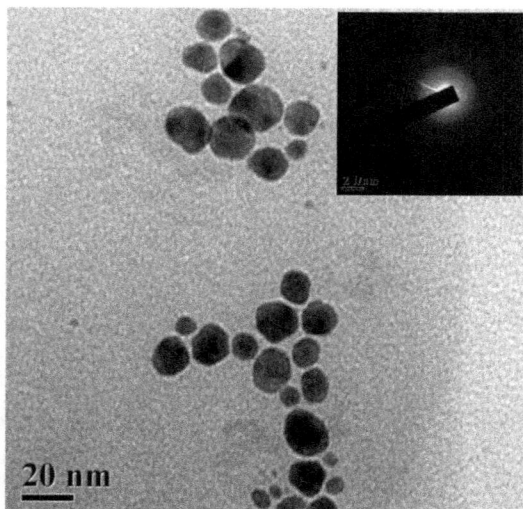

(set-d)

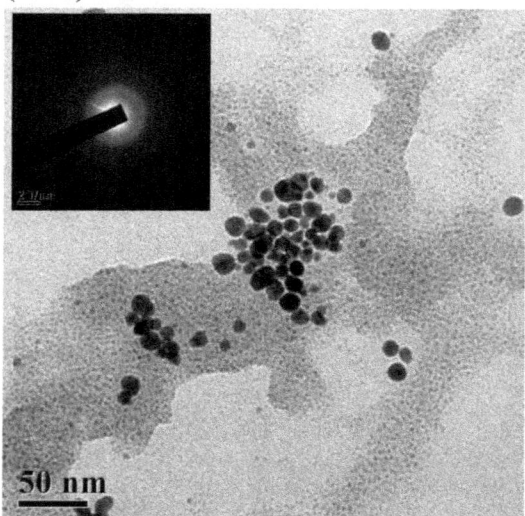
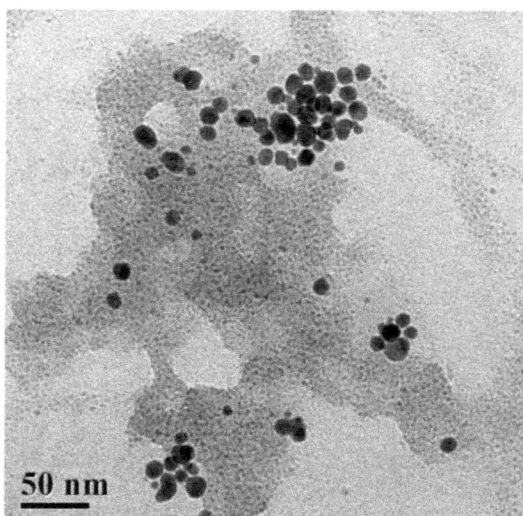

Fig. 2 continued

(set-e)

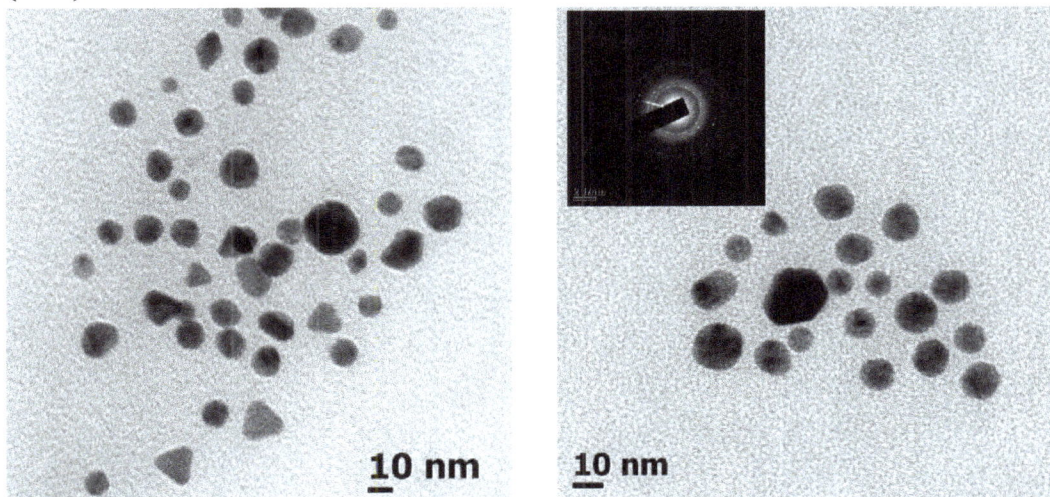

(iv)

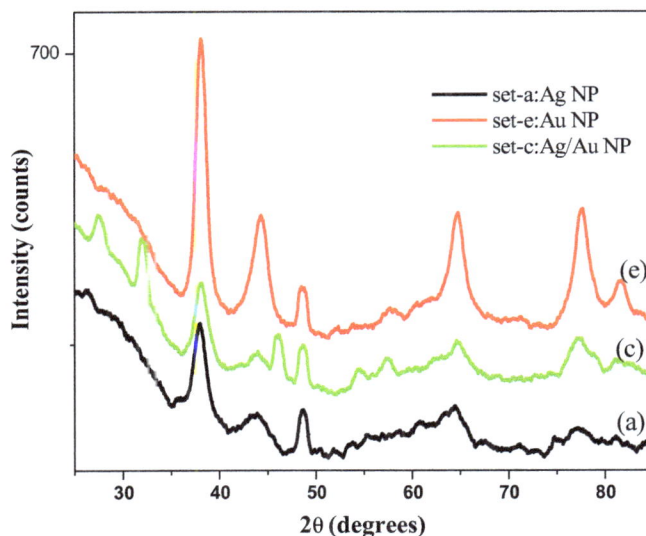

(v)

Fig. 2 continued

The same types of NPs were formed when quercetin was used as a reducing agent. Figure 3v shows the UV–Vis absorption spectra of Ag NP (set r), Au NP (set t), and Ag/Au NP (1:1) (set s) synthesized by quercetin, one of the constituents present in AMEL.

Morphology of monometallic (Ag and Au) and bimetallic (Ag/Au) NPs synthesized by AMEL and its chemical constituents as GMA

Morphology of the synthesized NPs was studied by the TEM. Figure 2iv [for sets (a) and (e)] shows the TEM images of the Ag and Au NPs synthesized by AMEL, respectively. In the case of Ag NPs, particles were found to

be well separated and mostly spheroidal in shape [Fig. 2-iv(set a)]. On the other hand, Au NPs of interesting morphologies, e.g., triangles and hexagons together with regular spheroidal shape NPs, were observed in the respective TEM images [Fig. 2iv(set e)].

Figure 2iv(sets b–d) shows the TEM images of the bimetallic Au/Ag NPs formed by the simultaneous reduction of Ag^+ and Au^{3+} ions in the reaction mixture having Ag^+:Au^{3+} concentration ratio 3:1 (set b), 1:1 (set c), and 1:3 (set d). The particles formed were predominantly spherical. In the close up view, occasional aggregations are quite visible. TEM images of the bimetallic core/shell-type structure usually show electron density banding with a dark Au core and a lighter Ag

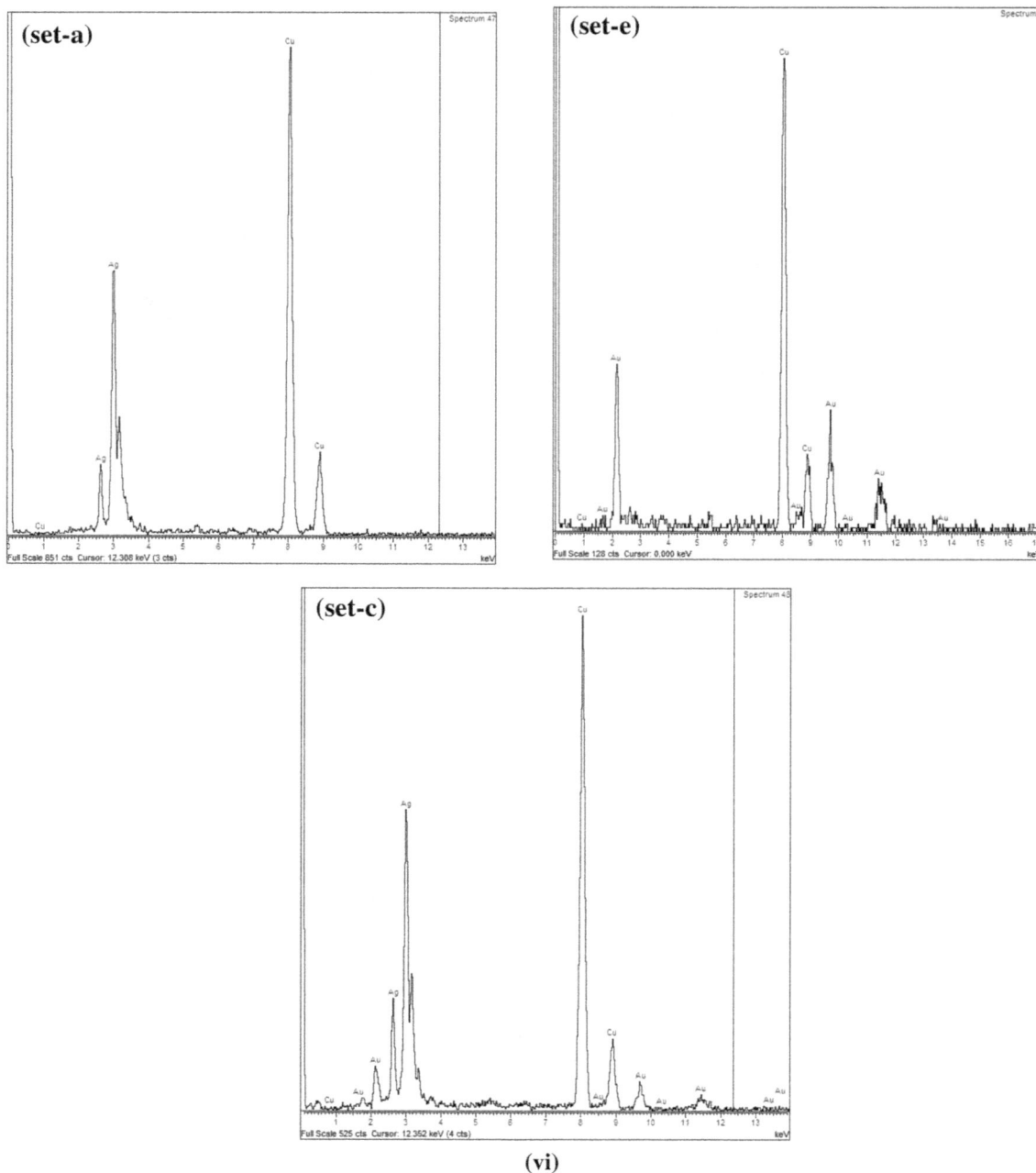

(vi)

Fig. 2 continued

shell [30] which was not observed in the present case. TEM images of the synthesized bimetallic Ag/Au (1:1) NP showed the uniform contrast for each NP. This suggested that the electron density was homogeneous within the volume of the particle [30]. Therefore, the bimetallic NPs synthesized presently were not core/shell type and these NPs closely resembled to the bimetallic alloy NPs. These results are in strong agreement with the UV–Vis spectroscopic data discussed previously and also the reported data [3, 30].

Compared to the mother extract (AMEL), more uniform morphologies of the NPs were observed in the case of NPs synthesized by its individual components (**A**, **B**, **C,** and **D**) (Fig. 3vi). There may be two possible reasons for this. First, only a single reducing component showed its activity. Second, carbazole alkaloids are not soluble in aqueous medium. Therefore, we had to use SDS to prevent their precipitation in aqueous medium. Thus, in addition to **A**, **B**, **C,** and **D,** SDS might also play a role in the stabilization of these NPs. However, like AMEL,

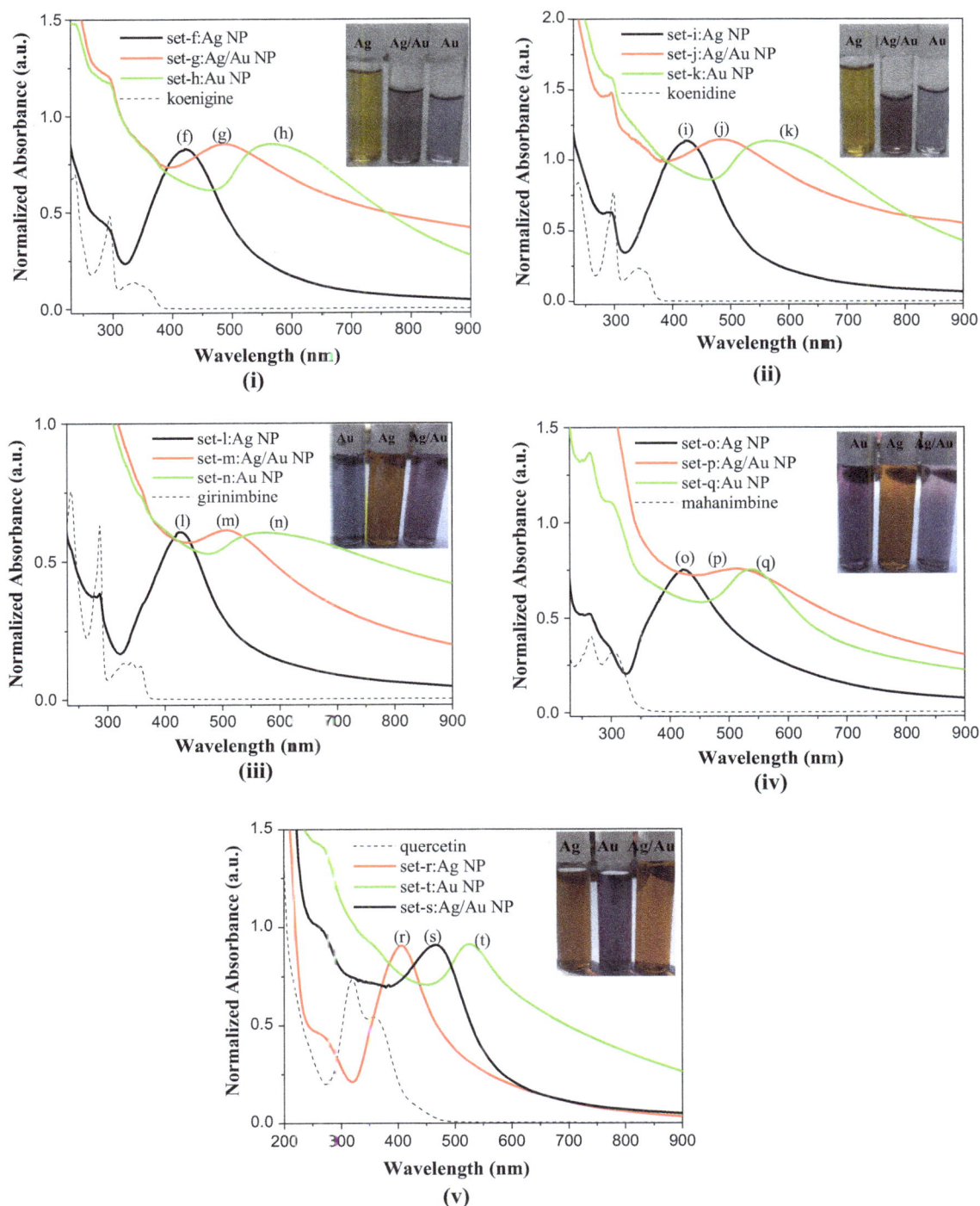

Fig. 3 i–v UV–Vis spectra and **vi** TEM image for the different sets of synthesized NPs: Ag NPs, Au NPs, and Ag/Au alloy (1:1) NPs prepared using koenigine (*sets f–h*), koenidine (*sets i–k*), girinimbine (*sets l–n*), mahanimbine (*sets o–q*), and quercetin (*sets r–t*), respectively. *Broken line* in **i–v** represents absorbance curve of koenigine, koenidine, girinimbine, mahanimbine, and quercetin, respectively. *Inset* in **i–v** shows the *color* of the corresponding NP solutions

in these four cases, also, Ag NPs were found to be mostly uniform in size and spherical in shape [Fig. 3-vi(sets f, i, l, and o)], while anisotropic structures were quite visible in the case of Au NPs [Fig. 3vi(sets h, k, n, and q)].

Interestingly, Ag and Au NPs synthesized by quercetin were found to be more uniform in size and NPs formed were smaller in size. In this case. Au NP showed a wide spectrum of morphology in addition to regular spherical shape [Fig. 3vi(sets r and t)].

(set-f)

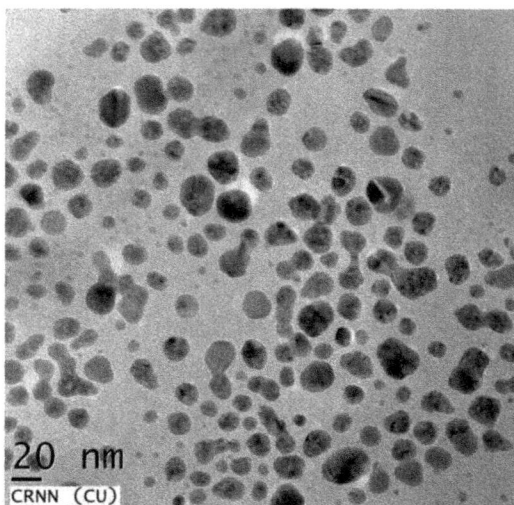 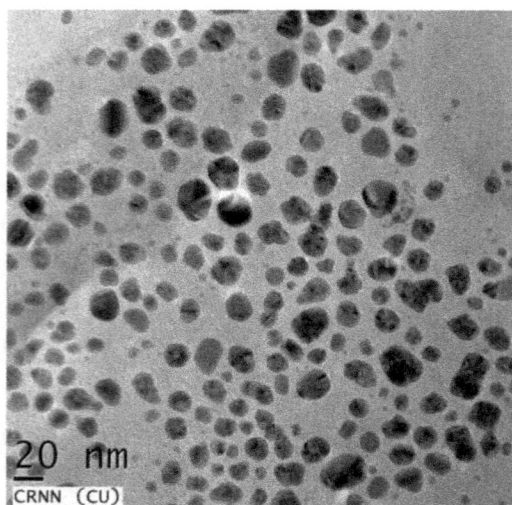

(set-g)

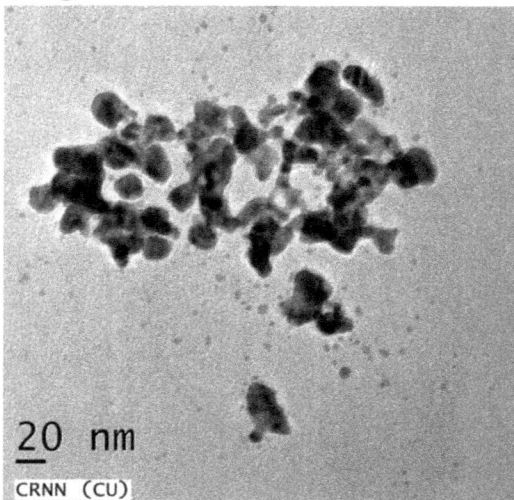 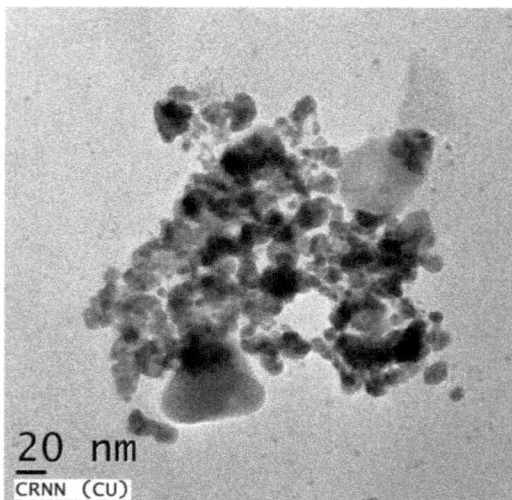

(set-h)

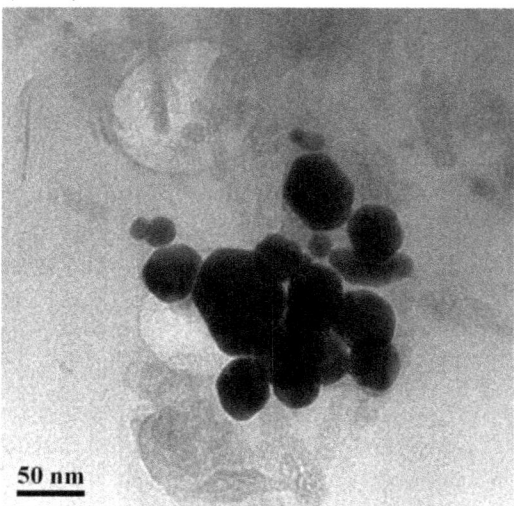 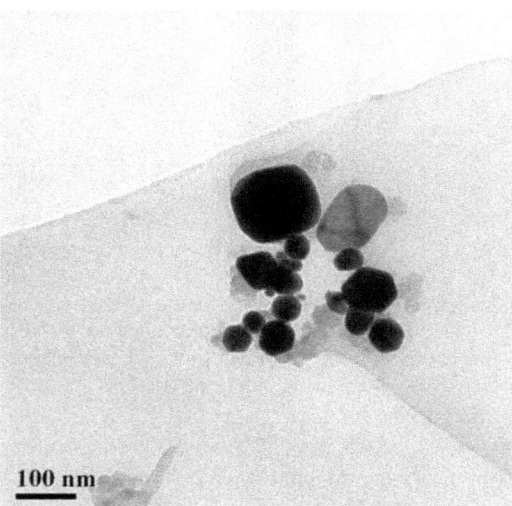

Fig. 3 continued

(set-i)

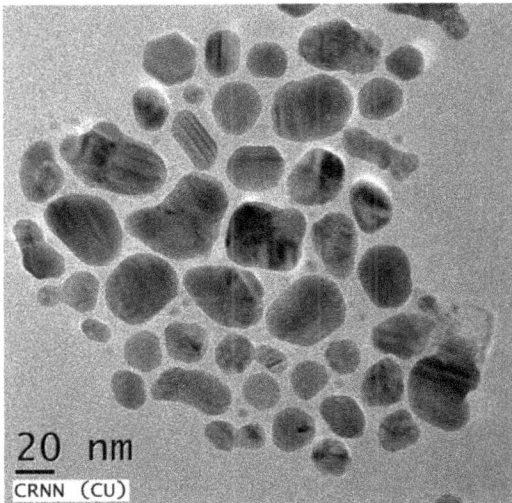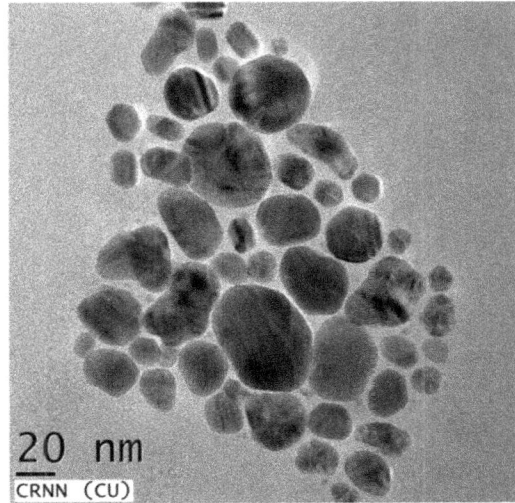

(set-j)

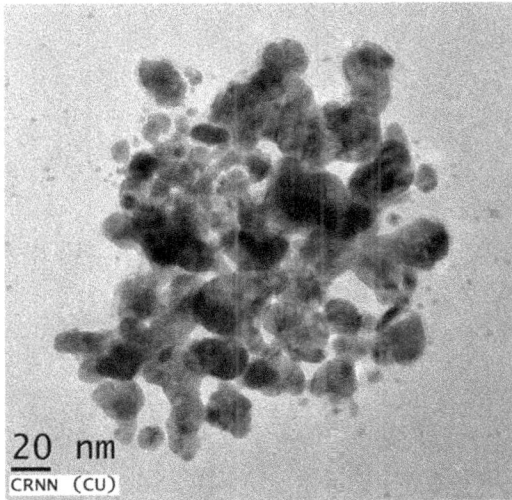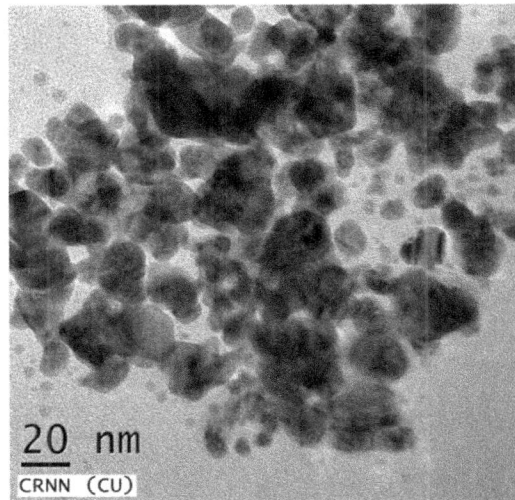

(set-k)

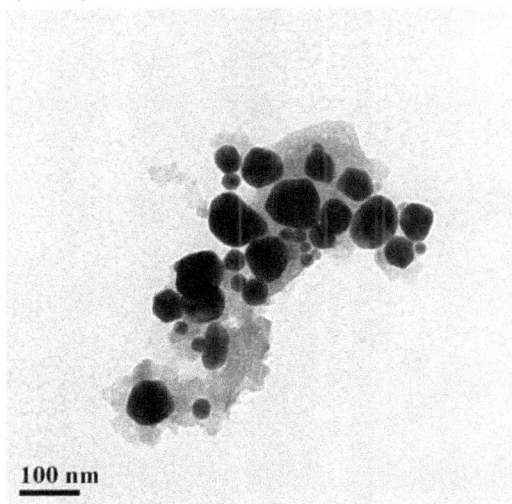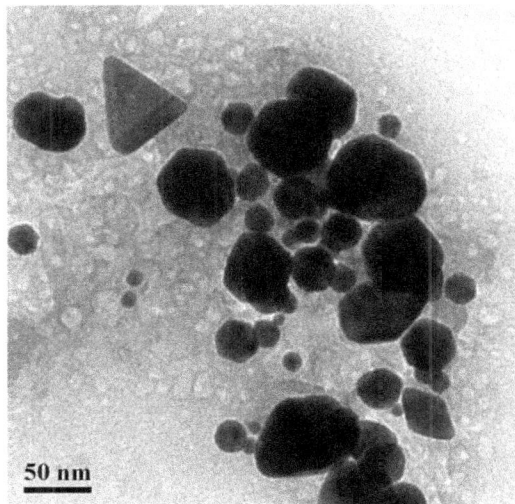

Fig. 3 continued

(set-l)

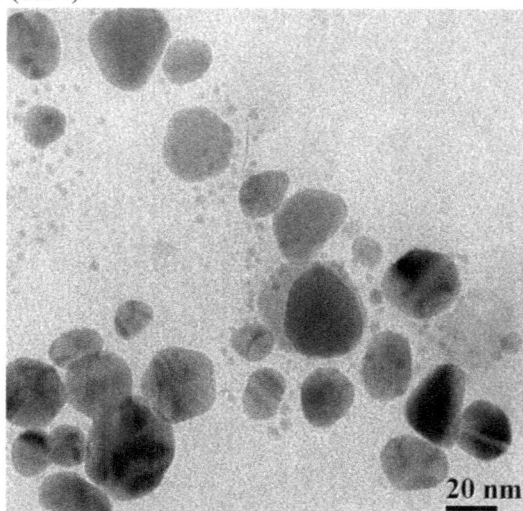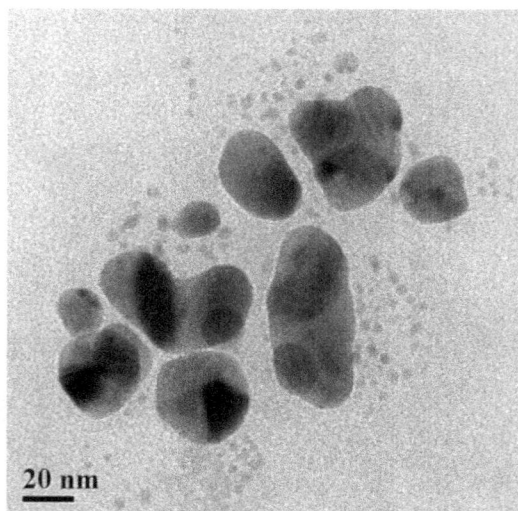

(set-m)

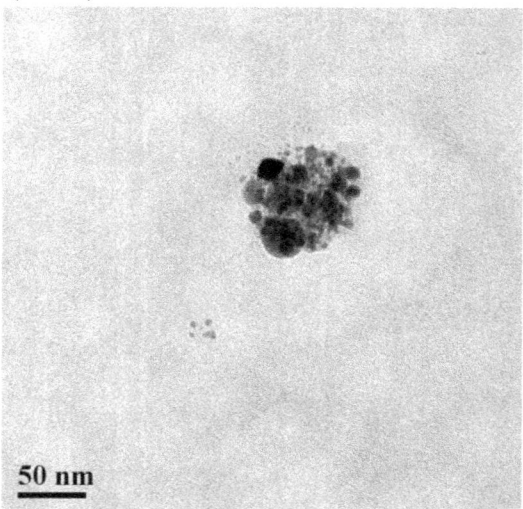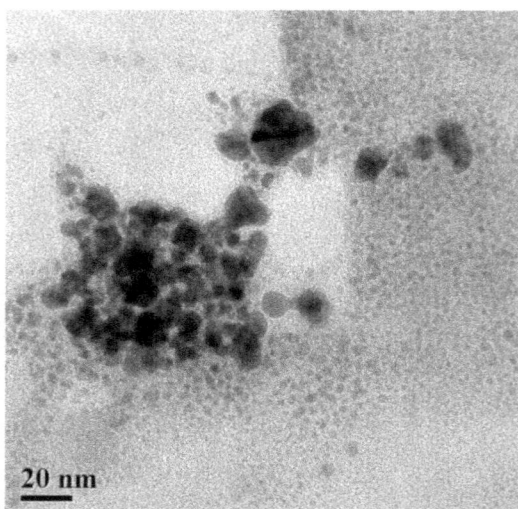

(set-n)

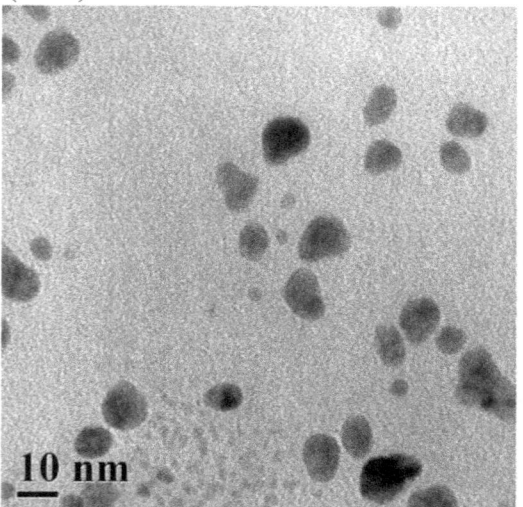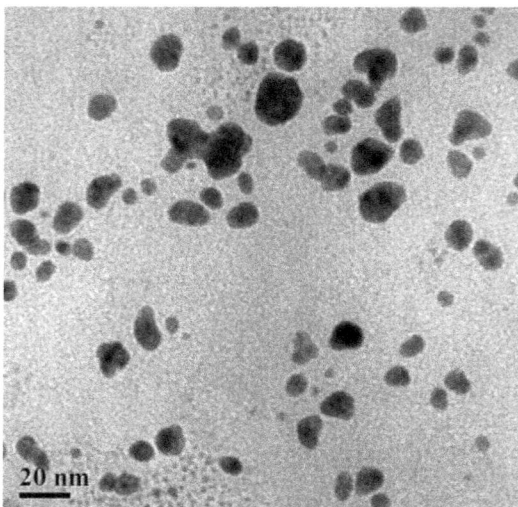

Fig. 3 continued

(set-o)

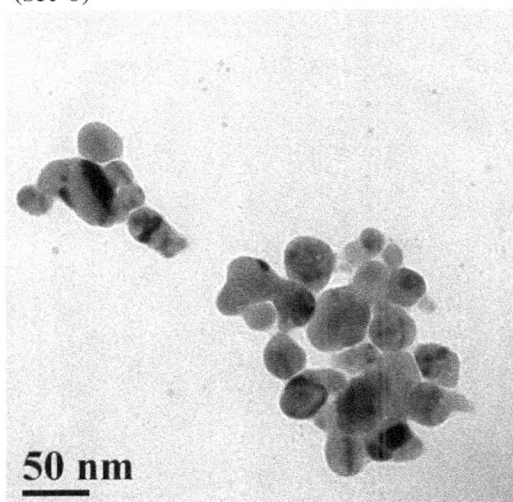 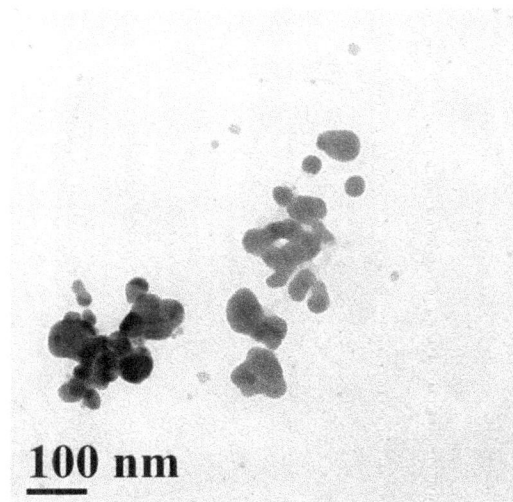

(set-p)

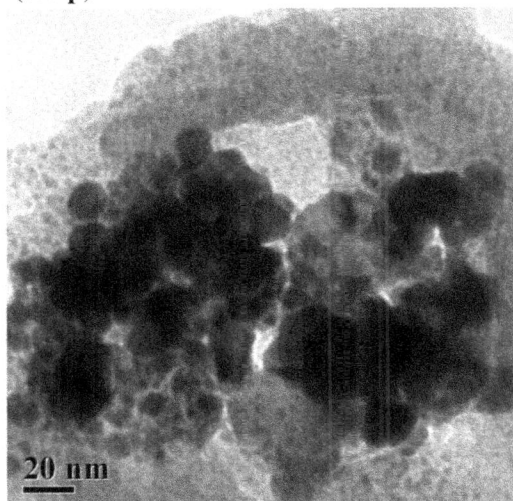 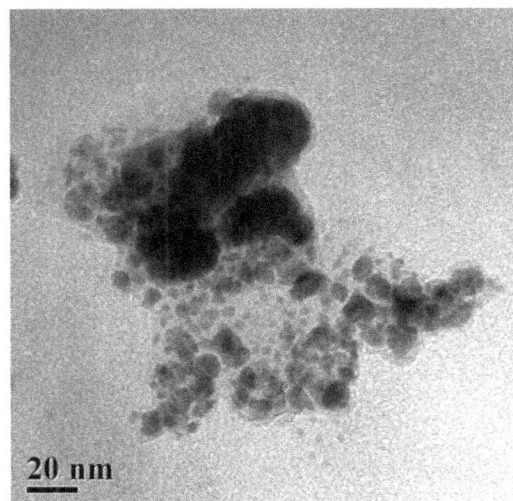

(set-q)

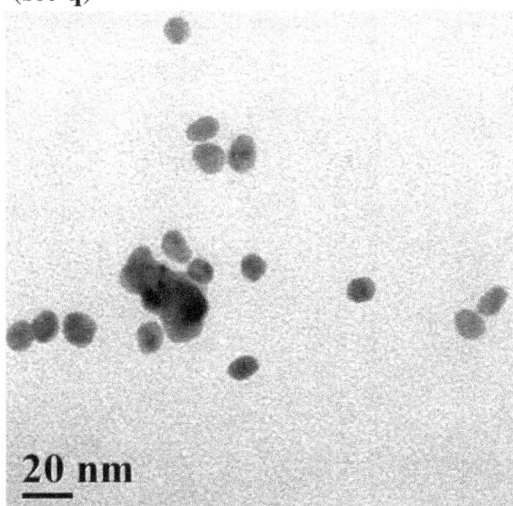 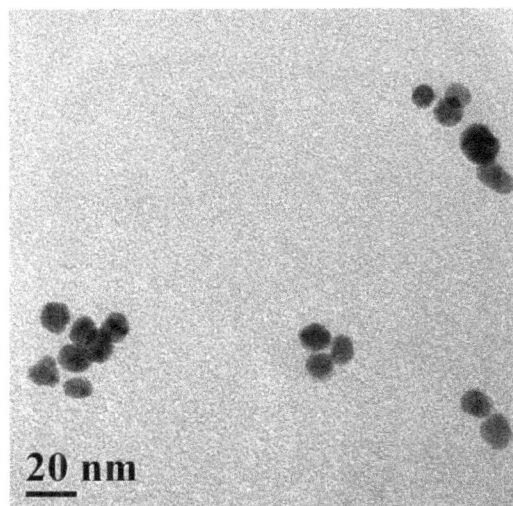

Fig. 3 continued

(set-r)

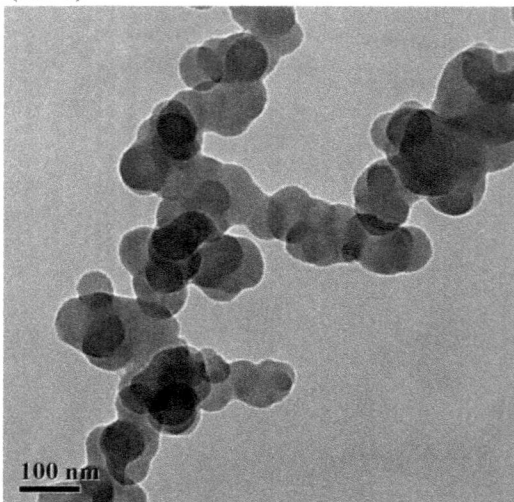 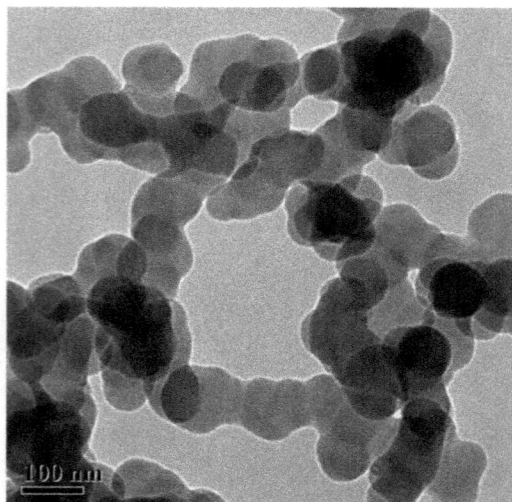

(set-s)

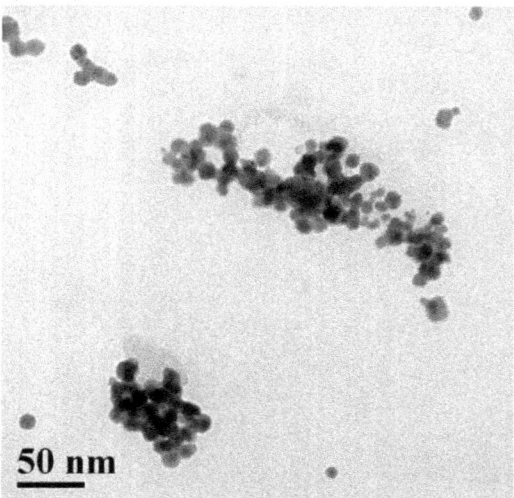 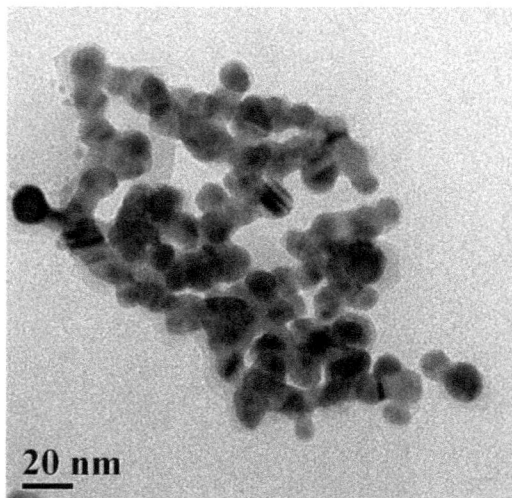

(set-t)

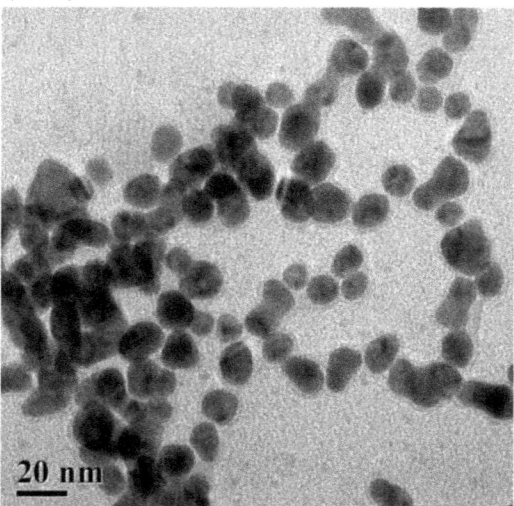 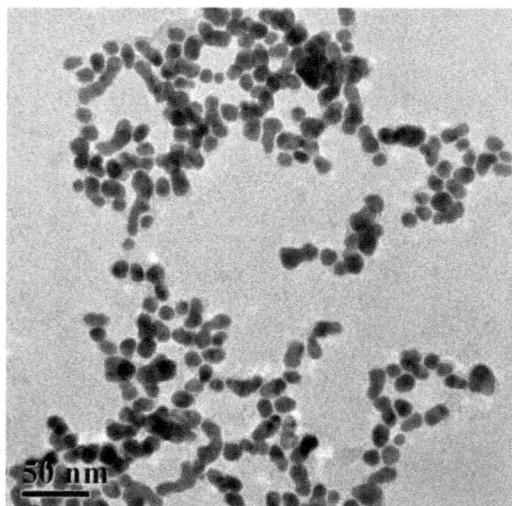

(vi)

Fig. 3 continued

In all the cases, TEM images [Fig. 3vi(sets g, j, m, p, and s)] clearly confirmed the formation of Ag/Au alloy NPs.

Selected area electron diffraction (SAED) patterns for Ag, Au, and Ag/Au [(1:1), (1:3), and (3:1)] alloy NPs synthesized by the present method are shown in the insets of Fig. 2iv. The diffraction rings in SAED correspond to the crystalline nature of the synthesized NPs. The crystalline nature of the synthesized NPs was further confirmed from their powder X-ray diffraction (XRD) patterns, as shown in Fig. 2v. The presence of four lattice planes viz, (111), (200), (220), and (311) confirmed the formation of fcc Ag NP, and these lattice planes were associated with the diffraction peaks at almost 33.06°, 43.78°, and 64.56° [Fig. 2v(set a)]. On the other hand, the presence of fcc Au NPs in the sample was indicated by the appearance of four lattice planes (111), (200), (220), and (311) corresponding to the diffraction peak at 38.18°, 44.39°, 64.83°, and 77.61° [Fig. 2v(set e)].

Four lattice planes (111), (200), (220), and (311) were associated with the diffraction peaks at 38.18°, 43.92°, 64.71°, and 77.26° for fcc Ag/Au (1:1) alloy NP [Fig. 2-v(set c)]. These results confirmed the formation of Ag (metallic silver, JCPDS 04-0783), Au (metallic gold, JCPDS 04-0784), and for bimetallic Ag/Au (1:1) alloy NPs in pure phase.

Energy dispersive X-ray (EDX) studies of the synthesized NPs [Ag NP, Au NP, and Ag/Au NP (1:1)] were done to detect their elemental composition, as shown in Fig. 2vi. In all these EDX spectra, strong signals for copper (Cu) were observed. This may be due to the presence of Cu in Cu-grid used for the experiment.

Understanding the roles of AMEL and its chemical constituents as GMA in controlling the surface structure of the synthesized monometallic (Ag and Au) and bimetallic (Ag/Au) NPs

TGA analysis of the monometallic Ag, Au, and bimetallic Ag/Au (1:1) alloy NPs synthesized by AMEL gave an idea about the thermal stability and surface adsorbed moieties of these NP sets. Results of TGA analysis are shown in Fig. 4i. The plot initially showed a decrease in weight up to 100 °C which may be due to the water molecules present with NPs. In all the cases, steady weight loss of NP samples was observed. Total 54% weight loss was observed for Ag NPs, while 34 and 36% weight losses were noticed in the case of Au and Ag/Au (1:1) NPs, respectively [Fig. 4i(sets a,e, and c)]. Therefore, after elimination of water from the NP surface, this weight loss might have occurred due to the surface desorption of the active chemical constituents of AMEL which acted as the reducing agent and adhered on the surface of the NPs to

give them stability and thus imparted a multifunctional activity to AMEL.

Elaborate IR spectroscopic measurements of each of the systems involved in the present protocol gave an idea about the surface chemistry of the synthesized NPs. In the case of AMEL itself (before the reduction of either Ag^+ to Ag^0 or Au^{3+} to Au^0), IR peaks observed were: 1711, 1613, 1393, 1216, 1142, and 1060 cm^{-1}. The IR peaks at 1711 and 1613 cm^{-1} may be associated with the stretching vibrations for –C=O (keto and ester) groups and for bending mode of vibration of N–H (amine) groups. On the other hand, IR peaks at 1393, 1216, 1142, and 1060 cm^{-1} can be associated with stretching vibrations for –C=C–[(in-ring) aromatic], C–O (ester, ether), C–O (polyol), and C–N (amine), respectively [Fig. 4ii (for AMEL, sets a, e, and c)]. These result further confirmed that the extract (AMEL) was rich in oxygen and nitrogen containing secondary metabolites which can be polyphenols, flavonoids (e.g., quercetin), and alkaloids, such as carbazole alkaloid [e.g., koenigine (A), koenidine (B), girinimbine (C), and mahanimbine (D) which we have also isolated from AMEL].

As a whole, IR spectra of Ag, Au, and Ag/Au (1:1) alloy NPs moderately resembled the IR spectrum of AMEL itself. However, some new peaks appeared with slight shifting in the case of AMEL itself. These were found to be at: 1709, 1600, 1385, and 1039 cm^{-1} (for Au NPs); 1729, 1614, 1450, 1383, and 1073 cm^{-1} (for Ag NPs), and 1726, 1623, 1444, 1342, 1210, and 1078 cm^{-1} (for Ag/Au (1:1) alloy NPs).

In the present case, we have not used any external reducing and stabilizing agent. Therefore, it is clear that AMEL itself played both these two roles and the chemical components of AMEL imparted these characteristics to AMEL. These chemical components not only reduced Ag^+ to Ag^0 or Au^{3+} to Au^0, but at the same time stabilized corresponding NPs either in their free form or in other oxidized form. Close resemblance of the FT-IR spectrum of the AMEL itself with NPs systems synthesized by it may be due to the fact that the unreacted chemical components of AMEL along with their reacted forms adhered on the surface of the NPs giving them stability. For further understanding of these facts, we have also done detailed IR spectroscopic studies of the individual chemical components [koenigine (A), koenidine (B), girinimbine (C), mahanimbine (D), and quercetin] and the NPs synthesized by these components of AMEL. The results are shown in Figs. S1–S6 of the Electronic Supplementary Material (pages S11–13). These results further confirmed the fact that the component actively took part in the reduction process as well as in the surface functionalization of these NPs.

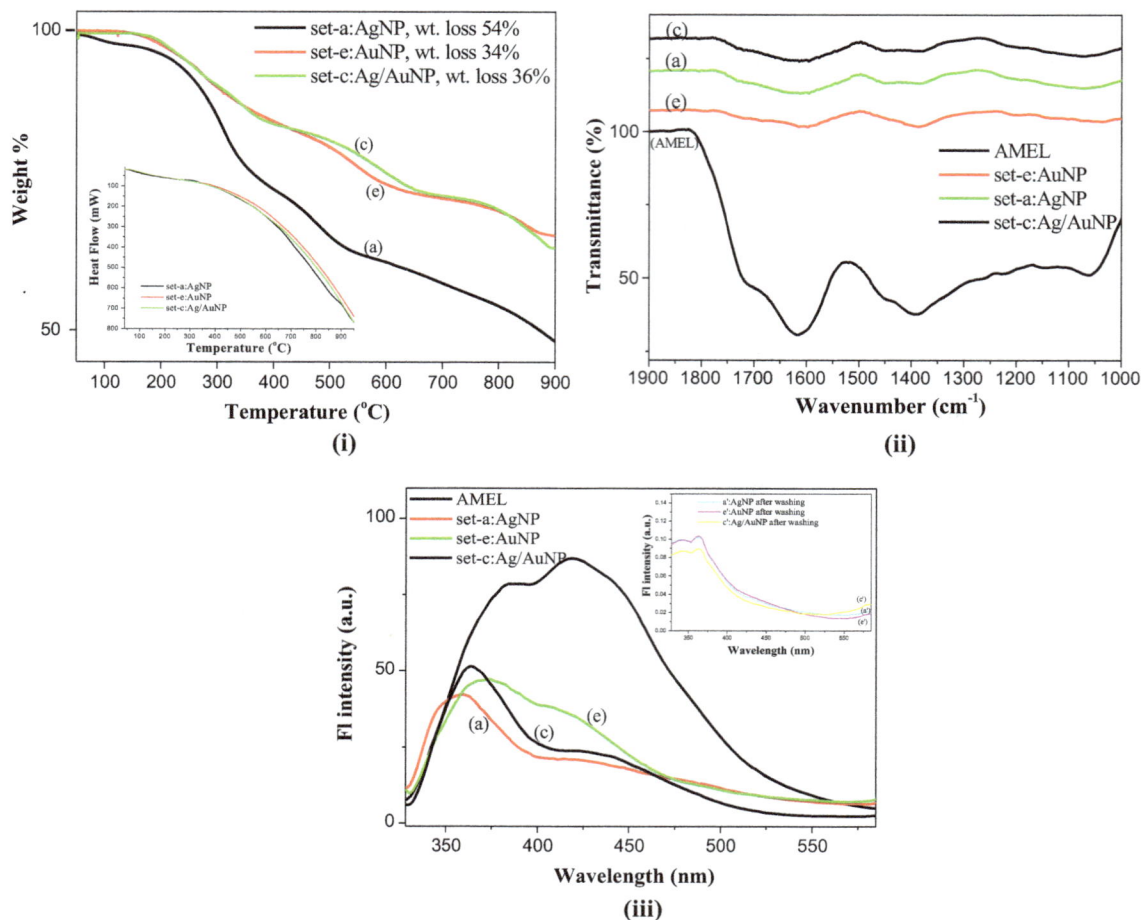

Fig. 4 **i** TGA plot for different sets of NPs synthesized by AMEL as GMA: Ag NP (*set a*), Ag/Au alloy NP (1:1) (*set c*), and Au NP (*set e*). *Inset* of **i** shows the corresponding DTA plots. **ii** FT-IR spectra of different sets of NPs and AMEL itself. **iii** Fluorescence spectra of AMEL before and after the formation of Ag NPs (*set a*), Ag/Au alloy NPs (1:1) (*set c*) and Au NPs (*set e*). *Inset* of **iii** shows the fluorescence spectra of NPs after repeated washing: Ag NP (*a′*), Ag/Au alloy NP (1:1) (*c′*), and Au NP (*e′*). Excitation wavelength was fixed at 308 nm

Role of AMEL and its chemical components on the surface functionalization of the synthesized NPs were further elaborated with the help of fluorescence spectroscopic measurements. The fluorescence spectra of Ag NP, Au NP, and Ag/Au (1:1) NP solution synthesized by AMEL in aqueous medium are shown in Fig. 4iii(sets a, e, and c). The fluorescence emission maximum [λ_{em} (max)] for Ag NP was observed at 350 nm, whereas λ_{em} (max) of Au NP was observed at 430 nm. However, in the case of Ag/Au (1:1) alloy NP, λ_{em} (max) was noticed at 364 nm with a broad hump at 430 nm [Fig. 4iii(sets a, e, and c)].

Similar type of fluorescence behavior was observed for the NP systems synthesized by the individual chemical components, as shown in Fig. S7 of the Electronic Supplementary Material (page S14).

In the case of aqueous solution of AMEL itself, the appearance of broad emission band [Fig. 4iii (AMEL)] indicated the presence of several fluorophoric moieties in it and their combined effect was reflected in the broad nature of the fluorescence emission band of this system. In the case of single fluorophoric system, such as koenigine (**A**), koenidine (**B**), girinimbine (**C**), and mahanimbine (**D**) along with quercetin and the NP sets synthesized by them, the respective emission band maxima were found to be comparatively sharper [Figs. S7 and S8 of the Electronic Supplementary Material (page S14)]. After repeated washing of these NPs, the fluorescence activity was found to be diminished [inset of Fig. 4iii and Fig. S9 of the Electronic Supplementary Material, page S15].

Active component, such as carbazole alkaloids and polyphenols, flavonoids of AMEL have strong fluorophoric

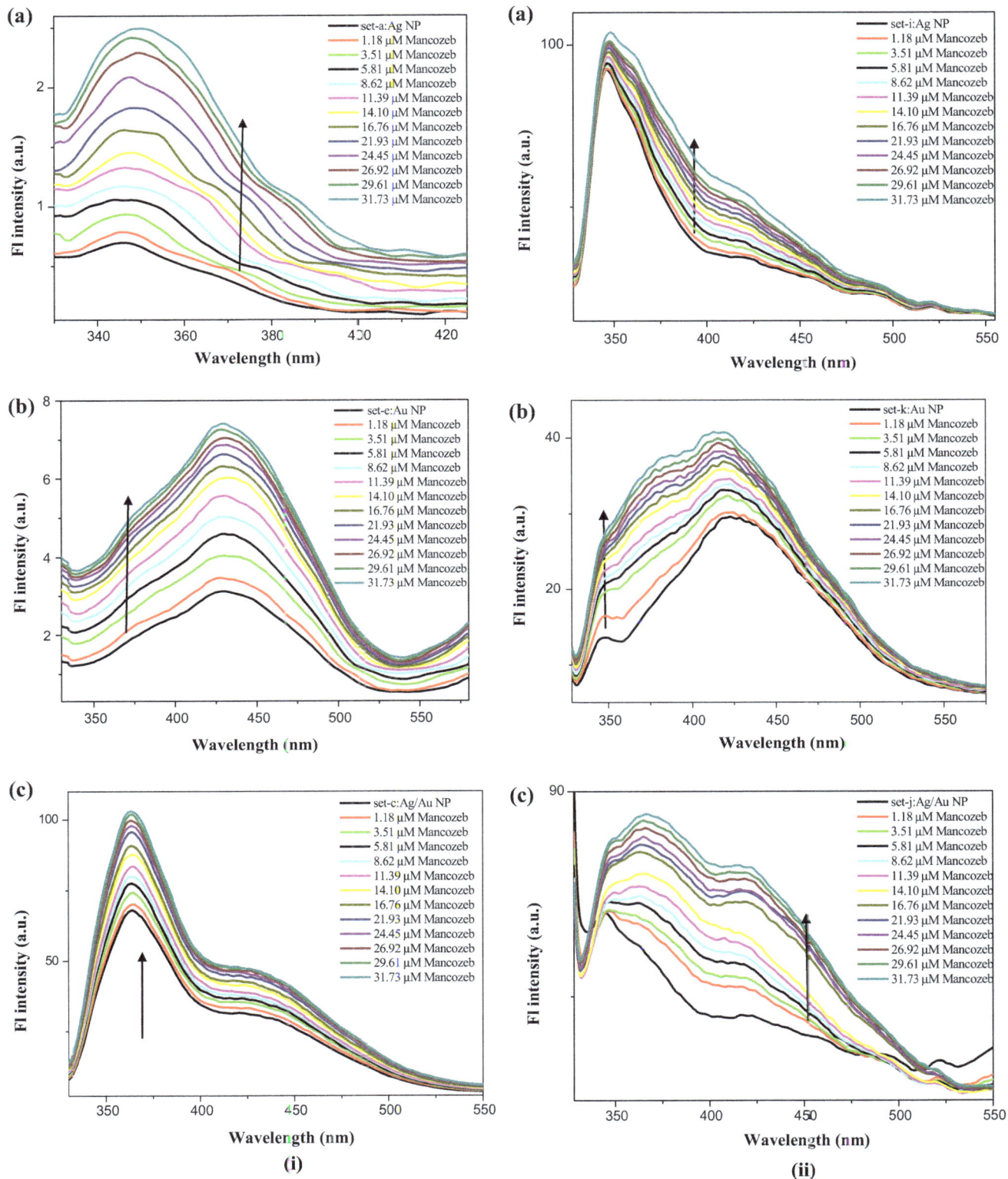

Fig. 5 Mancozeb sensing activity of monometallic Ag, Au and bimetallic Ag/Au (1:1) NPs synthesized by AMEL. Mancozeb sensing activity of different sets of NPs synthesized by **i** AMEL (*a–c*), **ii** koenidine (*a–c*), and **iii** quercetin (*a–c*). For studying this sensing activity, fluorescence spectra of different sets of synthesized NPs were recorded in the absence and presence of different concentrations (1.18–31.73 μM) of aqueous solution of mancozeb. Excitation wavelength was fixed at 308 nm

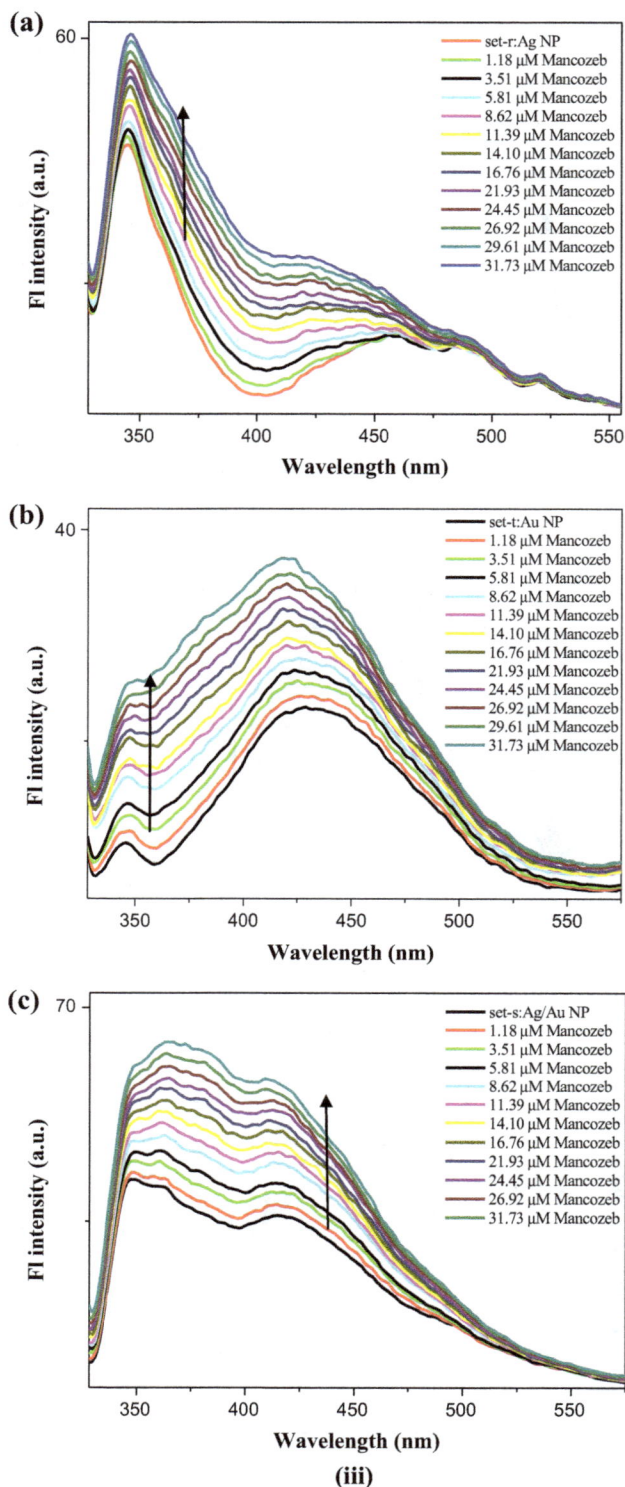

(a)

(b)

(c)

(iii)

Fig. 5 continued

Exploring the efficacy of the monometallic (Ag and Au) and bimetallic (Ag/Au) NPs synthesized by AMEL and its chemical constituents as GMA towards the fluorometric sensing of Mancozeb

We have tested the fluorescence sensing activity of the Ag, Au, and Ag/Au (1:1) alloy NPs synthesized by AMEL towards the sensing of Mancozeb in aqueous medium. During the addition of Mancozeb to NP solution, a significant enhancement of the fluorescence intensity of each of the NP sets was observed (Fig. 5i–iii). The fluorescence emission maximum [λ_{em} (max)] for Ag NP (set a), Au NP (set e), and Ag/Au (1:1) alloy NP (set c) systems (synthesized by AMEL) on the addition of Mancozeb solution were observed at 350, 430, and 364 nm with a broad hump at 430 nm, respectively [Fig. 5i(a–c)].

We have also tested the fluorometric sensing activity of Ag NP, Au NP, and Ag/Au (1:1) alloy NPs synthesized by koenidine (**B**) and quercetin towards Mancozeb in aqueous medium, as shown in Fig. 5ii and iii, respectively. In these cases also, rapid enhancement of fluorescence intensity was observed on the addition of Mancozeb solution. These two compounds represent two different classes of natural products (carbazole alkaloid and polyhydroxy flavonoid, respectively) present in AMEL. The fluorescence behavior of AMEL was found to be the combination of the fluorescence characteristics of both these two classes [Fig. 4iii (AMEL) and Fig. S8 (page S14)].

During the present study, we have found that neither AMEL itself nor its any chemical constituents selected for the present work showed fluorometric/colorimetric sensing activity towards Mancozeb [Fig. S10 of the Electronic Supplementary Material, (page S15)]. We have also tested the sensing activity of AMEL and its chemical constituents in the presence of Ag, Au, and Ag/Au (1:1) alloy NPs synthesized by chemical reduction method. However, in this case, negative results were obtained.

Therefore, it is very much clear that surface functionalization of these NP systems by AMEL or its chemical constituents was very much important for showing their fluorescence sensing activity towards Mancozeb. This enhancement of fluorescence emission intensity of the NP solution may be due to the fluorescence 'turn on' phenomenon [32]. Mancozeb itself was non-fluorescent. On gradual addition of Mancozeb to NP solution, the S-containing functional group of Mancozeb interacted more strongly with the NP surface by soft–soft interaction mode which helped Mancozeb moieties to remain anchored on the surface of the NPs more strongly than the fluorescent chemical components of AMEL and these strong fluorophoric moieties became free from the surface of NPs. Ultimately, this phenomenon increased the concentration of the fluorophoric moieties in the solution which rapidly enhanced or 'turned on' the fluorescence activity of the whole system (Fig. 5).

property [31] and these molecules imposed characteristics fluorescence activity to the synthesized NP systems and leaded in controlling the surface structure of the synthesized NPs.

Scheme 1 Schematic representation of fluorescence sensing study of GMA synthesized nanoparticles (NPs) towards Mancozeb, a dithiocarbamate pesticide

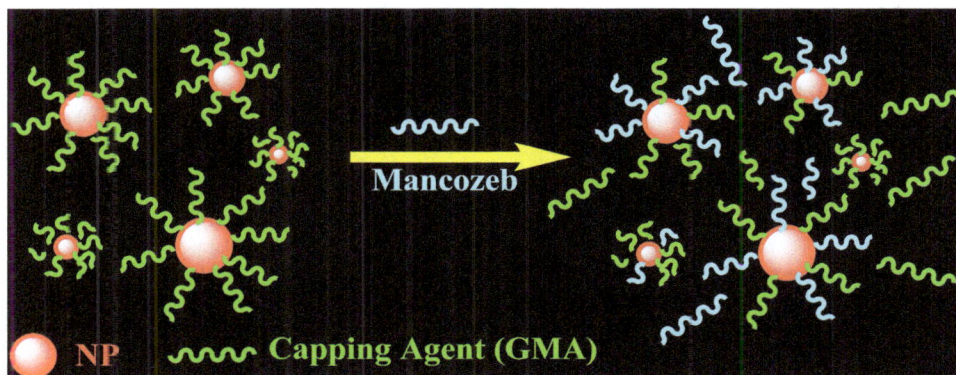

Table 3 Zeta-potential data of the synthesized NP and corresponding NP-Mancozeb systems

Reducing agents	Set of NP	NP composition	Zeta potentials (mV)	
			In absence of mancozeb	In presence of mancozeb
AMEL	a	Ag NP	−31.0	−21.3
	c	Ag/Au NP	−41.6	−17.4
	e	Au NP	−28.3	−19.4
Koenidine	i	Ag NP	−24.2	−15.4
	j	Ag/Au NP	−17.9	−15.5
	k	Au NP	−31.1	−13.7
Quercetin	r	Ag NP	−21.2	−16.9
	s	Ag/Au NP	−22.2	−19.3
	t	Au NP	−18.2	−13.7

Zeta potential of AMEL, koenidine, quercetin, and mancozeb is −19.5, −16.4, −21.3, and −6.52 mV, respectively

The sensing mechanism is shown in Scheme 1.

The attachment of Mancozeb on the NPs surface was also confirmed from the zeta-potential values of these NP solutions (Table 3). It was observed that after the addition of Mancozeb to the NP solutions, the increase (more positive) in zeta-potential value was observed. This is due to the attachment of Mancozeb molecules, which have more positive zeta-potential value (−6.52 mV) to the NP surface by replacing AMEL or its chemical compounds, such as koenidine or quercetin which have more negative zeta-potential (−19.5, −16.4, and −21.3 mV, respectively) (Table 3).

Conclusions

In conclusion, we have used an indigenous source (as leaves of Indian curry leaf plant, a well-known Indian medicinal plant along with its chemical components) as GMA to develop energy-efficient, economically viable and environment-friendly synthetic protocols for stable, crystalline, and fluorescent monometallic (Ag and Au) and also bimetallic (Ag/Au alloy) nanoparticles with wide spectrum of controlled and target specific morphologies under sonochemical condition. These NPs showed their extreme efficacy towards the fluorometric sensing of a dithiocarbamate pesticide, Mancozeb in aqueous medium. AMEL (aqueous methanol extract of leaves) itself along with its active chemical constituents having characteristic fluorescent activity controlled the surface chemistry of the synthesized NPs and also imposed specific fluorescent behavior (in situ) to them. This opened up their possible utility as the eco-friendly and easily synthesizable fluorometric sensors and detectors for the hazardous dithiocarbamate pesticides/fungicide, such as Mancozeb (extensively used in agriculture industries) in aqueous medium. Moreover, the use of non-toxic (leaves are edible) GMA may render these NPs biocompatibility, and thus, these NPs can be explored in the future for the in vivo detection of toxic fluorescent compounds.

Acknowledgements We thank SERB-DST [sanction order no. SR/SO/BB-0007/2011 dated 21.08.2012 to N. A. B.] for the financial support. M. N. A., S. B., and S. D. thanks SERB-DST, MANF-UGC, and CSIR, respectively, for their fellowships. We thank the Department of Chemistry, Siksha Bhavana, Visva-Bharati (Central University) and its DST-FIST and UGC-SAP (Phase-II) programmes for necessary infrastructural and instrumental facilities. We also thank the

Department of Physics, Siksha Bhavana, Visva-Bharati (Central University) for powder XRD analysis. We acknowledge Professor T. Basu in the Department of Bio-chemistry and Bio-physics, University of Kalyani, Kalyani-741235, W.B., India for his kind help in zeta-potential measurements. The acknowledgement is also due to the CRNN, University of Calcutta and IIT Kharagpur, W.B., India for the TEM facility.

Compliance with ethical standards

Conflict of interest The authors declare no competing financial interest.

References

1. Rao, C.N.R., Kulkarni, G.U., Thomas, P.J., Edwards, P.P.: Size-dependent chemistry: properties of nanocrystals. Chem. Eur. J. **8**, 28–35 (2002)
2. Sanchez-Ramirez, J.F., Pal, U., Nolasco-Hernandez, L., Mendoza-A' lvarez, J., Pescador-Rojas, J.A.: Synthesis and optical properties of Au–Ag nanoclusters with controlled composition. J. Nanomater. **2008**, 1–9 (2008)
3. Mondal, S., Roy, N., Laskar, R.A., Basu, S., Mandal, D., Begum, N.A.: Biogenic synthesis of Au, Ag and bimetallic Au/Ag nanopaticles using mahogany (Swietenia mahogany JACQ.) leaf extract. Colloids Surf. B Biointerf. **82**, 497–504 (2011)
4. Roy, N., Mondal, S., Laskar, R.A., Basu, S., Mandal, D., Begum, N.A.: Biogenic synthesis of Au and Ag nanoparticles by Indian propolis and its constituents. Colloids Surf. B Biointerf. **76**, 317–325 (2010)
5. Shiv Shankar, S., Rai, A., Ahmad, A., Sastry, M.: Rapid synthesis of A, Ag and bimetallic Au core-Ag shell nanoparticles using neem (*Azdirachta indica*) leaf broth. J. Colloid Interf. Sci. **275**, 496–502 (2004)
6. Alam, MdN, Chatterjee, A., Das, S., Batuta, S., Mandal, D., Begum, N.A.: Burmese grape fruit juice can trigger the "logic gate" like colorimetric sensing behavior of Ag nanoparticles towards toxic metal ions. RSC Adv. **5**, 23419–23430 (2015)
7. Thakkar, K.N., Mhatre, S.S., Parikh, R.Y.: Biological synthesis of metallic nanoparticles. Nanomedicine. **6**, 257–262 (2010)
8. Nadagouda, M.N., Varma, R.S.: Green and controlled synthesis of gold and platinum nanomaterials using vitamin B_2: density-assisted self-assembly of nanospheres, wares and rods. Green Chem. **8**, 516–518 (2006)
9. Anastas, P.T., Warner, J.C.: Green chemistry: theory and practice, p. 30. Oxford University Press, New York (1998)
10. Raveendran, P., Fu, J., Wallen, S.L.: Completely "Green" synthesis and stabilization of metal nanoparticles. J. Am. Chem. Soc. **125**, 13940–13941 (2003)
11. Shiv Shankar, S., Rai, A., Ahmad, A., Sastry, M.: Controlled the optical properties of lemongrass extract synthesized gold nano-triangles and potential application in infrared-absorbing optical coatings. Chem. Mater. **17**, 566–572 (2005)
12. Chandran, S.P., Chaudhary, M., Pasricha, R., Ahmad, A., Sastry, M.: Synthesis of gold triangles and silver nanoparticles using Aloe vera plant extract. Biotechnol. Prog. **22**, 577–583 (2006)
13. Huang, J., Li, Q., Sun, D., Lu, Y., Su, Y., Yang, X., Wang, H., Wang, Y., Chen, C.: Biosynthesis of silver and gold nanoparticles by novel sundried *Cinnamomum camphora* leaf. Nanotech. **18**, 105104 (2007)
14. Alam, M.N., Roy, N., Mandal, D., Begum, N.A.: Green chemistry for nanochemistry: exploring medicinal plants for the biogenic synthesis of metal NPs with fine-tuned properties. RSC Adv. **3**, 11935–11956 (2013)
15. Begum, N.A., Mondal, S., Basu, S., Laskar, R.A., Mandal, D.: Biogenic synthesis of Au and Ag nanoparticles using aqueous solutions of black tea leaf extracts. Colloids Surf. B Biointerf. **71**, 113 (2009)
16. Alam, MdN, Das, S., Batuta, S., Roy, N., Chatterjee, A., Mandal, D., Begum, N.A.: *Murraya koenegii* Spreng. leaf extract: an efficient green multifunctional agent for the controlled synthesis of Au nanoparticles. ACS Sustain. Chem. Eng. **2**, 652–664 (2014)
17. Alam, M.N., Batuta, S., Ahamed, G., Das, S., Mandal, D., Begum, N.A.: Tailoring the catalytic activity of Au nanoparticles synthesized by a naturally occurring green multifunctional agent. Arab. J. Chem. (2016). doi:10.1016/j.arabjc.2016.02.007
18. Kumar, B., Yadav, P.R., Goel, H.C., Rizvi, M.M.A.: Recent developments in cancer therapy by the use of nanotechnology. Digest J. Nanomat. Biostruct. **4**, 1–12 (2009)
19. Aragay, G., Pino, F., Merkoci, A.: Nanomaterials for sensing and destroying pesticides. Chem. Rev. **112**, 5317–5338 (2012)
20. Rohit, J.V., Solanki, J.N., Kailasa, S.K.: Surface modification of silver nanoparticles with dopamine dithiocarbamate for selective colorimetric sensing of mancozeb in environmental samples. Sens. Actuators B. **200**, 219–226 (2014)
21. Xiong, D., Li, H.: Colorimetric detection of pesticides based on calixarene modified silver nanoparticles in water. Nanotech. **19**, 465502–465507 (2008)
22. Menon, S.K., Modi, N.R., Pandya, A., Lodha, A.A.: Ultrasensitive and specific detection of dimethoate using p-sulphonato calix [4] resorcinarene functionalized silver nanoprobe in aqueous solution. RSC Adv. **3**, 10623–10627 (2013)
23. Jain, V., Momin, M., Laddha, K.: *Murraya koenigii*: an updated review. Int J Ayurvedic Herb Med **2**, 607–627 (2012)
24. Singh, A.P., Wilson, T., Luthria, D., Freeman, M.R., Scott, R.M., Bilenker, D., Shah, S., Somasundaram, S., Vorsa, N.: LCMS–MS characterisation of curry leaf flavonols and antioxidant activity. Food Chem. **127**, 80–85 (2011)
25. Chang, C., Yang, M., Wen, H., Chern, J.: Estimation of total flavonoids content in propolis by two complementary colorimetric methods. J. Food Drug Anal. **10**, 178–182 (2002)
26. Mcdonald, S., Prenzler, P.D., Autolovich, M., Robards, K.: Phenolic content and antioxidant activity of olive extracts. Food Chem. **73**, 73–84 (2001)
27. Roy, N., Alam, M.N., Mondal, S., Sk, I., Laskar, R.A., Das, S., Mandal, D., Begum, N.A.: Exploring Indian Rosewood as a promising biogenic tool for the synthesis of metal nanoparticles with tailor-made morphologies. Process Biochem. **47**, 1371–1380 (2012)
28. Link, S., Wang, Z.L., El-Sayed, M.A.: Alloy formation of gold–silver nanoparticles and the dependence of the plasmon absorption on their composition. J. Phys. Chem. B **103**, 3529–3533 (1999)
29. Ahmad, A., Mukherjee, P., Senapati, S., Mandal, D., Khan, M.I., Kumar, R., Sastry, M.: Extracellular biosynthesis of silver nanoparticles using the fungus *Fusarium oxysporum*. Coll. Surf. B Biointerfaces. **28**, 313–318 (2003)
30. Shankar, S.S., Ahmad, A., Khan, A.M.I., Sastry, M., Kumar, R.: Extracellular biosynthesis of bimetallic Au–Ag alloy nanoparticles. Small **1**, 517–520 (2005)
31. Begum, N.A., Roy, N., Mandal, S., Basu, S., Mandal, D.: Fluorescence spectroscopy of a naturally occurring carbazole alkaloid: murrayanine. J. Lumin. **129**, 158–163 (2009)
32. Senkbeil, S., Lafleur, J.P., Jensen, T.G., Kutter, J.P.: Gold nanoparticle-based fluorescent sensor for the analysis of dithiocarbamate pesticides in water. 16th international conference on miniaturized systems for chemistry and life sciences, Oct 28–Nov 1, Okinawa, Japan, pp 1423–1425 (2012)

10

Perforated ZnO nanoflakes as a new feature of ZnO achieved by the hydrothermal-assisted sol–gel technique

Zahra Khaghanpour[1] · Sanaz Naghibi[1]

Abstract The perforated ZnO nanoflakes with high degree of crystallinity and uniformity were synthesized via the hydrothermal-assisted sol–gel technique without any template. $ZnCl_2$ was used as a Zn-containing precursor, causing the oriented growth of particles. The observation of a hole on the facet of the as-synthesized particles was discussed in this work. XRD, TEM, and DRS were used to investigate the prepared powder and a simple mechanism was suggested to explain the hole formation on the surface of nanoflakes. As a result, the synthesized powder included pure ZnO with direct band gap energy of 3.24 eV. The range of particle size was within 1 μm in diameter and <50 nm in thickness. A circle hole with 300–500 nm in diameter was observed on the facet of the as-synthesized particles.

Keywords ZnO nanoflakes · Perforated flake · Hydrothermal · $ZnCl_2$ · TEM

Introduction

The hollow and perforated particles have recently attracted several scientists. For this reason, several methods and techniques have been presented and the improvement of the properties and the characteristics of the particles have been identified. Due to the unique morphology, the perforated particles have found specific applications in bioscience, energy storage, conversion, and adsorbents [1–4].

ZnO as a well-known ceramic material has become an interesting topic for several researches, and numerous preparation routs [5, 6], morphologies [7–10], and characteristics [11, 12] have been reported. The crystal structure of ZnO wurtzite is hexagonal. Zn and O atoms fill the crystallographic planes and form their unit cell. The anisotropic growth of ZnO crystallites is related to this structure, forming a large variety of features [13]. Morphology control of ZnO particles is an important parameter determining structure characteristics. Varieties of morphologies of ZnO particles have been synthesized with rod-like, hexagonal pyramid-like, truncated hexagonal conical, cauliflower-like, tubular, hourglass-like, flake-like, aggregate, and spherical shapes [7–10].

Nguyen et al. have been developed a template-based method for the preparation of hollow particles. In this approach, the ZnO hollow particles were synthesized via facilitating nucleation of ZnO crystallites under ultrasonic treatment. This process caused rising pressure and temperature at the interface between the cavitation bubbles (generated by ultrasonic vibrations) and the precursor solution. The solution temperature was measured up to 150 °C. In such circumstances, sol–gel stages (i.e., hydrolysis, pyrolysis, and condensation) occurred around the bubbles. In other word, the cavitation bubbles played important role in this process by providing uniform templates for the formation of hollow ZnO structure [14].

Zhang et al. have been synthesized ZnO hollow spheres by utilizing carbon microspheres as templates. In this process, Zn precursor was dissolved in dimethylformamide (DMF). Carbon microspheres as the sacrificial template and water as the hydrolysis agent were added to the solution. After a while, a puce precipitate was obtained. ZnO hollow spheres were obtained by heat treating the precipitated powder at 450 °C [15].

✉ Sanaz Naghibi
  naghibi@iaush.ac.ir

[1] Department of Materials Engineering, Shahreza Branch, Islamic Azad University, Pasdaran St., PO Box 86145-311, Shahreza, Isfahan Province, Iran

Jin et al. have been prepared hollow ZnO microsphere by a chemical method using polymethyl methacrylate (PMMA) as template. For this reason, PMMA powder, distilled water, and ammonia were mixed. After a while, Zn precursor was added to the solution, and then, the pH was adjusted to 8.5 using NaOH. The obtained precipitated powder was calcined at 350 °C. The synthesized microspheres were comprised of several layers of ZnO particles with hollow diameter of ∼10 μm. [16].

Zhu et al. have used electrospinning and subsequent heat treatment to achieve hollow ZnO nanospheres. For this reason, Zn precursor was dissolved in distilled water and then added to the solution of polyvinyl pyrrolidone and ethanol. The obtained solution was applied by electrospinning process. The hollow ZnO nanospheres with an average diameter of ∼250 nm were formed [17].

As can be seen, most of the researches have been focusing on using sacrificial template to achieve the hollow structures.

This paper focused on introducing a new morphology of ZnO particles. The hydrothermal-assisted sol–gel method was applied and zinc chloride was used as the Zn precursor. As a result, 2D ZnO hexagonal single crystals with dimension of >1 μm were obtained. Interestingly, there are micron-sized hole on the as-synthesized ZnO flaks. It should be mentioned that this procedure is a template-less method. This phenomenon and its plausible mechanism will be explained.

Materials and methods

In this work, zinc chloride ($ZnCl_2$, 99.8%, Merck, Germany), triethylamine (TEA, >99%, Merck), and distilled water were used as the precursor, pH adjusting reagent, and solvent, respectively. An adequate amount of $ZnCl_2$ was added to water to achieve a clear solution with the concentration of 5 g/L. Then, TEA was added to the stirring solution to make the pH reach 9. The precipitation occurred after 24 h and a white suspension achieved. This suspension was then treated by an autoclave for 2 h at 200 °C, and dried in an oven to obtain a white powder.

This powder sample was characterized using X-ray diffraction analysis, XRD (by a PANalytical's diffractometers, X'Pert Pro., The Netherlands), transmission electron microscopy, TEM (by a LEO equipment, Japan), as well as diffuse reflection spectroscopy, DRS (by a UV–visible scanning spectrophotometer, JASCO, Japan).

Based on the DRS results, the direct bandgap energy (E_g) was measured in accordance with the Tauc method explained elsewhere [18], via plotting $(\alpha h\upsilon)^2$ versus $h\upsilon$, where α and $h\upsilon$ are the photon energy and the absorption coefficient, respectively. The linear part of the curve extrapolated to $h\upsilon$ axis to achieve E_g value.

The photocatalytic behaviors of the synthesized sample as well as TiO_2-P25 were investigated via measuring the degradation of methylene blue as described previously [19]. For this reason, 100 mg of the powder samples were added to the methylene blue solution (30 mg/L), continuously stirred. The obtained suspensions were irradiated under UV lamp. The decomposition of methylene blue versus irradiation time was recorded by a UV–visible spectrophotometer. For this purpose, variation of the concentration, which is expressed as ln C_0/C (where C_0 refers to the initial concentration of methylene blue solution and C depicts the concentration of methylene blue after UV irradiation), was plotted versus time of UV exposure (30, 60, 90, and 120 min).

TiO_2-P25 (Degussa, Germany) is one of the well-known commercial semiconductors. Expression characteristics of this powder along with the obtained results (about the as-synthesized ZnO sample) could be useful for readers to give more detail on the interpretation of the characterizations.

Results and discussion

Figure 1 shows the XRD pattern of the sample, which can be attributed to the wurtzite [JCPDS: 75–576] with a space group of $P6_3mc$ and hexagonal crystal system without any

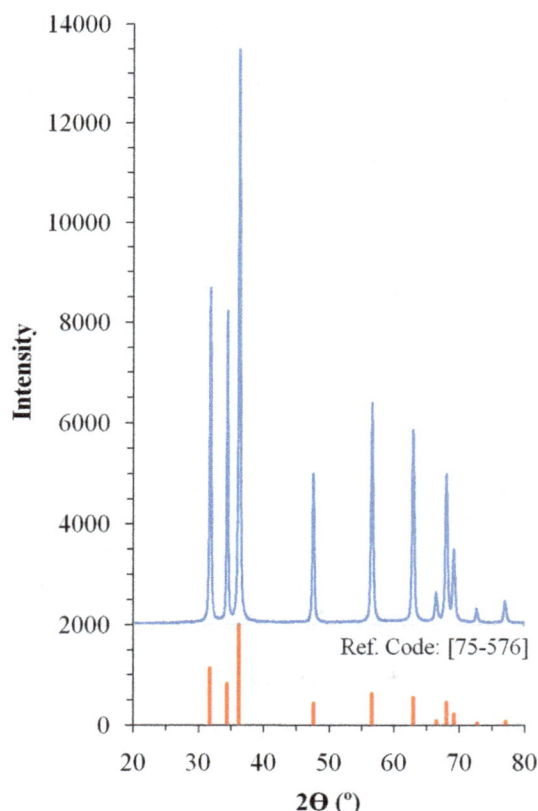

Fig. 1 XRD pattern of the as-synthesized powder along with the peaks of ZnO [JCPDS: 75–576]

Table 1 Experimentally acquired peaks position and intensity along with the standard parameters of ZnO [JCPDS: 75–576]

Peak no.	Observed position [°2Θ]	Standard position [°2Θ]	Observed intensity [cts, (%*)]	Standard intensity [cts]	Intensity changes** [%]	Observed d-spacing [Å]	Standard d-spacing [Å]	d-spacing changes*** [%]	Miller index
1	31.88	31.840	6395 (57)	56.1	+2	2.80,635	2.80,826	−0.07	100
2	34.53	34.503	5818 (52)	41.2	+26	2.59,717	2.59,740	−0.01	002
3	36.37	36.337	11,133 (100)	99.9	0	2.46,968	2.47,039	−0.03	101
4	47.65	47.653	2861 (26)	21.5	+21	1.9085	1.90,683	0.09	102
5	56.70	56.731	4192 (38)	30.9	+23	1.62,345	1.62,135	0.13	110
6	62.89	63.016	3492 (31)	27.2	+14	1.47,638	1.47,393	0.17	103
7	66.46	66.541	506 (5)	41.0	−88	1.40,562	1.40,413	0.11	200
8	67.98	68.120	2734 (25)	22.7	+10	1.37,779	1.37,538	0.18	112
9	69.20	69.261	1318 (12)	11.2	+7	1.35,648	1.35,549	0.07	201
10	72.59	72.759	258 (2)	18.0	−89	1.30,123	1.29,870	0.19	004
11	76.99	77.162	402 (4)	35.0	−89	1.23,751	1.23,520	0.19	202

* The observed height = (peaks height/11,133) × 100; These values can be compared to the standard intensities

** Intensity changes = [(Observed Intensity-Standard Intensity)/Standard Intensity] × 100

*** d-spacing changes = [(Observed d-spacing-Standard d-spacing)/Standard d-spacing] × 100

impurity. Table 1 presents all the experimentally acquired peaks position and intensity along with the standard parameters of this phase. The comparison between the intensity and d-spacing values of the standard ZnO as well as the observed pattern indicated that the as-synthesized particles may not be formed in regular shape. The orientated growth may occur in this process. By considering the intensity of the highest observed peak at $2\Theta = 36.37°$ as 100%, the relative intensities of the other peaks show that there are changes in some of the diffraction angles, such as 34.53°, 56.70°, 66.46°, 72.59°, and 76.99°. On the other hand, the interplanar spacing values show changes compared to the standard values in the orientation along $<10\bar{1}3>$, $<11\bar{2}2>$, $<0004>$, and $<20\bar{2}2>$ directions. To evaluate this assertion, TEM could be useful equipment.

Figure 2 shows the TEM image of the as-synthesized powder. The hexagonal flake-like particles are observed with approximately $\emptyset = 1$ µm and oriented in $<0001>$ direction [20]. This feature is due to the inherent structure of wurtzite. In addition, the thickness of the particles was <50 nm; therefore, these particles could be considered nanoflakes, which are similar to the other researchers' results [9, 20]. This irregularity is related to the crystal growth along c orientation. The ZnO crystallites are known as polar molecules, so that their polar faces are capable to absorb ions from the solution, in contrary ion adsorption along c axis are restricted, and then, the growth along this axis declines. This is the reason of the flaky particles formation. Since TEA acts as a catalyst in this system [21], $ZnCl_2$ ionizes into Zn^{2+} and Cl^-. The Zn^{2+} participates in the crystallization

Fig. 2 TEM image of the as-synthesized powder

reactions, whereas Cl^- ions determine the final feature of the particles. These ions adhere to the (0001) facet of the crystallites and hinder the proximity of Zn^{2+} and the crystallites surface [22].

The motivation for this work refers to the existence of a unique hole on the ZnO nanoflakes. This phenomenon has not been reported before, and this paper attempts to propose an impressive mechanism to investigate the preparation of perforated particles.

Figure 3 shows the schematic illustration of the suggested mechanism. In the early stages of the homogeneous nucleation, $Zn(OH)_2$ molecules have an important role. Increasing time and temperature of the hydrothermal vessel causes to provide the nucleation condition. When the initial ZnO crystallites appear in the solution (Fig. 3I), Zn^{2+} ions congregate around them (Fig. 3II). The Cl^- ions surround Zn^{2+} ions; therefore, Zn^{2+} ions

Fig. 3 Schematic illustration of the mechanism of the formation of holes

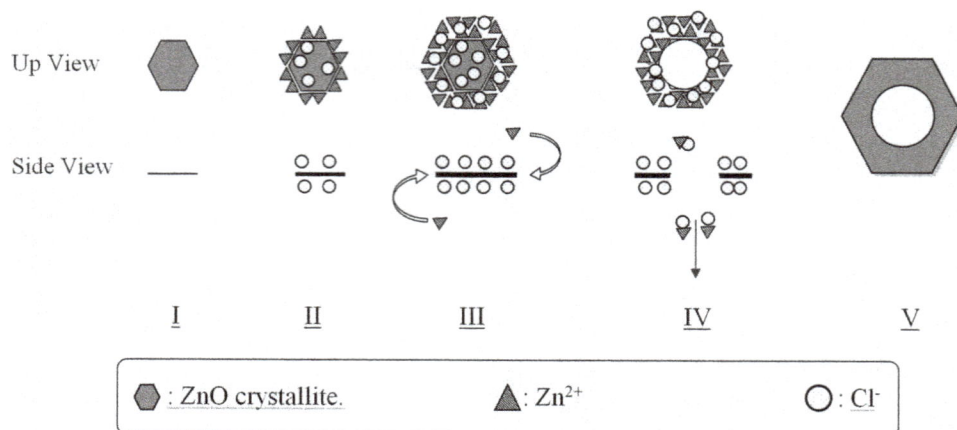

Fig. 4 Tauc results for ZnO powder

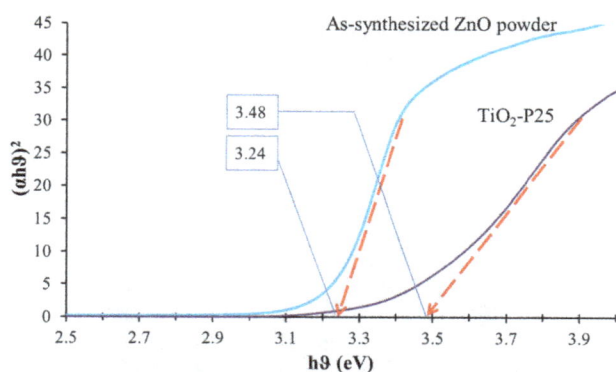

Fig. 4 Tauc results for ZnO powder

congregate on the edges and addition of the Zn^{2+} ions onto the facet become restricted (Fig. 3III). The conversion of intermediate product to the oxide form needs sufficient time and energy. In this period, Cl^- ions, which are accumulated close to the center of the crystallite, would probably participate in reaction to Zn^{2+} ions and remove them from the gathering place. This can led to form a hole through a particle (Fig. 3IV). This irregular feature could be considered the reason of the changes in the XRD results, which was previously explained (see Table 1).

As ZnO powder is known as a photocatalyst, the optical behavior of the as-synthesized powder should be studied.

Figure 4 represents the optical characteristics of the ZnO nanoflakes and TiO$_2$-P25 powder, providing Eg of nanoflakes and P25 as 3.24 and 3.48 eV, respectively. The E_g value of ZnO powder has been reported in the range of 3.26–3.30 eV [23–25]. This approved that the mentioned morphology cannot essentially affect E_g.

Figure 5 shows the kinetics of the photocatalytic degradation of methylene blue suspension containing the as-synthesized ZnO and TiO$_2$-P25 powders. By accordance with the ln (C_0/C) versus irradiation time, the first-order kinetic model is valid for both samples, but the

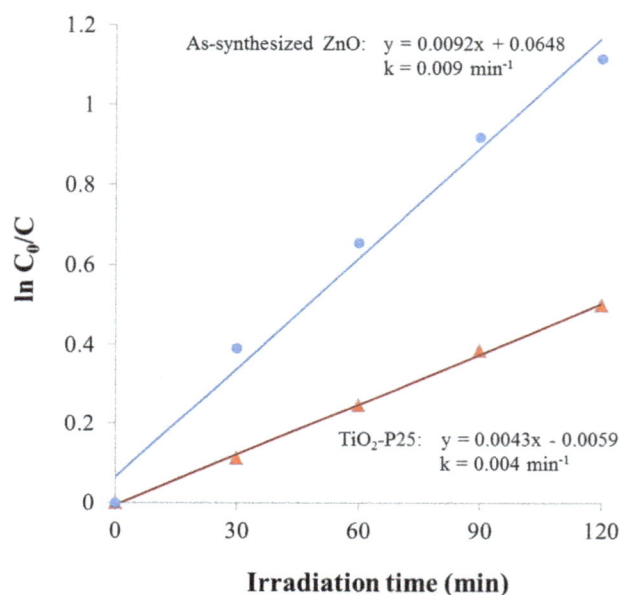

Fig. 5 Kinetics model of the photodegradation of dye solution containing as-synthesized ZnO powder (*filled circles*), and TiO$_2$-P25 powder (*filled triangle*)

lines slope show higher photoactivity of the as-synthesized powder in comparison with TiO$_2$-P25 nanoparticles. In other words, k_{ZnO} (0.009 min^{-1}) is more than k_{P25} (0.004 min^{-1}). Despite the smaller particle size of P25 and relatively similar band gap energy, as-synthesized ZnO powder indicates lower photoactivity. This characteristic is correlated to the higher specific surface area, larger pore volume, and higher degree of crystallinity [18]. The observed phenomenon can be referred to the higher surface area and lower E_g value of the as-synthesized powder. k_{ZnO} is more than two times greater than k_{P25}, whereas ZnO particle size is about 1 μm; 40 times greater than that of TiO$_2$-P25 (average particle size is about 25 nm). It may be related to the existence of the holes on the facet of ZnO particles.

Conclusions

Perforated ZnO nanoflakes were fabricated using the hydrothermal-assisted sol–gel technique without any template. TEA, water, and $ZnCl_2$ were used as catalyst, solvent, and Zn precursor, respectively. Cl^- ion was responsible for the oriented growth and the hole formation during the nucleation and growth of ZnO particles. They surrounded the crystallites of ZnO and prevented the addition of Zn^{2+} to the surface. On the other hand, they probably participate in reaction to Zn^{2+} and removed them from the vicinity of the crystallite, providing the condition for a hole formation. The synthesized powder consisted of ZnO wurtzite phase without any impurity and direct bandgap energy of 3.24 eV. The particles with 1 μm in diameter and less than 50 nm in thickness were observed. There were circle holes with 300–500 nm in diameter on the facet of the as-synthesized particles.

References

1. Andreyev, D.S., Arriaga, E.A.: Fabrication of perforated submicron silica shells. Scr. Mater. **57**(10), 957–959 (2007)
2. Chen, A., Li, Y., Yu, Y., Li, Y., Xia, K., Wang, Y., Li, S., Zhang, L.: Synthesis of hollow mesoporous carbon spheres via "dissolution-capture" method for effective phenol adsorption. Carbon **103**, 157–162 (2016)
3. Zhao, T., Luo, W., Deng, Y., Luo, Y., Xu, P., Liu, Y., Wang, L., Ren, Y., Jiang, W.: Monodisperse mesoporous TiO_2 microspheres for dye sensitized solar cells. Nano Energy **26**, 16–25 (2016)
4. Liu, J., Hui, A., Ma, J., Chen, Z., Peng, Y.: Fabrication and application of hollow ZnO nanospheres in antimicrobial casein-based coatings. Int. J. Appl. Ceram. Technol. (2016). doi:10.1111/ijac.12635
5. Gharagozlou, M., Baradaran, Z., Bayati, R.: A green chemical method for synthesis of ZnO nanoparticles from solid-state decomposition of Schiff-bases derived from amino acid alanine complexes. Ceram. Int. **41**(7), 8382–8387 (2015)
6. Gharagozlou, M., Naghibi, S.: Synthesis of ZnO nanoparticles based on Zn complex achieved from L-leucine. J. Chin. Chem. Soc. **63**(3), 290–297 (2016)
7. Wang, H., Xie, J., Yan, K., Duan. M.: Growth mechanism of different morphologies of ZnO crystals prepared by hydrothermal method. J. Mater. Sci. Technol. **27**(2), 153–158 (2011)
8. Xu, L., Hu, Y.-L., Pelligra, C., Chen, C.-H., Jin, L., Huang, H., Sithambaram, S., Aindow, M., Joesten, R., Suib, S.L.: ZnO with different morphologies synthesized by solvothermal methods for enhanced photocatalytic activity. Chem. Mater. **21**(13), 2875–2885 (2009)
9. Li, H., Jiao, S., Li, H., Li, L.: Growth and characterization of ZnO nanoflakes by hydrothermal method: effect of hexamine concentration. J. Mater. Sci. Mater. Electron. **25**(6), 2569–2573 (2014)
10. Liu, Y., Dong, J., Hesketh, P.J., Liu, M.: Synthesis and gas sensing properties of ZnO single crystal flakes. J. Mater. Chem. **15**(23), 2316–2320 (2005)
11. Lee, C.-P., Chen, P.-W., Li, C.-T., Huang, Y.-J., Li, S.-R., Chang, L.-Y., Chen, P.-Y., Lin, L.-Y., Vittal, R., Sun, S.-S., Lin, J.-J., Ho, K.-C.: ZnO double layer film with a novel organic sensitizer as an efficient photoelectrode for dye-sensitized solar cells. J. Power Sources **325**, 209–219 (2016)
12. Li, S.-Q., Zhou, P.-J., Zhang, W.-S., Chen, S., Peng, H.: Effective photocatalytic decolorization of methylene blue utilizing ZnO/rectorite nanocomposite under simulated solar irradiation. J. Alloys Compd. **616**, 227–234 (2014)
13. Zinc oxide (ZnO) crystal structure, lattice parameters. In: Madelung, O., Rössler, U., Schulz, M. (eds.) II–VI and I–VII Compounds; Semimagnetic Compounds, pp. 1–5. Springer, Heidelberg (1999). Book DOI: 10.1007/b71137, Chapter DOI: 10.1007/10681719_286
14. Nguyen, D.T., Kim, K.-S.: Structural evolution of highly porous/hollow ZnO nanoparticles in sonochemical process. Chem. Eng. J. **276**, 11–19 (2015)
15. Zhang, J., Wang, S., Wang, Y., Xu, M., Xia, H., Zhang, S., Huang, W., Guo, X., Wu, S.: ZnO hollow spheres: preparation, characterization, and gas sensing properties. Sens. Actuators B Chem. **139**(2), 411–417 (2009)
16. Jin, D., Liao, N., Xu, X., Yu, X., Wang, L., Wang, L.: Synthesis of hollow ZnO microspheres and its novel UV absorption. Mater. Chem. Phys. **123**(2–3), 363–366 (2010)
17. Zhu, C., Lu, B., Su, Q., Xie, E., Lan W.: A simple method for the preparation of hollow ZnO nanospheres for use as a high performance photocatalyst. Nanoscale **4**(10), 3060–3064 (2012)
18. Naghibi, S., Faghihi Sani, M.A., Madaah Hosseini, H.R.: Application of the statistical Taguchi method to optimize TiO_2 nanoparticles synthesis by the hydrothermal assisted sol–gel technique. Ceram. Int. **40**(3), 4193–4201 (2014)
19. Naghibi, S., Madaah Hosseini, H.R., Faghihi Sani, M.A., Shokrgozar, M.A., Mehrjoo, M.: Mortality response of folate receptor-activated, PEG–functionalized TiO_2 nanoparticles for doxorubicin loading with and without ultraviolet irradiation. Ceram. Int. **40**(4), 5481–5488 (2014)
20. Vabbina, P.K., Karabiyik, M., Al-Amin, C., Pala, N., Das, S., Choi, W., Saxena, T., Shur, M.: Controlled synthesis of single-crystalline ZnO nanoflakes on arbitrary substrates at ambient conditions. Part. Part. Syst. Charact. **31**(2), 190–194 (2014)
21. Ghasemzadeh, M.A., Safaei-Ghomi, J., Weaver, G.: Synthesis and characterization of ZnO nanoparticles: application to one-pot synthesis of benzo[b][1,5]diazepines. Cogent Chem. **1**(1), 1095060 (2015)
22. Sugunan, A., Warad, H.C., Boman, M., Dutta, J.: Zinc oxide nanowires in chemical bath on seeded substrates: role of hexamine. J. Sol–Gel Sci. Technol. **39**(1), 49–56 (2006)
23. Meyer, B.K.: ZnO band structure, energy gaps. In: Rössler, U. (eds.) New Data and Updates for IV–IV, III–V, II–VI and I–VII Compounds their Mixed Crystals and Diluted Magnetic Semiconductors, pp. 566–569. Springer, Heidelberg (2011). Book DOI: 10.1007/978-3-642-14148-5, Chapter DOI: 10.1007/978-3-642-14148-5_316
24. Deng, Z., Chen, M., Gu, G., Wu, L. A facile method to fabricate ZnO hollow spheres and their photocatalytic property. J. Phys. Chem. B **112**(1), 16–22 (2008)
25. Özgür, Ü., Alivov, Y.I., Liu, C., Teke, A., Reshchikov, M.A., Doğan, S., Avrutin, V., Cho, S.-J., Morkoç, H.: A comprehensive review of ZnO materials and devices. J. Appl. Phys. **98**(4), 041301 (2005). doi:10.1063/1.1992666

11

Synthesis and characterization of *Solanum nigrum*-mediated silver nanoparticles and its protective effect on alloxan-induced diabetic rats

Arumugam Sengottaiyan[1] · Adithan Aravinthan[2] · Chinnapan Sudhakar[1] ·
Kandasamy Selvam[1] · Palanisamy Srinivasan[1] · Muthusamy Govarthanan[1,3] ·
Koildhasan Manoharan[4] · Thangaswamy Selvankumar[1]

Abstract *Solanum nigrum*, a medicinal plant, traditionally used in treating diabetes mellitus. In this study, we
used the leaf extract of the plant to synthesize silver
nanoparticles (AgNPs), as a proposition to treat alloxan-
induced diabetic rats. The phytosynthesised AgNPs were
analyzed using UV–visible and Fourier transform infra-red
spectroscopy for their functional groups. Transmission
electron microscopy revealed that, the synthesized particles
are found to be 4–25 nm in size. Monodispersed and
spherical nature of synthesized AgNPs were shown by
scanning electron microscope and the presence of Ag in the
AgNPs was confirmed by energy dispersive spectrum. The
phytosynthesised AgNPs were evaluated for its antidiabetic
activity in alloxan-induced diabetic rats. AgNPs-treated
diabetic rats found to be significantly improved the dys-
lipidemic condition as seen in the diabetic control. Furthermore, it also reduced the blood glucose level over the
period of treatment. The improvement in body weight was
also found to be evidence for *S. nigrum* extract-mediated
AgNPs as a potential antidiabetic agent against alloxan-
induced diabetic rats.

Keywords Diabetes mellitus · *Solanum nigrum* · Silver
nanoparticles · Alloxan · Animal model

Introduction

Diabetes mellitus (DM) is a group of syndromes characterized by hyperglycemia; altered metabolism of lipids,
carbohydrates, and proteins; and an increased risk of
complications from vascular disease that affects 10 % of
the population [1, 2]. Despite, the availability and extensive utilization of hypoglycemic agents, diabetes and the
related complications continue to be major health con-
cerned worldwide [3]. According to International Diabetic
Federation the estimated diabetes prevalence in 2010 has
risen to 285 million, representing 6.4 % of the world's
adult population, with a prediction that by 2030, the
number of people with diabetes will have risen to 438
million, with this alarming concern, India has been
declared as the "Diabetic capital of world". Currently 40.9
million people in India suffering from diabetes and by 2030
there would be 79.44 million diabetics in India alone. It is
also estimated that by the year 2030, diabetes is likely to be
the seventh leading cause of death, accounting 3.3 % of
total deaths in the world [4].

Owing to the progressive nature of the disease, an
improved treatment strategy is required, which includes,
the discovery of new drugs [5]. In addition, drug treatment,
is not completely successful with diabetes and there is a
compelling need for better prevention and treatment

✉ Arumugam Sengottaiyan
  sengottaibiotech@rediffmail.com

✉ Muthusamy Govarthanan
  gova.muthu@gmail.com

✉ Thangaswamy Selvankumar
  t_selvankumar@yahoo.com

[1] Department of Biotechnology, Mahendra Arts and Science
    College (Autonomous), Kalippatti, Namakkal,
    Tamil Nadu 637501, India

[2] College of Veterinary Medicine, Biosafety Research Institute,
    Chonbuk National University, Iksan 570-752, South Korea

[3] Division of Biotechnology, Advanced Institute of
    Environment and Bioscience, College of Environmental and
    Bioresource Sciences, Chonbuk National University,
    Iksan 570-752, South Korea

[4] Raja Duraisingam Government Arts and Science College,
    Sivagangai, Tamil Nadu, India

strategies [6]. Nevertheless, traditionally people have been using many plants to treat diabetes empirically despite the lack of safety and efficacy [7]. At this juncture, it is necessary to test the herbal medicine as an alternative to synthetic agent [8].

In the search of new opportunities for treatment of DM, the current study has focused on the efficacy of *S. nigrum*. The plants belong to the Solanaceae family, are considered as poisonous to human, however, *S. nigrum* is an edible plant and possesses antioxidant and hepatoprotective activity [9]. Furthermore, many parts of this plant are used as a traditional medicine. Despite, it has been used as an antidiabetic agent in traditional medicine, it lacks scientific evidence [10]. In addition, this is the first of its kind reporting the efficiency of *S. nigrum*-mediated AgNPS as an anti-hyperglycemic agent.

Since, the last decade the applications of AgNPs have been increasing rapidly in various fields Originally, the silver metal was used as an anti-microbial agent, later in the 1990's it has been used in the medicinal field as silver colloids to treat various diseases [11]. AgNPs are usually a cluster of particles ranging between 1 and 100 nm in size and they exhibit new properties based on their size, distribution and morphology [12]. Nanoparticles are becoming the focus of intensive research, due to their wide range of applications in areas such as catalysis, optics, antimicrobials, and biomaterial production [13, 14].

Various approaches have been used for the synthesis of AgNPs. The phytosynthesis approach has many advantages over chemical, physical, and microbial synthesis [15–18], as there is no need of the elaborated process of culturing and maintaining the cell, hazardous chemicals and high-energy requirements.

In the present study, we have characterized the *S. nigrum* leaf extract-mediated AgNPs and assessed it, as an antidiabetic agent in alloxan-induced diabetic rat model.

Materials and methods

Collection of plants

The *S. nigrum* plant material was collected from Erode District, Tamil Nadu, India and authenticated by Dr. S. ArunPrakash, Department of Botany, Government Arts College, Namakkal, Tamil Nadu, India.

Preparation of aqueous extract

The leaves of the plants collected were washed and air dried in shade at room temperature for 7 days. The air dried plants were pulverized in an electric grinder and sieved using mesh to obtain uniform size for further study.

The extraction process was carried out with the help of Soxhlet apparatus. Fifteen grams of dry powder were subjected to soxhlet extraction with 300 mL methanol, the extraction was repeated up to 10 cycles at 45 °C to ensure the complete recovery.

Synthesis of AgNPs

AgNPs synthesis was carried out according to Aravinthan et al. [19]. Briefly, 4 mL of the aqueous extract was mixed with 96 mL of 1 mM $AgNO_3$ solution and the resulting greenish white mixture was incubated for 8 h in a rotary shaker (200 rpm) at 26 °C. Reduction of Ag^+ ions to Ag nanocrystals was monitored by a change in color of the reaction mixture from greenish white to dark brown [19].

Characterization of AgNPs

Surface Plasmon resonance (SPR) bands of the synthesized AgNPs were characterized using UV–Vis spectroscopy (Shimadzu UV-2450) in the range of 200–600 nm. Morphological analysis and the crystalline nature of the particles were investigated by transmission electron microscopy (TEM) using FEI Tecnai TF 20 high-resolution TEM instrument operated at an accelerating voltage of 200 kV [20]. The characterization of the synthesized AgNPs was conducted with X-ray diffractometer (XPERT-Pro diffractometer using Cu K_α radiation), operated at 2θ from 30 to 80° at 0.041°/min with a time constant of 2 s. AgNPs synthesis and the elemental composition were further confirmed by SEM–EDS (SEM–EDS; JEOL-64000, Japan). The chemical characterization of changes in the surface and surface composition was performed by Fourier transform infrared spectroscopy (Shimadzu) within the mid IR region of frequency 4000–400 cm^{-1} [19].

Induction of diabetes

Diabetes was induced in male Wistar albino rats aged more than 8 weeks (140–160 g body weight) by single intraperitoneal administration of alloxan (single dose of 200 mg/kg body weight). Within 48 h after alloxan administration, blood glucose concentrations were measured via tail clip sampling. Animals with a blood glucose concentration 200 mg/dL were considered to be diabetic [21].

Toxicity assessment in rat

Acute toxicity test on AgNPs was performed in experimental rats with a graded dose levels of AgNPs (10 and 20 mg/kg body weight) [22]. The rats were observed continuously for 2 h for behavioral, neurological and

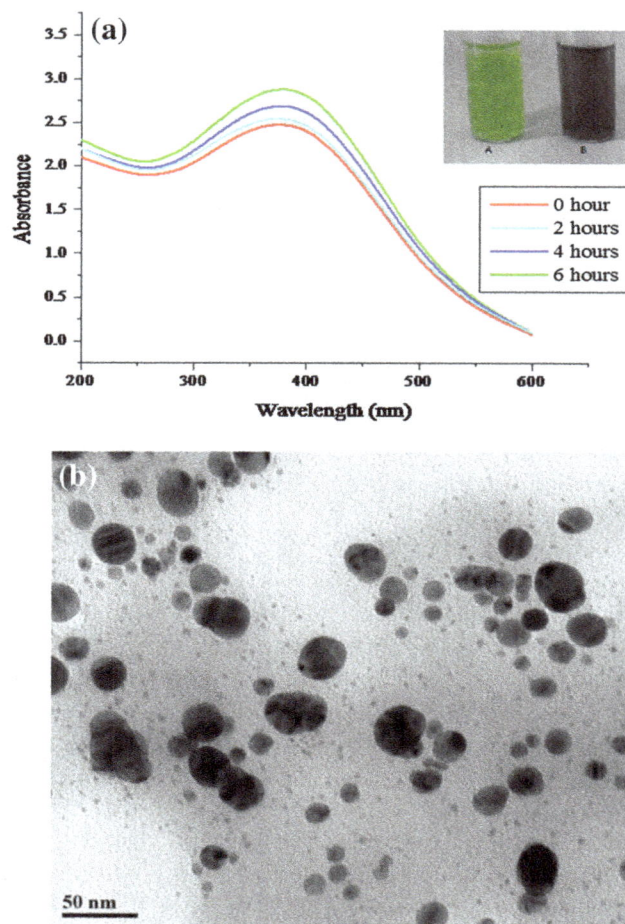

Fig. 1 a UV–visible spectrum of green synthesized AgNPs at different time intervals, showed peak at 420 nm. **b** TEM image of AgNPs

autonomic profiles and after a period of 24 and 72 h for any lethality or death. Further experiments were carried out using 10 mg/kg body weight AgNPs.

Experimental design

The animals were divided into five groups and each group consisted of 5 rats. Group I: Untreated normal rats; (Normal control, received only distilled water) Group II: Diabetic control, alloxan 200 mg/kg single dose, received no treatment; Group III: Diabetic rats, received glibenclamide 0.5 mg/kg for 21 days; Group IV: Diabetic rats, received *S. nigrum* Methanol extract 10 mg/kg for 21 days; Group V: Diabetic rats, received phytosynthesised AgNPs 10 mg/kg for 21 days.

Oral glucose tolerance test

The oral glucose tolerance test was performed after 21 days administration of AgNPs and plant extract to the

respective groups. The rats were fasted overnight, and 2 g/kg glucose was administered orally and the blood was collected from the tail vein prior and the post administration of glucose at 0,30,60,90, and 120 min and fasting blood glucose level was measured using a glucometer (Accu-Chek, Roche Diagnostic, Manheim, Germany).

Biochemical analysis

Blood samples were collected by retro orbital puncture and centrifuged at $1000 \times g$ for 15 min and the collected serum was stored at -80 °C until analysis. Total cholesterol and Triglycerides were estimated using (Biovision, Milpitas, CA, USA) kit as manufacture's instruction.

Statistical analysis

All the data were expressed as mean ± SD, and the statistical significance between the groups was analyzed using one-way ANOVA followed by Tukey's post hoc test. If $P \leq 0.05$ were considered significant.

Results and discussion

Synthesis and characterization of AgNPs

The *S. nigrum* leaf extract was mixed in the aqueous solution of the silver ion complex, and became yellowish brown due to the reduction of silver ion, indicating the completion of the reaction. Similar changes in color have also been reported earlier [23, 24]. The AgNps exhibit yellowish brown color in aqueous solution owing to excitation of surface plasmon vibrations in AgNPs [25, 26]. The plant leaf extract known to posses many active compounds such as, sugar [27] caffeine and theophylline [28] and antioxidants [29] are reported to involve in the formation of AgNPs. As *S. nigrum* also contains high antioxidants, we speculate this AgNPs formation due to the presence of high amount of antioxidants in the extract [9]. The Fig. 1a represents the absorbance spectrum of AgNPs formed at various times (0, 2, 4, 6 h) and the maximum absorption peak was observed in the range of 370–420 nm. It is reported that the absorption spectrum of spherical AgNPs present at 420 nm, and maximum absorption peak shift towards the blue region directly correlates with smaller particle size, and providing the clue for the size of the synthesized nanoparticle in the range of ≤ 25 nm [30]. Figure 1b shows a representative TEM image of phytosynthesized AgNPs. The AgNPs were seen to be spherical and monodispersed with the size range of 4–25 nm. This result is well agreed with the UV–Vis spectra result, in which the maximum peak shifts towards the blue region (λ

Fig. 2 **a** X-ray diffraction for crystalline nature of *S. nigrum* leaf extract-mediated silver nanoparticles. **b** SEM EDX spectrum of AgNPs, after formation of AgNPs with different X-ray emission peaks labeled

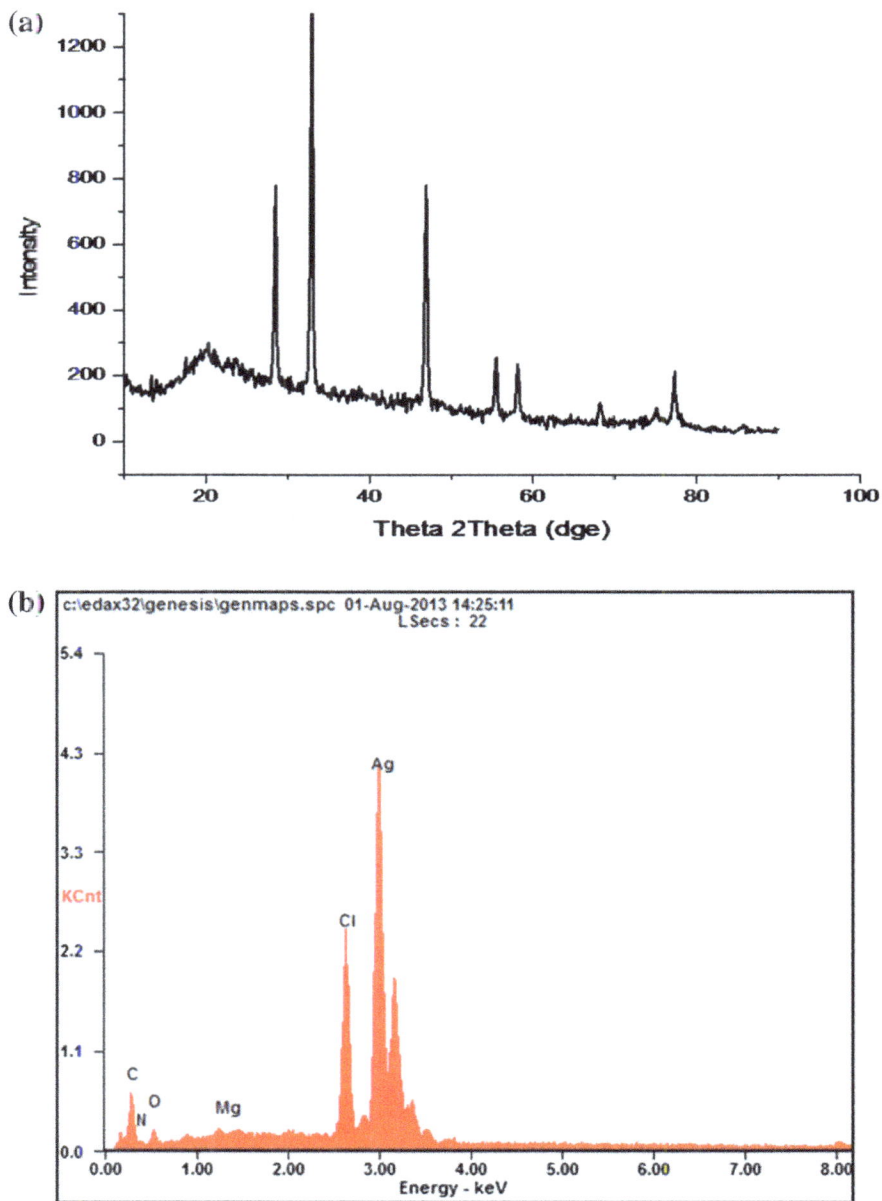

Fig. 3 FTIR spectrum of phytosynthesized AgNPs

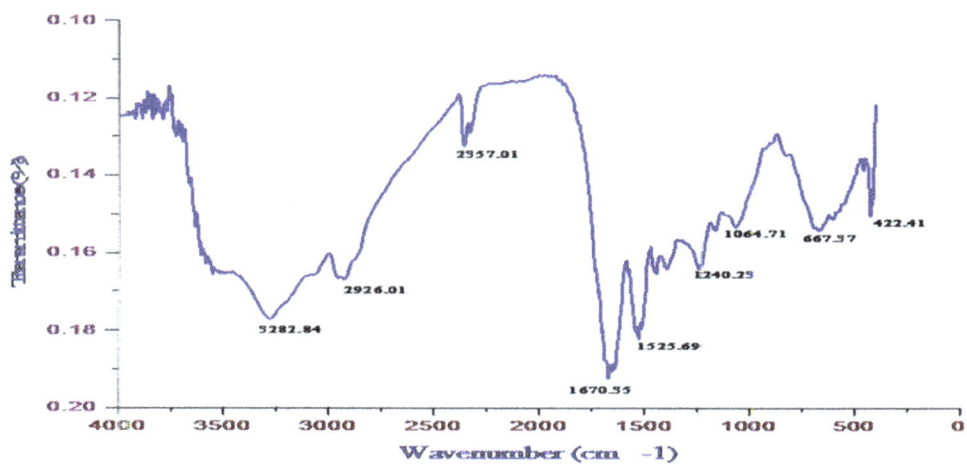

max at 420 nm), and providing the direct evidence for size [30].

The crystalline nature of the silver nanoparticles of *S. nigrum*, the XRD analysis was undertaken, Fig. 2a revealing six peaks at degree (2θ) 10.00, 11.51, 12.00, 27.75, 32.15, and 45.94 corresponding to six diffraction facets of silver. The broadening of X-ray peaks observed can be attributed to the organic content of leaf extract. Analysis of AgNPs through Energy dispersive X-ray (EDX) spectrometers confirmed the presence of elemental

silver signal at 3 keV along (Fig. 2b) with other weak signals corresponding to oxygen and carbon, which could have essentially derived from the plant extract and consistent with previous studies [31, 32]. FTIR measurement was carried out to identify the possible biomolecules in *S. nigrum* leaf extract responsible for capping leading to efficient stabilization of the silver nanoparticles (Fig. 3). The IR spectrum of silver nanoparticles manifests prominent absorption bands located at 3282, 2926, 2357, 1670, 1523, 1240, 1064, 667, and 422.41 cm^{-1}. The bands seen

Fig. 4 **a** Effect of oral administration of *S. nigrum* leaf extract and AgNPs at a dose of 10 mg/kg, in Blood glucose level of alloxan-induced diabetic rats before and after treatment. *Each column* represents mean ± SD for five rats. *$P < 0.05$; **$P < 0.01$; ***$P < 0.001$ different from diabetic control. **b** Effect of *S. nigrum* leaf extract and AgNPs in oral glucose tolerance test after 21 days of treatment. *Each column* represents mean ± SD ($n = 5$). All the values were found to be statistically significant compared to diabetic control at *$P < 0.05$

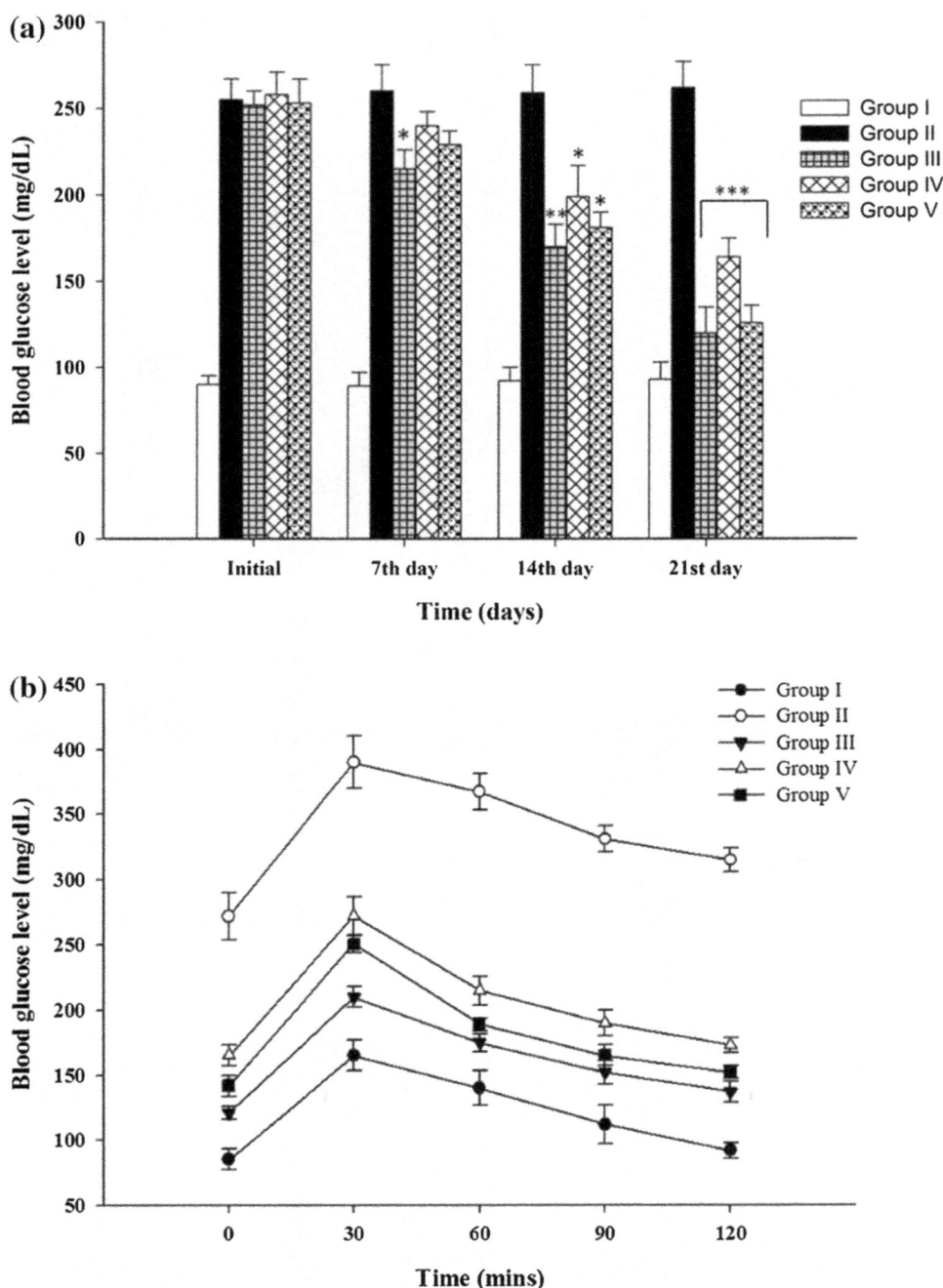

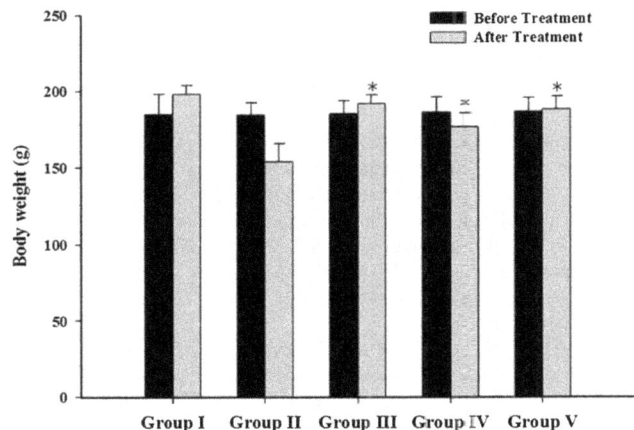

Fig. 5 Effect of methanolic extract of *S. nigrum* and AgNPs on body weight before and after treatment. *Each column* represents mean ± SD (*n* = 5). All the values were found to be significant when compared to diabetic control at *P < 0.05

in 3282 and 2926 cm^{-1} represent the stretching vibrations of primary and secondary amines, respectively. The peak at 2357 could be attributed to H bonded OH stretching of a secondary metabolites present in the extract which might act as a reducing agent to synthesize AgNPs [33].

The antidiabetic activity of phytosynthesized AgNPs

An acute toxicity study revealed the non-toxic nature of the methanolic extract and green synthesized AgNPs. The rats treated with different doses of *S. nigrum* did not show any drug-induced physical signs of toxicity during the whole experimental period, and no deaths were observed. Based on the acute toxicity studies, with a varying concentration (10, 20 mg/kg) of AgNPs, 10 mg/kg body weight was fixed

as an optimal concentration, and thus used in the treatment of alloxan-induced diabetic rats.

The blood glucose level of each group has been estimated during the course of treatment (Fig. 4a), to evaluate anti-hyperglycemic effect of plant extract and phytosynthesized AgNPs. At 14 and 21 days the treated groups (extract/AgNPs) pronouncedly decrease the blood glucose level compared to that of diabetic control (Group II). AgNPs were shown to have reduced blood glucose level higher than plant extract alone. On the other hand glibenclamide-treated groups significantly reduced the blood glucose level throughout the period of treatment compared to group II, IV and V.

The glucose tolerance level of *S. nigrum* leaf extract-mediated AgNPs-treated group was evaluated in diabetic rats (Fig. 4b), by estimating its efficacy in reducing hyperglycemic condition in blood followed an administration of glucose orally, and compared it with standard drug (Glibenclamide 0.5 mg/kg). The phytosynthesized AgNPs-treated animals shown to have a significant improvement in lowering blood glucose compared to Group II. Among the treated groups the efficiency of compounds in lowering the blood glucose level over a period of 2 h has followed with an order, group III > group V > group IV. Phytosynthesized AgNPs treatment is shown to improve the blood glucose level which is comparable to glibenclamide-treated groups. Alloxan-induced diabetic rat model has been widely used in several studies [34, 35]. In addition, the mode of action of alloxan is well documented [36]. In the treatment of diabetes, blood glucose concentration is considered as a routine and major biochemical marker to monitor the improvement in the disease condition, the results obtained for blood glucose concentration of treated groups showed a smaller total area

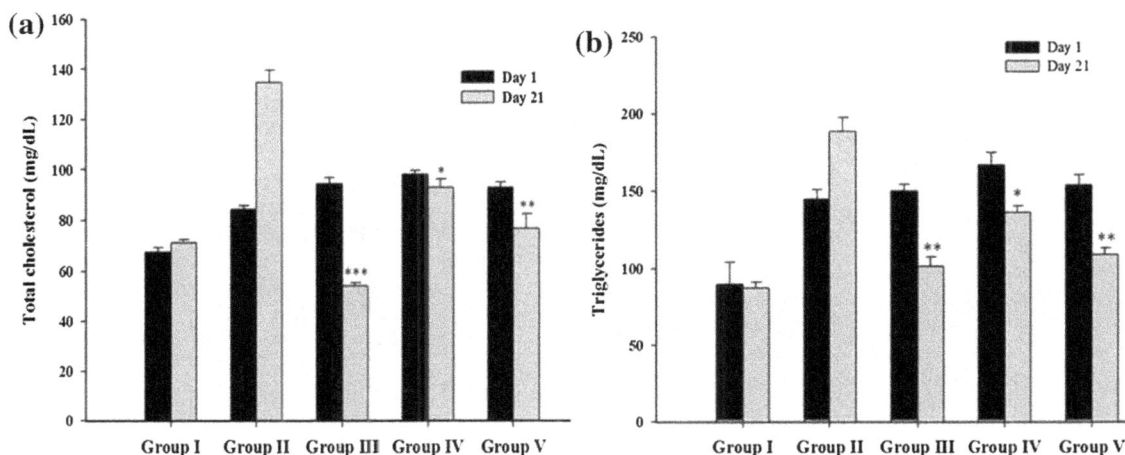

Fig. 6 **a** Effect of oral administration of methanolic extract of *S. nigrum* and AgNPs on serum cholesterol. **b** Triglycerides level before and after treatment. All the values were found to be significant with diabetic control at *P < 0.05; **P < 0.01; ***P < 0.001

under the curve which is significantly lesser than the diabetic control group, and these findings are in accordance with previous research [36]. Figure 5 demonstrates the level of changes in body weight of experimental rats at before and after treatment. Weight loss is one of the major syndrome associated with diabetes, probably due to muscle wasting [37]. In our study the diabetic-induced rat group (Group II) showed significant weight loss (Fig. 5) Compared to group I (Normal control). The total cholesterol (TC) and triglycerides (TG) levels were also elevated in diabetic rats (Fig. 6a, b). It is well documented that DM tends to reduce body weight as a result of increased muscle wasting, dehydration, and fat catabolism. Whereas, the treatment with plant extract as well as phytosynthesized AgNPs significantly improved the body weight loss. The TC and TG levels were retrieved comparable to normal group over a period of 21 days administration of AgNPs, the mechanism of preventing the muscle loss could probably attribute to reversal of antagonism [38].

Conclusion

In this present study, AgNPs were rapidly synthesized using aqueous leaf extract of *S. nigrum* as a bio-reductant. The efficacy of phytosynthesized AgNPs as an anti-hyperlipidemic agent has been evaluated. This study revealed that the AgNPs are found to be safe and environmentally friendly, hence, these AgNPs can be considered in treating diabetes associated syndrome.

References

1. Alvin, C.P., David Allesio, D.D.: Endocrine pancreas and pharmacotherapy of diabetes mellitus and hypoglycaemia. In: Brunton, L., Chabner, B., Knollman, B. (eds.) Goodman and Gilman's the Pharmacological Basis of Therapeutics, 12th edn, p. 1237. McGraw-Hill, New York (2011)
2. Foster, D.W.: Diabetes Mellitus Harrison's Principles of Internal Medicine, pp. 1979–1981. McGraw Hill, USA (1994)
3. El-Amrani, F., Rhallab, A., Alaoui, T., El-Badaoui, K., Chaki, S.: Hypoglycaemic effect of *Thymelaea hirsuta* in normal and streptozotocin-induced diabetic rats. J. Med. Plants Res. **3**(9), 625–629 (2009)
4. Singh, U., Kochhar, A., Singh, S.: Blood glucose lowering potential of some herbal plants. J. Med. Plants Res. **5**(19), 4691–4695 (2011)
5. Modi, P.: Diabetes beyond insulin: review of new drugs for treatment of diabetes mellitus. Curr. Drug Discov. Technol. **4**, 39–47 (2007)
6. Baynes, J.W., Thorpe, S.R.: The role of oxidative stress in diabetic complications. Endocrinol. J. **3**, 277–284 (1996)
7. Alarcon-Aguilara, F.J., Roman-Ramos, R., Perez-Gutierrez, S., Aguilar-Contreras, A., Contreras-Webar, C.C., Felores-Saenz, J.L.: Study of the anti-hyperglycemic effect of plants used as antidiabetics. J. Ethnopharmacol. **61**, 101–110 (1998)

8. Grover, J.K., Yadav, S., Vats, V.: Medicinal plants of India with antidiabetic potential. J. Ethanopharmacol. **81**, 81–100 (2002)
9. Jain, R., Sharma, A., Gupta, S., Sarethy, I.P., Gabrani, R.: *Solanum nigrum*: current perspectives on therapeutic properties. Altern. Med. Rev. **6**, 78–85 (2011)
10. Sohrabipour, S., Kharazmi, F., Soltani, N., Kamalinejad, M.: Effect of the administration of *Solanum nigrum* fruit on blood glucose, lipid profiles, and sensitivity of the vascular mesenteric bed to phenylephrine in streptozotocin-induced diabetic rats. Med. Sci. Monit. Basic Res. **19**, 133–140 (2013)
11. Kenneth, K.Y., Liu, X.: Silver nanoparticles—the real "silver bulltet" in clinical medicine? Med. Chem. Comm. **1**, 125–131 (2010)
12. Satyavani, K., Gurudeeban, S., Ramanathan, T., Balasubramanian, T.: Biomedical potential of silver nanoparticles synthesized from calli cells of *Citrullus colocynthis* (L.). Schrad. J. Nanobiotechnol. **9**, 43 (2011)
13. Kalimuthu, K., Babu, R.S., Venkataraman, D., Bilal, M., Gurunathan, S.: Biosynthesis of silver nanocrystals by *Bacillus licheniformis*. Colloid Surf. B. **65**, 150–153 (2008)
14. Smitha, S.L., Nissamudeen, K.M., Philip, D., Gopchandran, K.G.: Studies on surface plasmon resonance and photoluminescence of silver nanoparticles. Spectrochim. Acta Mol. Biomol. Spectros. **71**, 186–190 (2008)
15. Liu, Y.C., Lin, L.H.: New pathway for the synthesis of ultrafine silver nanoparticles from bulk silver substrates in aqueous solution by sonoelectrochemical methods. Electrochem. Commun. **6**, 78–86 (2004)
16. Bae, C.H., Nam, S.H., Park, S.M.: Formation of silver nanoparticles by laser ablation of a silver target in NaCl solution. Appl. Surf. Sci. **197**, 628–634 (2002)
17. Basavaraja, S., Balaji, D.F., Lagashetty, A., Rajasab, A.H., Venkataraman, A.: Extracellular biosynthesis of silver nanoparticles using the fungus *Fusarium semitectum*. Mater. Res. Bull. **43**, 1164–1170 (2008)
18. Jha, A.K., Prasad, K.: Green synthesis of silver nanoparticles using Cycas leaf. Int. J. Green Nanotechnol.: Phys. Chem. **1**, 110–117 (2010)
19. Aravinthan, A., Govarthanan, M., Selvam, K., et al.: Sunroot mediated synthesis and characterization of silver nanoparticles and evaluation of its antibacterial and rat splenocyte cytotoxic effects. Int. J. Nanomed. **10**, 1977–1983 (2015)
20. Mukunthan, K.S., Elumalai, E.K., Patel, T.N., Murty, V.R.: *Catharanthus roseus*: a natural source for the synthesis of silver nanoparticles. Asian Pac. J. Trop. Biomed. **1**(4), 270–274 (2011)
21. Tanquilut, N.C., Tanquilut, M.A.C., Torres, E.B., Rosario, J.C., Reyes, B.A.S.: Hypoglycemic effect of *Lagerstroemia speciosa* (L.) Pers. on alloxan-induced diabetic mice. J. Med. Plants Res. **3**, 1066–1071 (2009)
22. Ghosh, M.N.: Fundamentals of Experimental Pharmacology, 2nd edn, pp. 153–158. Scientific Book Agency, Culcutta (1984)
23. Singhal, G., Bhavesh, R., Kasariya, K., Sharma, A.R., Singh, R.P.: Biosynthesis of silver nanoparticles using *Ocimum sanctum* (Tulsi) leaf extract and screening its antimicrobial activity. J. Nanoparticle Res. **13**, 2981–2988 (2011)
24. Banerjee, P., Satapathy, M., Mukhopahaya, A., Das, P.: Leaf extract mediated green synthesis of silver nanoparticles from widely available Indian plants: synthesis, characterization, antimicrobial property and toxicity analysis. Bioresour. Bioprocess. **1**, 3 (2014)
25. Maiti, S., Barman, G., Konar, L.J.: Synthesis of silver nanoparticles having different morphologies and its application in estimation of chlorpyrifos. Adv. Sci. Focus. **1**, 145–150 (2013)
26. Barman, G., Samanta, A., Maiti, S., Konar, L.J.: Detection of Cu^{+2} ion by the synthesis of bio-mass-silver nanoparticle nanocomposite. Int. J. Sci. Eng. Res. **6**, 1086–1097 (2014)

27. Shankar, S.S., Rai, A., Ahmad, A., Sastry, M.: Rapid synthesis of Au, Ag, and bimetallic Au core–Ag shell nanoparticles using Neem (*Azadirachta indica*) leaf broth. J. Colloid Interface Sci. **275**, 496–502 (2004)

28. Krishnaraj, C., Jagan, E.G., Rajasekar, S., Selvakumar, P., Kalaichelvan, P.T., Mohan, N.: Synthesis of silver nanoparticles using *Acalypha indica* leaf extracts and its antibacterial activity against water borne pathogens. Colloids Surf. B **76**, 50–56 (2010)

29. Ramteke, C., Chakrabarti, T., Saranki, B.K., Pandey, R.A.: Synthesis of silver nanoparticles from the aqueous extract of leaves of *Ocimum sanctum* for enhanced antibacterial activity. J. Chem. (2013). doi:10.1155/2013/278925

30. Martinez-Castanon, G.A., Nino-Martinez, N., Martinez-Gutierrez, F., Martinez-Mendoza, J.R., Ruiz, F.: Synthesis and antibacterial activity of silver nanoparticles with different sizes. J. Nanopart. Res. **10**, 1343–1348 (2008)

31. Govarthanan, M., Selvankumar, T., Manoharan, K., et al.: Biosynthesis and characterization of silver nanoparticles using panchakavya, an Indian traditional farming formulating agent. Int. J. Nanomed. **9**, 1593–1599 (2014)

32. Lee, K.J., Park, S.H., Govarthanan, M., et al.: Synthesis of silver nanoparticles using cow milk and their antifungal activity against phytopathogens. Mater. Lett. **105**, 128–131 (2013)

33. Dubey, S.F., Lahtinen, M., Särkkä, E., Sillanpää, M.: Bioprospective of *Sorbus aucuparia* leaf extract in development of silver and gold nanocolloids. Colloids Surf. B **80**, 26–33 (2010)

34. Attanayake, A.P., Jayatilaka, K.A.P.W., Pathirana, C., Mudduwa, L.K.B.: Study of antihyperglycaemic activity of medicinal plant extracts in alloxan induced diabetic rats. Anc. Sci. Life **32**, 193–198 (2013)

35. Ragini, V., Prasad, K.V., Bharathi, K.: Antidiabetic activity of *Shorea tumbuggaia* rox. Int. J. Innov. Pharm. Res. **2**, 113–121 (2011)

36. Fröde, T.S., Medeiros, Y.S.: Animal models to test drugs with potential antidiabetic activity. J. Ethnopharmacol. **115**, 173–183 (2008)

37. Swanston-Flatt, S.K., Day, C., Bailey, C.J., Flatt, P.R.: Traditional plant treatment for diabetes: studies in normal and streptozotocin diabetic mice. Diabetologia **33**, 462–464 (1990)

38. Whitton, P.D., Hems, D.A.: Glycogen synthesis in perfused liver of streptozotocin diabetic rats. Biochem. J. **21**, 150–153 (1975)

12

Streptomycin loaded TiO$_2$ nanoparticles: preparation, characterization and antibacterial applications

S. Kalaiarasi[1] · M. Jose[1]

Abstract We report a facile synthesis of Titanium dioxide (TiO$_2$) nanoparticles by sol gel technique assisted by biogenic route using the rind of Aloevera and demonstrate the antibacterial assessment against human pathogens causing urinary tract infection. The synthesized nanoparticles were characterized by powder XRD analysis, FT-IR analysis and scanning electron microscopic analysis. The XRD spectrum confirmed that the synthesized TiO$_2$ nanoparticles exhibit anatase phase. Average grain size was calculated using Debye–Scherrer formula and it was found to decrease from 13 to 8 nm with increasing template concentration. FTIR spectrum showed characteristic bands at 1626, 1056 and 1074 cm^{-1} revealing C–N stretching of amino groups present in the protein cages of Aloevera which assist in the formation of TiO$_2$ nanoparticles. Morphological characterization analyzed by SEM showed nanocoral network and all the networks displayed excellent invitro bioactivity against *Staphylococcus aureus*, *Escherichia coli*, *Klebsiella pneumoniae*, *Salmonella typhi* and *Proteus mirabilis*. Drug delivery was assessed by the release of anti-inflammatory streptomycin which is evidenced by the release profile suggesting that it has the potential to provide better deliveries.

Keywords X-ray diffraction · Electron microscopy · Antibacterial activity · Human pathogens and Minimal inhibitory concentration

✉ M. Jose
  mjosh1231@gmail.com; jose@shctpt.edu

[1] Department of Physics, Sacred Heart College (Autonomous), Tirupattur 635601, India

Introduction

Nanostructured materials have captured the attention of researchers across the world because of the transitional physical and chemical attributes [1]. Metal oxide nanoparticles play a critical and dominant part in diverse areas that are capable of forming diversified oxide compounds adopting many geometrical structures with unique properties due to size and edge corner sites [2]. Titanium dioxide (TiO$_2$) is one of the fascinating biologically compact material, which can live in three nanostructured forms, anatase (tetragonal), brookite (rhombohedral) and rutile (tetragonal) with six coordinate titanium ions possessing high refractive index, low absorption emission in visible and near infrared spectral region [3]. At the bulk state, rutile is thermodynamically stable, whereas anatase and brookite are capable of transforming to rutile by thermal treatment. All the crystalline modifications have different adsorption properties and responses due to different band structures. In nanocrystalline regime, the anatase TiO$_2$ structure is dominant and industrially relevant in many applications such as photo catalysis, photovoltaic cells, sensors and biomedical applications [4–8].

Moreover, TiO$_2$ at anatase phase with its increased thermal stability up to the range of sintering temperature shows enhanced photo catalytic and antimicrobial performances [7]. As the demand for drugs is increasing exponentially worldwide, various mechanisms were adopted to synthesize anatase phase TiO$_2$ nanoparticles for antibiotic resistance against human and animal pathogens. Various synthesis routes such as sol gel, hydrothermal and solvothermal have been adopted to prepare anatase phase TiO$_2$ nanoparticles [9]. Texturing nanocoral morphology seems to be challenging because of the synthesize protocol but it could be achieved by template technique [10].

Instead of employing various biotemplating materials like gelatin, starch, gum, rice straw, egg shell membrane, bamboo membrane, east pollen, etc., It is seen that lignocellulosic material play a vital role as templating agent and it is proved that soft biotemplates enhances the formation of TiO_2 nanoparticles successfully [11–13].

Aloe barbadensis is a medicinal plant and a lignocellulosic material whose pulp contains 0.5% solid materials, which is composed of water-soluble compounds, fat-soluble vitamins, minerals, enzymes, polysacrides, organic acids and 99.5% of water [14–16]. In the extracellular synthesis of nanoparticles using plants, the aggregation of nanoparticles is done by stabilization due to the protein-nanoparticles interaction, and therefore, morphology can be influenced by the carbonyl compounds present in Aloevera. Very recently, enhanced textural properties and coral morphology of TiO_2 nanoparticles have been discussed and reported while using water storage cells of Aloevera as template medium [17]. However, analysing the related literature reveals that the bactericidal activity of TiO_2 against human pathogens has not yet been studied. In this work, Aloevera templated bioactive anatase phase TiO_2 nanocorals were studied to evaluate the level of adsorption and release of streptomycin nanocorals, to counter the human pathogens causing urinary tract infection (UTI). Subsequently, we prove that the percentage of templating agent and the calcination temperature enhance the effectiveness of structural and morphological attributes. Interpretation from the antimicrobial activity of pathogens loaded with streptomycin drug suggest that TiO_2 assisted by Aloevera could be a better antibacterial agent for biomedical and pharmaceutical applications.

Materials and methods

All chemicals used in this work were purchased from Merck and were used without any further purification. Titanium (IV) isopropoxide and glacial acetic acid were used as main precursor and chelating agent, respectively, and deionized water was used throughout the experiment. Aloevera leaves used in the experiment were washed and the rind parts were finely chopped, shadow dried, ground into fine powder using agate mortar.

Synthesis of TiO_2 nanoparticles

TiO_2 solution was prepared by adding 2 mL of titanium (IV) isopropoxide in 200 mL of deionized water under continuous magnetic stirring at room temperature. During the preparation of precursor solution, 20 mL of glacial acetic acid acting as the chelating agent was added in drops, which prevent titanium isopropoxide from the

nucleophilic attacks by water. The resulting solution was stirred for few hours and then Aloevera rind powder with 0.4, 0.8, 1.2 and 1.6 g concentrations were introduced into the solution (the ratios of titanium (IV) isopropoxide to Aloevera rind powder were chosen at 1 : 0, 1:0.4, 1:0.8, 1:1.2 and 1:1.6 w/w%). The solution was heated at 80 °C, until the xerogel was completely dried and then cooled naturally to room temperature. The dried gel was crushed into fine powder and finally, the powdered samples were calcinated at 500 °C for 5 h in a muffle furnace. Before calcination, FTIR spectra were obtained to study the functional group vibrations.

Result and discussion

Powder XRD analysis

The structure and phase of the synthesized TiO_2 nanoparticles were determined using Bruker D8 advance Powder XRD diffractometer and the recorded XRD pattern is shown in the Fig. 1. The predominant peaks observed for various template concentrations of Aloevera in the following ratios of 1:0, 1:0.4, 1:0.8, 1:1.2 and 1:1.6 g calcinated at 500 °C were investigated.

All the diffraction peaks agreed very well with the tetragonal structure, with lattice parameters $a = b = 3.784$ Å and $c = 9.514$ Å (JCPDS File No. 78-2486) and the appearance of diffraction planes of (101), (200), (004), (211) and (204) confirms that the exhibited phase was anatase. As the template concentration exceeds the precursor concentration at the ratios from 1:1.2 to 1:1.6 g, a slight transformation starts, which was confirmed by the appearance of additional diffraction peaks at 31.6°

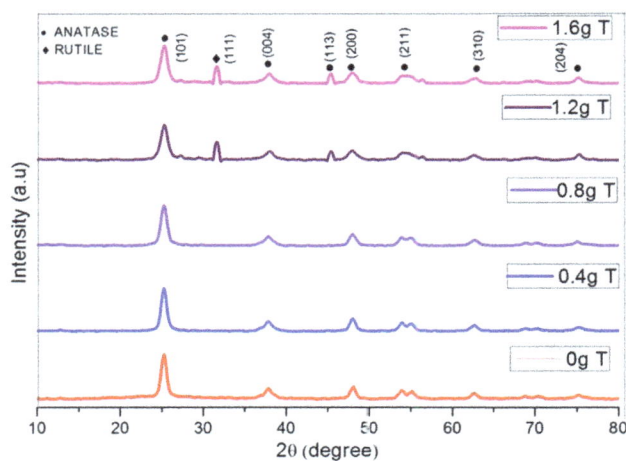

Fig. 1 X-ray diffraction pattern of synthesized TiO_2 nanoparticles calcinated at 500 °C at varying template concentration of Aloevera (0, 0.4, 0.8, 1.2, 1.6 g T)

and 45.4° corresponding to (111) and (113) diffraction planes indicating that the percentage of Aloevera composition hinders the grain growth of the nanoparticles. Analysis of prominent diffraction peaks using Debye–Scherer's equation showed that the average grain size decreased from 13 to 8 nm with increasing template concentration with cell parameters $a = b = 3.7856$ Å and $c = 9.514$ Å, which agrees very well with the anatase phase of the TiO$_2$ nanoparticles. The XRD pattern shows that the main phase is still anatase, however, when the template concentration is increased, minor peaks of the rutile phase appear and the crystallite size decreases to 8 nm.

FTIR spectroscopy analysis

FT-IR spectra of the prepared TiO$_2$ nanoparticles calcinated at 500 °C have been recorded using BRUKER RFS 27 spectrophotometer ranging from 400 to 4000 cm^{-1} at room temperature (Fig. 2). The broad band appeared at 3406 cm^{-1} which was assigned to hydrogen bonded OH stretching vibration. The absorption bands in the range of 3000–3500 and 1600 cm^{-1} in all the spectra indicate the presence of hydroxyl stretching vibration groups. The peaks observed at 1023, 1056 and 1074 cm^{-1} correspond to Ti–O–C bond which may be due to the increased intensity of template concentration indicating the interaction between organic and inorganic components present in the precursor and the template during condensation and chelation process. The bands that appeared at 1630 and 1626 cm^{-1} are attributed to amide groups present in Aloevera. A slight shift in the spectrum occurring at 1528 and 1626 cm^{-1} indicates binding of proteins with the surface of TiO$_2$ nanoparticles at high Aloevera concentrations [15].

Similarly the bands appeared at 1427 and 1074 cm^{-1} are associated with carboxylic acid and C–N stretching vibration of amine group, respectively. The bands at 649, 652 and 621 cm^{-1} are attributed to the stretching vibration of TiO$_2$. The broad peak at 564 cm^{-1} indicates Ti–O–Ti stretching vibration.

SEM analysis

Scanning electron microscopy (SEM) analysis was performed using FEI quanta 200 electron microscope to study the morphology of the synthesized TiO$_2$ nano powder at different template concentration and the images are shown in Fig. 3a, e. It clearly indicates the step wise transition of TiO$_2$ nanoparticles while increasing the template concentration from relatively undefined agglomeration. SEM image of TiO$_2$ nanoparticles without template (Fig. 3a) depicts the constitution of some coral shaped TiO$_2$ nanoparticles associated with agglomeration and scraggy surface, which might be due to the process of simultaneous hydrolysis and condensation. At this initial stage, it is interesting to note that the nanoparticles appear to be agglomerated completely with no sign of well developed faces despite the appearance of anatase phase. It is evident from Fig. 3b, d that the average size of the TiO$_2$ nanoparticles synthesized with varying template concentration varies from 30 to 15 nm, which can be associated with the shrinkage of mesostructure upon protein cages present in the Aloevera. On the contrary, both SEM analysis and XRD analysis suggest that the size of the nanoparticles varies significantly upon the increasing template concentration. However, the increased concentration of Aloevera seems to have decreased agglomeration which might be due to the presence of organic ligands at the surface of the nanoparticles implying that the organic complexes present in Alovera effectively controls the size and morphology of nanoparticles. It is apparent from the Fig. 3b–e that the impact of template is clearly observed on the control of agglomeration and reduction of particle size. However, from Fig. 3d, e it is observed that there is no significant change in morphology which could be attributed to the decreased difference in concentration percentages. Uniformity of particles point out the association of bioconstituents with TiO$_2$ nanoparticles during the synthesis process. The elemental composition of the synthesized TiO$_2$ nanoparticles was confirmed by the energy dispersive X-Ray analysis under SEM (Fig. 3f). It reveals the presence of Ti and O signals with an atomic ratio 2:1. The less significant percentage of oxygen composition is attributed due to the adsorption of oxygen from the Aloevera constituents during the formation of nanoparticles.

The fabrication procedure of TiO$_2$ nanostructures is schematically illustrated in Fig. 4. The results indicate the

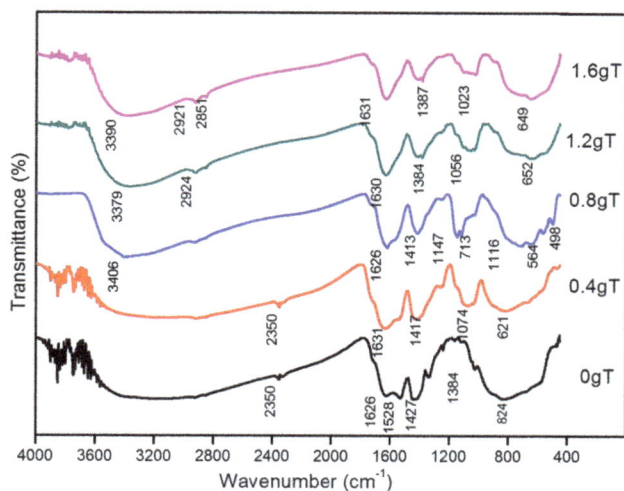

Fig. 2 FTIR spectra of TiO$_2$-NPs synthesized at different wt% of Aloevera (0, 0.4, 0.8, 1.2, 1.6 g T)

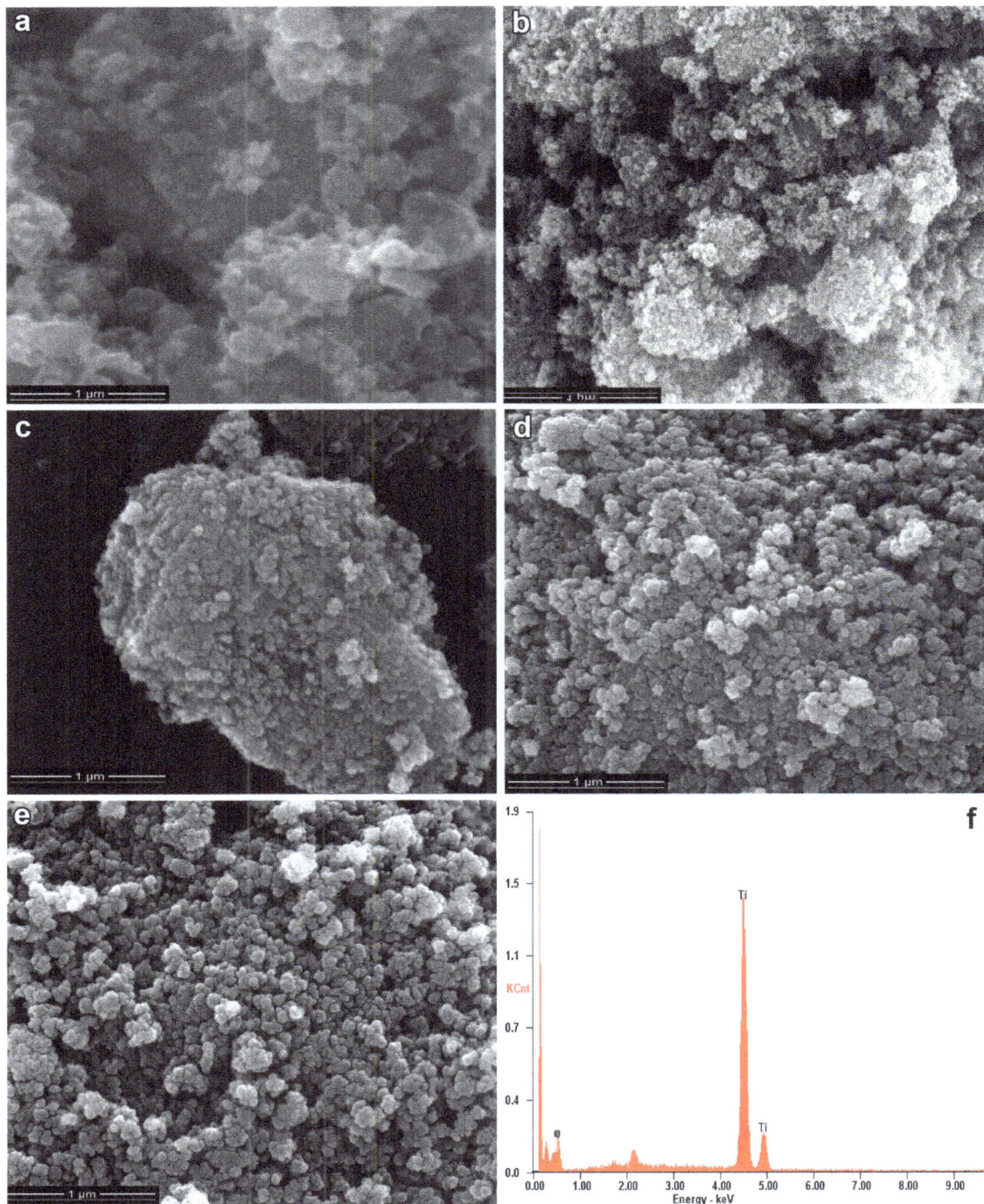

Fig. 3 a–e Scanning electron microscopic images of synthesized TiO_2 nanoparticles calcinated at 500 °C for varying template concentration, **f** EDS spectrum of calcinated TiO_2 nanoparticles

presence of nucleation-seed-growth step in the formation of nanoparticles. Addition of Aloevera rind powder tends to change the nature of precursor to slurry. This observation is supported by Fig. 3b. Furthermore, the solution was suspended into the template medium and kept static for 48 h. Ti ions are absorbed on the surface of the template, especially on the pores of protein constituents that are

present in the phosynthetic cells of the rind part of Aloevera. The protein constituent in the Aloevera functions both as the capping agent and structure directing agent such that the chemisorbed Ti hydroxides are dehydrated to form TiO_2 nanocomposites during calcinations. At the calcinations temperature, removal of bioconstituents, as well as

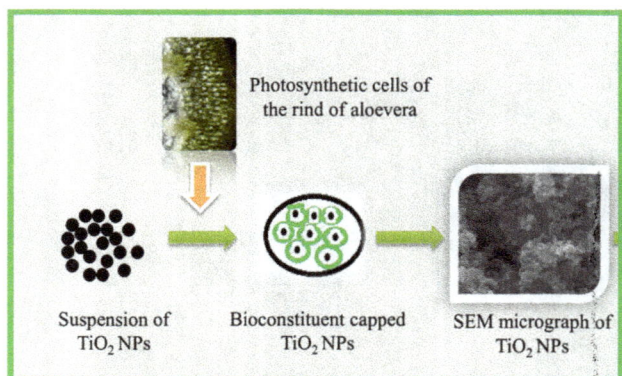

Fig. 4 Schematic representation of growth mechanism of TiO$_2$ nanoparticles

spontaneous aggregation takes place that results in the formation of nanocoral morphology.

Antibacterial activity

Various bacterial strains (*Staphylococcus aureus, Escherichia coli, Klebsiella pneumoniae, Salmonella typhi* and *Proteus mirabilis*) that cause UTI were maintained on Nutrient agar slants at 4 °C. Antibacterial activity of synthesized TiO$_2$ nanoparticles by Agar well diffusion technique. The pathogens to be tested were spread on the Mueller–Hinton Agar plates with sterile cotton swap. Subsequently, wells were punctured into the agar plates of 7 mm diameter each and filled with 100 mL of synthesized TiO$_2$ nanoparticles loaded with and without streptomycin drug dissolved in 10% DMSO and allowed to diffuse at room temperature for 2 h. Then, the plates were incubated at 37 °C for 48 h. The standard antibiotic disc of streptomycin was used as positive control and the experiments

were performed in triplicate each week during a 2 months period. The results were expressed in mean standard deviation. After incubation, the growth incubation zone diameters were measured in mm. Inhibitory concentration of synthesized nanoparticles were maintained constant (10 µL) for all test pathogens and bactericidal activity of synthesized TiO$_2$ nanoparticles loaded streptomycin at 6:4% concentration is shown in (Table 1).

The synthesized TiO$_2$ nanoparticles exhibit high antibacterial activity compared to standard drugs and demonstrates its capability of deactivating cellular enzymes. It is evidenced from Table 1 that synthesized nanoparticles exhibit zone of inhibition for all the test pathogens. However, *P. mirabilis* does not show antibacterial activity and *E. Coli* shows no activity for nanoparticles templated at 1.2 and 1.6% probably because the negative carrier of *P. mirabilis* and *E.coli* at specific concentration show less affinity towards titanium metal ions. In addition, inhibition of streptomycin loaded nanoparticles at all template concentration exhibit relatively larger zone in comparison with the test control. The statistical interpretation from Table 1 shows that *P. mirabilis* and *E. coli* show a remarkable antibacterial profile when, tested with TiO$_2$ nanoparticles loaded with streptomycin. Surprisingly zone width for control is less compared to zone width observed for TiO$_2$ loaded streptomycin and hence the affinity of titanium metal ions in association with the drug against *P. mirabilis* and *E. coli* implicates the significant efficacy of drug release.

From Fig. 5f, the dissolution profiles of TiO$_2$ loaded with streptomycin nanoparticles showed a superior increase in the release/dissolution rate that reflected inhibition zone width larger in correlation with synthesized TiO$_2$ nanoparticles and the positive control of

Table 1 Inhibition growth diameter obtained by well diffusion method using different template concentrations of synthesized TiO$_2$ NPs against selected human pathogens

S. no	Solvent	% of Aloevera concentration	Zone of inhibition (mm)				
			K. pneumoniae	S. typhi	P. mirabilis	E. coli	S. aureus
1	5 ml of DMSO in 0.1 g of TiO$_2$	0	18 ± 0.062	10 ± 0.051	N$_Z$	16 ± 0.061	16 ± 0.040
2		0.4	20 ± 0.081	12 ± 0.080	N$_Z$	08 ± 0.040	12 ± 0.073
3		0.8	13 ± 0.022	08 ± 0.053	N$_Z$	N$_Z$	12 ± 0.032
4		1.2	08 ± 0.056	10 ± 0.039	N$_Z$	N$_Z$	12 ± 0.009
5		1.6	12 ± 0.021	08 ± 0.06	N$_Z$	06 ± 0.005	17 ± 0.036
6	5 ml of DMSO in 0.07 g of TiO$_2$:0.03 g of streptomycin	0	30 ± 0.025	16 ± 0.052	26 ± 0.032	31 ± 0.091	20 ± 0.042
7		0.4	31 ± 0.051	17 ± 0.020	31 ± 0.054	39 ± 0.060	20 ± 0.021
8		0.8	39 ± 0.053	18 ± 0.015	31 ± 0.026	32 ± 0.081	23 ± 0.042
9		1.2	30 ± 0.022	17 ± 0.066	30 ± 0.011	32 ± 0.024	24 ± 0.060
10		1.6	29 ± 0.072	18 ± 0.023	28 ± 0.036	36 ± 0.049	23 ± 0.091
11	Positive control streptomycin	–	17 ± 0.001	15 ± 0.002	20 ± 0.001	20 ± 0.002	15 ± 0.005

Fig. 5 a–e Inhibition zone diameter histogram of TiO$_2$ nanoparticles assisted with different template concentrations against selected human pathogens at 0, 0.4, 0.8, 1.2, 1.4 and f release kinetics of TiO$_2$ loaded with streptomycin

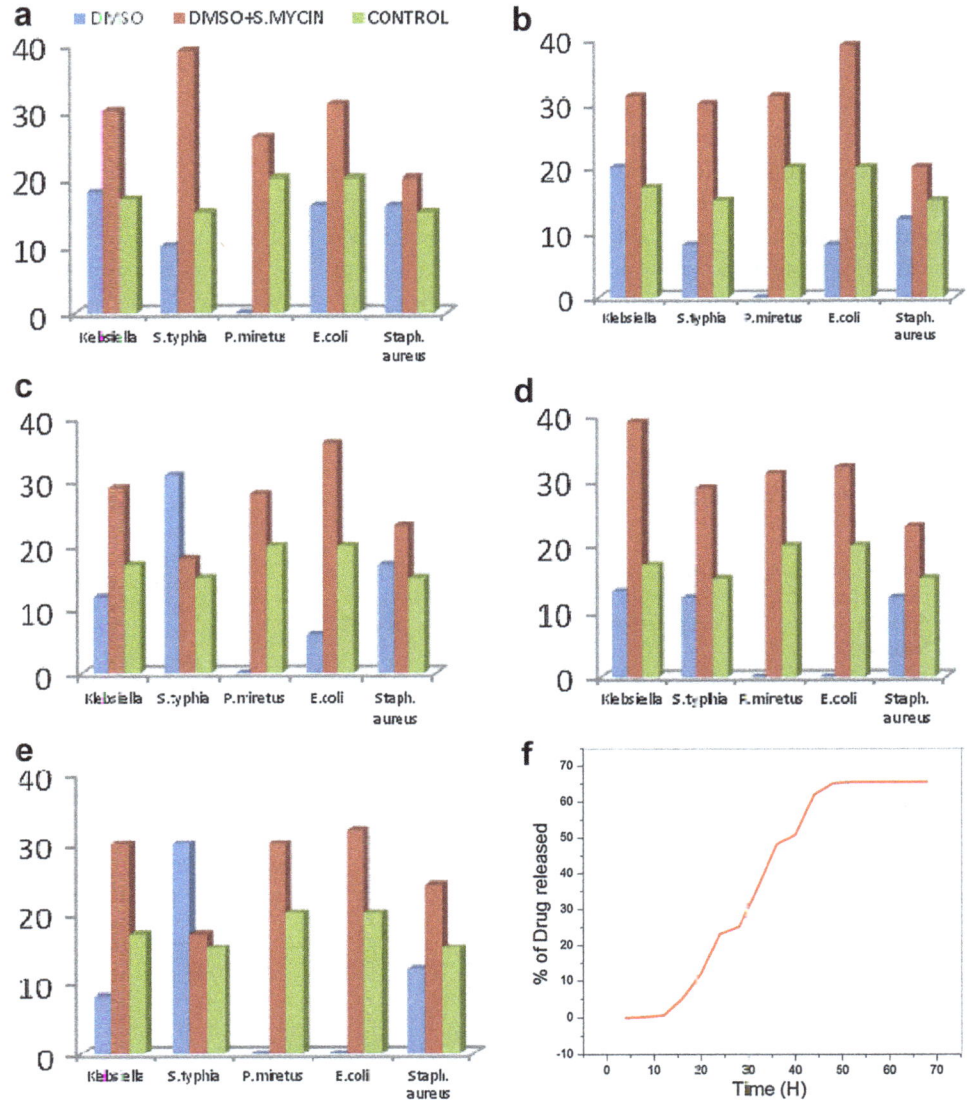

streptomycin against human pathogens that are responsible for causing UTI. Diffusion controlled release of TiO$_2$ nanoparticles loaded with streptomycin were studied by dialysis method wherein 10 mg of streptomycin dissolved in 2 μL PBS solution at pH 7.4 (release media) was poured into the inner tube of the dialyzer. The dialyzer tube was suspended into a 200 mL glass beaker containing release media, which was magnetically stirred continually at 500 rpm at room temperature. The percentage of drug released from TiO$_2$ nanocorals at different time intervals is shown in Fig. 5f. It is also evident that increase in the pore size of the TiO$_2$ microspheres on the surface might be due to the release of entrapped protein and deformation in structure, leading to the sustained release of drug. The initial release of streptomycin happens after 60 min. After the first stage, the protonation takes place on the surface of TiO$_2$ nanoparticles, which promotes the release of chemisorbed drug in the medium, so that the release percentage increases to 45 and then to 65% [18].

Conclusion

The study concludes that Aloevera is an effective template for the synthesis of anatase phase TiO$_2$ nanoparticles with an average grain size ranging from 8 to 13 nm. Powder XRD analysis was employed to elucidate the structure and phase of TiO$_2$ nanoparticles. The presence of Ti–O and the biological molecules responsible for the formation of TiO$_2$ nanoparticles were analyzed by FTIR spectroscopic technique. SEM results rationalized the dependence of template concentration on particle size and nanoscale morphology of TiO$_2$ nanoparticles. Interpretation from the antimicrobial activity of pathogens causing UTI loaded with strepto-mycin drug shows that TiO_2 assisted by Aloevera could be a potential antibacterial agent for practical applications. Future perspective of the work could focus on the construction of polymeric nanoarchitectures to improve loading capacity of drugs. Although invitro experiment disclose that TiO_2 networks exhibit good response towards improving the efficacy of the drug delivery system, further investigations have to be systematically carried out to expedite the process and broaden the possibility of practical applications in drug delivery and transport mechanism.

References

1. Kaviyarasu, K., Premanand, D.: Synthesis of Mg doped TiO_2 nanocrystals prepared by wet-chemical method: optical and microscopic studies. Int. J. Nanosci. **12**(5), 13500331–13500336 (2013)
2. Gratzel, M.: Photochemical cells. Nature **414**, 338–344 (2001)
3. Gopal, M., Moberly Chan, W.J., De Jonghe, L.C.: Room temperature synthesis of crystalline metal oxides. J. Mater. Sci. **32**, 6001–6008 (1997)
4. Fujishima, A., Honda, K.: Electrochemical photolysis of water at a semiconductor electrode. Nature **238**, 37–38 (1972)
5. Yang, S., Gao, L.: Preparation of titanium dioxide nanocrystallite with high photocatalytic activities. J. Am. Ceram. Soc. **88**, 968–970 (2005)
6. Addamo, M., Augugliaro, V., Di Paola, A., Garcıa-López, E., Loddo, V., Marcı', G., Molinari, R., Palmisano, L., Schiavello, M.: Preparation, characterization, and photoactivity of polycrystalline nanostructured TiO_2 catalysts. J. Phys. Chem. **108**, 3303–3310 (2004)
7. Chun, H., Lan, Y., Jiuhui, Q., Xuexiang, H., Wang, A.: Ag/AgBr/TiO2 visible light photocatalyst for destruction of azodyes and bacteria. J. Phys. Chem. B **110**, 4066–4072 (2006)
8. Sakatani, Y., Grosso, D., Nicole, L., Boissie're, C., de AA Soler-Illiab, G.J., Sanchez, C.: Optimised photocatalytic activity of grid-like mesoporous TiO_2 films: effect of crystallinity, pore size distribution and pore accessibility. J. Mater. Chem. **16**, 77–82 (2006)
9. Rajesh Kumar, S., Pillai, S.C., Hareesh, U.S., Mukundan, P., Warrier, K.G.K.: Synthesis of thermally stable high surface area anatase-alumin a mixed oxides. Mater. Lett. **43**, 286–290 (2000)
10. Malia, S.S., Betty, C.A., Bhosale, P.N., Patila, P.S.: Sensitized solar cells (DSSCs) based on novel nanocoral TiO_2: a comparative study. Electrochim. Acta. **59**, 113–120 (2012)
11. Eshun, K., He, Q.: *Aloe vera*: a valuable ingredient for the food, pharmaceutical and cosmetic industries—a review. Crit. Rev. Food Sci. Nutr. **44**, 91–96 (2004)
12. Boudreau, M.D., Beland, F.A.: An evaluation of the biological and toxicological properties of Aloe barbadensis (miller). Aloe vera J. Environ. Sci. Health. **24**, 103–154 (2006)
13. Kumar, S., Upadhyaya, J.S., Negi, Y.S.: Preparation of nanoparticles from corn cobs by chemical treatment methods. Bio. Res. **5**, 1292–1300 (2010)
14. Hamman, J.H.: Composition and applications of *Aloe vera* leaf gel. Molecules **13**, 1599–1616 (2008)
15. Venkatesh, K.S., Krishnamoorthi, S.R., Palani, N.S., Thirumal, V., Jose, S.P., Wang, F.M., Ilangovan, R.: Facile one step synthesis of novel TiO2 nanocoral by sol–gel method using *Aloe vera* plant extract. Indian J. Phys. **89**, 445 (2014)
16. Chekin, F., Bagheri, S., Abd Hamid, S.B.: Synthesis of Pt doped TiO_2 nanoparticles: characterization and application for electrocatalytic oxidation of l-methionine. Sen. Actuators B Chem. **177**, 898–903 (2013)
17. Shankar, S.S., Ahmad, A., Sastry, M.: *Geranium* leaf assisted biosynthesis of silver nanoparticles. Biotechnology **19**, 1627–1631 (2003)
18. Niwa, T., Takeuchi, H., Hino, T., Kunou, N., Kawashima, Y.: In vitro drug release behavior of D, L-lactide/glycolide copolymer (PLGA) nanospheres with nafarelin acetate prepared by a novel spontaneous emulsification solvent diffusion method. J. Pharm. Sci. **83**(5), 727–732 (1994)

13

Gold nanoparticles Wells–Dawson heteropolyacid nanocomposite film as an effective nanocatalyst in photocatalytic removal of azo dyes from wastewaters

Neda Rohani[1] · Fatemeh F. Bamoharram[2] · Azam Marjani[1] · Majid M. Heravi[3]

Abstract A novel nanophotochromic film consisting of Au nanoparticles and Wells–Dawson-type heteropoly-acid, $[H_6P_2W_{18}O_{62}]$, was prepared by combination of sol–gel and photoreduction method. This film was characterized by UV–visible spectroscopy, particle size distribution, Field emission scanning electron microscopy, and energy dispersive spectroscopy analysis. For the preparation of Au nanoparticles, heteropolyacid, $[H_6P_2W_{18}O_{62}]$ was used in form of composite film as a green reductant and stabilizer. Au nanoparticles with particle size in the range of 10–20 nm were synthesized and monodispersed in the nanocomposite using dip coating method. The photocatalytic activity of this composite film was studied in the decolorization of methyl orange (MeO) and methyl red (MR) as carcinogenic pollutant dyes using UV irradiation. The pseudo-first order rate constants was established and calculated for these reactions. Comparison of this composite film with the prepared composite by Preyssler heteropolyacid disclosed that the structure of heteropolyacid can affect loading amount of Au nanoparticles.

Keywords Decolorization · Gold nanoparticles · Heteropolyacid · Nanocomposite · Wells–Dawson

Introduction

Au nanoparticles have attracted much attention nowadays mainly due to their exceptional physical and chemical properties. Thus, much effort has been made into their synthesis and characterization [1]. Polyoxometalates (POMs) have been extensively used in the preparation of Au nanoparticles reported in various reports [2–5]. The chemistry of POMs and their salts has been extensively reviewed [6, 7]. They have been defined as a sub-class of inorganic metal oxide clusters, which possess fascinating structures and various properties [8]. Interestingly, when they are submitted to stepwise and multi-electron redox reactions, their structures remain intact. In addition, in their presence it is possible to reduce a compound, electrochemically or photochemically using a suitable reducing agent [9, 10]. Mandal et al. [2] performed a simple process and successfully observed that several metal nanoparticles were formed in the presences of photochemically reduced Keggin heteropolyanions as photocatalysts. Mandal and co-workers demonstrated that the reduction of Ag^+ on the surface of gold nanoparticles can be facilitated in the presence of photochemically generated $[PW_{12}O_{40}]^{4-}$ under UV irradiation. In addition, Au nanoparticles were readily prepared via a simple photoreduction mediated by transition metal monosubstituted Keggin heteropolyanions [5]. Although Keggin and its derivatives have been used in the synthesis of Au nanoparticles, the role of Wells–Dawson heteropolyacids (HPAs) has been largely disregarded. A heteropolyacid is a class

✉ Fatemeh F. Bamoharram
abamoharram@yahoo.com

[1] Department of Chemistry, Arak Branch, Islamic Azad University, Arak 38361-1-9131, Iran

[2] Research Center for Animal Development Applied Biology - Department of Nanobiotechnology, Mashhad Branch, Islamic Azad University, Mashhad 91865-397, Iran

[3] Department of Chemistry, Alzahra University, Tehran 1993891176, Iran

of acid made up of a particular combination of hydrogen and oxygen with certain metals and non-metals. This type of acid is a common re-usable acid catalyst in chemical reactions [5]. Wells–Dawson-type heteropolyacid $[H_6P_2W_{18}O_{62}]$ is remarkable due to its high acidity and outstanding stability both in solution and in the solid state [11]. Literature survey revealed that in spite of these interesting properties the synthesis of nanoparticles using this heteropolyacid in nanocomposite films has been unstudied. We are interested in the chemistry of HPAs [12]. Armed with these experiences and due to the recent interest in the synthesis of nanoparticles in solid matrices [13], we were encouraged to study the preparation of Au nanoparticles incorporated in Wells–Dawson heteropolyacid nanocomposite film. Recently, we have introduced the Preyssler-type heteropolyacid ($H_{14-}[NaP_5W_{30}O_{110}]$), as an unique reducing agent and stabilizer for the synthesis of Au nanoparticles in an organic–inorganic nanocomposite film under UV irradiation [13]. To extend our research to other HPAs, we investigated the capability of Wells–Dawson heteropolyacid in the preparation of Au nanoparticles in an organic–inorganic nanocomposite film.

Experimental

Chemicals and instruments

For the synthesis of Dawson acid, at first potassium salt was synthesized and then Wells–Dawson acid, $[H_6P_2W_{18}O_{62}]$, was prepared by passage of potassium salt solution in water through a column of Dowex resin (H^+ form). The elute was evaporated to dryness under reduced pressure [14].

All chemicals were obtained from commercial sources and used as received. Spin Coater, (S.C.S.86, Japan), Field Emission Scanning Electron Microscope (FESEM) (VEGA\\TESCAN-XMU, Czech Republic), Energy Dispersive Spectrometry (EDS) (Mira 3-XMU, Czech Republic), and UV–visible spectrophotometer, (Optizen UV3220, Germany) were used for characterization of samples. The average particle size of Au nanoparticles in the film was measured by SPSS software using Image J program.

Preparation of nanocomposite

Polyvinylalcohol (PVA) 30 wt% was dissolved in deionized water and stirred in an oil bath for 10 min. Then tetraethylorthosilicate 20 wt% and Wells–Dawson acid solutions were added. This mixture was refluxed at 353 K for 6 h and a clear viscous gel is formed. At the next stage, 250 μL of the obtained transparent gel was used to make films by spin coating method (500 rpm/min) and after reducing, it was dipped in different times including 5, 10, and 30 min into $HAuCl_4$ solution (25 mL, 3 mM) [13].

Photodegradation experiments

The photo reactor was designed in our laboratory. In a typical reaction, a 250-mL Pyrex glass was equipped with a magnetic stirrer, azo dye solution, and nanocomposite film. The mixture was stirred and purged with nitrogen for 1 h, and then it was irradiated under the high-pressure mercury lamp (Philips, 125 W, wavelength 254 nm) as UV light source. The temperature in the glass reactor was set to 25 ± 2 °C by the circulating water. The experimental procedure is described as follows: a series of 10 mL solutions containing 30 mg L^{-1} of azo dye solution were prepared and then sample was placed under light irradiation in the presence of nanocomposite film. The progress of reaction was investigated using a UV–visible spectrophotometer.

Results and discussion

Heteropolyacids are well-established as powerful oxidants for the oxidation of a wide range of organic compounds. In this process, the excitation of the O → M charge transfer can occur under UV irradiation [10, 15, 16]. Interestingly, the reduced form of heteropolyacid can act as a strong reducing reagent which can be readily re-oxidized by various chemicals especially as metal ions [3, 15].

In this research, Wells–Dawson acid was used in synthesis of Au nanoparticles through a simple sol–gel and photoreduction process. In this process, Wells–Dawson acid played a dual role (a) photocatalytic reducing agent and (b) stabilizer [13].

The photoreduction of nanocomposite film was confirmed by UV–visible spectroscopy (Fig. 1). The formation of reduced Wells–Dawson in the composite film was confirmed by observation of deep blue color (Fig. 1). In UV–visible spectra (Fig. 1), the reduced Wells–Dawson acid showed absorption peaks at about 300 and 690 nm, which the observed peak in 690 nm is related to the formation of single electron-reduced Wells–Dawson ion. This figure also shows that there is an increase in the intensity up to 60 min, but after that, there is no change. This is a proof that all the Wells–

Fig. 1 UV–Visible spectra of composite film under irradiation at various time intervals: 0 min (*a*), 10 min (*b*), 20 min (*c*), 30 min (*d*), 40 min (*e*), 50 min (*f*), 60 min (*g*)

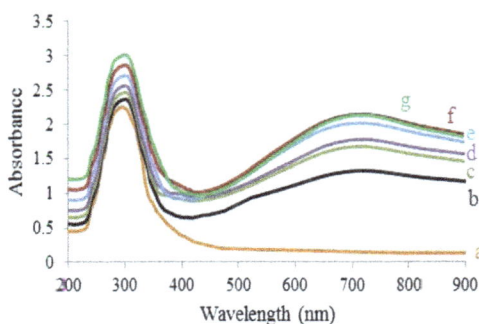

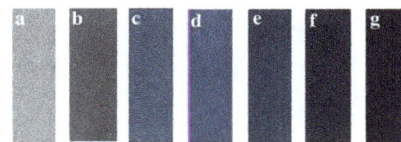

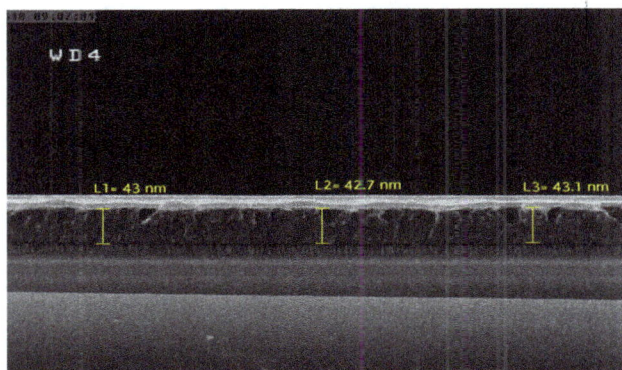

Fig. 2 Thickness of the synthesized nanocomposite film (FESEM image)

Figure 4 shows the FESEM images of reduced composite film contacted with $HAuCl_4$ solution at different times. This figure shows when the time of dipping is increased from 5 to 30 min, the average particle size is also increased. The average size of nanoparticles is 10.54, 16.57, and 19.61 in Fig. 3a, b, d, respectively.

The EDS analysis is shown in Fig. 5. It shows the elemental composition of Au nanoparticles and Wells–Dawson in the generated nanofilms. The elemental constitution to control the loading amount of Au nanoparticles confirmed 9.91% w/w in the nanocomposite film. The quantitative results are shown in Tables 1 and 2.

Dawson ions in the composite have been reduced. The reduced film was used as reducing medium and host for the formation of Au nanoparticles [17].

With dipping of the reduced nanocomposite film with thickness 43 nm (Fig. 2) into $HAuCl_4$ solution, a band at 530 nm was appeared which confirms the formation of Au nanoparticles in the composite film (Fig. 3). The formation of Au nanoparticles was further approved by changing the blue film to pink and violet color, respectively.

Interestingly, comparison our results with the obtained results in preparation of Au nanoparticles using Preyssler HPA [13], $H_{14}[NaP_5W_{30}O_{110}]$, showed a higher loading of gold nanoparticles (22%) under similar reaction conditions. Thus, the structure of heteropolyacid can affect the loading amount in this process. The difference can be attributed to the difference between oval and spherical structure of Preyssler and Dawson, respectively [18]. It is suggested that the larger number of H^+ and W atoms in Preyssler may lower the activation barrier, and provides many "sites" on the oval-shaped structure. They are likely to render the loading effectiveness.

Fig. 3 UV–Visible spectra of Au-embedded composite film at different dipping time intervals: 0 min (*a*), 5 min (*b*), 10 min (*c*), 20 min (*d*), 30 min (*e*)

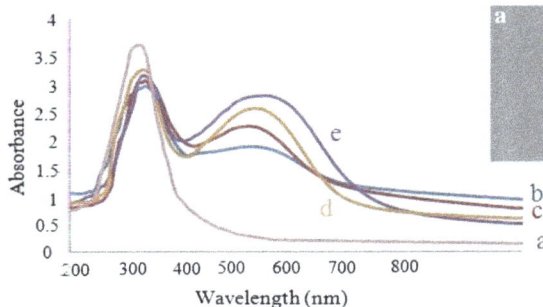

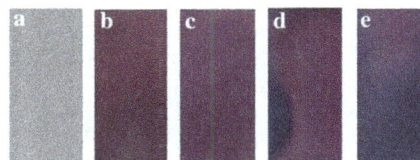

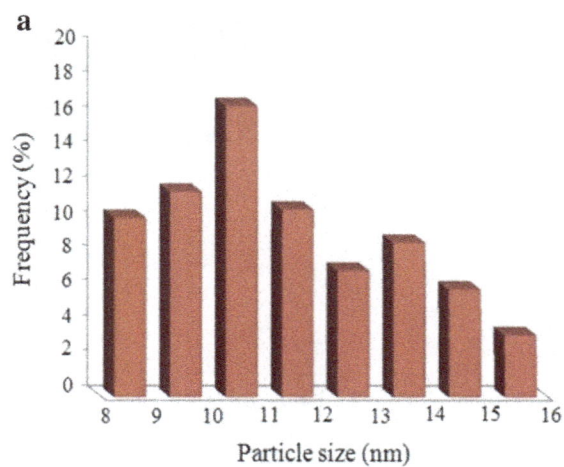

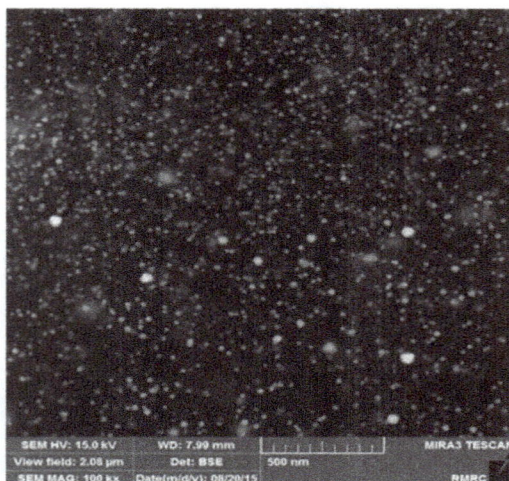

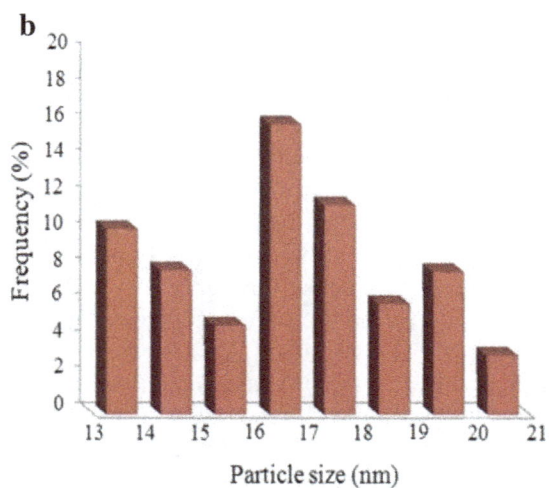

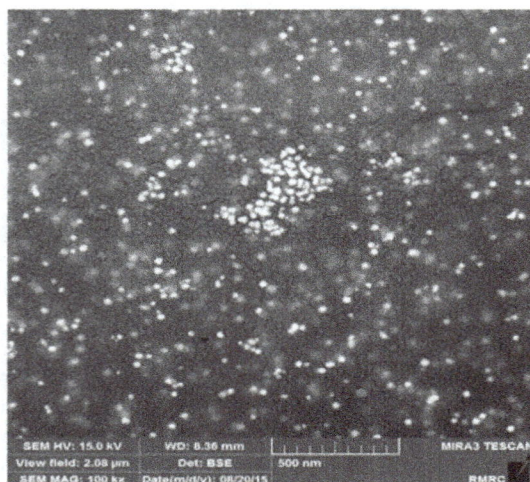

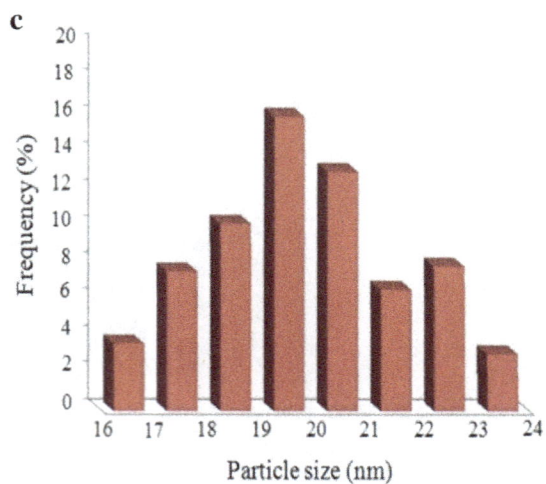

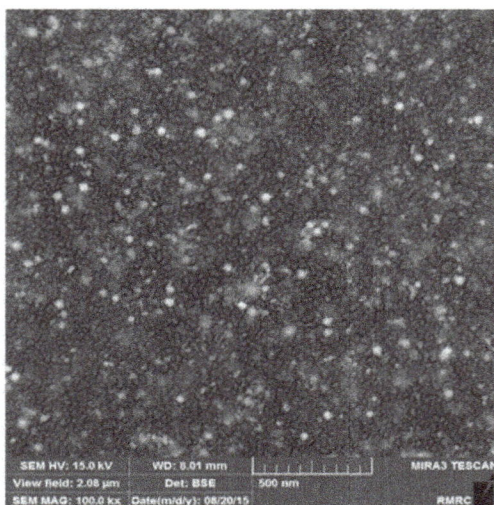

Fig. 4 FESEM images of synthesized gold nanoparticles at different dipping time intervals in 3 mM of HAuCl$_4$: 5 min (**a**), 10 min (**b**), 30 min (**c**)

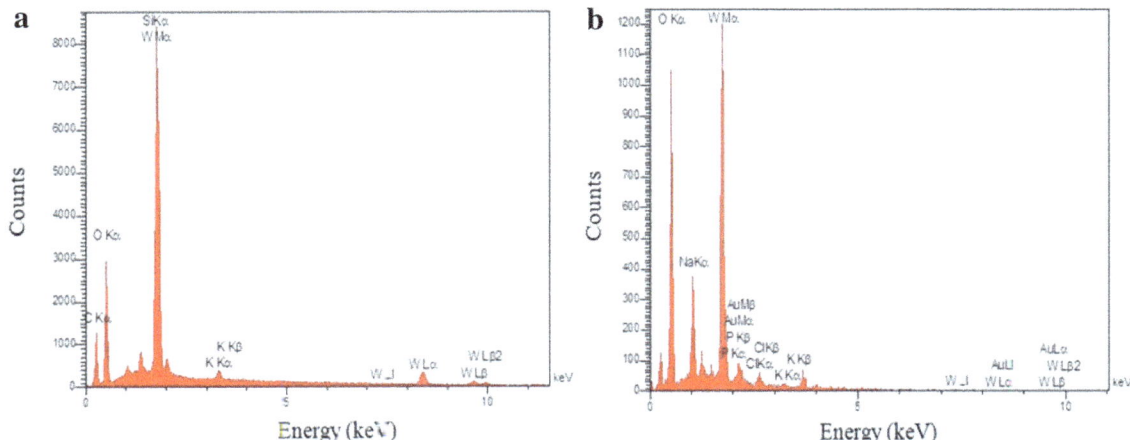

Fig. 5 EDS spectrum of nanocomposite film: before (**a**), after (**b**) dipping in HAuCl$_4$ solution

Table 1 Atomic percentage for Fig. 5a

Elt	Line	Int	Error	K	Kr	W%	A%	ZAF	Formula	Ox%	Pk/Bg
Nanocomposite film											
C	Ka	105.2	41.5241	0.1744	0.0851	23.44	45.82	0.3629		0.00	48.85
O	Ka	263.0	13.8659	0.2172	0.1060	32.84	48.20	0.3227		0.00	56.63
Si	Ka	0.0	0.0000	0.0000	0.0000	0.00	0.00	0.8872		0.00	24.35
K	Ka	27.6	1.0019	0.0144	0.0070	0.83	0.50	0.8443		0.00	3.10
W	La	51.0	0.6128	0.5940	0.2898	42.89	5.48	0.6757		0.00	6.40
				1.0000	0.4879	100.00	100.00			0.00	

Table 2 Atomic percentage for Fig. 5b

Elt	Line	Int	Error	K	Kr	W%	A%	ZAF	Formula	Ox%	Pk/Bg
Nanocomposite film + Au nanoparticles											
O	Ka	162.7	53.9753	0.7160	0.4303	68.07	83.52	0.6320		0.00	275.91
Na	Ka	71.7	13.1083	0.1206	0.0725	15.88	13.56	0.4563		0.00	22.23
P	Ka	2.3	3.1933	0.0042	0.0025	0.32	0.20	0.7737		0.00	3.83
Cl	Ka	11.1	3.1933	0.0255	0.0153	1.91	1.06	0.8019		0.00	4.85
K	Ka	3.3	5.1739	0.0092	0.0055	0.65	0.33	0.8481		0.00	2.97
W	La	0.5	0.7890	0.0324	0.0195	3.26	0.35	0.5983		0.00	2.85
Au	La	0.5	0.7890	0.0922	0.0554	9.91	0.99	0.5591		0.00	4.06
				1.0000	0.6010	100.00	100.00			0.00	

The catalytic activity of the synthesized nanocomposite film was studied in the decolorization of MeO and MR azo dyes. The results are reported in Figs. 6 and 7, and a comparison is reported in Table 3. A significant decrease in the absorbance bands can be observed with a decolorization degree of 96 and 87.2% after 36 and 43 min for MeO and MR, respectively.

The following equation was used to calculate the degree of azo dye decolorization [19–21]:

$$C = (C_0 - C_e)/C_0 \times 100.$$

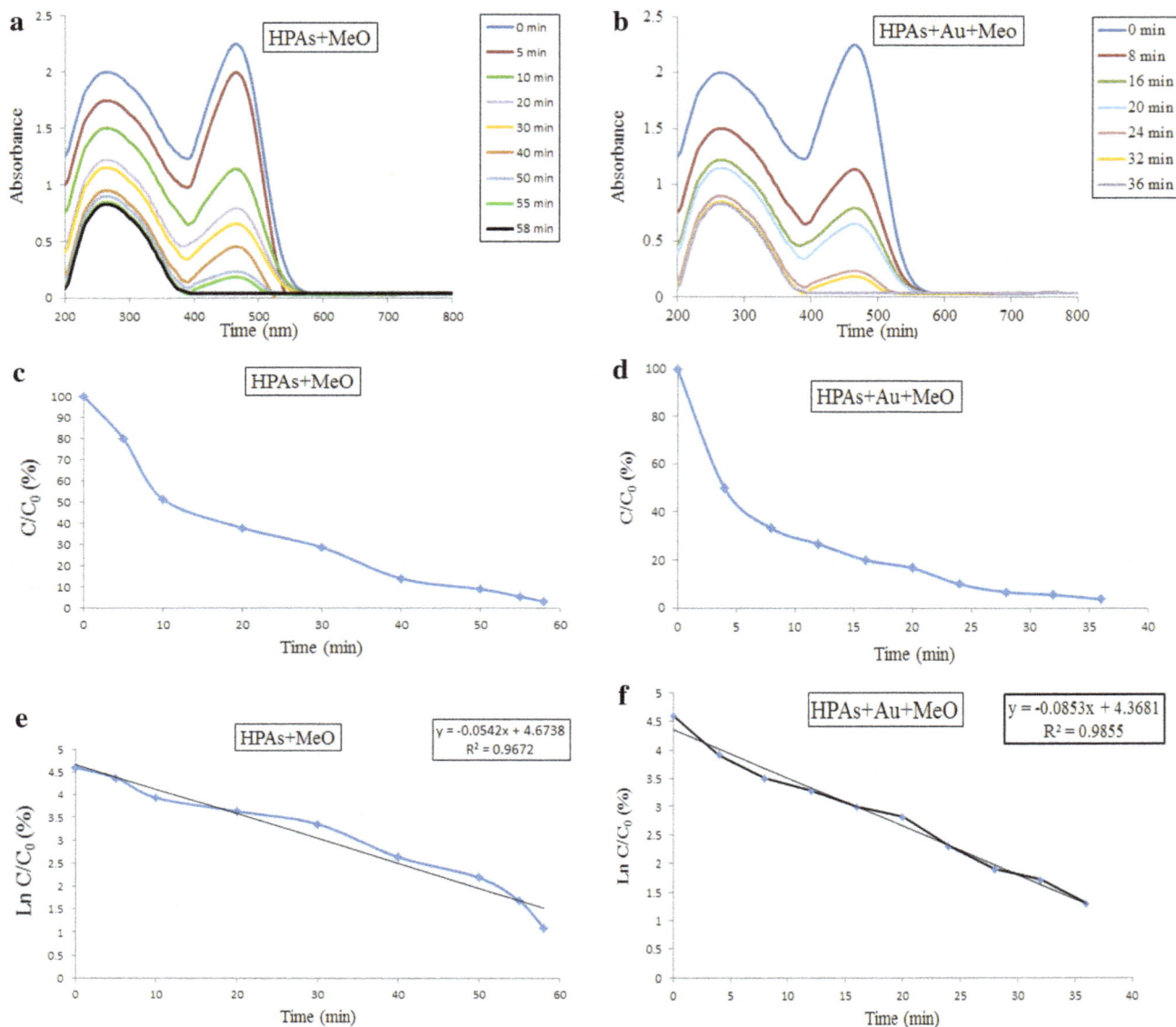

Fig. 6 Catalytic decolorization of MeO: in the absence of Au nanoparticles (**a, c, e**) and in the presence of Au nanoparticles (**b, d, f**)

In this equation, C and C_0 are the decolorization degree and the initial absorbance of dyes solution, respectively. C_e stands for the absorbance of the dye solution after photocatalysis. The pseudo-first order rate constants are given by the plot of $\ln(C/C_0)$ versus time.

Conclusions

Dawson acid, $[H_6P_2W_{18}O_{62}]$, can be used as a superior and green-reducing reagent and stabilizer in the preparation of Au nanoparticles in a nanophotochromic film, via a simple

and fast sol–gel procedure and photolysis. The synthesized nano film showed a significant catalytic activity in the decolorization process of MeO with pseudo-first order kinetic behavior.

Interestingly, our findings showed that heteropolyacid structure can control the loading amount of embedded Au nanoparticles up to twice. Further investigation in this field will provide a great opportunity for researches and scientists to develop techniques for the green and controlled synthesis of various nanoparticles in the presence of other heteropolyacids.

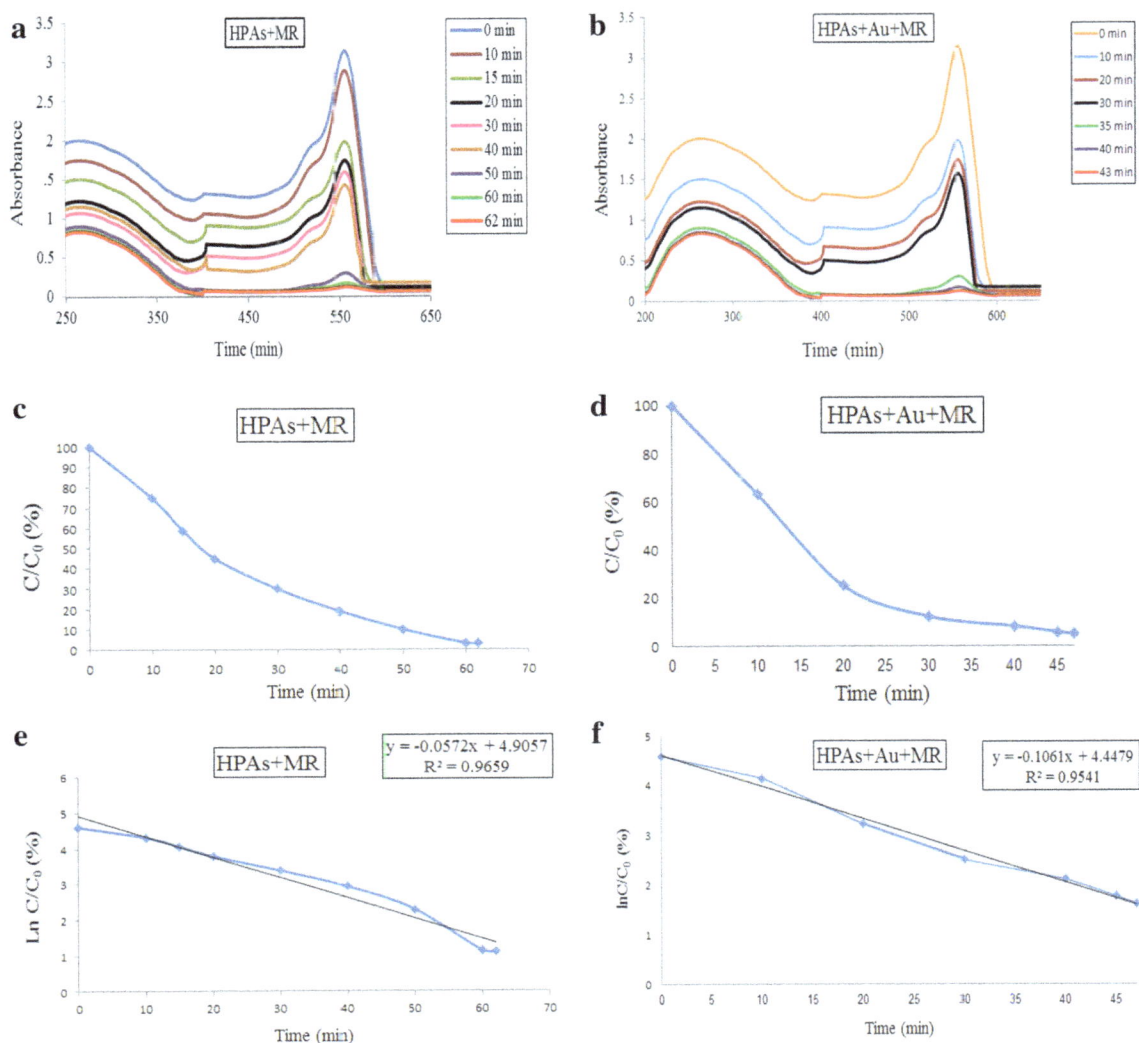

Fig. 7 Catalytic decolorization of MR: in the absence of Au nanoparticles (**a**, **c**, **e**) and in the presence of Au nanoparticles (**b**, **d**, **f**)

Table 3 Comparison for decolorization of MeO and MR

Catalyst	Time for decolorization of Azo dyes (min)	Yield (%)
Dawson film/MeO	58	91.2
Dawson film/Au/MeO	36	96
Dawson film/MR	62	84.4
Dawson film/Au/MR	43	87.2

Acknowledgements The authors would like to thank Research Center for Animal Development Applied Biology-Department of Nanobiotechnology, Mashhad Branch, Islamic Azad University, Mashhad, Iran and Department of Chemistry. Arak Branch, Islamic Azad University, Arak, Iran.

References

1. Ayati, A., Ahmadpour, A., Bamoharram, F.F., Tanhaei, B., Mänttäri, M., Lahtinen, M., Sillanpää, M.: Novel Au NPs/Preyssler acid/TiO_2 nanocomposite for the photocatalytic removal of azo dye. Sep. Purif. Technol. **133**, 415–420 (2014)
2. Mandal, S.. Das, A., Srivastava, R., Sastry, M.: Keggin ion mediated synthesis of hydrophobized Pd nanoparticles for multifunctional catalysis. Langmuir **21**(6), 2408–2413 (2005)
3. Wang, J., Lu, X., Fan, Sh, Zhao, W., Li, W.: In situ growth of gold nanoparticles on SiO_2/lanthanide–polyoxometalates composite spheres: an efficient catalytic and luminescent system. J. Alloys Compd. **632**, 87–93 (2015)
4. Mandal, S., Selvakannan, R.P., Pasricha, R., Sastry, M.: Keggin ions as UV-switchable reducing agents in the synthesis of Au core-Ag shell nanoparticles. J. Am. Chem. Soc. **125**(28), 8440–8441 (2005)

5. Niu, C., Wu, Y., Wang, Z., Li, Z., Li, R.: Synthesis and shapes of gold nanoparticles by using transition metal monosubstituted heteropolyanions as photocatalysts and stabilizers. Front. Chem. China **4**(1), 44–47 (2009)

6. Sadjadi, S., Heravi, M.M.: Recent advances in applications of POMs and their hybrids in catalysis. Curr. Org. Chem. **20**(13), 1404–1444 (2016)

7. Heravi, M.M., Sadjadi, S.: Recent developments in use of heteropolyacids, their salts and polyoxometalates in organic synthesis. J. Iran. Chem. Soc. **6**(1), 1–54 (2009)

8. Heravi, M.M., Bamoharram, F.F., Rajabzadeh, G., Seifi, N., Khatami, M.: Preyssler heteropolyacid $[NaP_5W_{30}O_{110}]^{14-}$, as a new, green, and recyclable catalyst for the synthesis of [1,2,4] triazino [4,3-b][1,2,4,5]tetrazines. J. Mol. Catal. A Chem. **25**(1), 213–217 (2006)

9. He, T., Yao, J.: Photochromism in composite and hybrid materials based on transition-metal oxides and polyoxometalates. Prog. Mater. Sci. **51**, 810–879 (2006)

10. Xing, X., Wang, M., Liu, R., Zhang, Sh, Zhang, K., Li, B., Zhang, G.: Highly efficient electrochemically driven water oxidation by graphene-supported mixed-valent Mn16-containing polyoxometalate. Green Energy Environ. **1**, 138–143 (2016)

11. Zhang, Y., Bo, X., Nsabimana, A., Munyentwali, A., Han, C., Li, M., Guo, L.: Green and facile synthesis of an Au nanoparticles@polyoxometalate/ordered mesoporous carbon tri-component nanocomposite and its electrochemical applications. Biosens. Bioelectron. **66**, 191–197 (2015)

12. Bamoharram, F.F.: Role of polyoxometalates as green compounds in recent developments of nanoscience. Synth. React. Inorg. Met. Org., Nano Met. Chem. **41**(8), 893–922 (2011)

13. Rohani, N., Bamoharram, F.F., Marjani, A., Heravi, M.M.: A novel photochromic film based on Preyssler heteropolyacid and gold nanoparticles as a green and recyclable nanocatalyst for removal of azo dye from wastewaters. Curr. Nanosci. **12**, 605–610 (2016)

14. Sheshmani, S., Arab, F.M., Mirzaei, M., Abedi Rad, B., Nouri, G.S., Yousefi, M.: Preparation, characterization and catalytic application of some polyoxometalates with Keggin, Wells–Dawson and Preyssler structures. Indian J. Chem. **50 A**, 1725–1729 (2011)

15. Triantis, T., Troupis, A., Gkika, E., Alexakos, G., Boukos, N.: Photocatalytic synthesis of Se nanoparticles using polyoxometalates. Catal. Today **144**(1–2), 2–6 (2009)

16. Liu, ChG, Zheng, T., Liu, Sh, Zhang, H.Y.: Photodegradation of malachite green dye catalyzed by Keggin-type polyoxometalates under Visible-light irradiation: transition metal substituted effects. J. Mol. Struct. **1110**, 44–52 (2016)

17. Ayati, A., Tanhaei, B., Bamoharram, F.F., Ahmadpour, A., Maydannik, Ph, Sillanpää, M.: Photocatalytic degradation of nitrobenzene by gold nanoparticles decorated polyoxometalate immobilized TiO_2 nanotubes. Sep. Purif. Technol. **171**, 62–68 (2016)

18. Rohani, M., Bamoharram, F.F., Khosravi, M., Baharara, J., Heravi, M.M.: Preparation and characterization of Preyssler heteropolyacid-cellulose acetate hybrid nanofibers: a new, green and recyclable nanocatalyst for photodegradation of methyl orange as the model dye. J. Exp. Nanosci. 1–13 (2016). doi:10.1080/17458080.2016.1246754

19. Chen, Zh, Liao, J., Chen, Y., Zhang, J., Fan, W., Huang, Y.: Synthesis of oxygen deficient BiOI for photocatalytic degradation of methyl orange. Inorg. Chem. Commun. **74**, 39–41 (2016)

20. Kumar, D.P., Reddy, N.L., Karthikeyan, M., Chinnaiah, N., Bramhaiah, V., Durga Kumari, V., Shankar, M.V.: Synergistic effect of nanocavities in anatase TiO_2 nanobelts for photocatalytic degradation of methyl orange dye in aqueous solution. J. Colloid Interface Sci. **477**, 201–208 (2016)

21. Ayati, A., Ahmadpour, A., Bamoharram, F.F., Heravi, M.M., Rashidi, H.: Photocatalytic synthesis of gold nanoparticles using Preyssler acid and their photocatalytic activity. Chin. J. Catal. **32**(6), 978–982 (2011)

14

Antioxidant, antimicrobial and cytotoxic activities of silver and gold nanoparticles synthesized using *Plumbago zeylanica* bark

S. Priya Velammal[1] · T. Akkini Devi[1] · T. Peter Amaladhas[1]

Abstract Utilization of bioresources for the synthesis of metal nanoparticles is the latest field in green chemistry. The present work reports the utilization of the aqueous bark extract of *Plumbago zeylanica* for the biosynthesis of Ag and Au NPs. The Ag and Au NPs thus obtained were characterized by UV–Vis, FT-IR, TEM, XRD, and EDAX analysis. The water-soluble components of the extract were responsible for the reduction of Ag^+ and Au^{3+} ions. FT-IR spectra revealed that the –OH and $>C=O$ groups present in the biomolecules were responsible for reduction and stabilization of nanoparticles. TEM images showed the existence of spherical Ag and Au NPs with average size of 28.47 and 16.89 nm, respectively, which was further substantiated by XRD analysis. The presence of elemental Ag and Au along with C and O from the attached biomolecules was proved by EDAX analysis. The antimicrobial, antioxidant, and in vitro cytotoxic activities of the synthesized nanoparticles were studied by disc diffusion, DPPH, and MTT assay methods, respectively. The free radical inhibition was found to be 78.17 and 87.34 % for Ag and Au nanoparticles, respectively. The Ag and Au NPs showed 61.56 and 65.61 % toxicity against DLA cell line, respectively. The DNA binding ability of Ag and Au NPs were investigated using CT-DNA. The hyperchromism shift inferred the groove binding of nanoparticles with CT-DNA.

Keywords Biosynthesis · Silver nanoparticles · Gold nanoparticles · Antioxidant · Cytotoxicity · DLA cell line

Introduction

Nanoscience and nanotechnology, the two current fields of research have strong and beautiful footprints in ancient times. Nano mercury and its sulfide of sizes finer than 10 nm have been used as medicines in India since long back [1]. Silver and copper nanoparticles have been used to produce iridescent metallic effects on ancient ceramic objects [2]. The ancient Vedic rite of Agnihotra has used the nanoparticles to cleanse the environment. In Agnihotra, wood of trees like *Butea frondosa* (palasa), *Mangifera indica* (mango), *Ficus religiosa* (pipal), medicinal herbs, ghee, cloves, cardamom, and camphor are offered to the fire god by throwing them into the fire. This has been reported to fill the atmosphere with nutrients in nano size and medicinal properties, which in turn eliminate pathogenic bacteria or germs [3]. In Ayurvedha, herbo-mineral, and herbo-metallic drugs have been prepared by prolonged heating which significantly reduces the particle size. The medicinal use of colloidal gold (Swarnabhasma) and mercury remain popular till this day and the Swarnabhasmas (gold ash) has been characterized as globular particles of gold with the average size between 56 and 57 nm. It has been used to treat arthiritis, asthma, diabetes, and diseases of nervous system [4].

✉ T. Peter Amaladhas
peteramaldhast@yahoo.co.in; peteramaladhas@gmail.com

S. Priya Velammal
subra a1950@gmail.com

T. Akkini Devi
t.deviagni@gmail.com

[1] PG and Research Department of Chemistry, V.O. Chidambaram College, Tuticorin 628008, Tamil Nadu, India

Metal and metal oxide nanoparticles have been synthesized using many methods in the past, such as chemical reduction [5], electrochemical reaction [6], microwave method [7], reverse micelles [8], sonochemical method [9], and biosynthesis. Biosynthesis of nanoparticles has received great attention recently as the synthesized nanoparticles are non-toxic and can be used for biomedical applications. Different fungi like *Fusarium oxysporum* [10], *Cladosporium cladosporioides* [11], *Fusarium semitactum* [12], plants like *Syzygium aromaticum* [13], *Medicago sativa* [14], *Azadirachta indica* [15], *Terminalia cuneata* [16], and *Trignellafoenum graecum* seeds [17] have been employed to synthesize metal nanoparticles. Recently, sunlight-induced rapid synthesis of silver nanoparticles has been reported by our group [18, 19]. Biosynthesis of silver nanoparticles using bark of *Cinnamon zeylanicum* [20], *Piper nigrum* [21], *Breynia rhamnoides* [22], *Avicennia marina* [23], *Callicarpa maingayi* [24], *Cissus quadrangularis* [25], *Artocarpus elasticus* [26], and *Saraca indica* [27] has also been reported. The use of plant barks for the synthesis of nanoparticles avoids the problems that arise due to seasonal variation of biochemicals in the leaves.

Plumbago zeylanica is a medicinal herb belonging to the family of *Plumbaginaceae*. It has been used as a remedy for skin diseases, ringworm, leprosy, sores, and ulcers. The root or leaves are used as counter-irritant and vesicant. The powdered bark is used to treat leprosy, spleen and liver diseases and also for plaque [28]. It contains a variety of important chemical compounds, such as naphthaquinones, alkaloids, glycosides, steroids, triterpenoids, tannins, phenolic compounds, flavanoids, saponins, coumarins, and carbohydrates [29, 30]. Plumbagin, 5-hydroxy-2-methyl-1,4-naphthoquinone is the principal active compound of *P. zeylanica* [31]. In this work, biosynthesis of silver (Ag) and gold (Au) nanoparticles (NPs) was achieved using the bark extract of *P. zeylanica*. The water-soluble biocomponents in the extract not only act as reductants to reduce Ag^+ and Au^{3+} to Ag^o and Au^o, respectively, but also stabilize them by attaching onto the nanoparticles. The nanoparticles stabilized by active biomolecules may have the potential to treat various diseases. The results of antimicrobial, antioxidant, in vitro cytotoxicity of Ag and Au NPs and their ability to bind DNA are reported here.

Methods

Preparation of bark extract: the reducing agent

P. zeylanica bark was purchased from the local market from Madurai, Tamil Nadu, India. The bark was finely cut, washed with double-distilled water to remove any impurities and air-dried. About 2 g of bark was weighed and boiled for 10 min with 100 mL distilled water in Erlenmeyer flask. After cooling to room temperature, it was filtered through Whatman no. 41 filter paper. The light yellow colored extract (pH 6) was utilized for the synthesis of Ag and Au NPs.

Biosynthesis of Ag and Au NPs

Analar grade silver nitrate ($AgNO_3$) with 99.9 % purity was purchased from Spectrochem, India and 99.999 % pure hydrogen tetrachloroaurate(III) trihydrate ($HAuCl_4 \cdot 3H_2O$) was purchased from Sigma-Aldrich, USA. To prepare AgNPs, about 1 mL of 20-mM $AgNO_3$ was diluted to 20 mL with double-distilled water and treated with 10 mL of bark extract, and the solution was exposed to bright sunlight. There was a visible color change from pale yellow to brown in 2 min, and this reaction was monitored with time using UV–Vis spectrophotometer. In the case of AuNPs, 5 mL of 1-mM $HAuCl_4$ was mixed with 5 mL of the extract, and the intensity of pink color was monitored with time using UV–Vis spectrophotometer. In both the cases, 1 mL of the aliquot was diluted to 10 mL with double-distilled water before measuring the absorbance.

Characterization of Ag and Au NPs

UV–Vis spectral analysis and DNA binding studies were performed on a JASCO, V-650 spectrophotometer. FT-IR spectra for the dry powder of bark extract and nanoparticles were recorded in the range 4000–400 cm^{-1} with Thermo scientific, Nicolet iS5 spectrometer. The size of the nanoparticles was assessed using PHILIPS, CM 200, TEM microscope operated at 200 kV with a resolution of 2.4 Å. The diffraction pattern of nanoparticles was recorded using a PANalytical X'Pert Powder X'Celerator diffractometer, with CuKα monochromatic filter. To ascertain the elemental composition of Ag and Au NPs, energy dispersive X-ray (EDAX) analysis was performed on a JEOL, Model JED-2300 microscope. Ultrasonic probe sonicator, Enertech make, model: ENUP-500A was used to disperse CT-DNA in Tris–HCl buffer solution.

Antimicrobial activity

Kirby–Bauer disc diffusion method was employed to assess the antimicrobial activity of biosynthesized Ag and Au NPs. The bacteria *Escherichia coli*, *Pseudomonas aeruginosa*, *Bacillus subtilis*, and *Staphylococcus aureus*, and the fungi *Candida tropicalis* and *Aspergillus flaves* were chosen for study. For disc diffusion method, the bacterial inoculums were standardized to a density equivalent to barium sulfate standard of 0.5 McFarland units and swabbed into petridish of 4 mm depth About 2 mm loopful of extract, Ag and Au NPs were taken and lowered carefully onto the paper disc. The moistened disc was then placed on the surface of inoculated plate and incubated for 16–18 h at 35–37 °C. Netilmicin and fluconazole were used as controls for bacteria and fungi, respectively.

Antioxidant activity

Antioxidant activity of the synthesized Ag and Au NPs was assessed using DPPH assay. From the stock solution of the nanoparticles (1 mg/mL), samples containing 50, 100, 150, and 200 μg were prepared and mixed with 0.1 % DPPH. The reaction mixture was incubated for 30 min at room temperature. When DPPH reacts with any antioxidant, it gets reduced, and the intensity of color decreases; and the decrease in absorbance at 517 nm was recorded, and butylated hydroxytoluene (BHT) was used as a standard control. The experiment was performed in triplicate, and the scavenging activity was calculated as % inhibition according to the following formula, where, A is absorbance.

$$\% \text{ Inhibition} = 100 \times \frac{(A \text{ of standard control} - A \text{ of sample})}{A \text{ of standard control}}$$

In vitro cytotoxicity of Ag and Au NPs

Cytotoxic effect was determined by tetrazolium dye (3-(4,5-Dimethylthiazol-2-yl)-2,5-diphenyltetrazolium bromide)-based microtitration assay (MTT assay) against DLA (Dalton Lymphoma Ascites) cell lines. Cells were maintained in Dulbecco's modified Eagle's medium (DMEM) supplemented with 10 % fetal bovine serum (FBS) at 37 °C in humidified atmosphere with 5 % CO_2. The cells were seeded in 96-well flat-bottom tissue culture plates at a density of approximately 1.2×10^4 cells per well and were allowed to attach overnight at 37 °C. Cells were exposed to plant extract, Ag and Au NPs over a range of concentration for 24 h. After the incubation, medium was discarded; the MTT (10 μL of 5 mg/mL) was added to 100 μL of fresh medium and was incubated for additional 4 h. MTT assay was based on the measurement of the mitochondrial activity of viable cells by the reduction of the tetrazolium salt (MTT) to form a blue water-insoluble product, formazan. After 4 h of incubation, the medium was discarded, and 100 μL of DMSO was added to dissolve the formazan crystals, which were formed by the reduction of tetrazolium salt only by metabolically active cells [32]. The absorbance assay plates were measured at 570 nm in a micro-titer plate reader. Since the absorbance

was directly correlated with the number of viable cells, cell survival was calculated by the following formula:

$$\% \text{ viability} = \frac{A_t}{A_c} \times 100$$

where, A_t is the absorbance of sample and A_c is the absorbance of control. Percentage cytotoxicity is calculated by the formula: cytotoxicity $\% = 100 - \%$ viability.

DNA binding studies

In every organism, deoxyribonucleic acid (DNA) is an important genetic substance. The study of interaction of drug and DNA has acquired great significance for designing and synthesizing new drugs targeted at DNA. Presuming biomolecules capped nanoparticles as drug, the ability to bind DNA was investigated. UV–Visible spectral studies were employed to study the binding of Ag and Au NPs with Calf-thymus DNA (CT-DNA). DNA solutions were prepared using the buffer 5 mM Tris–HCl/50 mM NaCl at pH = 7.2 in water and sonicated for 30 cycles, where each cycle consisted of 1 min with 30 s interval; which gave ratio of absorbance at 260 and 280 nm, A260/A280 as 1.87. This ratio indicates that the DNA was sufficiently free of protein. Absorption spectra were recorded at fixed-concentration Ag and Au NPs for varying concentration of the CT-DNA (2–10 μg/mL) or vice versa. Stock solutions of DNA were stored at 4 °C and used in 4 days.

Results and discussion

Parameters influencing the formation of AgNPs

The effect of time, volume of extract, concentration of metal precursors and pH, which influence the formation of AgNPs were studied in detail, and the results are presented here.

Effect of time

The mixture of the extract and silver nitrate solution was exposed to sunlight, and the SPR band was monitored in regular intervals using UV–Vis spectrophotometer. The color of the solution changed from yellow to dark brown in 2 min, the intensity of the color increased with time and finally attained a constant value in 40 min implying that the formation of AgNPs is almost complete (Fig. 1a). The decrease in peak width and the shift in λ_{max} from 426 to 416 nm with time infer that the formation of monodispersed and small-sized nanoparticles, respectively [33].

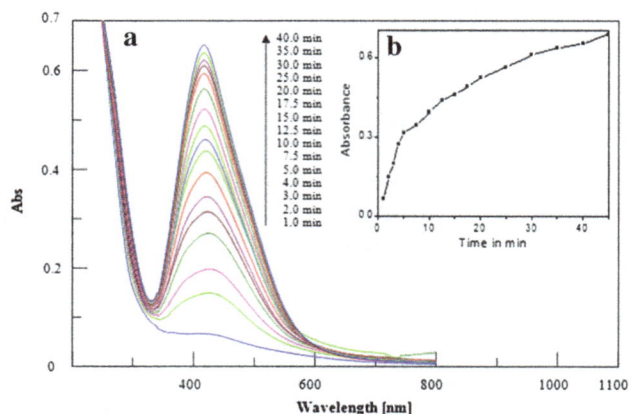

Fig. 1 a UV–Vis spectra of the mixtures of the extract and AgNO$_3$ with time **b** Variation of absorbance with time

Influence of extract volume

To assess the effect of concentration of extract on the formation of AgNPs, the volume was varied from 1 to 15 mL by maintaining the total volume of the reaction mixture as constant (20 mL). The solutions were then exposed to sunlight for 10 min, and the UV–Visible spectra were recorded. The intensity of the SPR band around 449 nm increased with increase in the volume of extract up to 10 mL, accompanied by a blue shift in λ_{max} from 449 to 417 nm (Fig. 2a). This infers that size of nanoparticles decreases with increase in concentration of extract (reducing agent). When the volume of extract exceeds 10 mL, decrease in the absorbance was noticed (Fig. 2b), maybe due to the aggregation of nanoparticles. Other researchers [34] have also reported this trend in biosynthesis of AgNPs. Narrow peaks were obtained for higher volume of extract implying the formation of smaller sized AgNPs.

Influence of concentration of silver nitrate

To optimize the concentration of AgNO$_3$ required for the synthesis of stable AgNPs, it was varied from 1 to 10 mM, and after exposing the solutions to sunlight for 10 min, the SPR band was monitored. With increase in concentration of AgNO$_3$, there was a red shift in SPR band from 418 to 442 nm (Fig. 3a) indicating the increase of particle size. At further high concentrations (5–8 mM), the metallic silver got deposited on the sides of the flask as the silver mirror and the absorbance of the reaction mixture had decreased consequently (Fig. 3b). On further increase of Ag$^+$ ion concentration (9 and 10 mM), a black precipitate was obtained, maybe due to formation of Ag$_2$O (results not shown).

Effect of pH

The AgNPs were synthesized over a wide pH range (pH 2–11) using 0.1-M sulphuric acid or 0.1-M sodium hydroxide. In acidic pH, large-sized nanoparticles were formed, while in alkaline pH, small-sized nanoparticles were produced, as indicated by blue shift in λ_{max} from 420 to 413 nm (Fig. 4a). In alkaline pH, many functional groups are ionized and are available for reduction and facilitate the formation of small size nanoparticles. The pH of the extract was 6, and stable nanoparticles were formed in this condition, and hence, all experiments were carried out at this pH unless otherwise mentioned.

UV–Visible spectral analysis of AuNPs

The AuNPs synthesized using *P. zeylanica* were analyzed by UV–Visible spectrophotometer, where SPR band was observed at 542 nm. The formation of AuNPs was so quick; the color of solution turns red with pink tinge within

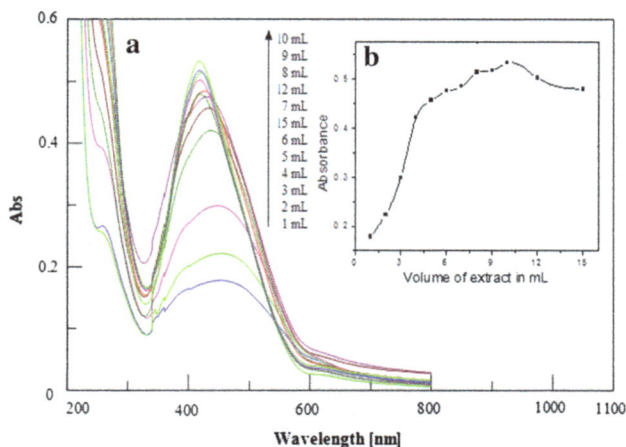

Fig. 2 a UV–Visible spectra of reaction mixture at various volume of extract **b** Change in absorbance with volume of extract

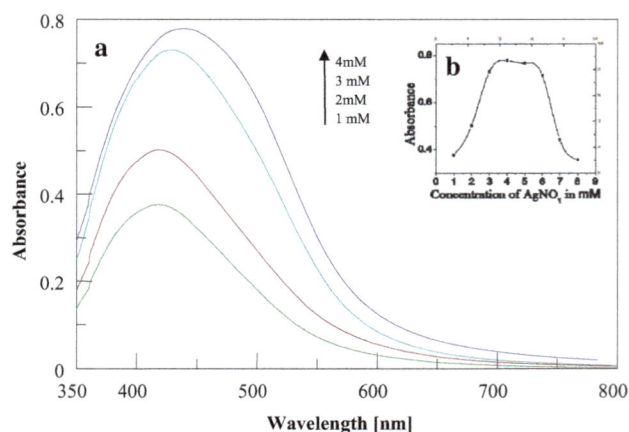

Fig. 3 a UV–Visible spectra of AgNPs at varying concentration of AgNO$_3$ **b** Variation of absorbance with concentration of silver nitrate

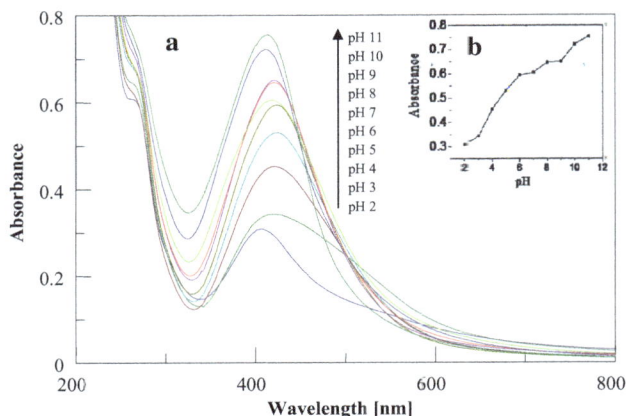

Fig. 4 a UV–Visible spectra of AgNPs synthesized at various pH **b** Variation of absorbance with pH

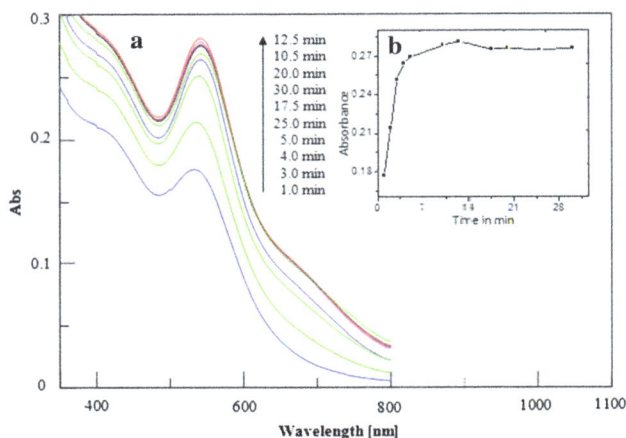

Fig. 5 a UV–Visible spectra of the synthesized AuNPs as a function of time **b** Variation of absorbance with time

the time of mixing, and the absorbance became constant within 10 min indicating the completion of the reaction (Fig. 5b). No appreciable shift in the SPR band was observed with time, and the narrow SPR band indicates the formation of monodispersed AuNPs.

FT-IR studies

FT-IR spectra of Ag and Au NPs were recorded to identify the possible biomolecules responsible for the reduction and capping. The Ag and Au nanosolutions were centrifuged at 13000 rpm for 15 min and air-dried. This dried powder was made as pellets with KBr, and the spectra were recorded. Figure 6 shows the FT-IR spectra of the bark extract, Ag and Au NPs. The spectrum of extract (powder) has peaks at 1063, 1384, and 1631 cm^{-1}. The AgNPs have peaks at 1033, 1384, and 1615 cm^{-1}, while AuNPs exhibits peaks at 1032, 1383, and 1602 cm^{-1}. The main peaks present in the extract are also present in both Ag and Au

NPs inferring that the attachment of biomolecules which are present in the extract to nanoparticles.

Different plant parts of *P. zeylanica* contain naphthaquinones, alkaloids, glycosides, steroids, triterpenoids, tannins, phenolic compounds, flavanoids, coumarins, carbohydrates, saponins, proteins, oils, and fats [29]. Of all the chemical constituents, plumbagin is the principle active compound. Plumbagin (5-hydroxy-2-methyl-1,4-naphthoquinone $-C_{11}H_8O_3$) is primarily present in roots in higher amounts. The structures of plumbagin, glycoside and derivatives of glycoside are represented below [35]. The major water-soluble components present in *P. zeylanica* extract are glycosides [36].

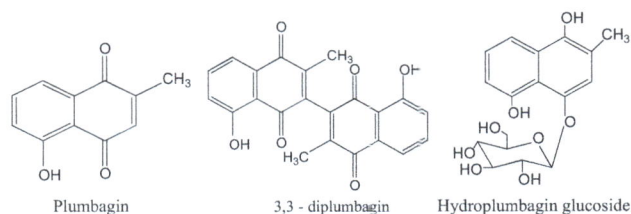

Plumbagin   3,3 - diplumbagin   Hydroplumbagin glucoside

The peak at 1631 cm^{-1} in extract gets shifted to 1615 and 1602 cm^{-1} in the spectra of Ag and Au NPs, respectively. This peak corresponds to chelated >C=O group [37], inferring that the binding of biomolecules with the nanoparticles through this group. The peaks at 3424, 3420, and 3384 cm^{-1} in the spectra of the extract, Ag and Au NPs are due to –OH groups in the biocomponents, and they may be responsible for bioreduction. The peaks near 2912 cm^{-1} in all the three spectra are due to –C–H stretching of aromatic ring. Further, the peaks at 1033 cm^{-1} and 1032 cm^{-1} in the spectra of Ag and Au NPs, respectively, which appears as shoulder in extract, correspond to C–O stretching frequency of biocomponents. These data indicate the involvement of >C=O and –OH functional groups in the reduction and stabilization of nanoparticles.

Mechanism of formation and stabilization of AgNPs

The phytoconstituents like plumbagin and hydroplumbagin glucoside contain many hydroxyl groups. On addition of bark extracts to the $AgNO_3$ solution, the Ag^+—phytoconstituent complex is formed as an intermediate. The nascent hydrogens produced during this process reduce Ag^+ to Ag^o followed by coalescence, cluster formation, and growth of clusters, finally yielding AgNPs [38]. The phenolic groups present in the bark extract subsequently undergo oxidation and get converted to their quinone forms. The electrochemical potential difference between Ag^+ and phytoconstituents drives the reaction. The formed AgNPs are stabilized through the lone pair of electrons and π electrons of quinone structures (Scheme 1).

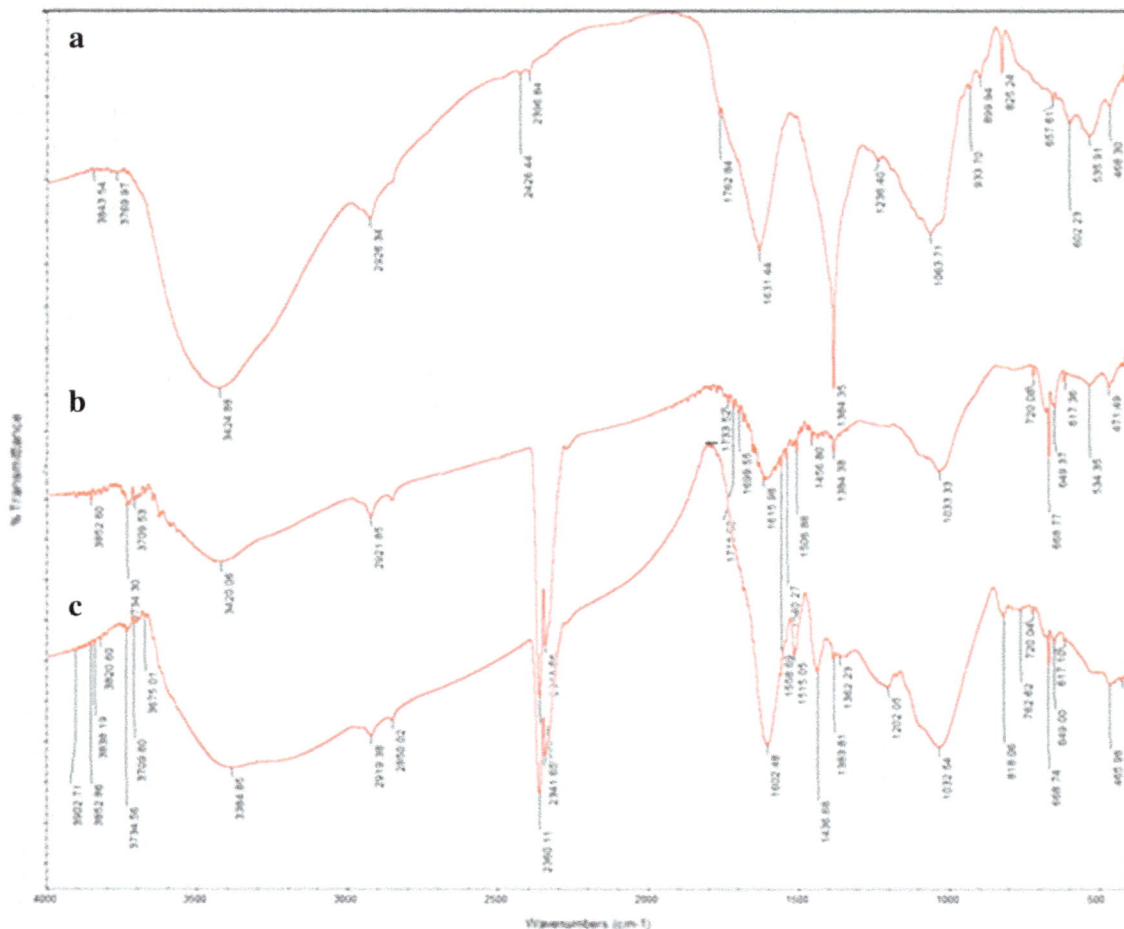

Fig. 6 FT-IR spectra of **a** *Plumbago zeylanica* bark extract, **b** AgNPs and **c** AuNPs

TEM studies

TEM analysis reveals the size and morphology of biosynthesized NPs. TEM images of the Ag and Au NPs at different magnification are depicted in Figs. 7 and 8, respectively. The TEM images clearly show that the nanoparticles are predominantly spherical in shape. The size is in the range of 2.76–51.57 nm (average size 28.47 nm) for AgNPs and 10–25 nm (average size 16.89 nm) for AuNPs. Figures 7d and 8d show the SAED patterns (circular rings) of Ag and Au NPs, that reveal the crystalline nature of nanoparticles. The diffraction rings arise due to reflections of planes of fcc crystals which is evinced from broad Bragg's reflection observed in the XRD spectra.

XRD studies

X-ray diffraction patterns of the Ag and Au NPs are shown in Fig. 9a, b, respectively. A comparison of XRD pattern of AgNPs with the standard JCPDS file no. 89-3722 [39]

confirms the crystalline face-centered cubic (fcc) nature, as evidenced by the peaks at 2θ values of 37.99°, 64.52°, and 76.69°, which can be indexed to 111, 220, and 311 planes, respectively. AuNPs show similar characteristic peaks of metallic fcc gold (JCPDS: 4-0784) [27]. The particle sizes of Ag and Au NPs were computed by Debye–Scherrer equation from which the average size was found to be 17.26 and 12.84 nm, respectively.

$$D = k\lambda/\beta\cos\theta \tag{1}$$

where, k is the Scherrer constant, λ is the wavelength of the X-ray, β and θ are the half-width of the peak and half of the Bragg angle, respectively.

EDAX analysis

EDAX (Energy Dispersive X-ray) analysis was performed to confirm the elemental form of silver and gold. Figure 10a represents the EDAX profile of AgNPs. The EDAX spectrum shows a strong signal at 3 keV, which corresponds to metallic nanosilver [40]; peak

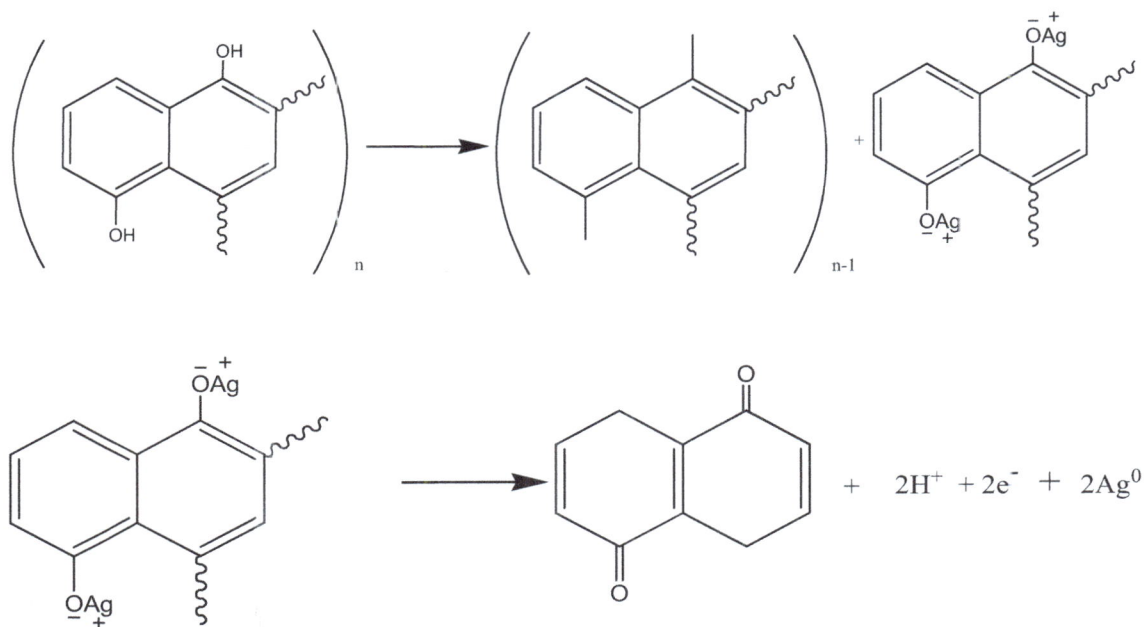

Scheme 1 Mechanism of reduction of Ag^+ to Ag^0

Fig. 7 TEM images (a–c) AgNPs (d) SAED pattern

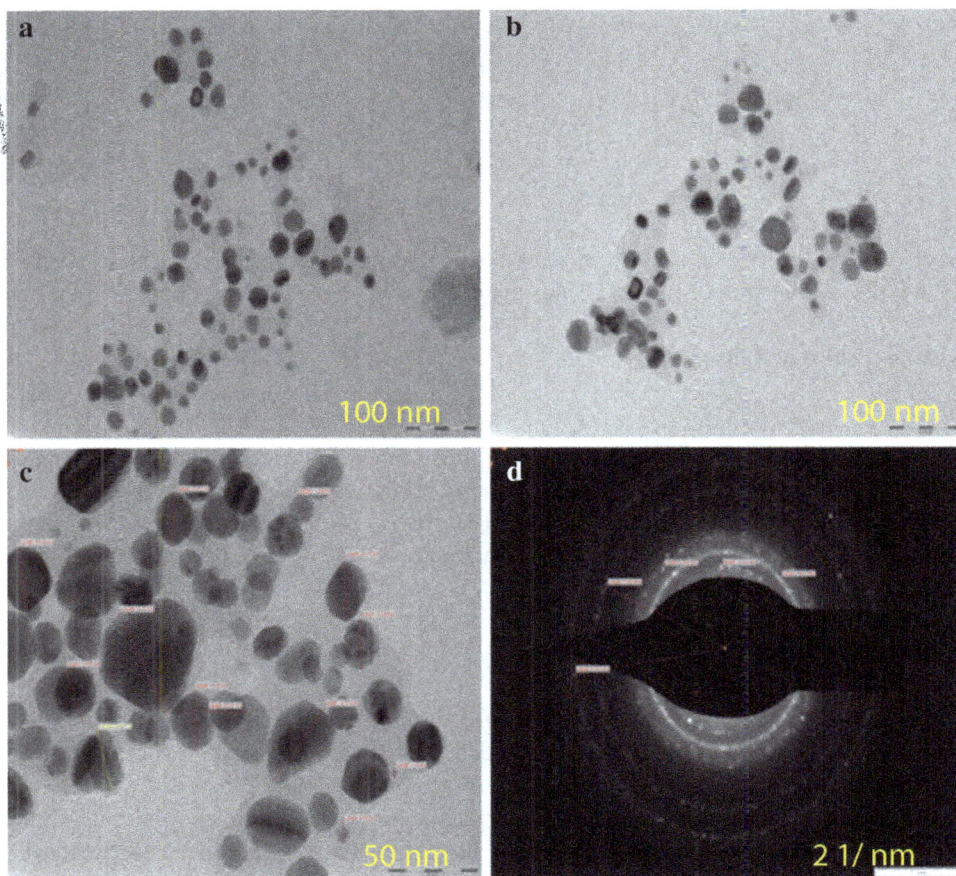

corresponding to Cl^- impurity is also present in the spectrum. There are also weak signals for carbon and oxygen, possibly due to the biocomponents attached to AgNPs. Figure 10b represents EDAX profile of AuNPs, which has strong signals around 2 and 10 keV, which correspond to metallic nanogold [41]. There are also

Fig. 8 TEM images of (**a–c**) AuNPs **d** SAED pattern

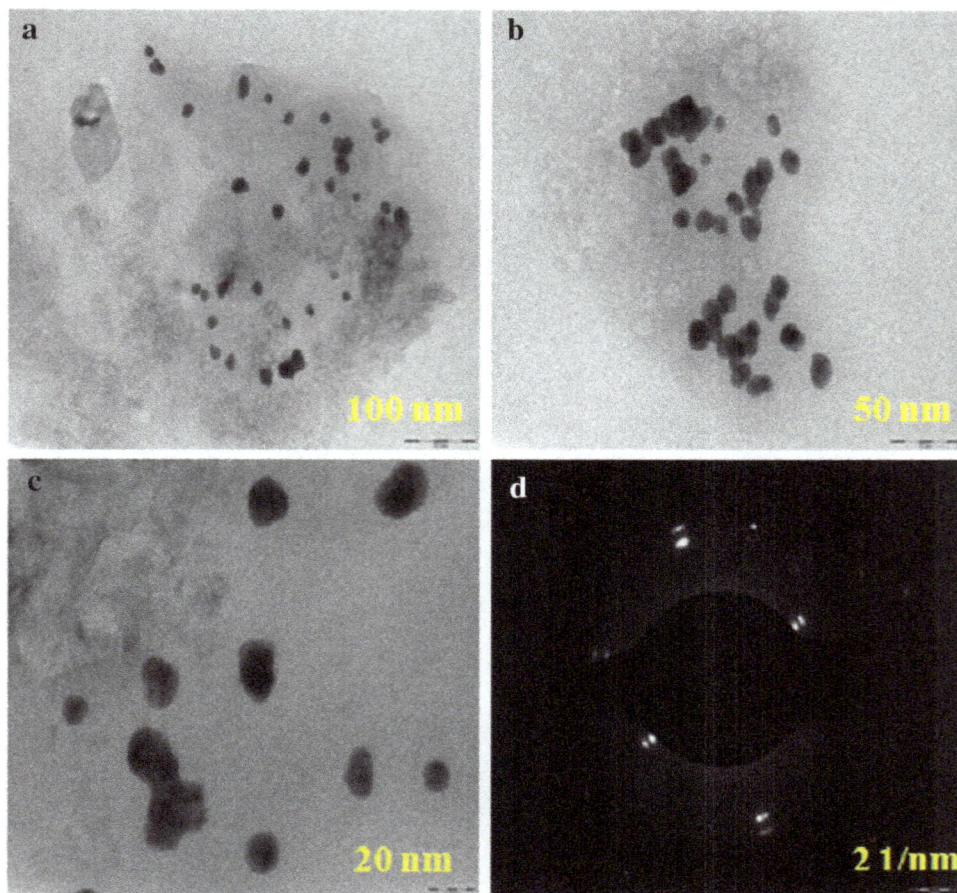

weak signals for carbon and oxygen, indicating capping of AuNPs by biocomponents from the plant extract.

Antimicrobial activity

The AgNPs have desirable activity against *Bacillus subtilis* (12 mm) and *Pseudomonas aeruginosa* (11 mm), but the zone of inhibition for *Staphylococcus aureus, Candida tropicalis, E.Coli* was 8 mm each; for *Aspergillus flaves* no inhibition was observed (Fig. 11). The extract and AuNPs have no effect on any of the species.

Antioxidant activity using DPPH assay

Antioxidants protect cells against damaging effects of reactive oxygen species, which can neutralize free radicals before they can do harm and undo some damage already caused to specific cells. Medicinal plants contain many free radical scavenging molecules like phenolic compounds and vitamins. *P. zeylanica* contains a variety of important chemical compounds and, hence, expected to be a potential antioxidant. DPPH (2,2-diphenyl-1-picryl-

hydrazyl) assay method was adopted to study the antioxidant activity of the synthesized Ag and Au NPs. Substances which are capable of reducing DPPH free radical are radical scavengers and considered to be antioxidants [42]. DPPH is a stable nitrogen-centered free radical with a purple color. When it reacts with an antioxidant, which can donate an electron to DPPH radical, it changes to a stable compound (DPPH-H) with yellow color [43]. As a consequence, purple color decays can be monitored through UV spectrophotometer at 517 nm.

The percentage inhibition of free radicals is presented in Fig. 12. Free radical scavenging activity increases with increasing the concentration of extract, Ag and Au NPs. At 200 µg/mL concentration, extract inhibits 71.16 % of free radicals, whereas, AuNPs, AgNPs, and standard BHT exhibit 87.34, 78.17, and 74.88 % inhibition, respectively. At this concentration, the antioxidant activity decreases in the order: AuNPs > AgNPs > extract > standard BHT. The results are appreciable when compared with the previous report [44]. The IC$_{50}$ values for Au and Ag NPs are higher than standard BHT (Table S1).

Fig. 9 XRD pattern of **a** AgNPs **b** AuNPs

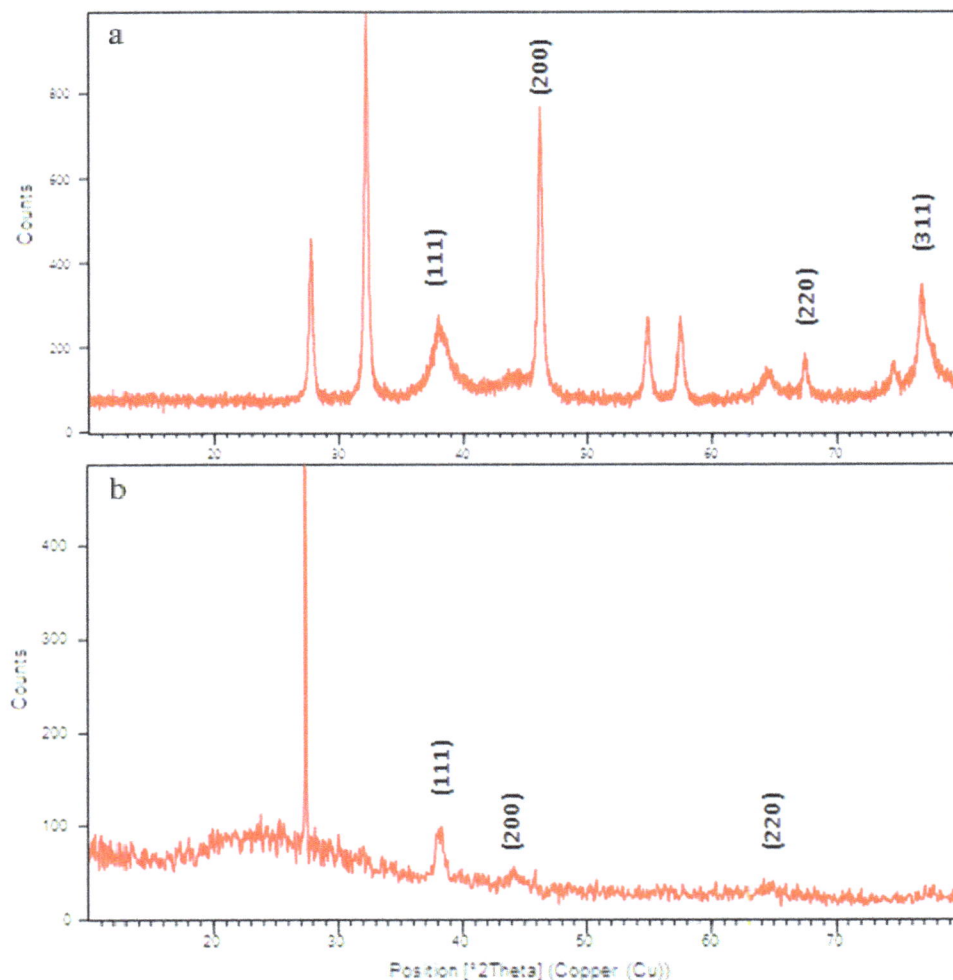

In vitro cytotoxicity against DLA cell line

Only live cells reduce the tetrazolium dye but not the dead cells. When MTT is reduced, the quaternary amine is converted into a tertiary amine by opening the tetrazolium ring. A second tertiary amine, which is formed, binds to a hydrogen atom of the quarternary amine. The absorption spectrum of the molecule changes as a result of the displacement of this hydrogen atom at high pH. Subsequently, the cell numbers can be estimated. Thus, absorbance is directly correlated to number of viable cells [45].

The percentage cell toxicity of extract, Ag and Au NPs, against DLA cell line is given in the Fig. 13. The data reveal that the increase in concentration of extract increases the toxicity. While comparing the results with other studies, it is obvious that the higher toxicity is observed in aqueous extract, whereas in previous reports, other solvent extracts were found to have higher toxicity [46]. At 150 µg/mL concentration, the extract, Au and Ag NPs exhibit 58.35, 61.56, and 65.61 % toxicity, respectively (Table S2). One interesting observation is that the extract

itself shows considerable toxicity against DLA cell line. This may be due to the presence of plumbagin, which is reported to be a potent anticancer drug [47].

DNA binding studies

Nucleic acids show a strong absorbance in the region of 240–275 nm. It originates from the $\pi \to \pi^*$ transitions of the pyrimidine and purine ring systems of the nucleobases. The bases can be protonated, and therefore, the spectra of DNA and RNA are sensitive to pH. At neutral pH, the absorption maxima range from 253 nm (for guanosine) to 271 nm (for cytidine); as a consequence, polymeric DNA and RNA show a broad and strong absorbance near 260 nm.

Electronic spectroscopy is one of the useful techniques to study the binding of DNA with small molecules. Small molecules can react with DNA via covalent or non-covalent interactions. For non-covalent interactions, generally there are three modes for binding viz. electrostatic attractions with the anionic sugar–phosphate backbone of DNA,

Fig. 10 EDAX spectrum of
a AgNPs **b** AuNPs

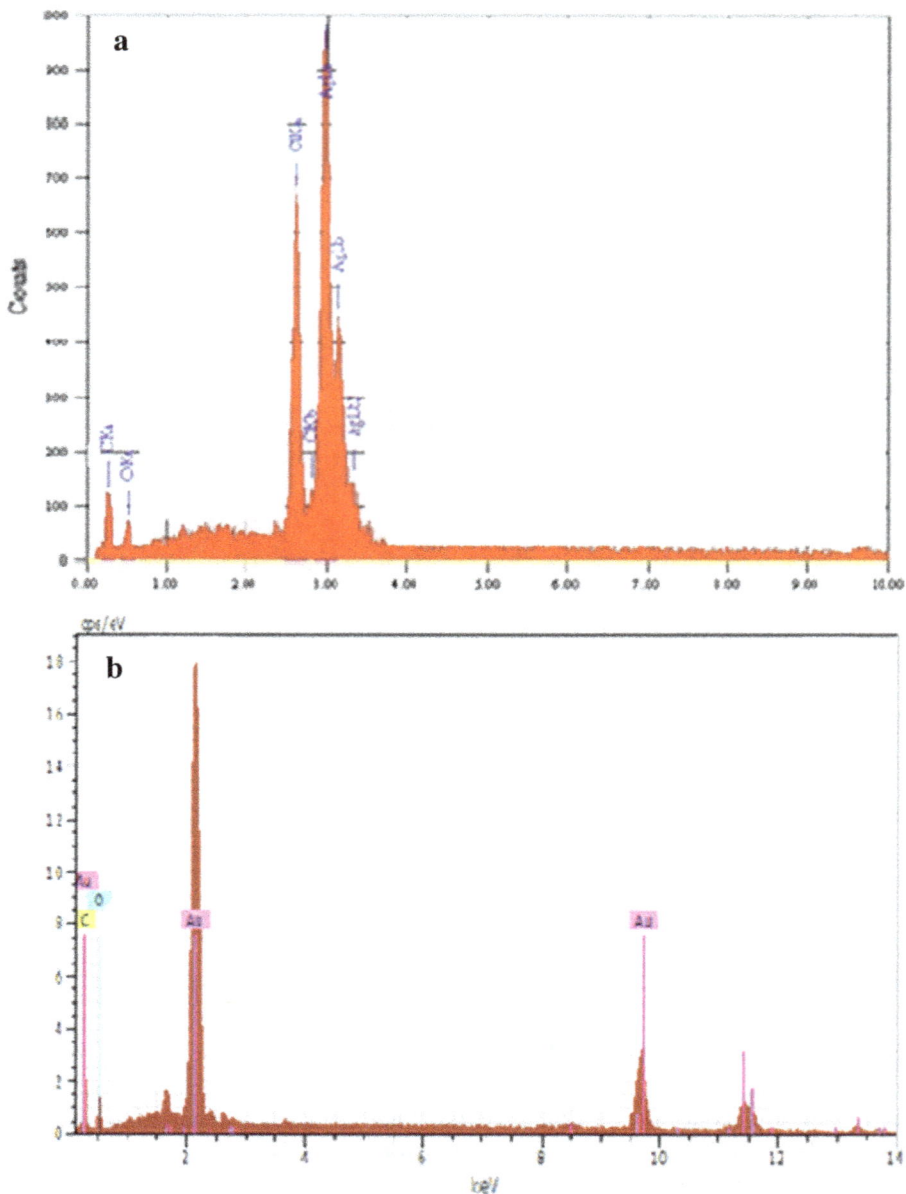

interactions with the DNA groove, and intercalation between base pairs through DNA groove [48]. The absorption spectra of CT-DNA with the increase in concentration of Ag and Au NPs are given in Figs. 14 and 15. The increase of absorption intensity (hyperchromism) at 258 nm (due to $\pi \rightarrow \pi^*$ transition of sugar phosphate in DNA) with increase in concentration of Ag and Au NPs indicates the interaction of the nanoparticles with DNA. In general, DNA exhibits both hyper and hypochromism in the electronic spectrum. Hypochromism suggests the interaction of small molecules with DNA through intercalation, and hyperchromism suggests that the interaction is through groove binding; moreover, hyperchromism is an indication of the breakage of secondary structure of DNA

[49]. The values of the binding constant, K_{app} are obtained from absorbance values as reported in the literature [50]. For weak binding affinities, the data are examined by linear reciprocal plots based on the following equation:

$$\frac{1}{A_{obs} - A_o} = \frac{1}{A_c - A_o} + \frac{1}{K_{app}(A_c - A_0)[AgNPs]}$$

where A_o is the absorbance of CT-DNA in the absence of AgNPs, and A_c is the recorded absorbance of CT-DNA for different concentration of AgNPs. The double reciprocal plot of $1/A_{obs} - A_o$ vs $1/[AgNPs]$ is linear (Fig. 14a), and the binding constant (K_{app}) can be calculated from the ratio of intercept to slope. K_{app} is found to be 7.03×10^7 and 1.15×10^7 for Ag and Au NPs using the above equation.

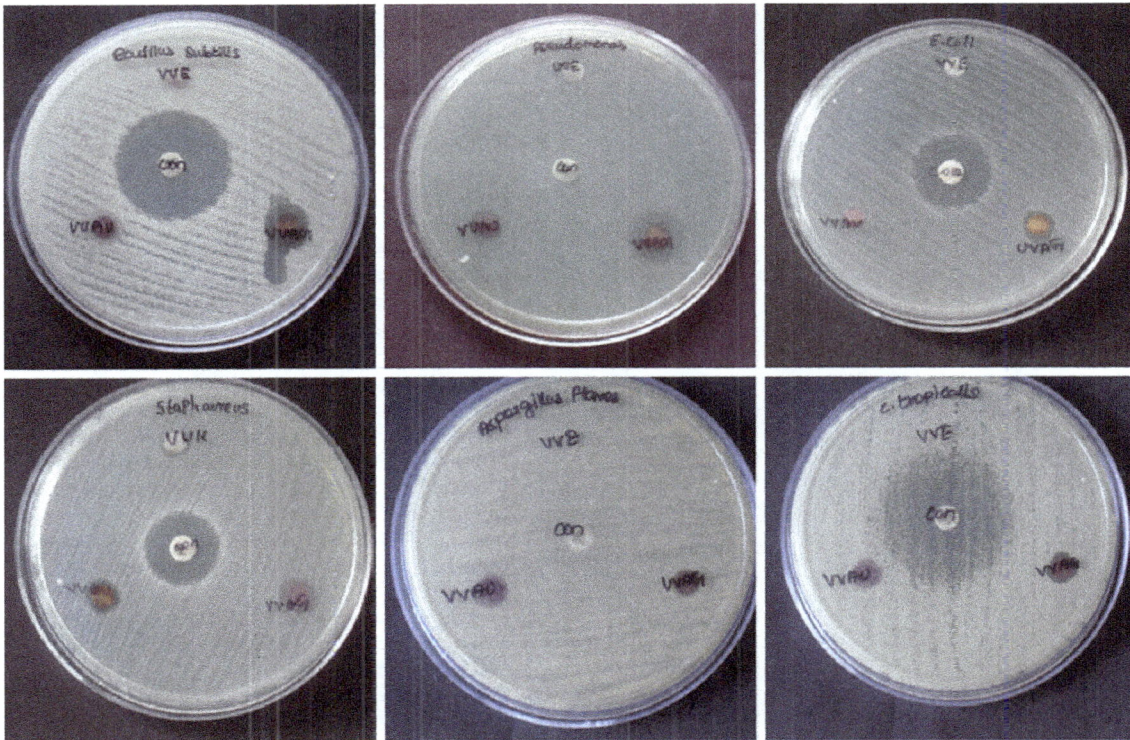

Fig. 11 Antibacterial activity of *Plumbago zeylanica* extract (VVE), AgNPs (VVAG), and AuNPs (VVAU)

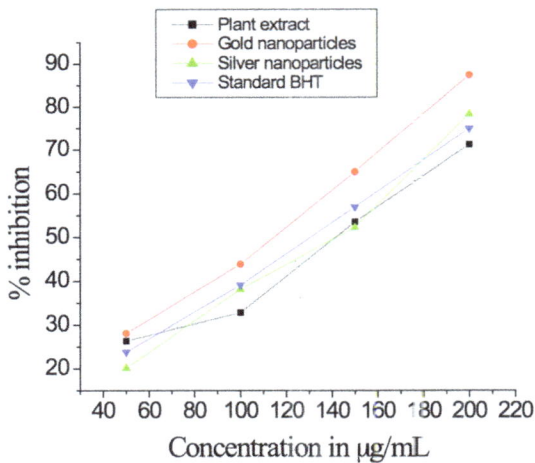

Fig. 12 Antioxidant activity of plant extract and that of Ag and Au NPs

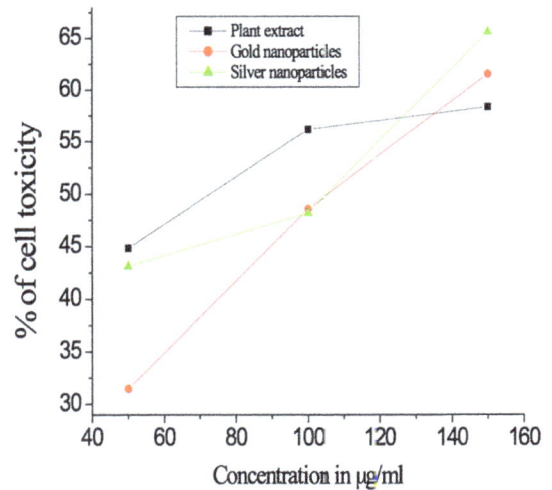

Fig. 13 In vitro cytotoxicity of extract. Ag and Au NPs

The binding constant values are high when compared to normal intercalators [51], which further support groove binding. The intensity of the SPR band also increased with the increase in concentration of Ag and Au NPs.

Electronic spectra were also recorded for different concentration of CT-DNA at fixed concentration of Ag and Au NPs (Figs. S3 and S4). No appreciable increase in absorbance was noticed with varying concentration of CT-DNA.

Conclusions

Bioreductive synthesis of Ag and Au NPs using a medicinally important plant *Plumbago zeylanica,* yielded spherical nanoparticles with average size of 28.47 and 16.89 nm, respectively. AgNPs have desirable antibacterial activity against *Bacillus subtilis* and *Pseudomonas aeruginosa* in comparison to plant extract and AuNPs. Ag and Au NPs

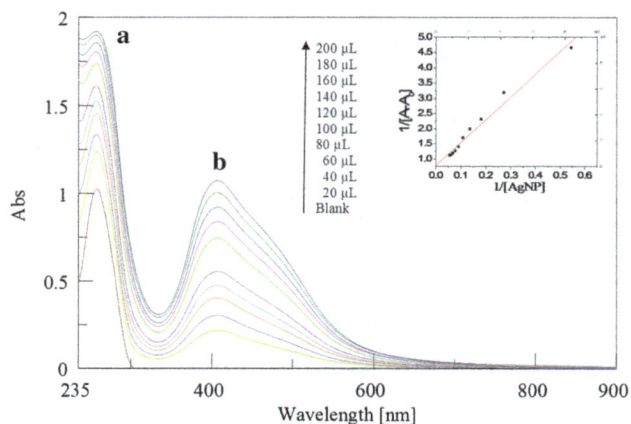

Fig. 14 Absorption spectra of (**a**) CT-DNA with varying concentration of (**b**) AgNPs; Insert: *Plot* of $1/[A_{obs} - A_0]$ vs $1/[AgNP]$

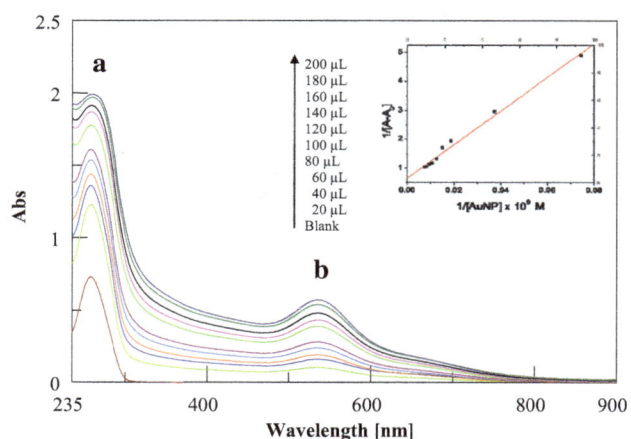

Fig. 15 Absorption spectra of (**a**) CT-DNA with varying concentration of (**b**) AuNPs; Insert: *Plot* of $1/[A - A_0]$ vs $1/[AuNPs]$

have potential antioxidant activity, and the percentage of inhibition of free radicals is found to be 78 and 87 %, respectively. In vitro cytotoxicity studies showed that Ag and Au NPs have 65 and 61 % of toxicity against DLA cell lines. The aqueous extract shows enhanced cytotoxicity in comparison with previous reports. Binding of Ag and Au NPs with CT-DNA is demonstrated by UV–Visible spectrophotometry. The hyperchromism in absorbance suggests the groove binding of these NPs with CT-DNA. The data presented in this study may contribute to the development of nanomaterials as an alternative drug to treat cancer.

Acknowledgments This work is Sponsored by University Grants Commission (UGC), New Delhi. The authors, S. Priya @ Velammal thanks UGC for Minor Research Project and T. Akkini Devi thanks UGC for Project Fellowship. Dr. M.A. Neelakantan, Dean of National Engineering College, Kovilpatti, Tamil Nadu is acknowledged for his help in DNA binding studies. The authors thank Dr. Samasun fertility lab services, Tirupur, Tamil Nadu and Dr. V. R. Mohan, PG and Research Department of Botany, V.O.C. College for their help in antioxidant and cytotoxic studies. SAIF, IIT Bombay, India is acknowledged for TEM analysis.

References

1. Padhi, P., Sahoo, G., Gosh, S., Das, K., Panigrahi, S.C.: Synthesis of black and red mercury sulfide nano-powder by traditional indian method for biomedical application. AIP Conf. Proc. **1063**, 431–438 (2008)
2. Fermo, P., Padeleeti, G.: The use of nanoparticles to produce iridescent metallic effects on ancient ceramic objects. J. Nanosci. Nanotechnol. **12**, 8764–8769 (2012)
3. Satish Kapoor, K.: Nanoconcept in Indian tradition. Dav's Ayurveda Holist. Health Mag. **1**, 22 (2012)
4. Arvizo, R.R., Bhattacharyya, S., Kudgus, R.A., Giri, K., Bhattacharya, R., Mukherjee, P.: Intrinsic therapeutic applications of noble metal nanoparticles: past, present and future. Chem. Soc. Rev. **41**, 2943–2970 (2012)
5. Nikhil, R.J., Latha, G., Catherine, J.M.: Wet chemical synthesis of high aspect ratio cylindrical gold nanorods. J. Phys. Chem. B. **105**, 4065–4067 (2001)
6. Rodriguez, S.L., Blanco, M.C., Lopez Quintela, M.A.: Electrochemical synthesis of silver nanoparticles. J. Phys. Chem. B **104**, 9683–9686 (2009)
7. Pastoriza, S., Liz Marzan, L.M.: Preparation of PVP-protected metal nanoparticles in DMF. Langmuir **18**, 2888–2895 (2002)
8. Noritomi, H., Umezawa, Y., Miyagawa, S., Kato, S.: Preparation of highly concentrated silver nanoparticles in reverse micelles of sucrose fatty acid esters through solid-liquid extraction method. Adv Chem Eng Sci **1**, 299–304 (2011)
9. Zhu, J.J., Liu, S.W., Palchik, O., Koltypin, Y., Gedanken, A.: Shape-controlled synthesis of silver nanoparticles by pulse sonoelectrochemical methods. Langmuir **16**, 6396–6399 (2000)
10. Ahmad, A., Mukherjee, P., Senapati, S., Mandal, D., Islam Khan, M., Kumar, R., Sastry, M.: Extracellular biosynthesis of silver nanoparticles using the fungus *Fusarium oxysporum*. Colloids Surf. B **28**, 313–318 (2003)
11. Balaji, D.S., Basavaraja, S., Raghunandan, D., Mahesh, B., Prabhakar, B.K., Venkataraman, A.: Extracellular biosynthesis of functionalized silver nanoparticles by strains of *Cladosporium cladosporioides* fungus. Colloids Surf. B **68**, 88–92 (2009)
12. Basavaraja, S., Balaji, D.S., Lagashetty, A., Rajasab, A.H., Venkataraman, A.: Extracellular biosynthesis of silver nanoparticles using the fungus *Fusarium semitectum*. Mater. Res. Bull. **43**, 1164–1170 (2008)
13. Raghunandan, D., Basavaraja, S., Mahesh, B., Balaji, S., Manjunath, S.Y., Venkataraman, A.: Rapid biosynthesis of irregular shaped gold nanoparticles from macerated aqueous extracellular dried clove buds (*Syzygium aromaticum*) Solution. Colloids Surf. B **79**, 235–240 (2010)
14. Gardea Torresdey, L.J., Gomez, E., Peralta Videa, R.J., Parsons, J.G., Troiani, H., Jose Yacaman, M.: Alfalfa sprouts: a natural source for the synthesis of silver nanoparticles. Langmuir **19**, 1357–1361 (2003)
15. Shankar, S.S., Rai, A., Ahmad, A., Sastry, M.: Rapid synthesis of Au, Ag, and bimetallic Au Core–Ag shell nanoparticles using neem (*Azadirachta Indica*) leaf broth. J. Colloid Interface Sci. **275**, 496–502 (2004)
16. Jebakumar, T., Lee, Y.R., Sethuraman, M.G.: Green synthesis of silver nanoparticles using *Terminalia cuneata* and its catalytic action in reduction of direct yellow-12 dye. Spectrochim. Acta Part A Mol. Biomol. Spectrosc. **161**, 122–129 (2016)

17. Vidhu, V.K., Philip, D.: Catalytic degradation of organic dyes using biosynthesized silver nanoparticles. Micron **56**, 54–62 (2014)

18. Peter Amaladhas, T., Usha, M., Naveen, S.: Sunlight induced rapid synthesis and kinetics of silver nanoparticles using leaf extract of *Achyranthes aspera* L. and their antimicrobial application. Adv. Mat. Lett. **4**, 779–785 (2013)

19. Akkini Devi, T., Ananthi, N., Peter Amaladhas, T.: Photobiological synthesis of noble metal nanoparticles using *Hydrocotyle asiatica* and application as catalyst for the photodegradation of cationic dyes. J. Nanostruct. Chem. **6**, 75–92 (2016)

20. Sathishkumar, M., Sneha, K., Won, S.W., Cho, C.W., Kim, S., Yun, Y.S.: *Cinnamon zeylanica* bark extract and powder mediated green synthesis of nano-crystalline silver particles and its bactericidal activity. Colloids Surf B Biointerfaces **73**, 332–338 (2009)

21. Paulkumar, K., Gnanajobitha, G., Vanaja, M., Rajeshkumar, S., Malarkodi, C., Pandian, K., Annadurai, G.: *Piper nigrum* leaf and stem assisted green synthesis of silver nanoparticles and evaluation of its antibacterial activity against agricultural plant pathogens. Sci World J **2014**, 1–9 (2014)

22. Gangula, A., Podila, R., Ramakrishna, M., Karanam, L., Janardhana, C., Rao, A.M.: Catalytic reduction of 4-Nitrophenol using biogenic gold and silver nanoparticles derived from *Breynia rhamnoides*. Langmuir **27**, 15268–15274 (2011)

23. Gnanadesigan, M., Anand, M., Ravikumar, S., Maruthupandy, M., Syed Ali, M., Vijayakumar, V., Kumaraguru, K.: Antibacterial potential of biosynthesised silver nanoparticles using *Avicennia marina* mangrove plant. Appl. Nanosci. **2**, 143–147 (2012)

24. Kamyar, S., Mansor Bin, A., Emad, A.J., Ibrahim, N.A., Parvaneh, S., Abdolhossein, R., Yadollah, A., Samira, B., Sanaz, A., Muhammad, S.U., Mohammed, Z.: Green biosynthesis of silver nanoparticles using *Callicarpa maingayi* stem bark extraction. Molecules **17**, 8506–8517 (2012)

25. Santhoshkumar, T., Rahuman, A.A., Bagavan, A., Marimuthu, S., Jayaseelan, C., Kirthi, A.V., Kamaraj, C., Rajakumar, G., Zahir, A.A., Elango, G., Velayutham, K., Iyappan, M., Siva, C., Karthick, L., Bhaskara Rao, K.V.: Evaluation of stem aqueous extract and synthesized silver nanoparticles using *Cissus quadrangularis* against *Hippobosca maculate* and *Rhipicephalus* (*Boophilus*) *microplus*. Exp. Parasitol. **132**, 156–165 (2012)

26. Abdullah, N.I.S.B., Ahmad, M.B., Shameli, K.: Biosynthesis of silver nanoparticles using *Artocarpus elasticus* stem bark extract. Chem. Cent. J. **9**, 61 (2015)

27. Shib Shankar, D., Rakhi, M., Arun Kanti, S., Braja Gopal, B., Biplab Kumar, P.: *Saraca indica* bark extract mediated green synthesis of polyshaped gold nanoparticles and its application in catalytic reduction. Appl. Nanosci. **4**, 485–490 (2014)

28. Tilak, J.C., Adhikari, S., Devasagayam, T.P.A.: Antioxidant properties of *Plumbago zeylanica,* an Indian medicinal plant and its active ingredient, plumbagin. Redox Rep. **9**, 219–227 (2004)

29. Ming, Y., Wang, J., Yang, J., Liu, W.: Chemical constituents of *Plumbago zeylanica*. Adv. Mater. Res. **308**, 1662–1664 (2011)

30. Kapoor, L.D.: Handbook of Ayurvedic medicinal plants. CRC Press, London (1990)

31. Krishnaswamy, M., Purusothaman, K.K.: Plumbagin: a study of its anticancer, antibacterial and antifungal properties. Indian J. Exp. Biol. **18**, 876–877 (1980)

32. Sarah Brown, D., Paola, N., Jo-Ann, S., David, S., Paul, E.R., Balaji, V., David Flint, J., Jane Plumb, A., Duncan, G., NialWheate, J.: Gold nanoparticles for the improved anticancer drug delivery of the active component of oxaplatin. J. Am. Chem. Soc. **132**, 4678–4684 (2010)

33. Sheny, D.S., Mathew, J., Philip, D.: Phytosynthesis of Au, Ag and Au-Ag bimetallic nanoparticles using aqueous extract and dried leaf of *Anacardium occidentale*. Spectrochemica Acta Part A **79**, 254–262 (2011)

34. Muhammad, A., Farooq, A., Muhammad Ramzan, S.A.J., Muhammad, A.I., Umer, R.: Green synthesis of silver nanoparticles through reduction with *Solanum Xanthocarpum L.* berry extract: characterization, antimicrobial and urease inhibitory activities against *Helicobacter pylori*. Int. J. Mol. Sci. **13**, 9923–9941 (2012)

35. Zhang, O.R., Mei, Z.N., Yang, G.Z., Xiao, Y.X.: Chemical constituents from aerial parts of *Plumbago zeylanica* Linn. J. Chin. Med. Mater. **30**, 558–560 (2007)

36. Navneet, K., Bhuwan Mishra, B., Vinod Tiwari, K., Vyasji, T.: An account of phytochemicals from *Plumbago zeylanica* (Family: Plumbaginaceae): A natural gift to human being. Chron. Young Sci. **3**, 178 (2012)

37. Lie Chwen, L., Ling Lang, Y., Cheng Jen, C.: Cytotoxic naphthaquinones and plumbagic acid glucosides from *Plumbago zeylanica*. Phytochemistry **62**, 619–622 (2003)

38. Emrah, B., Mahmut, O.: Rapid, facile synthesis of silver nanostructure using hydrolysable tannin. Ind. Eng. Chem. Res. **48**, 5686–5690 (2009)

39. Kasi, G., Shanmugam, G., Ayyakannu, A.: Phytosynthesis of silver nanoparticles using *Pterocarpus santalinus* leaf exract and their antibacterial properties. J. Nanostruct. Chem. **3**, 68 (2013)

40. Nyoman Rupiasih, N., Avinash, A., Suresh, G., Vidyasagar, P.B.: Green synthesis of silver nanoparticles using latex extract of *Thevetia peruviana*: a novel approach towards poisonous plant utilization. J. Phys. Conf. Ser. **423**, 12032–12039 (2013). doi:10.1088/1742-6596/423/1/012032

41. Narayanan, K.B., Sakthivel, N.: Phytosynthesis of gold nanoparticles using leaf extract of *Coleus amboinicus* Lour. Mater. Charact. **61**, 1232–1238 (2010)

42. Sharad, M., Prachi, B., Man Mohan, S.: Enhanced antioxidant activity of gold nanoparticle embedded 3,6-dihydroxyflavone: a combinational study. Appl. Nanosc. **4**, 153–161 (2014)

43. Elena, H.N., Miron, C.T., Gabriela, P., Mihaela, H., Titus, C., Alexandru, B.T.: Reaction of 2,2-diphenyl-1-picrylhydrazyl with HO, O_2^-, HO^- and HOO^- radicals and anions. Int. J. Mol. Sci. **7**, 130–143 (2006)

44. Veerapandian, S., Sawant, S.N., Dobie, M.: Antibacterial and Antioxidant activity of protein capped silver and gold nanoparticles synthesized with *Escherichia coli*. J. Biomed. Nanotechnol. **8**, 140–148 (2012)

45. Berridge, M.V., Herst, P.M., Tan, A.S.: Tetrazolium dyes as tools in cell biology: new insights into their cellular reduction. Biotechnol. Annu. Rev. **11**, 127–152 (2005)

46. Krishnaraj, C., Muthukumaran, P., Ramachandran, R., Balakumaran, M.D., Kalaichelvan, P.T.: *Acalypa indica* Linn: biogenic synthesis of silver and gold nanoparticles and their cytotoxic effects against MDA-MB-231, human breast cancer cells. Biotechnol. Rep. **4**, 42–49 (2014)

47. Jordan Sand, M., BilalBin, H., Mohammad, S.J., Olya, W., Emily, M.S., Joseph, F., Ajit, K.V.: Plumbagin (5-hydroxy-2-methyl-1,4-naphthoquinone), isolated from *Plumbago zeylanica*, inhibits ultraviolet radiation-induced development of squamous cell carcinomas. Carcinogenesis **33**, 184–190 (2011)

48. Yunhua, W., Guangzhong, Y.: Interaction between Garcigenrin and DNA by spectrophotometry and fluorescence spectroscopy. Spectrosc. Lett. **43**, 28–35 (2010)

49. Nahid, S., Saba, H.: Spectroscopic studies on the interaction of calf thymus DNA with the drug levetiracetam. Spectrochim. Acta Part A Mol. Biomol. Spectrosc. **96**, 278–283 (2012)

50. Swarup, R., Ratan, S., Utpal, G., Tapan, K.D.: Interaction studies between biosynthesized silver nanoparticle with calf thymus DNA and cytotoxicity of silver nanoparticles. Spectrochim. Acta Part A Mol. Biomol. Spectrosc. **141**, 176–184 (2015)

Green synthesis of silver nanoparticles using *phlomis* leaf extract and investigation of their antibacterial activity

A. R. Allafchian[1] · S. Z. Mirahmadi-Zare[2] · S. A. H. Jalali[3,4] · S. S. Hashemi[1] · M. R. Vahabi[4]

Abstract In recent years, the green synthesis of silver nanoparticles using various plant extracts has attracted great attention. This is because, these methods are simple, inexpensive and, eco-friendly. In this study, it was observed that silver ions were reduced by *phlomis* leaf extract after 5 min, leading to the formation of crystalline silver nanoparticles. *Phlomis* species is known as a rich source of flavonoids, phenylpropanoids and other phenolic compounds. The silver nanoparticles produced by the *phlomis* extract were characterized by different techniques including UV–vis spectrophotometry, X-ray diffraction, scanning electron microscopy (SEM), transmission electron microscopy (TEM), and FT-IR. The SEM and TEM results indicated that AgNPs were predominantly spherical in shape with an average particle size of 25 nm. In addition, the antibacterial activity of biologically synthesized nanopartilcles against Gram-positive (*Staphyloccocus aureus* and *Bacillus cereus*) and Gram-negative (*Salmonella typhimurium* and *Escherichia coli*) bacteria was proved. This study, therefore, showed that the phlomis leaf extract could be used for the green synthesis of silver nanoparticles with the appropriate antibacterial activity.

Keywords Silver nanoparticles · *Phlomis* · Green synthesis · Antibactrial activity · Plant extract

Introduction

One of the most important fields of research in nanotechnology is the synthesis of different nanoparticles such as silver, gold, iron, etc. [1–4]. There are various chemical and physical methods for the synthesis of metallic nanoparticles for example, one can mention reduction of solutions, photochemical reactions in reverse micelles, electrochemical reduction, heat evaporation and radiation assisted methods, among others. Physical and chemical methods have usually been successful in the synthesis of nanomaterials in large quantities in short periods of time, as well for specific size and shape. However, most of these methods are extremely expensive and they also involve the use of toxic, hazardous chemicals as the stabilizers which may pose potential environmental and biological risks [5–8].

In recent years, the use of biological methods for the synthesis of metallic nanoparticles has received considerable attention, because these are inexpensive and eco–friendly; also, they can be carried out in one step [9]. Green synthesis methods utilize miscellaneous biological natural substances such as microorganisms, whole plants, plant tissues and fruits, plant extracts, marine algae and micro–fluids for the reduction and stabilization of nanoparticles. Synthesis of nanoparticles using plant extracts has several advantages over other environmentally green synthesis methods, because plants are broadly distributed, readily scalable, easily available, safe to handle and less expensive [2].

Among diverse nanoparticles, silver nanoparticles due to various properties such as catalysis, electrochemical

✉ A. R. Allafchian
    Allafchian@cc.iut.ac.ir

[1] Nanotechnology and Advanced Materials Institute, Isfahan University of Technology, Isfahan 84156–83111, Iran

[2] Department of Molecular Biotechnology at Cell Science Research Center, Royan Institute for Biotechnology, ACECR, Isfahan, Iran

[3] Institute of Biotechnology and Bioengineering, Isfahan University of Technology, Isfahan 84156–83111, Iran

[4] Department of Natural Resources, Isfahan University of Technology, Isfahan 84156–83111, Iran

conductivity and antimicrobial activity, can be used in different applications like biomedicine, agriculture, photo chemicals and food chemistry [10, 11]. Since ancient times, the silver has been known to be efficient against a wide range of microorganisms [12]. Nowadays, the most important applications of silver nanoparticles in biotechnology science correspond to their antibacterial and antifungal activities [13].

Phlomis is a perennial native herb plant belonging to Lamiaceae family with more than 100 species in the world. This genus plant is mostly grown in Turkey, Iran, Asia, North Africa and Europe. About 17 species of this plant are in Iran, of which 10 are endemic. A review of several studies shows that *phlomis* species have unique aromatic compounds, medicinal properties and antimicrobial activity [14, 15].

A large number of plant leaf extracts like *Cinnamomum camphora* [16], *Chenopodium album* [17], *Stevia rebaudiana* [18], *Murraya koenigii* [19], *Annona squamosal* [20], *Desmodium gangeticum* [21], *Pulicaria glutinosa* [22], *Brucea javanica* [23], *Artocarpus heterophyllus* [24], *Caesalpinia coriaria* [25], *Abutilon indicum* [26], *Rosa indica* [27], *Tephrosia tinctoria* [28] have been employed by several researchers for the synthesis of silver nanoparticles.

In this research, the leaf extract of *phlomis* was used for rapid, simple and biosynthetic synthesis of silver nanoparticles. Furthermore, green synthesis Ag nanoparticles were characterized by FT-IR, XRD, TEM, and SEM techniques. Finally, its antibacterial activity was investigated by the disk diffusion method.

Materials and methods

Materials

Silver nitrate was obtained from Merck Company for this study. All glassware were washed in dilute HNO_3 acid and rinsed with distilled water and dried in a hot air oven before use. During the period of flowering and the vegetative phase, samples of *phlomis* Arial parts consisting of leaves, stems and flower buds were collected. By comparing the collected voucher specimen with that of a known identity available in the herbarium of the Department of Natural Resources, Isfahan University of Technology, Iran, the taxonomic identity of the plant was confirmed.

Microorganisms and growth conditions

In this study, Gram-negative bacteria like *Escherichia coli* (ATCC 35218) and *Salmonella typhimurium* (ATCC 14028) and Gram-positive bacteria like *Staphylococcus*

aureus (ATCC 29213) and *Bacillus cereus* (ATCC 14579) were employed as bacteria strains for the antimicrobial experiments. These bacteria strains were cultured at 37 °C on Luria-Bertani (LB) agar.

Preparation of the plant extract

The *phlomis* leaf extract was employed to prepare silver nanoparticles. The aerial parts of the fresh collected plant were washed several times with distilled water, to remove dust particles and any dirt. Leaves of the plants were dried in the shade and at room temperature. About 40 g of its leaves was subsequently ground into fine powder and boiled for 2 h in 250 ml of distilled water. The extracts were cooled to room temperature and filtered through whatman filter paper No. 1. The filtered extract was stored in refrigerator at 4 °C until it was used as the reducing and stabilizing agent.

Synthesis of silver nanoparticles

To synthesize AgNPs, 5.0 ml of the *phlomis* plant extract was added into 45.0 ml of 0.01 M of silver nitrate solution at room temperature. The change from pale yellow and dark brown indicated the formation of colloidal silver nanoparticles.

Characterization of silver nanoparticles

UV–visible absorption analysis was carried out using a SUV–S2100 spectrophotometer. FT–IR spectra of silver nanoparticles were performed using a JASCO FTIR (680 plus, Japan) spectrometer with KBr pellet in the range of $4000–400\ cm^{-1}$. The crystalline nature of silver nanoparticles was investigated by XRD analysis. X-ray diffraction data of AgNPs were obtained using a Philips–X'Pert Pro MPD with Cu kα radiation ($\lambda = 1.54$ Å) in the 2θ range of $20°$ to $80°$, and with a steps size of $0.02°$ at 40 kV and 30 mA. The morphology of the AgNPs was examined by TEM (cm30–Philphs). Furthermore, SEM (HITACHI S–4160) study was carried out to investigate the shape, size and the surface area of the AgNPs.

Antibacterial assay

The antibacterial activity of bio–synthesized AgNPs was tested against various positive and negative bacteria by the standard agare well diffusion method. To examine the antibacterial activity of biosynthesized AgNPs, Muller-Hinton agar plates were sterilized and allowed to solidify. After solidification, 30 µl of each bacterial suspension was inoculated on the petriplates by a strile glass rod. Then, 0.1 g of synthesized–AgNPs powder were dissolved in

1.0 ml (100 ppm) autoclaved distillated water to provide the suspension of AgNPs. Sterilized paper disks (diameter 6.4 mm) were impregnated with 30 µl of suspension and placed on Muller Hinton agar plates. The negative (distilled water) and positive (AgNO₃) controls, and the *phlomis* leaf extract were also employed for the antibacterial assay. The plates were incubated at 37 °C for 24 h. After the incubation period, the zones of inhibitions were observed around the discs. Antibacterial activity was investigated by measuring the diameter of the zones of inhibition after using the plant extract.

Results and discussion

UV–Vis spectral studies and FT–IR analysis

The *phlomis* plant extract was employed for the green synthesis of AgNPs. After the addition of the plant leaf extract to the silver nitrate solution, it was observed that the color of the reaction mixture was gradually changed from light yellow to dark brown, indicating the formation of silver nanoparticles. UV–Vis absorption spectroscopy is an important method to detect the formation and stability of metal NPs in the reaction mixture. Figure 1 shows the UV–

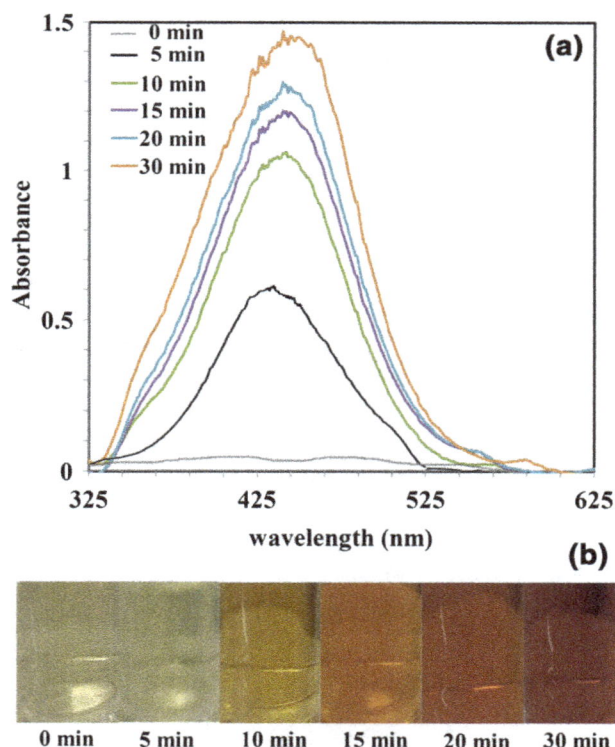

Fig. 1 Formation of silver nanoparticles (AgNPs), **a** UV–Vis spectra of synthesized AgNPs using *phlomis* leaf extract; and **b** photograph showing the color change of AgNPs aqueous solution during nanoparticle synthesis

Visible spectra recorded at different times of reaction. No change in absorbance was observed after 30 min, indicating the complete conversion of Ag⁺ to Ag. In this work, the UV–vis spectra of silver nanoparticles synthesized displayed a strong broad peak around 440 nm due to the formation of AgNPs. This peak corresponded to the surface plasmon resonance of the synthesized AgNPs [29]. UV–VIS absorption measurements for silver NPs further were confiris in the range of 450–500 nm [2].

To determine the possible biomolecules and functional groups involved in reduction, capping and efficient stabilization of newly synthesized Ag nanoparticles, FTIR spectroscopy was employed. The FTIR spectrum of stabilized silver nanoparticles is depicted in Fig. 2. The spectra showed absorption bands at 3419, 1749, 2927, 1625, 1383, 1235, 1069, 832 and 602 cm⁻¹. The strong peaks at 3419 cm⁻¹ corresponded to –OH stretching due to phenolic compounds present in the *phlomis* leaf extract. The band at 2927 cm⁻¹ was attributed to alkane C–H stretching vibration. C=O stretching possibly due to the presence of carbonil group. The peak at 1625 cm⁻¹ corresponded to C=C stretching vibration of aromatic rings. The peak at 1383 cm⁻¹ was corresponded to C–H bending. Further, peaks assigned at 1235 and 1069 cm⁻¹ were attributed to C–N stretching possibly due to the presence of amines group. Another intense band at 832 cm⁻¹ was characteristic of the aromatic ring [30–33].

The precipitation stage during the reduction process usually involves fast chemical reactions and nucleation kinetics, thus the type of reducer a significant factor in the particle size distribution of the product [34]. The results previously reported indicated that the phlomis leaf extract containing glycosides such as flavonoids, iridoids, diterpenoids, phenylpropanoids, triterpenoids and other phenolic compounds possibility contributed to the process of nanoparticle synthesis [15]. Furthermore, based on the above results, it was clear that the functional groups like –OH (hydroxyl), –C=O (Carbonil) and C–N (amine) present in the leaf extract were involved in synthesis of AgNPs.

XRD analysis

The X-ray diffraction analyses were carried out to determine the known phase of the silver nanoparticles. Figure 3 demonstrates the XRD pattern of the dried synthesized Ag nanoparticles by the *phlomis* extract. The spectrum exhibited five distinct separate peaks at $2\theta = 38.10°$, 44.37°, 64.17°, 77.57° and 81.67° that could be indexed to (111) (200), (220), (311) and (222) planes, respectively. The diffraction peaks data obtained were in accordance with the reports of FCC structure from Joint Committee Powder Diffraction Standards (JCPDS) file No. 04–0783. The mean grain crystalline size of green synthesized

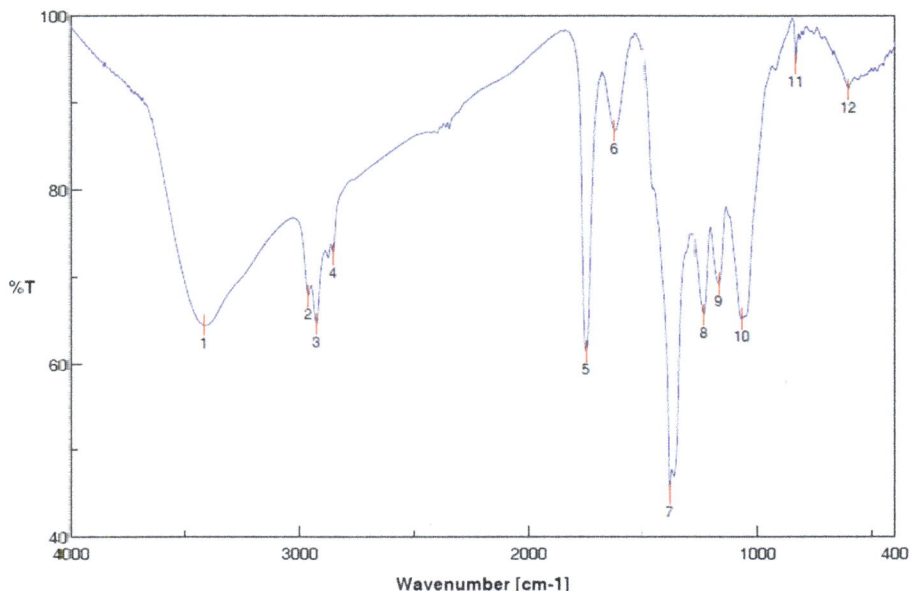

Fig. 2 The FT–IR spectra of *green* synthesized AgNPs from *phlomis* leaf extract

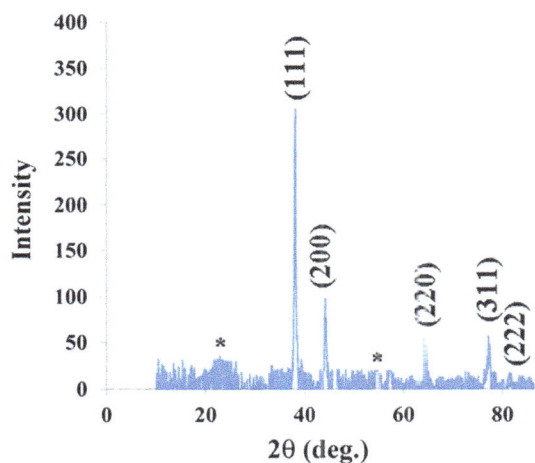

Fig. 3 X–ray diffraction patterns of synthesized AgNPs using *phlomis* leaf extract

AgNPs was calculated by employing Debye–Scherrer formula.

$$D = \frac{K\lambda}{\beta \cos\theta} \tag{1}$$

where D is the average crystalline diameter size (Å), K is a constant (0.9), 'λ' is the X-ray wavelength used ($\lambda = 1.54$ Å), 'β' is the angular line width at the half maximum of diffraction (radians) and 'θ' is the braggs angle (degrees) [11]. The average grain crystalline size of AgNPs was estimated to be approximately 27 nm. A small number of unassigned peaks (marked with stars) were also recorded that might be due to the crystallization of bioorganic phases present in *phlomis* extract on the surface of the silver

nanoparticles. Similar results were also obtained for AgNPs synthesized using the *beetroot* extract and *Ixora coccinea* leaves extract [35, 36].

SEM and TEM analysis

SEM technique was employed to determine the surface morphology and the topography of synthesized silver nanoparticles. Figure 4 shows the size of silver nanoparticles from 19 to 30 nm, with an average size 25 nm. SEM image exhibited that the biosynthesized silver nanoparticles were mostly spherical in shape. The shape and size of the biosynthesized AgNPs were further analyzed by TEM. TEM image (Fig. 5) demonstrated that the most AgNPs were obviously spherical in shape and well dispersed, with an average size around 25 nm. The obtained results from TEM image were in a good agreement with the SEM data.

Antibacterial activity

The antibacterial activity of biosynthesized silver nanoparticles, $AgNO_3$, *phlomis* leaf extract and distilled water was studied against Gram-positive (*S. aureus* and *B. cereus*) and Gram-negative (*S. typhimurium* and *E. coli*) bacteria using the agar well diffusion assay, and the zone of inhibition was tabulated as shown in Table 1 and Fig. 6. The synthesized AgNPs displayed efficient antibacterial activity against both Gram-negative and Gram-positive bacteria. The silver nanoparticles synthesized by *phlomis* leaf extracts showed the maximum zone of inhibition around 15 mm for *S. typhimurium* and *E. coli*, which were followed by *S. aureus* (14.7 mm) and *B. cereus* (12.1 mm).

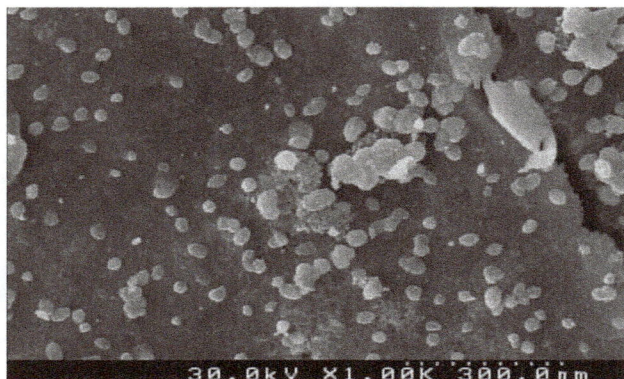

Fig. 4 Scanning electron microscopic analysis of AgNPs synthesized using *phlomis* leaf extract

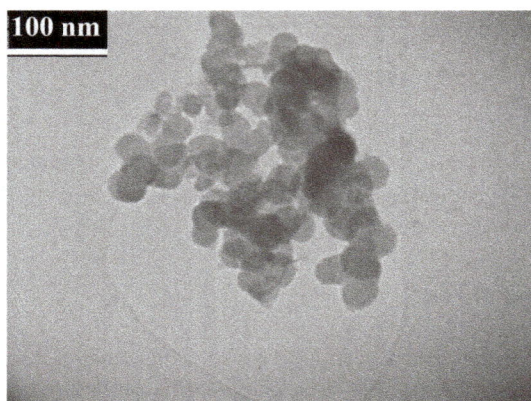

Fig. 5 TEM image of synthesized AgNPs using *phlomis* leaf extract

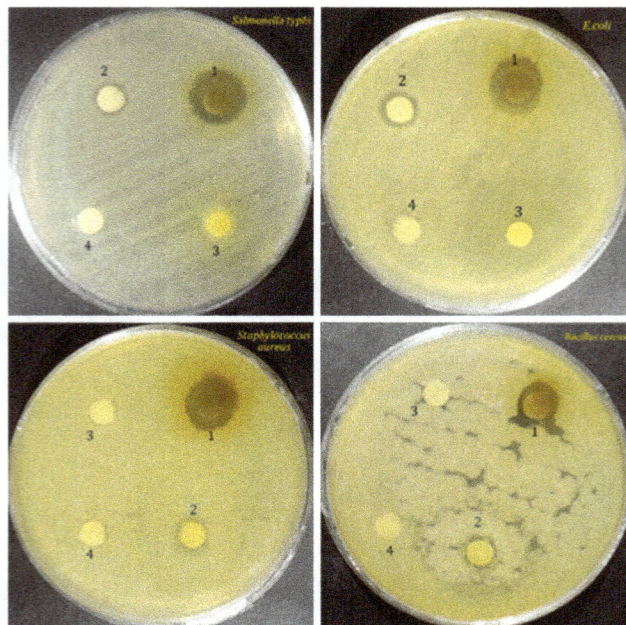

Fig. 6 Antimicrobial activity of *1* silver nanoparticles, *2* AgNO₃, *3* *phlomis* leaf extract and *4* distilled water against Gram-positive (*Staphyloccocus aureus* and *Bacillus cereus*) and Gram-negative (*Salmonella typhimurium* and *Escherichia coli*) pathogenic bacteria

On the other hand, the negative control (distilled water) and the leaf extract did not exhibit any zone of inhibition. The positive control (AgNO₃) displayed antimicrobial activity against all tested microorganisms, Gram-negative bacteria, *E. coli* (11.3 mm) and *S. typhimurium* (9.5 mm), and Gram-positive bacteria, *S. aureus* and *B. cereus* (9.4 mm).

The mechanisms of efficient antibacterial activity by silver nanoparticles against various pathogenic bacteria, are still unclear and required further investigation [30, 35]. AgNPs might have been attached to the surface of the cell membrane of microorganisms, leading to the disturbance of its functions like permeability and respiration. It is obvious, therefore, that the binding of particles to the microorganism depends on the surface area available for intraction. In general, small nanoparticles have a larger surface area for intraction with bacteria, as compared to that of bigger particles, due to greater antibacterial activity [35, 37, 38]. In our results, the Gram-positive bacteria showed the lower zone of inhibition, as compared to Gram-negative bacteria. This could be due to the cell wall of Gram-positive bacteria composed of a rigid thicker multiple layer of peptidoglycan, as it prevented the nanoparticles from entering into cell wall [39].

Table 1 Zone of inhibition (mm) of AgNO₃, Ag NPs, *phlomis* leaf extract and deionized water against tested bacteria

Components	Zone of inhibition (mm)			
	Staphylococcus aureus	*Bacillus cereus*	*Escherichia coli*	*Salmonella typhimurium*
AgNPs	14.7	12.1	15.1	14.9
Plant extract	NA	NA	NA	NA
AgNO₃	9.4	9.4	11.3	9.5
Distillate water	NA	NA	NA	NA

The disc's diameter was 6.4 mm

NA No activity

Conclusion

In the present study, we showed, for the first time, the biosynthesis of stable and nearly spherical silver nanoparticles using the *phlomis* plant extract as a reducing and capping agent. The biosynthesized nanoparticles presented strong antimicrobial activity against Gram-positive (*S. aureus* and *B. cereus*) and Gram-negative (*S. typhimurium* and *E. coli*) bacteria. Also, we confirmed the formation of silver nanoparticles using the *phlomis* plant extract from UV–Vis, XRD, SEM, and FTIR. UV–Vis peak was observed for AgNPs at 450 nm. The synthesized AgNPs were found to have a crystalline structure as investigated by XRD method. The particle size of the silver nanoparticles ranged in size from 19 to 30 nm, with an average size of 25 nm. Finally, the green synthesis of silver nanoparticles using plant material was found to be the most eco-friendly and conventional method, in comparison to chemical and physical synthesis methods.

Acknowledgements The authors wish to thank Isfahan University of Technology (IUT) Research Council, Center of Excellency in Applied Nanotechnology and the Iranian Nanotechnology Council for supporting this work.

References

1. Allafchian, A.R., Majidian, Z., Ielbeigi, V., Tabrizchi, M.: A novel method for the determination of three volatile organic compounds in exhaled breath by solid-phase microextraction–ion mobility spectrometry. Anal. Bioanal. Chem. **408**, 839–847 (2016)
2. Mittal, A.K., Chisti, Y., Banerjee, U.C.: Synthesis of metallic nanoparticles using plant extracts. Biotechnol. Adv. **31**, 346–356 (2013)
3. Muthoosamy, K., Bai, R.G., Abubakar, I.B., Sudheer, S.M., Lim, H.N., Loh, H.S., Huang, N.M., Chia, Ch.H., Manickam, S.: Exceedingly biocompatible and thin-layered reduced graphene oxide nanosheets using an eco-friendly mushroom extract strategy. Int. J. Nanomedicine. **10**, 1505–1519 (2015)
4. Ng, C.M., Chen, P.C., Manickam, S.: Green high-gravitational synthesis of silver nanoparticles using a rotating packed bed reactor (RPBR). Ind. Eng. Chem. Res. **51**, 5375–5381 (2012)
5. Kuppusamy, P., Ichwan, S.J., Parine, N.R., Yusoff, M.M., Maniam, G.P., Govindan, N.: Intracellular biosynthesis of Au and Ag nanoparticles using ethanolic extract of *Brassica oleracea* L. and studies on their physicochemical and biological properties. J. Environ. Sci. **29**, 151–157 (2015)
6. Prakash, P., Gnanaprakasam, P., Emmanuel, R., Arokiyaraj, S., Saravanan, M.: Green synthesis of silver nanoparticles from leaf extract of *Mimusops elengi*, Linn. for enhanced antibacterial activity against multi drug resistant clinical isolates. Colloids Surf. B **108**, 255–259 (2013)
7. Orbaek, A.W., McHale, M.M., Barron, A.R.: Synthesis and characterization of silver nanoparticles for an undergraduate laboratory. J. Chem. Edu. **92**, 339–344 (2014)
8. Bar, H., Bhui, D.K., Sahoo, G.P., Sarkar, P., Pyne, S., Misra, A.: Green synthesis of silver nanoparticles using seed extract of *Jatropha curcas*. Colloids Surf. A **348**, 212–216 (2009)
9. Otari, S., Patil, R., Nadaf, N., Ghosh, S., Pawar, S.: Green biosynthesis of silver nanoparticles from an actinobacteria *Rhodococcus* sp. Mater. Lett. **72**, 92–94 (2012)
10. Allafchian, A.R., Bahramian, H., Jalali, S.A.H., Ahmadvand, H.: Synthesis, characterization and antibacterial effect of new magnetically core–shell nanocomposites. J. Magn. Magn. Mater. **394**, 318–324 (2015)
11. Allafchian, A.R., Jalali, S.A.H.: Synthesis, characterization and antibacterial effect of poly (acrylonitrile/maleic acid)–silver nanocomposite. J. Taiwan Inst. Chem. Eng. **57**, 154–159 (2015)
12. Kalishwaralal, K., Deepak, V., Pandian, S.R.K., Kottaisamy, M., BarathManiKanth, S., Kartikeyan, B., et al.: Biosynthesis of silver and gold nanoparticles using *Brevibacterium casei*. Colloids Surf. B **77**, 257–262 (2010)
13. Allafchian, A., Jalali, S.A.H., Bahramian, H., Ahmadvand, H.: Preparation, characterization, and antibacterial activity of NiFe$_2$O$_4$/PAMA/Ag–TiO$_2$ nanocomposite. J. Magn. Magn. Mater. **404**, 14–20 (2016)
14. Sarikurkcu, C., Uren, M.C., Tepe, B., Cengiz, M., Kocak, M.S.: Phenolic content, enzyme inhibitory and antioxidative activity potentials of *Phlomis nissolii* and *P. pungens* var. *pungens*. Ind. Crop. Prod. **62**, 333–340 (2014)
15. Sarkhail, P., Rahmanipour, S., Fadyevatan, S., Mohammadirad, A., Dehghan, G., Amin, G., et al.: Antidiabetic effect of *Phlomis anisodonta*: effects on hepatic cells lipid peroxidation and antioxidant enzymes in experimental diabetes. Pharmacol. Res. **56**, 261–266 (2007)
16. Huang, J., Li, Q., Sun, D., Lu, Y., Su, Y., Yang, X., et al.: Biosynthesis of silver and gold nanoparticles by novel sundried *Cinnamomum camphora* leaf. Nanotech. **18**, 105104 (2007)
17. Dwivedi, A.D., Gopal, K.: Biosynthesis of silver and gold nanoparticles using *Chenopodium album* leaf extract. Colloids Surf. A **369**, 27–33 (2010)
18. Yilmaz, M., Turkdemir, H., Kilic, M.A., Bayram, E., Cicek, A., Mete, A., et al.: Biosynthesis of silver nanoparticles using leaves of *Stevia rebaudiana*. Mater. Chem. Phys. **130**, 1195–1202 (2011)
19. Philip, D., Unni, C., Aromal, S.A., Vidhu, V.: *Murraya koenigii* leaf–assisted rapid green synthesis of silver and gold nanoparticles. Spectrochim. Acta A **78**, 899–904 (2011)
20. Vivek, R., Thangam, R., Muthuchelian, K., Gunasekaran, P., Kaveri, K., Kannan, S.: Green biosynthesis of silver nanoparticles from *Annona squamosa* leaf extract and its in vitro cytotoxic effect on MCF–7 cells. Process Biochem. **47**, 2405–2410 (2012)
21. Thirunavoukkarasu, M., Balaji, U., Behera, S., Panda, P., Mishra, B.: Biosynthesis of silver nanoparticle from leaf extract of *Desmodium gangeticum* (L.) DC. and its biomedical potential. Spectrochim. Acta A **116**, 424–427 (2013)
22. Khan, M., Khan, M., Adil, S.F., Tahir, M.N., Tremel, W., Alkhathlan, H.Z., et al.: Green synthesis of silver nanoparticles mediated by *Pulicaria glutinosa* extract. Int. J. nanomedicine. **8**, 1507–1516 (2013)
23. Notriawan, D., Angasa, E., Suharto, T.E., Hendri, J., Nishina, Y.: Green synthesis of silver nanoparticles using aqueous rinds extract of *Brucea javanica* (L.) Merr at ambient temperature. Mater. Lett. **97**, 181–183 (2013)
24. Jagtap, U.B., Bapat, V.A.: Green synthesis of silver nanoparticles using *Artocarpus heterophyllus* Lam. seed extract and its antibacterial activity. Ind. Crop. Prod. **46**, 132–137 (2013)
25. Jeeva, K., Thiyagarajan, M., Elangovan, V., Geetha, N., Venkatachalam, P.: *Caesalpinia coriaria* leaf extracts mediated biosynthesis of metallic silver nanoparticles and their antibacterial activity against clinically isolated pathogens. Ind. Crop. Prod. **52**, 714–720 (2014)
26. Ashokkumar, S., Ravi, S., Kathiravan, V., Velmurugan, S.: Synthesis of silver nanoparticles using *A. indicum* leaf extract and their antibacterial activity. Spectrochim. Acta A **134**, 34–39 (2015)

27. Manikandan, R., Manikandan, B., Raman, T., Arunagirinathan, K., Prabhu, N.M., Basu, M.J., et al.: Biosynthesis of silver nanoparticles using ethanolic petals extract of *Rosa indica* and characterization of its antibacterial, anticancer and anti–inflammatory activities. Spectrochim. Acta A **138**, 120–129 (2015)
28. Rajaram, K., Aiswarya, D., Sureshkumar, P.: Green synthesis of silver nanoparticle using *Tephrosia tinctoria* and its antidiabetic activity. Mater. Lett. **138**, 251–254 (2015)
29. Velusamy, P., Das, J., Pachaiappan, R., Vaseeharan, B., Pandian, K.: Greener approach for synthesis of antibacterial silver nanoparticles using aqueous solution of neem gum (*Azadirachta indica* L.). Ind. Crop. Prod. **66**, 103–109 (2015)
30. Ahmed, M.J., Murtaza, G., Mehmood, A., Bhatti, T.M.: Green synthesis of silver nanoparticles using leaves extract of *Skimmia laureola*: characterization and antibacterial activity. Mater. Lett. **153**, 10–13 (2015)
31. Joseph, S., Mathew, B.: Microwave–assisted green synthesis of silver nanoparticles and the study on catalytic activity in the degradation of dyes. J. Mol. Liq. **204**, 184–191 (2015)
32. Kumar, P.V., Pammi, S., Kollu, P., Satyanarayana, K., Shameem, U.: Green synthesis and characterization of silver nanoparticles using *Boerhaavia diffusa* plant extract and their anti bacterial activity. Ind. Crop. Prod. **52**, 562–566 (2014)
33. Kumar, S., Singh, M., Halder, D., Mitra, A.: Mechanistic study of antibacterial activity of biologically synthesized silver nanocolloids. Colloids Surf. A **449**, 82–86 (2014)
34. Skoog, D., West, D., Holler, F., Crouch, S.: Fundamentals of analytical chemistry. 9th edn. Brooks Cole (2013)
35. Bindhu, M., Umadevi, M.: Silver and gold nanoparticles for sensor and antibacterial applications. Spectrochim. Acta A. **128**, 37–45 (2014)
36. Karuppiah, M., Rajmohan, R.: Green synthesis of silver nanoparticles using *Ixora coccinea* leaves extract. Mater. Lett. **97**, 141–143 (2013)
37. Mata, R., Bhaskaran, A., Sadras, S.R.: Green–synthesized gold nanoparticles from *Plumeria alba* flower extract to augment catalytic degradation of organic dyes and inhibit bacterial growth. Particuology. (2015). doi:10.1016/j.partic.2014.12.014
38. Naraginti, S., Sivakumar, A.: Eco–friendly synthesis of silver and gold nanoparticles with enhanced bactericidal activity and study of silver catalyzed reduction of 4–nitrophenol. Spectrochim. Acta A. **128**, 357–362 (2014)
39. Ahmed, S., Ahmad, M., Swami, B.L., Ikram, S.: A review on plants extract mediated synthesis of silver nanoparticles for antimicrobial applications: a green expertise. J. Adv. Res. (2015). doi:10.1016/j.jare.2015.02.007

16

Ionic liquid attached to colloidal silica nanoparticles: as high performance catalyst for the preparation of dihydrofurans under microwave irradiation

Javad Safaei-Ghomi[1] · Hossein Shahbazi-Alavi[1,2] · Seyed Hadi Nazemzadeh[1]

Abstract The preparation of *trans*-dihydrofurans has been achieved by a one-pot condensation reaction of 4-bromophenacyl bromide, aromatic aldehydes and 5,5-dimethyl-1,3-cyclohexanedione using bis (1(3-trimethoxysilylpropyl)-3-methyl-imidazolium) copper tetrachloride tethered to colloidal silica nanoparticles under microwave irradiation. Ionic liquid tethered to colloidal silica nanoparticles have been characterized by H NMR spectroscopy, scanning electron microscope, energy dispersive spectroscopy, thermogravimetric analysis and dynamic light scattering.

Keywords Diastereoselective synthesis · Ionic liquid · Colloidal silica nanoparticles · Heterogeneous catalysts

Introduction

Dihydrofurans exhibit biological activities including anti-influenza [1], anti-HSV-1 [2], analgesic [3], anticancer [4], and anti-Alzheimer's disease [5]. These features make dihydrofurans useful targets in organic synthesis. A number of methods have been developed for the synthesis of dihydrofurans in the presence of diverse catalysts, such as manganese(III) acetate [6], potassium carbonate [7], piperidine [8], cerium(IV) ammonium nitrate (CAN) [9],

copper(I) triflate [10], triphenylphosphine (Ph$_3$P) [11], [BMIm]OH [12], *N*-methyl imidazole [13], NaOH [14], and Et$_3$N [15]. Some of the reported procedures endure drawbacks such as long reaction times, and undesirable reaction conditions. Hence, to avoid these disadvantages, the finding of an effective method for the preparation of dihydrofurans is still favored.

In recent years, synthesis and immobilization of nanoparticles in ionic liquids (ILs) have been widely studied [16–22]. Ionic liquids can be considered as valuable key precursor compounds for catalysts [23–29]. The nature of cation–anion interactions in ambient temperature ionic liquids is an issue of increasing interest [30, 31]. The structures of 1-ethyl-3-methylimidazolium (Emim) and 1-butyl-3-methylimidazolium (Bmim) with transition metal chloride anions including $NiCl_4^{2-}$, $CoCl_4^{2-}$, and $PdCl_4^{2-}$ were investigated [32–35]. Herein, we report the use of bis (1(3-trimethoxysilylpropyl)-3-methyl-imidazolium) copper tetrachloride tethered to colloidal silica nanoparticles as an efficient catalyst for the diastereoselective synthesis of dihydrofurans by reaction of 4-romophenacyl bromide, aromatic aldehydes and 5,5-dimethyl-1,3-cyclohexanedione under microwave irradiation (Scheme 1).

Experimental

Materials and characterization

DLS was carried out using a Malvern Zetasizer Nano-S. The thermogravimetric analysis (TGA) curves are recorded using a V5.1A DUPONT 2000. To study the morphology and particle size of NPs, FE-SEM analysis and EDS spectrum of the products was visualized by a Sigma ZEISS.

✉ Javad Safaei-Ghomi
safaei@kashanu.ac.ir

[1] Department of Organic Chemistry, Faculty of Chemistry, University of Kashan, P.O. Box 87317-51167, Kashan, Islamic Republic of Iran

[2] Young Researchers and Elite Club, Islamic Azad University, Kashan Branch, Kashan, Islamic Republic of Iran

Scheme 1 The preparation of dihydrofurans

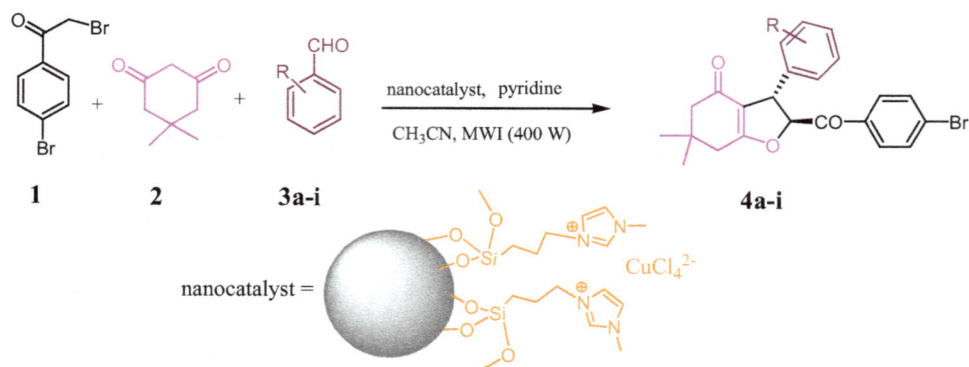

Table 1 Optimization of reaction condition using different catalysts

Entry	Solvent (MWI)	Catalyst (amount)	Time (min)	Yield[a] (%)
1	CH$_3$CN (500 W)	–	30	12
2	EtOH (500 W)	Piperidine (10 mol %)	25	52
3	CH$_3$CN (400 W)	Piperidine (10 mol %)	25	60
4	CH$_3$CN (500 W)	ZrO$_2$ (3 mol %)	25	30
5	CH$_3$CN (500 W)	p-TSA (10 mol %)	30	25
6	CH$_3$CN (500 W)	InCl$_3$ (5 mol %)	30	20
7	CH$_3$CN (400 W)	Et$_3$N (10 mol %)	25	62
8	CH$_3$CN (400 W)	Nano-ZnO (8)	25	38
9	CH$_3$CN (400 W)	IL/nano-colloidal silica (12 mg)	15	86
10	CH$_3$CN (400 W)	IL/nano-colloidal silica (14 mg)	15	90
11	CH$_3$CN (400 W)	IL/nano-colloidal silica (16 mg)	15	90
12	CH$_3$CN (300 W)	IL/nano-colloidal silica (14 mg)	15	84
13	CH$_3$CN (500 W)	IL/nano-colloidal silica (14 mg)	15	90
14	EtOH (400 W)	IL/nano-colloidal silica (16 mg)	15	74
15	DMF (400 W)	IL/nano-colloidal silica (16 mg)	15	66
16	H$_2$O (500 W)	IL/nano-colloidal silica (16 mg)	15	52
17	CH$_2$Cl$_2$ (500 W)	IL/nano-colloidal silica (16 mg)	15	35

Reaction conditions: a mixture of 4-bromophenacyl bromide (1 mmol), 5,5-dimethyl-1,3-cyclohexanedione (1 mmol), benzaldehyde 1 (R=H) (1 mmol), in the presence of pyridine (1 mmol) and catalyst for various times

[a] Isolated yield

Preparation of ionic liquid/nano-colloidal silica

In a typical procedure, 0.098 mL of colloidal silica nanoparticles (LUDOX SM colloidal silica 30 wt% suspension in H$_2$O) was diluted in 3 mL of deionized water, and 1.8 mmol of 1-(3-trimethoxysilylpropyl)-3-methylimidazolium chloride (IL) was added slowly with continuous stirring during 1 h. Then, 0.15 g of CuCl$_2$·2H$_2$O was added and refluxed for 24 h. After 24 h, IL functionalized silica nanoparticles was separated by centrifugation and washed with acetone and methanol for four times, then IL/Cu^{2+}/SiO$_2$ was dried by lyophilization/freeze-drying. The purity of the resultant IL/Cu^{2+}/SiO$_2$ was confirmed using ^{1}H NMR spectrum. The Cu loading was measured using XRF to be 4.7 wt%.

General procedure for the preparation of trans-2,3-dihydrofuran (4a–i)

1 mmol of pyridine and 1 mmol of 4-bromophenacyl bromide were stirred for 2 min. Then, aldehyde (1 mmol), 5,5-dimethyl-1,3-cyclohexanedione (1 mmol) and bis (1(3-trimethoxysilylpropyl)-3-methyl imidazolium) copper tetrachloride tethered to silica nanoparticles (nanocatalyst) (14 mg) in 15 mL of acetonitrile were added and the mixture was irradiated in microwave oven at 50 °C and 400 W (Table 1). After ending of the reaction (TLC), CHCl$_3$ was added. The catalyst was insoluble in CHCl$_3$ and it could therefore be recycled by an easy filtration. The solvent was evaporated and the solid obtained recrystallized from ethanol to afford the trans-2,3-dihydrofuran.

Fig. 1 a ^{1}H NMR spectrum of 1(3-trimethoxysilylpropyl)-3-methyl-imidazolium chloride and **b** bis (1(3-trimethoxysilylpropyl)-3-methyl-imidazolium) copper tetrachloride tethered to silica nanoparticles (nanocatalyst) in dimethyl sulfoxide (DMSO)

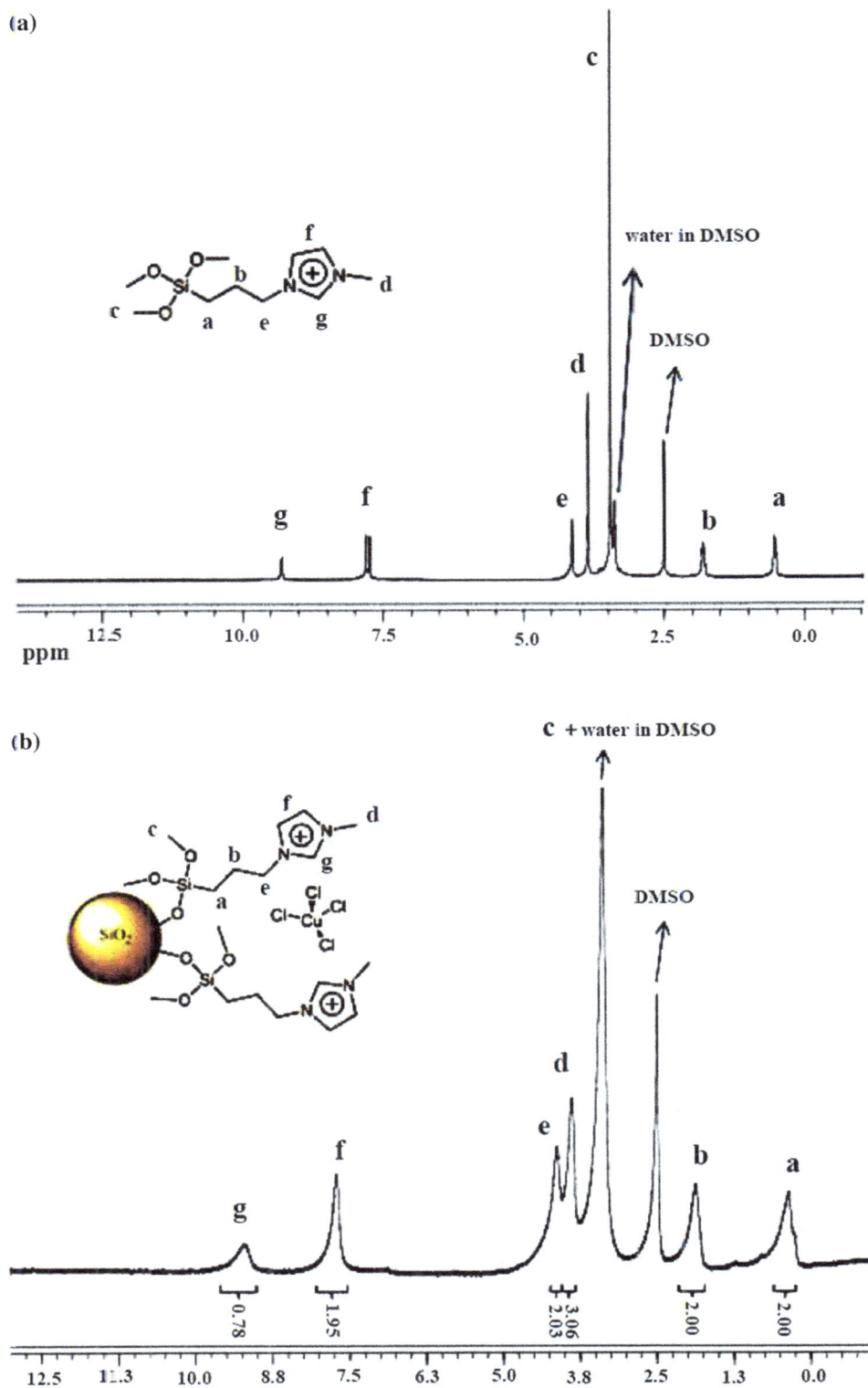

Fig. 1 a ^{1}H NMR spectrum of 1(3-trimethoxysilylpropyl)-3-methyl-imidazolium chloride and **b** bis (1(3-trimethoxysilylpropyl)-3-methyl-imidazolium) copper tetrachloride tethered to silica nanoparticles (nanocatalyst) in dimethyl sulfoxide (DMSO)

trans-2-(4'-bromobenzoyl)-3-phenyl-6,6-dimethyl-3,5,6, 7-tetrahydro-2H-benzofuran-4-one (**4a**): White solid, m.p. 152–154 °C, IR (KBr): v_{max} = 3056.2, 3032.5, 2966.7, 2894.2, 1697.5, 1634.6, 1586.3, 1224.7 cm^{-1}; ^{1}H NMR (400 MHz, CDCl$_3$): δ 1.123 (s, 3H, CH$_3$), 1.179 (s, 3H, CH$_3$), 2.219 (ABq, J = 16.0 Hz, 2H), 2.338 (ABq, J = 18.0 Hz, 2H), 4.702 (d, J = 4.8 Hz, 1H), 5.625 (d, J = 4.8 Hz, 1H), 7.125 (m, 3H, ArH), 7.227 (m, 2H, ArH),

7.613 (d, $J = 8.0$ Hz, 2H, ArH), 7.684 (d, $J = 8.0$ Hz, 2H, ArH) ppm; ^{13}C NMR (100 MHz, CDCl$_3$): δ 28.9, 29.6, 34.8, 37.8, 49.4, 51.5, 91.5, 115.4, 126.5, 127.4, 127.7, 129.5, 133.6, 135.5, 136.6, 141.5, 176.7, 192.5, 194.4 ppm; Anal. Calcd. for C$_{23}$H$_{21}$BrO$_3$: C, 64.95; H, 4.98; Found C, 64.76; H, 4.82.

trans-2-(4'-bromobenzoyl)-3-(4-chloro-phenyl)-6,6-dimethyl-3,5,6,7-tetrahydro-2H-benzo furan-4-one (**4b**): White solid, m.p. 194–196 °C, IR (KBr): $\nu_{max} = 3055.4$, 3027.7, 2956.2, 2883.5, 1694.7, 1646.3, 1578.5, 1235.2 cm^{-1}; ^{1}H NMR (400 MHz, CDCl$_3$): δ 1.163 (s, 3H, CH$_3$), 1.247 (s, 3H, CH$_3$), 2.235 (ABq, $J = 18.0$ Hz, 2H), 2.254 (ABq, $J = 20.0$ Hz, 2H), 4.406 (d, $J = 4.4$ Hz, 1H), 5.765 (d, $J = 4.4$ Hz, 1H), 7.173 (d, $J = 8.4$ Hz, 2H), 7.327 (d, $J = 8.4$ Hz, 2H), 7.614 (d, $J = 8.0$ Hz, 2H), 7.685 (d, $J = 8.0$ Hz, 2H) ppm; ^{13}C NMR (100 MHz, CDCl$_3$): δ 28.9, 29.6, 34.4, 37.8, 49.2, 51.5, 89.7, 115.8, 127.5, 128.3, 128.8, 130.2, 131.7, 133.5, 136.3, 137.5, 176.8, 193.3, 193.8 ppm; Anal. Calcd. for C$_{23}$H$_{20}$BrClO$_3$: C, 60.08; H, 4.38; Found: C, 59.85; H, 4.43.

Results and discussion

Figure 1a, b exhibit the ^{1}H NMR spectra for the 1(3-trimethoxysilylpropyl)-3-methyl-imidazolium chloride and bis (1(3-trimethoxysilylpropyl)-3-methyl-imidazolium) copper tetrachloride tethered to silica nanoparticles in dimethyl sulfoxide (DMSO), respectively. The NMR spectra of both materials are consistent with expected results for untethered and silica-tethered ionic liquids.

Figure 2 shows FE-SEM image of bis (1(3-trimethoxysilylpropyl)-3-methyl-imidazolium) copper tetrachloride tethered to silica nanoparticles (nanocatalyst). It is observed that the particles are strongly aggregated and glued with very large and continuous aggregates easily observed.

To investigate the size distribution of nanocatalysts [36, 37], dynamic light scattering (DLS) measurements of the nanoparticles were showed in Fig. 3. The size distribution is centered at a value of 43.8 nm. The dispersion for DLS analysis (2.5 g of nanocatalyst at 50 mL of ethanol) was performed using an ultrasonic bath (60 W) for 30 min. This analysis is in accordance with the previous SEM picture.

The elemental compositions of the nanocatalyst were studied by energy dispersive spectroscopy (EDS). EDS confirmed the attendance of Si, O, N, C, Cl and Cu in the compound (Fig. 4).

Thermogravimetric analysis (TGA) considers the thermal stability of the ionic liquid of untethered to SiO$_2$ (pure ionic liquid) and silica-tethered ionic liquids. The curve shows a weight loss about 36.8% for ionic liquid@nano-

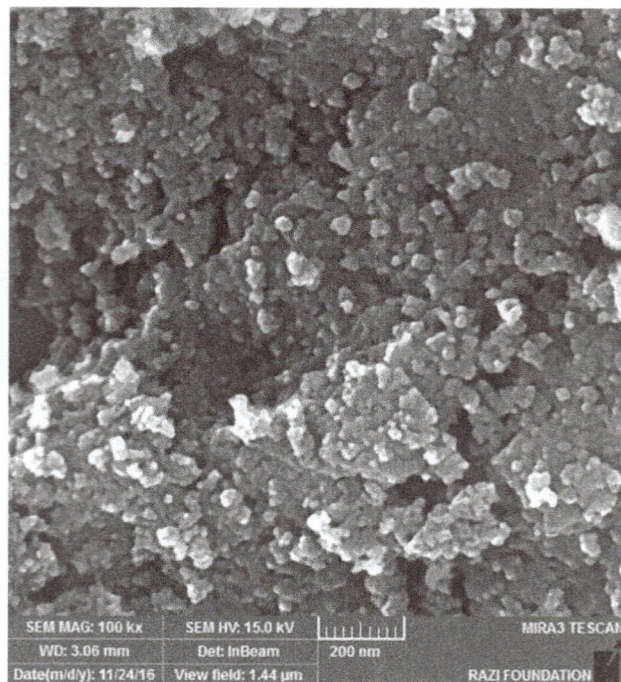

Fig. 2 FE-SEM image of nanocatalyst

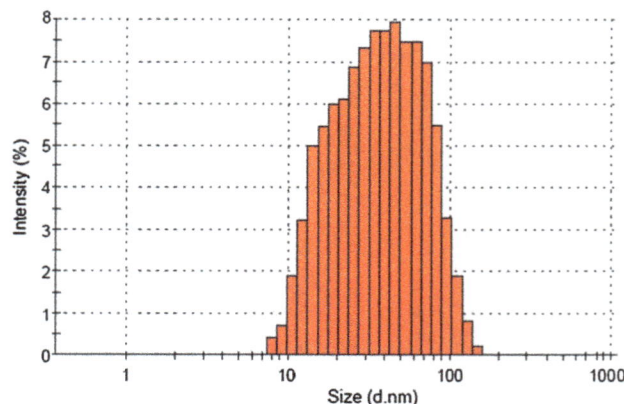

Fig. 3 DLS of nanocatalyst

colloidal silica from 240 to 550 °C, resulting from the destruction of organic spacer attaching to the nanoparticles. Hence, the nanocatalyst was stable up to 240 °C (Fig. 5).

We commenced our investigation by testing the reaction of 4-bromophenacyl bromide **1** (1 mmol), 5,5-dimethyl-1,3-cyclohexanedione **2** (1 mmol) benzaldehyde **3a** (R=H) (1 mmol) in the presence of pyridine (1 mmol) for the synthesis of dihydrofuran derivative **4a**. To obtain the ideal reaction conditions for the synthesis of compound **4a**, we studied some other catalysts, and solvents which are shown in Table 1. Screening of different catalysts such as piperidine, ZrO$_2$, *p*-TSA, InCl$_3$, Et$_3$N, nano-ZnO and IL/nano-colloidal silica revealed IL/nano-colloidal silica (14 mg) as the most effective catalyst to perform this reaction under

Fig. 4 EDS of nanocatalyst

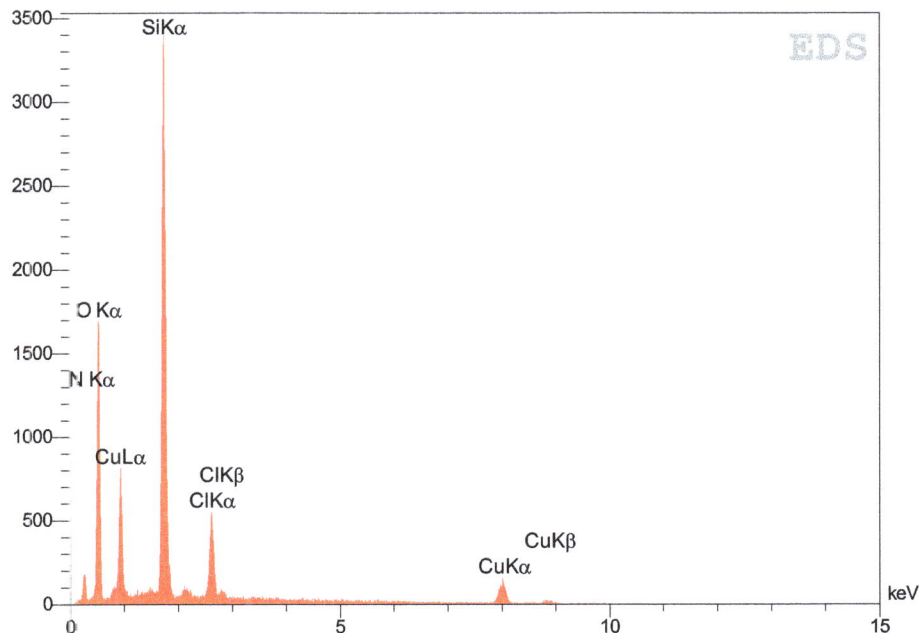

Fig. 5 TGA of IL and nanocatalyst

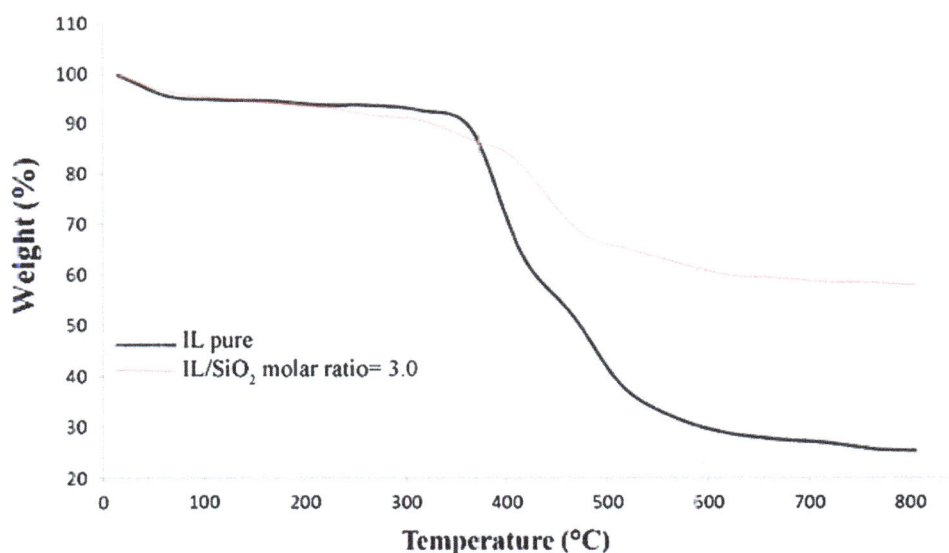

microwave irradiation (400 W) (Table 1). In this study, microwave irradiation is utilized as a rapid procedure for the synthesis of dihydrofurans. The reaction is heated from the inside and the microwave energy is transmitted straightly to the substrates and catalyst [38, 39]. The nanocatalysts absorb microwave irradiation, hence they can utilize as an internal heat origin for the reactions.

We explored the feasibility of the reaction by choosing some representative substrates (Table 2). It has been considered that better yields are achieved with substrates having electron-withdrawing groups.

We also considered reusability of the IL/nano-colloidal silica as catalyst under microwave irradiation in acetonitrile for the synthesis of product **4a** and it was found that product yields reduced to a small extent on each reuse (run 1, 90%; run 2, 90%; run 3, 39%; run 4, 89%; run 5, 88%).

A mechanism for the preparation of dihydrofurans **4a–i** using IL/nano-colloidal silica is proposed in Scheme 2. The reaction occurs via a Knoevenagel condensation between arylaldehydes **3a–i** and 5,5-dimethyl-1,3-cyclo-hexanedione **2**, forming the intermediate **I** on the active sites of IL/nano-colloidal silica. Afterwards, the Michael addition of pyridinium ylide with enone **I** affords the zwitterionic intermediate that undergoes cyclization to the title product. This mechanism is supported by literatures

Table 2 Yields of a series of dihydrofurans derivatives **4a–i** (R = various)

Entry	R	Product	Time (min)	Yield[a] (%)	m.p./°C found
1	H	**4a**	15	90	152–154
2	p-Cl	**4b**	15	94	194–196
3	o-Cl	**4c**	15	92	174—176
4	m-Cl	**4d**	15	91	155–157
5	p-CH$_3$	**4e**	20	85	182–185
6	p-OCH$_3$	**4f**	20	80	175–177
7	m-NO$_2$	**4g**	15	92	204–206
8	p-NO$_2$	**4h**	20	95	215–217
9	p-Br	**4i**	20	94	135–137

Reaction conditions: a mixture of 4-bromophenacyl bromide **1** (1 mmol), 5,5-dimethyl-1,3-cyclohexanedione **2** (1 mmol), arylaldehyde **3a–i** (R = various) (1 mmol), in the presence of pyridine (1 mmol), and IL/nano-colloidal silica (14 mg) under microwave irradiation (400 W) in acetonitrile (15 mL)

[a] Isolated yield

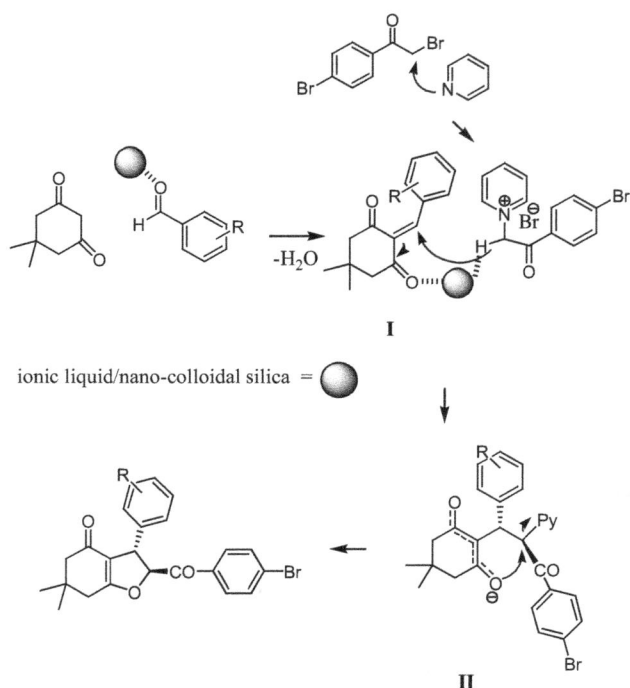

ionic liquid/nano-colloidal silica =

Scheme 2 Possible mechanism for the synthesis of *trans*-2,3-dihydrofuran using IL/nano-colloidal silica

[12, 13, 15]. In this mechanism, the cation of ionic liquid activates the C=O group for better reaction with nucleophiles. Also, anion of ionic liquid acts as a weak base to promote removal of acidic protons.

Conclusion

In this study, we described the preparation of *trans*-2,3-dihydrofuran using ionic liquid attached to colloidal silica nanoparticles as a reusable and efficient catalyst. The

advantages of this method are reusability of the catalyst, diastereoselective synthesis and utilizing of microwave as clean procedure.

Acknowledgements The authors are grateful to the University of Kashan for supporting this work under Grant No. 159148/XII.

References

1. Matsuya, Y., Sasaki, K., Ochiai, H., Nemoto, H.: Synthesis and biological evaluation of dihydrofuran-fused perhydrophenanthrenes as a new anti-influenza agent having novel structural characteristic. Bioorg. Med. Chem. **15**, 424–432 (2007)
2. Scala, A., Cordaro, M., Risitano, F., Colao, I., Venuti, A., Sciortino, M.T., Primerano, P., Grassi, G.: Diastereoselective multicomponent synthesis and anti-HSV-1 evaluation of dihydrofuran-fused derivatives. Mol. Divers. **16**, 325–333 (2012)
3. Salat, K., Moniczewski, A., Salat, R., Janaszek, M., Filipek, B., Malawska, B., Wieckowski, K.: Analgesic, anticonvulsant and antioxidant activities of 3-[4-(3-trifluoromethylphenyl)-piperazin-1-yl]-dihydrofuran-2-one dihydrochloride in mice. Pharmacol. Biochem. Behav. **101**, 138–147 (2012)
4. Zhang, Y., Zhong, H., Wang, T., Geng, D., Zhang, M., Li, K.: Synthesis of novel 2,5 dihydrofuran derivatives and evaluation of their anticancer activity. Eur. J. Med. Chem. **48**, 69–80 (2012)
5. Sugimoto, K., Tamura, K., Tohda, C., Toyooka, N., Nemoto, H., Matsuya, Y.: Structure–activity relationship studies on dihydrofuran-fused perhydrophenanthrenes as an anti-Alzheimer's disease agent. Bioorg. Med. Chem. **21**, 4459–4471 (2013)
6. Garzino, F., Meou, A., Brun, P.: Asymmetric Mn(III)-based radical synthesis of functionalized 2,3-dihydrofurans. Tetrahedron Lett. **41**, 9803–9807 (2000)
7. Cao, W., Ding, W., Chen, J., Chen, Y., Zang, Q., Chen, G.: A highly stereoselective synthesis of 2,3,4,5-tetrasubstituted-trans-2,3-dihydrofurans. Synth. Commun. **34**, 1599–1608 (2004)
8. Xing, C., Zhu, S.: Unexpected formation of tetrasubstituted 2,3-dihydrofurans from the reactions of β-Keto polyfluoroalkanesulfones with aldehydes. J. Org. Chem. **69**, 6486–6488 (2004)
9. Maiti, S., Perumal, P.T., Menendez, J.C.: CAN-promoted, diastereoselective synthesis of fused 2,3-dihydrofurans and their

transformation into tetrahydroindoles. Tetrahedron **66**, 9512–9518 (2010)

10. Son, S., Fu, G.C.: Copper-catalyzed asymmetric [4 + 1] cycloadditions of enones with diazo compounds to form dihydrofurans. J. Am. Chem. Soc. **129**, 1046–1047 (2007)

11. Yavari, I., Amiri, R., Haghdadi, M.: Triphenylphosphine-mediated synthesis of 5-oxo-2,5-dihydrofurans through the reaction of dialkyl acetylenedicarboxylates and butane-2,3-dione. J. Chem. Res. **11**, 766–767 (2004)

12. Rajesh, S.M., Perumal, S., Menendez, J.C., Pandian, S., Murugesan, R.: Facile ionic liquid-mediated, three-component sequential reactions for the green, regio- and diastereoselective synthesis of furocoumarins. Tetrahedron **68**, 5631–5636 (2012)

13. Kumar, A., Srivastava, S., Gupta, G.: Cascade [4 + 1] annulation via more environmentally friendly nitrogen ylides in water: synthesis of bicyclic and tricyclic fused dihydrofurans. Green Chem. **14**, 3269–3272 (2012)

14. Khan, A.T., Lal, M., Basha, R.S.: Regio- and diasteroselective synthesis of trans-2,3-dihydrofuran derivatives in a aqueous medium. Synthesis **45**, 406–412 (2013)

15. Wang, Q.F., Hou, H., Hui, L., Yan, C.G.: Diastereoselective Synthesis of trans-2,3-dihydrofurans with Pyridinium ylide assisted tandem reaction. J. Org. Chem. **74**, 7403–7406 (2009)

16. Moganty, S.S., Srivastava, S., Lu, Y., Schaefer, J.L., Rizvi, S.A., Archer, L.A.: Ionic liquid-tethered nanoparticle suspensions: a novel class of ionogels. Chem. Mater. **24**, 1386–1392 (2012)

17. Carvalho, A.P.A., Soares, B.G., Livi, S.: Organically modified silica (ORMOSIL) bearing imidazolium—based ionic liquid prepared by hydrolysis/co-condensation of silane precursors: synthesis, characterization and use in epoxy networks. Eur. Polym. J. **83**, 311–322 (2016)

18. Delacroix, S., Sauvage, F., Reynaud, M., Deschamps, M., Bruyère, S., Becuwe, M., Postel, D., Tarascon, J.M., Nhien, A.N.V.: SiO2/ionic liquid hybrid nanoparticles for solid-state lithium ion conduction. Chem. Mater. **27**, 7926–7933 (2015)

19. Čepin, M., Jovanovski, V., Podlogar, M., Orel, Z.C.: Amino- and ionic liquid-functionalised nanocrystalline ZnO via silane anchoring—an antimicrobial synergy. J. Mater. Chem. B **3**, 1059–1067 (2015)

20. He, Z., Alexandridis, P.: Nanoparticles in ionic liquids: interactions and organization. Phys. Chem. Chem. Phys. **17**, 18238–18261 (2015)

21. Lu, Y., Moganty, S.S., Schaefer, J.L., Archer, L.A.: Ionic liquid–nanoparticle hybrid electrolytes. J. Mater. Chem. **22**, 4066–4072 (2012)

22. Zarezadeh-Mehrizi, M., Badiei, A., Mehrabadi, A.R.: Ionic liquid functionalized nanoporous silica for removal of anionic dye. J. Mol. Liq. **180**, 95–100 (2013)

23. Welton, T.: Ionic liquids in catalysis. Coord. Chem. Rev. **248**, 2459–2477 (2004)

24. Olivier-Bourbigou, H., Magna, L., Morvan, D.: Ionic liquids and catalysis: recent progress from knowledge to applications. Appl. Catal. A **373**, 1–56 (2010)

25. Fatemi, S.M., Foroutan, M.: Recent findings about ionic liquids mixtures obtained by molecular dynamics simulation. J. Nanostruct. Chem. **5**, 243–253 (2015)

26. Shen, J.C., Jin, R.Z., Yuan, K., Zhang, M.M., Wang, X.S.: A green synthesis of fused polycyclic 5H-chromeno[3,2-c]quinoline-6,8(7H,9H)-dione derivatives catalyzed by TsOH in ionic liquids. Polycycl. Aromat. Compd. **36**, 758–772 (2016)

27. Valizadeh, H., Amiri, M., Shomali, A., Hosseinzadeh, F.: Ionic liquid 1-(3-trimethoxysilylpropyl)-3-methylimidazolium nitrite as a new reagent for the efficient diazotization of aniline derivatives and In Situ synthesis of azo dyes. J. Iran. Chem. Soc. **8**, 495–501 (2011)

28. Khedkar, M.V., Sasaki, T., Bhanage, B.M.: Immobilized Palladium metal-containing ionic liquid-catalyzed alkoxycarbonylation, phenoxycarbonylation, and aminocarbonylation reactions. ACS Catal. **3**, 287–293 (2013)

29. Jung, J.Y., Kim, J.B., Taher, A., Jin, M.J.: Pd(OAc)2 immobilized on Fe3O4 as magnetically separable heterogeneous catalyst for Suzuki reaction in water. Bull. Korean Chem. Soc. **30**, 3082–3084 (2009)

30. Safaei-Ghomi, J., Sadeghzadeh, R., Shahbazi-Alavi, H.: A pseudo six-component process for the synthesis of tetrahydrodipyrazolo pyridines using an ionic liquid immobilized on a FeNi3 nanocatalyst. RSC Adv. **6**, 33676–33685 (2016)

31. Armand, M., Endres, F., MacFarlane, D.R., Ohno, H., Scrosati, B.: Ionic-liquid materials for the electrochemical challenges of the future. Nat. Mater. **8**, 621–629 (2009)

32. Dullius, J.E.L., Suarez, P.A.Z., Einloft, S., de Souza, R.F., Dupont, J.: Selective catalytic hydrodimerization of 1,3-butadiene by palladium compounds dissolved in ionic liquids. Organometallics **17**, 815–819 (1998)

33. Angell, C.A., Byrne, N., Belieres, J.P.: Parallel developments in aprotic and protic ionic liquids: physical chemistry and applications. Acc. Chem. Res. **40**, 1228–1236 (2007)

34. Zhong, C., Sasaki, T., Tada, M., Iwasawa, Y.: Ni ion-containing ionic liquid salt and Ni ion-containing immobilized ionic liquid on silica: application to Suzuki cross-coupling reactions between chloroarenes and arylboronic acids. J. Catal. **242**, 357–364 (2006)

35. Carmichael, A.J., Hardacre, C., Holbrey, J.D., Nieuwenhuyzen, M., Seddon, K.R.: A method for studying the structure of low-temperature ionic liquids by XAFS Anal. Chem. **71**, 4572–4574 (1999)

36. Bootz, A., Vogel, V., Schubert, D., Kreuter, J.: Comparison of scanning electron microscopy, dynamic light scattering and analytical ultracentrifugation for the sizing of poly(butylcyanoacrylate) nanoparticles. Eur. J. Pharm. Biopharm. **57**, 369–375 (2004)

37. Na, K., Zhang, Q., Somorjai, G.A.: Colloidal metal nanocatalysts: synthesis, characterization, and catalytic applications. J. Clust. Sci. **25**, 83–114 (2014)

38. Bruckmann, A., Krebs, A., Bolm, C.: Organocatalytic reactions: effects of ball milling, microwave and ultrasound irradiation. Green Chem. **10**, 1131–1141 (2008)

39. Mohaqeq, M., Safaei-Ghomi, J., Shahbazi-Alavi, H., Teymuri, R.: ZnAl2O4 nanoparticles as efficient and reusable heterogeneous catalyst for the synthesis of 12-phenyl-8,12-dihydro-8,10-dimethyl-9H-naphtho[1′,2′:5,6] pyrano[2,3-d]pyrimidine-9,11-(10H)-diones under microwave irradiation. Polycycl. Aromat. Compd. **37**, 52–62 (2017)

17

The use of palladium nanoparticles supported on active carbon for synthesis of disproportionate rosin (DPR)

Ramin Mostafalu[1] · Akbar Heydari[1] · Abbas Banaei[2] · Fatemeh Ghorbani[2] · Marzban Arefi[1]

Abstract Disproportionate rosin (DPR) is a mixture of rosin acids with dehydro-abietic acid as its major component. Alkaline salts of DPR are used as emulsifier surfactant in emulsion polymerization reactions. In this work, synthesis of DPR by the use of palladium nanoparticles loaded on activated carbon was studied. The nanocatalyst was characterized by TEM, SEM, XRD, N_2 adsorption–desorption and AAS. The reusability of the prepared nanocatalyst was successfully examined three times with only a very slight loss of catalytic activity.

Keywords Disproportionated rosin · DPR · Gum rosin · Palladium nanoparticles · Palladium–carbon

Introduction

Disproportionated rosin (DPR) is a mixture of rosin acids with dehydro-abietic acid (DAA) (**2a**) as its major component [1]. Because of its low brittleness, high thermal stability, good oxidation resistance and light color, DPR is widely used in the production of butadiene and chloroprene rubber [2]. Alkaline salts of DPR are used as emulsifier surfactant in emulsion polymerization reactions. This reaction is used for polystyrene and styrene butadiene rubber (SBR) preparation in petrochemical industries [3, 4]. The main constituent of the DPR (i.e., DAA) is also of potential value in the pharmaceutical industry and a number of its derivatives have many biological activities such as anti-cancer effects [5–7].

Disproportionation of rosin is described as a hydrogen exchange between molecules of resin [8, 9]. The main product of the reaction is DAA, and reaction can be viewed as the conversion of abietic acid (AA) to DAA (Fig. 1). At temperatures between 250 and 280 °C, the reaction is very slow and the addition of a catalyst increases the reaction rate.

Iodide, sulfur, lithium and iron salt, have traditionally been used to promote or catalyze disproportionation of rosin [9–11]. At the present time, palladium catalysts have been of great interest, in part because of the good properties that the DPR feature regarding light color, low odor, medium softening-point, and excellent resistance to oxidation [1, 2, 12–15]. In the continuation of our research [16–21], the disproportionation of rosin with Pd nanoparticles supported on activated carbon (AC) was studied in this work.

Experimental section

Materials and methods

The reagents were purchased from the Merck, Sigma-Aldrich and Daejung companies and were used without further purification. The starting gum rosin was commercially available. For TEM studies, samples were placed on copper grids covered with carbon film and examined with a

---

✉ Akbar Heydari
  heydar_a@modares.ac.ir

[1] Chemistry Department, Tarbiat Modares University, P.O. Box 14155-4838, Tehran, Iran

[2] Research and Development, Padideh Shimi Jam Co., Eshtehard Industrial Town, Karaj, Iran

Fig. 1 Disproportionation of rosin

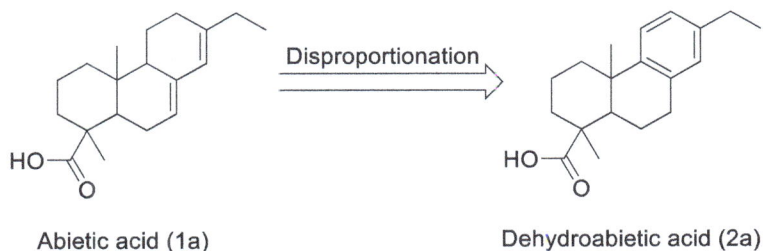

Abietic acid (1a)   Dehydroabietic acid (2a)

300 keV transmission electron microscope (TEM) JEM-3010 UHR (Jeol Ltd., Japan), equipped with a retractable high-resolution slow scan CCD-Camera (Gatan Inc., USA) with GOS phosphorous scintillator and lanthanum hexaboride cathode as the electron source. The X-ray powder patterns were recorded with a D8 ADVANCE (Bruker, Germany) diffractometer (CuK-radiation). Pd atomic absorption spectroscopy (AAS) was performed on an Atomic Absorption Spectrometer Varian SpectrAA 110. Prior to analysis, the sample was added to hydrochloric acid and H_2O_2 and the reaction was carried out for 180 min at 90 °C. The solutions were then diluted, and analyzed by AAS.

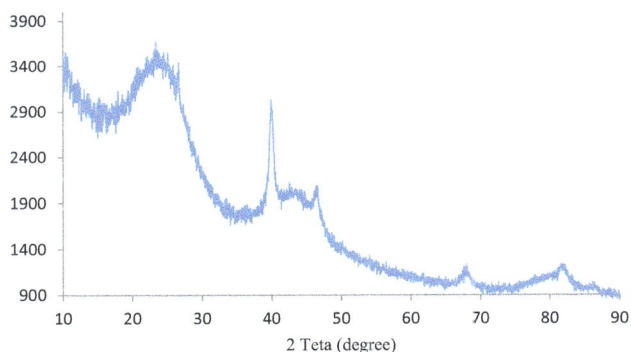

Fig. 2 XRD pattern of Pd-NP-AC catalyst

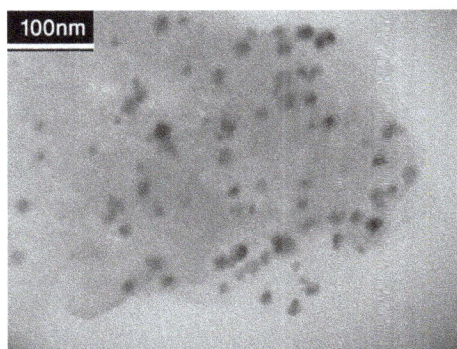

Fig. 3 TEM image of Pd-NP-AC catalyst

Preparation of palladium chloride

Palladium chloride is prepared by dissolving 1 g palladium metal in 4 ml freshly prepared aqua regia (mixture of nitric acid (99%) and hydrochloric acid (37%) optimally in a volume ratio of (1:3) for 4 h at 80 °C. After 4 h a blood-red solution was obtained.

Preparation of palladium nanoparticles loaded on activated carbon (Pd-NP-AC)

Pd nanoparticles were loaded on AC through a liquid phase reduction method. 19 g AC was suspended in 50 ml of water. Prepared palladium chloride was added and the reaction was continued for 2 h at 80 °C. Subsequently, the mixture was filtered in vacuum and rinsed using Millipore water. Prepared active carbon–palladium chloride was suspended in 60 ml of water. The pH of solution was adjusted to 9 by the use of NaOH and the suspension was stirred for 2 h. After 2 h, 60 ml of formalin (37%) reductant was added dropwise to the solution. The obtained suspension was magnetically stirred for two additional hours at 80 °C. Subsequently, the mixture was filtered in vacuum and rinsed using Millipore water. The resultant product was dried in a furnace at 105 °C overnight. The final amount of Pd loaded in sample was determined by atomic absorption.

Disproportionation of rosin by Pd-NP-AC catalyst

100 g of rosin were inserted into a 250-ml three-neck round-bottom flask equipped with a mechanical stirrer, temperature sensor and condenser. The reaction was run at 280 °C in an N_2 atmosphere to avoid oxidation. Once the reaction temperature was reached, the Pd-NP-AC catalyst (0.05% w/w) was added to the reaction. At the reaction temperature, a zero time sample was withdrawn before the catalyst was added, and more samples were taken during the first 6 h following the addition of the catalyst. A quantitative GC–FID analysis of the withdrawn samples was performed. Samples were methylated with tetramethylammonium hydroxide solution (10%) and analyzed on an Agilent model 7890A gas chromatograph with a flame

Fig. 4 Elemental maps of the Pd-NP-AC catalyst with carbon on the *left* and palladium on the *right*

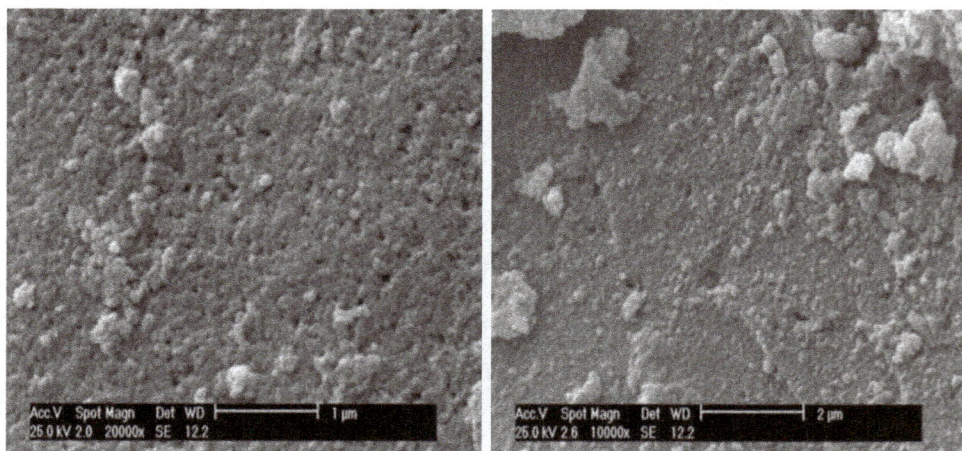

Fig. 5 SEM images of the Pd-NP-AC catalyst

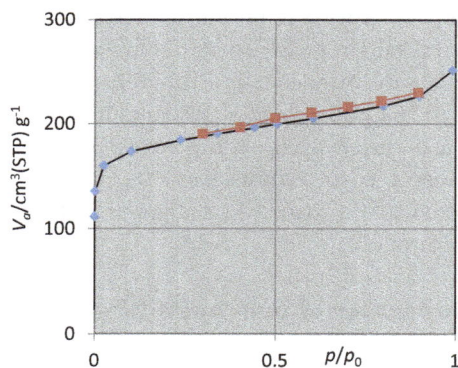

Fig. 6 Adsorption/desorption isotherm of Pd-NP-AC catalyst

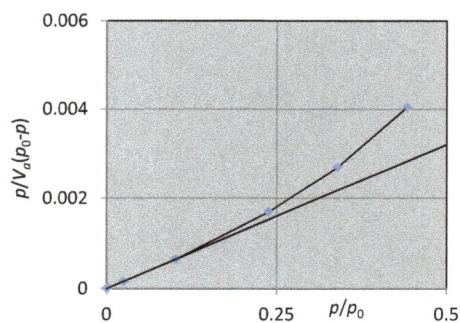

V_m	155	$[cm^3(STP)\ g^{-1}]$
$a_{s,BET}$	678	$[m^2\ g^{-1}]$
C	102830	
Total pore volume(p/p_0=0.990)	0.39	$[cm^3\ g^{-1}]$
Average pore diameter	2.29	[nm]

Fig. 7 BET plot of Pd-NP-AC catalyst

ionization detector. The instrument conditions are as follows:

Inlet: heater = 300 °C; pressure = 10.8 psi; total flow = 14 ml/min;
Septum purge: flow = 3 ml/min; split = 10:1;
Analytical column: DB 1701 (60 m);
Oven: initial = 100 °C, 5 min; ramp 1:2 °C/min; 270 °C 40 min; run time 130 min;
Detector: FID; heater = 300 °C; H_2 flow = 30 ml/min; air flow = 300 ml/min; make up flow = 40 ml/min.

Results and discussion

Pd-NP-AC was prepared by the method described by Mamlouk et al. [22] with some modifications. The structure of prepared compounds was characterized with various techniques, including TEM, BET, XRD and AAS. The XRD pattern of the Pd-NP-AC sample is shown in Fig. 2.

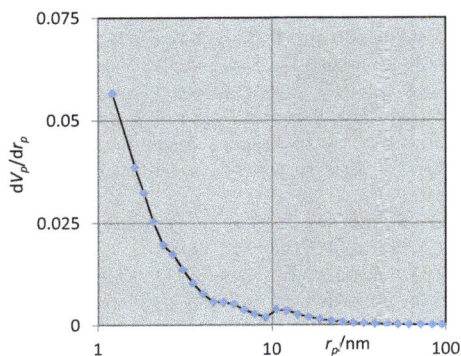

Fig. 8 BJH plot of Pd-NP-AC catalyst

Table 1 Comparison of different catalysts with our catalyst in the disproportionation of rosin

Catalyst	Wt% of catalyst	AA%	DAA%	Reaction time[a] (h)
Pd-NP-AC	0.05	0.07	70	3
FeCl$_3$	1	35	19	2
FeCl$_3$–I$_2$	1	0.42	51	6
S–I$_2$	1	2	72	6
S	1	56	24	2
KI	1	11	50	3
Fe–LiI	1	4.3	45	6
Fe–I$_3$	1	22.44	30	5

[a] They are the time it takes to reach maximum conversion. All reactions were carried out at 280 °C

Table 2 Recyclability test of Pd-NP-AC catalyst

DAA% 1st run	DAA% 2nd run	DAA% 3rd run
70	70	68

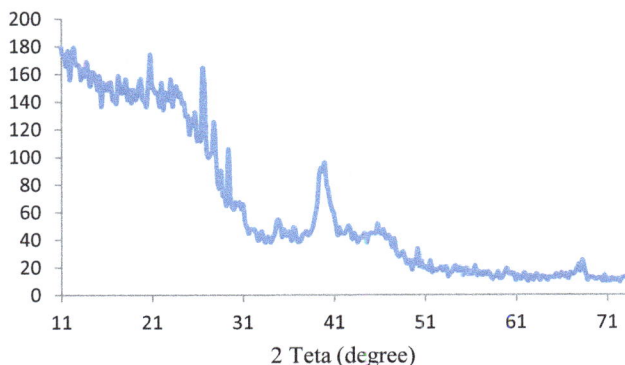

2 Teta (degree)

Fig. 9 The XRD pattern of the recovered catalyst after the third run

In the spectra the sharp and narrow peaks at $2\theta = 40°$, 46.6°, 68°, 82° and 87°, which correspond to (111), (200), (220), (311) and (222) crystalline planes of Pd, were attributed to the presence of crystalline palladium and indicating that palladium element exists in the form of Pd(0). All diffraction peaks and positions for palladium match well with those from the JCPDS card no. 05-0681. The crystallite size of palladium nanoparticles was evaluated using Scherrer equation for the (111) peak at $2\theta = 40°$ and was found to be 15 nm in size. The assignments are concordant with the Sarioglan [23], Drelinkiewicz et al. [24] and Zamani and Hossieni [25]. These crystalline palladium peaks were well separated from the broad peak of AC at around $2\theta = 26°$ which corresponds to the peak of graphite [26].

A TEM was used to obtain direct information about the structure and morphology of the palladium nanoparticles. Figure 3 shows the TEM images of the Pd-NP-AC. The mean diameter of palladium nanoparticles is about 10–45 nm with a mostly spherical shape.

Figure 4 shows the chemical maps of the Pd-NP-AC catalyst. It can be seen from the figure that the maps for palladium and carbon clearly reveal their presence in the structure of the catalyst. The amount of palladium (4.65%) of Pd-NP-AC was determined by atomic absorption analysis.

AC is a highly porous substance and has an extremely large surface area. The SEM of the Pd-NP-AC catalyst (Fig. 5) shows the porous characteristics of AC. To obtain detailed information about the pore volume, pore size distribution, and specific surface area, the N$_2$ adsorption and desorption isotherms at 77 K are performed on the samples (Fig. 6). BET indicated that surface area of the Pd-NP-AC is 678 m^2/g while total pore volume is 0.39 cm^3/g (Fig. 7). The average pore size diameter was calculated to be 2.29 nm using BJH methods (Fig. 8).

Disproportionation of rosin involves dehydrogenation, isomerisation and hydrogenation reactions of abietic-type acids so that the mixture of acids evolves to a final composition that is more stable from a thermodynamical viewpoint. The catalytic behavior of the Pd-NP-AC nanoparticles was studied for disproportionation of gum rosin and the progress of reaction was monitored by GC. In the GC spectrums, the peak at 98.6 min was from AA and the peak at 96.5 min was from DAA. The progress of reaction is monitored as an increase in DAA peak and decrease in AA peak. When the reaction was carried out with 0.1% (w/w) of catalyst at 280 °C for 5 h, DPR was obtained with a 72.4% yield. After evaluation of the catalytic efficiency, the optimization of time and temperature showed that the best result was obtained after 3 h at 280 °C using 0.05% (w/w) of catalyst in nitrogen. Under this condition, DPR was obtained with a 70% yield. A comparison with other reported efficient catalysts for the disproportion of rosin demonstrated that our present catalytic system exhibited a higher conversion and yield (see supplementary material) (Table 1).

An important issue related to solid catalysts is reusability. The reusability of the Pd-NP-AC catalyst was also studied. To recycle the catalyst, the catalyst was filtered from the reaction, washed with hot 2-propanol and dried for the next cycle at 80 °C. The Pd-NP-AC catalyst showed good stability for at least three runs in terms of DAA% (Table 2). The XRD pattern of the recovered catalyst after the 3rd run is shown in Fig. 9. In the spectra, the sharp and narrow peaks at $2\theta = 40°$, 46.6° and 68° were attributed to the presence of crystalline palladium [23–26].

Conclusions

In this paper, the catalytic disproportionation of Gum rosin over palladium nanoparticles loaded on AC was investigated. The catalyst was characterized by TEM, XRD, BET and AAS. Compared with other reported efficient catalysts in the literature, Pd-NP-AC is among the best catalysts for the disproportion of rosin.

Acknowledgements We acknowledge Tarbiat Modares University, Iran National Science Foundation (INSF) and Padideh Shimi Jam Co. for support of this work.

References

1. Souto, J.C., Yustos, P., Ladero, M., Garcia-Ochoa, F.: Disproportionation of rosin on an industrial Pd/C catalyst: reaction pathway and kinetic model discrimination. Bioresour. Technol. **102**, 3504–3511 (2011)
2. Wang, L., Chen, X., Sun, W., Liang, J., Xu, X., Tong, Z.: Kinetic model for the catalytic disproportionation of pine oleoresin over Pd/C catalyst. Ind. Crop. Prod. **49**, 1–9 (2013)
3. Mayer, M.J.J., Meuldijk, J., Thoenes, D.: Influence of disproportionated rosin acid soap on the emulsion polymerization kinetics of styrene. J. Appl. Polym. Sci. **56**, 119–126 (1995)
4. Gonzlez, M.A., Pérez-Guaita, D., Correa-Royero, J., Zapata, B., Agudelo, L., Mesa-Arango, A., Betancur-Galvis, L.: Synthesis and biological evaluation of dehydroabietic acid derivatives. Eur. J. Med. Chem. **45**, 811–816 (2010)
5. Tanaka, R., Tokuda, H., Ezaki, Y.: Cancer chemopreventive activity of rosin constituents of Pinus spez. and their derivatives in two-stage mouse skin carcinogenesis test. Phytomedicine **15**, 985–992 (2008)
6. Fonseca, T., Gigante, B., Marques, M.M., Gilchrist, T.L., De Clercq, E.: Synthesis and antiviral evaluation of benzimidazoles, quinoxalines and indoles from dehydroabietic acid. Bioorg. Med. Chem. **12**, 103–112 (2004)
7. Häkkinen, S.T., Lackman, P., Nygrén, H., Oksman-Caldentey, K.M., Maaheimo, H., Rischer, H.: Differential patterns of dehydroabietic acid biotransformation by *Nicotiana tabacum* and *Catharanthus roseus* cells. J. Biotechnol. **157**, 287–294 (2012)
8. Brites, M.J., Guerreiro, A., Gigante, B., Marcelo-Curto, M.J.: Quantitative determination of dehydroabietic acid methyl ester in disproportionated rosin. J. Chromatogr. **641**, 199–202 (1993)
9. Pinghui, Z., Zhendong, Z., Liangwu, B., Yanju, L., Dongmei, L.: Review on colorless disproportionated rosin and its catalysts. J. Bioprocess. Eng. Biorefin. **1**, 140–147 (2012)
10. Jadhav, J.: Process to Produce Disproportionate rosin Based Emulsifier for Emulsion Polymerization. US Patent 6087318 (2000)
11. Zhao, G., Rouge, B.: Method of Producing Disproportionated Rosin. US Patent 0097061 A1 (2008)
12. Song, Z.Q., Zavarin, E., Zinkel, D.F.: On the palladium-on-charcoal disproportionation of rosin. J. Wood Chem. Technol. **5**, 535–542 (1985)
13. Fleck, E.E., Palkin, S.: Catalytic isomerization of the acids of pine oleoresin and rosin. J. Am. Chem. Soc. **59**, 1593–1595 (1937)
14. Fleck, E.E., Palkin, S.: On the nature of pyroabietic acids. J. Am. Chem. Soc. **60**, 921–925 (1938)
15. Enos, H.I., Harris, G.C., Hedrich, G.W.: Rosin and rosin derivatives. In: Mark, H.F., McKetta Jr., J.J., Othmer, D.F. (eds.) Kirk–Othmer Encyclopedia of Chemical Technology, vol. 17, 2nd edn, p. 475. Wiley, New York (1968)
16. Mostafalu, R., Banaei, A., Riazi, M.H., Ghorbani, F.: A modified method for the determination of N-nitrosodiethanolamine in coconut diethanolamide using HPLC with dual-wavelength UV–Vis detector. J. Surfactants Deterg. **19**, 431–435 (2016)
17. Mostafalu, R., Banaei, A., Ghorbani, F.: An inaccuracy in the determination of cocoamidopropyl betaine by the potentiometric method. J. Surfactants Deterg. **18**, 919–922 (2015)
18. Kaboudin, B., Mostafalu, R., Yokomatsu, T.: Fe₃O₄ nanoparticle-supported Cu(II)-β-cyclodextrin complex as a magnetically recoverable and reusable catalyst for the synthesis of symmetrical biaryls and 1,2,3-triazoles from aryl boronic acids. Green Chem. **15**, 2266–2274 (2014)
19. Mostafalu, R., Kaboudin, B., Kazemi, F., Yokomatsu, T.: N-arylation of amines: C–N coupling of amines with arylboronic acids using Fe₃O₄ magnetic nanoparticles-supported EDTA–Cu(II) complex in water. RSC Adv. **4**, 49273–49279 (2014)
20. Arefi, M., Heydari, A.: Transamidation of primary carboxamides, phthalimide, urea and thiourea with amines using Fe(OH)₃@-Fe₃O₄ magnetic nanoparticles as an efficient recyclable catalyst. RSC Adv. **6**, 24684–24689 (2016)
21. Arefi, M., Saberi, D., Karimi, M., Heydari, A.: Superparamagnetic Fe(OH)₃@Fe₃O₄ nanoparticles: an efficient and recoverable catalyst for tandem oxidative amidation of alcohols with amine hydrochloride salts. ACS Comb. Sci. **17**, 341–347 (2015)
22. Alvarez, G.F., Mamlouk, M., Senthil-Kumar, S.M., Scott, K.: Preparation and characterization of carbon-supported palladium nanoparticles for oxygen reduction in low temperature PEM fuel cells. J. Appl. Electrochem. **41**, 925–937 (2011)
23. Sarioglan, S.: Recovery of palladium from spent activated carbon-supported palladium catalysts. Platin. Met. Rev. **57**, 289–296 (2013)
24. Drelinkiewicz, A., Hasik, M., Kloc, M.: Pd/polyaniline as the catalysts for 2-ethylanthraquinone hydrogenation. The effect of palladium dispersion. Catal. Lett. **64**, 41–47 (2000)
25. Zamani, F., Hosseini, S.M.: Palladium nanoparticles supported on Fe₃O₄/amino acid nanocomposite: highly active magnetic catalyst for solvent-free aerobic oxidation of alcohols. Catal. Commun. **43**, 164–168 (2014)
26. Gupta, A.K., Ganeshan, K., Sekhar, K.: Adsorptive removal of water poisons from contaminated water by adsorbents. J. Hazard. Mater. **137**, 396–400 (2006)

18

Theoretical study of geometry, stability and properties of Al and AlSi nanoclusters

Ali Arab[1] · Mohaddeseh Habibzadeh[1]

Abstract Geometry, stability, and properties of Al_n ($n = 1$–13) and Al_nSi_m ($n + m = 5$–7) nanoclusters were investigated by density functional theory. We found that while geometry of some clusters change significantly by substituting of Al atom(s) with Si atom(s) the geometry of some others remain without significant variation. The relative stability of clusters was discussed on the basis of binding energy per atom, fragmentation energy, and second-order difference of cluster energies. Our results reveal that Al_7 is the most stable cluster among pure clusters. For Al_nSi_m clusters, it is observed that Al_2Si_3 (60 % Si), and Al_4Si_2 (33.33 % Si) are the most stable clusters. The reactivity of Al_n and Al_nSi_m nanoclusters was also investigated on the basis of chemical hardness. The most important feature of chemical hardness is its oscillating behavior as a function of atomic percentage of Si indicating that the reactivity of Al_nSi_m clusters strongly depends on the composition of cluster.

Keywords DFT · AlSi nanoclusters · Geometry · Stability · Chemical hardness

Introduction

The study of nanoclusters is an important subject of research due to their various applications in many fields [1–6]. By studying the properties of nanoclusters as a function of size, one hopes to learn how the bulk properties evolve and it can be providing important clues to the understanding of the mechanism of catalysis and other chemical properties [2, 3]. In both theoretical and experimental aspects, more studies on the clusters improve the processes of understanding their structures, electronic properties, and catalysis [4]. From experimental aspect, the development of laser vaporization technique has enabled experimentalists to produce and characterize atomic clusters of specific size and composition [2]. But synthesis of clusters is sometimes difficult because of (1) size selected clusters cannot be produced in sufficient quantities and (2) these are metastable, they would coalesce when brought in the vicinity of each other [2]. In addition, there is no experimental technique that can provide direct information on cluster geometry [2]. From theoretical aspect, on the other hand, geometry and many other properties of clusters have been successfully predicted and there are abundant reports in the literature that investigate the properties of clusters theoretically [7–11].

The aluminum nanoclusters have simple electronic structure and high electrical conductivity and, therefore, are important in catalytic processes such as hydrogenation and dehydrogenation reactions as well as fuel cell technology [12, 13]. In recent years, Al and Al-based nanoclusters are the subject of numerous experimental and theoretical studies [14–20]. The reaction of Al clusters, Al_n ($n = 7$–24), formed by laser vaporization, with oxygen and ammonia has been studied using a fast flow reactor [14]. Reber et al. [15] have shown that some of the Al clusters

✉ Ali Arab
a.arab@semnan.ac.ir

[1] Department of Chemistry, Semnan University, P.O. Box 35131-19111, Semnan, Iran

are reactive toward even less reactive hydrocarbons. The interaction between Al clusters and several molecules such as H_2O [4, 16], H_2 [17, 18] NH_2 [19] and O_2 [20] was theoretically investigated.

Real catalysts mainly consist of a heterometallic or bimetallic system, which can enhance reactivity and selectivity [13]. Therefore, in theoretical study of clusters, much attention should be paid to the study of bimetallic clusters and their application in catalytic reactions. Al clusters are the most important lightweight materials for hydrogen adsorption and binding of hydrogen to the Al clusters can be significantly improved via doping. Kumar et al. [21] investigated the most stable structures and physical properties of $Al_{12}Si$, $Al_{18}Si$, and $Al_{22}Si$ clusters. They reported that Si impurity makes these clusters electronically closed shell and leads to a large gain in the binding energy. Majumder et al. [22] studied the equilibrium geometry and energetics of Si_n and $Si_{n-1}Al$ clusters using a combination of the density functional theory and molecular dynamics simulation under the local spin density (LSD) approximation. Their results revealed that clusters with $n = 4, 6, 10$ show higher stability as compared to its neighboring clusters. Effect of aluminum impurity atoms on the structures and stabilities of neutral and ionic Si_n($n = 2$–21) clusters have been investigated using full-potential linear-muffin-tin-orbital molecular-dynamics [23]. They found that most of the ground-state structures for neutral and ionic Si_nAl ($n = 1$–20) clusters can be obtained by substituting one Si atom of their corresponding Si clusters with an Al atom.

In this study, we performed a comprehensive and systematic DFT study on the properties of Al_n($n = 1$–13) and Al_nSi_m($n + m = 5$–7) nanoclusters. Binding energy per atom, second-order difference of cluster energies, fragmentation energy, dipole moment (μ), Al–Al bond distance, and chemical hardness of nanoclusters were calculated and discussed. Variations of geometry, stability and reactivity of nanoclusters were also analyzed as a function of cluster size as well as cluster composition.

Computational methods

The geometry of all clusters including Al_n ($n = 1$–13) and Al_nSi_m ($n = 4$–6, $m = 1$–3, $n + m = 5$–7) were optimized without any constraint using the functionals of Becke' three-parameter hybrid exchange functional [24] and the Lee–Yang–Parr correlation functional [25] (B3LYP) and 6–$31 + G^*$ basis set. Effect of basis set was also studied on the electronic properties of most stable structures considering 6–$31 + G^{**}$ and 6–$311 + G^*$ basis sets. All calculations were done by the Gaussian03 package [26]. The B3LYP hybrid density functional has been used abundantly

for the study of Al clusters. For example, B3LYP functional has been used for the study of structure and electronic properties of Al_nAs clusters [7]. Guo et al. [13] employed B3LYP functional to investigate the structure of Al_nV clusters and their interaction with molecular hydrogen. Dissociation of hydrogen on small Al clusters has been studied through B3LYP functional [17]. In addition, all calculations involved the determination of vibrational frequencies at the same level of theory and basis set for the validation of the local energy minima of each optimized structure as well as zero point energy correction.

Results and discussion

Al_n ($n = 1$–13) clusters

To find the most stable structure of Al_n clusters we optimized each cluster, initiating with different possible structures as input file, at three spin multiplicities of, 1, 3, and 5 for clusters with even electrons number and 2, 4, and 6 for clusters with odd electrons number. Figure 1 presents the most stable structure of Al_n clusters while the corresponding spin multiplicities (Ms) and dipole moments (μ) are summarized in Table 1. Other higher energy structures along with the corresponding spin multiplicities as well as relative energies (E_r) are collected as supporting information in table S.1. The relative energies are calculated by subtracting the energy of the most stable structure from the energy of a structure in the series. The calculated vibrational frequencies of most stable structures of Al_n clusters computed at B3LYP/6–$31 + G^*$ level of theory are collected as supporting information in table S.2. The structures reported in Fig. 1 are in agreement with those reported in the literature [2, 4, 7, 17]. It has been shown that for Al_n clusters up to five atoms the structures are planar and for clusters containing 6–13 atoms the structures become three dimensional [2, 4, 7, 17]. For Al_6 cluster we found that the most stable structure is trigonal prism geometry with Ms = 3 while the octahedral geometry with Ms = 1 is only 0.01 eV less stable. Pino et al. [17] obtained different geometries as the most stable structure of Al_6 depending on the method used for optimization. With B3LYP they obtained trigonal prism geometry with Ms = 3 as the most stable structure while the octahedral geometry with Ms = 1 is only 0.006 eV less stable. With CCSD(T), on the other hand, they obtained octahedral geometry with Ms = 1 as the most stable structure while the trigonal prism geometry with Ms = 3 is 0.162 eV less stable. For Al_{13} cluster, we obtained distorted icosahedron geometry with Ms = 2 as the most stable structure in agreement with some reported results [2, 7]. Some authors

Fig. 1 The most stable structures of Al$_n$ ($n = 2$–13) clusters computed at B3LYP/6–31 + G* level of theory

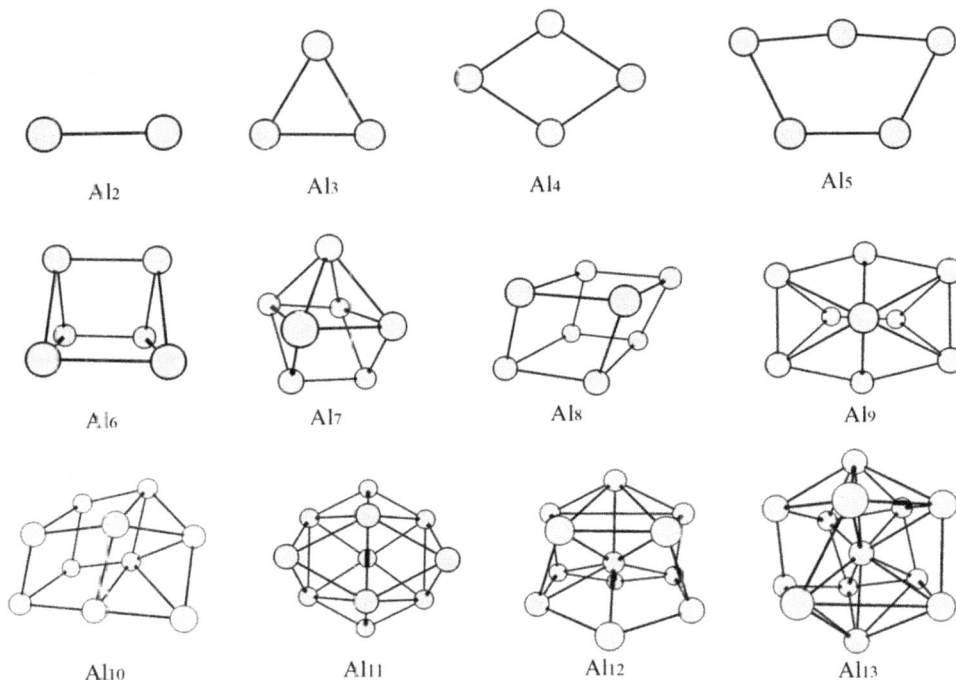

Table 1 The spin multiplicity and dipole moment of most stable Al$_n$ clusters presented in Fig. 1

Cluster	Al$_2$	Al$_3$	Al$_4$	Al$_5$	Al$_6$	Al$_7$	Al$_8$	Al$_9$	Al$_{10}$	Al$_{11}$	Al$_{12}$	Al$_{13}$
Ms	3	2	3	2	3	2	1	2	1	2	1	2
μ/Debye	0.000	0.000	0.002	0.520	0.004	0.420	0.001	0.180	0.420	0.104	0.081	0.440

reported regular icosahedron geometry (I_h symmetry) as the most stable structure for Al$_{13}$ cluster [4, 12].

According to the results of Table 1 the doublet spin state is the lowest in energy (most stable state) for all Al$_n$ clusters with odd number of atoms. For Al$_n$ clusters with even number of atoms, on the other hand, the triplet spin state is the lowest in energy for Al$_2$, Al$_4$, Al$_6$ and the singlet spin state is the lowest in energy for Al$_8$, Al$_{10}$ and Al$_{12}$. Cox et al. [27] experimentally found that the spin multiplicity of small Al clusters ($n < 10$) is 2 for clusters with odd number of atoms and 3 for clusters with even number of atoms. Interestingly, we observed that increasing the spin multiplicity of clusters to 4, 5, and 6 do not affect final structure of the clusters; however, clusters with higher spin multiplicity are very unstable energetically (in some cases more than 2 eV). Results of Table 1 show that the dipole moment of Al clusters strongly depend on the size and structural symmetry of the clusters which is in agreement with the results of references [2, 28, 29]. It has been reported that clusters with odd number of atoms are less symmetric and have larger values of dipole moment while clusters with even number of atoms are more symmetric and have smaller values of dipole moment [2]. Among Al$_n$ clusters; Al$_5$, Al$_7$, Al$_{10}$, and Al$_{13}$ have the highest values of dipole moment.

For the most stable Al$_n$ clusters we investigate some important properties as a function of cluster size. The relative stability of the Al$_n$ clusters has been investigated according to the binding energy per atom, second-order difference of cluster energies, and fragmentation energy.

The binding energy per atom of clusters was calculated according to the Eq. 1.

$$E_b = \frac{nE(\text{Al}) - E(\text{Al}_n)}{n} \qquad (1)$$

where in this equation n is the number of Al atoms in cluster, $E(\text{Al})$ is the energy of Al atom in the most stable state and $E(\text{Al}_n)$ is the energy of the most stable Al$_n$ cluster. Variation of binding energy per atom as a function of cluster size is shown in Fig. 2a. According to this figure the binding energy per atom increases monotonically to the maximum value of 2.01 eV for Al$_{13}$ cluster while is smaller than the Al bulk cohesive energy of 3.39 eV [2]. Distinctive regions which separated from each other at $n = 3$, 7, and 10 are also observed in Fig. 2a.

To gain better insight about the relative stability of clusters, it is more instructive to analyze the second-order difference of cluster energies ($\Delta_2 E$) and fragmentation

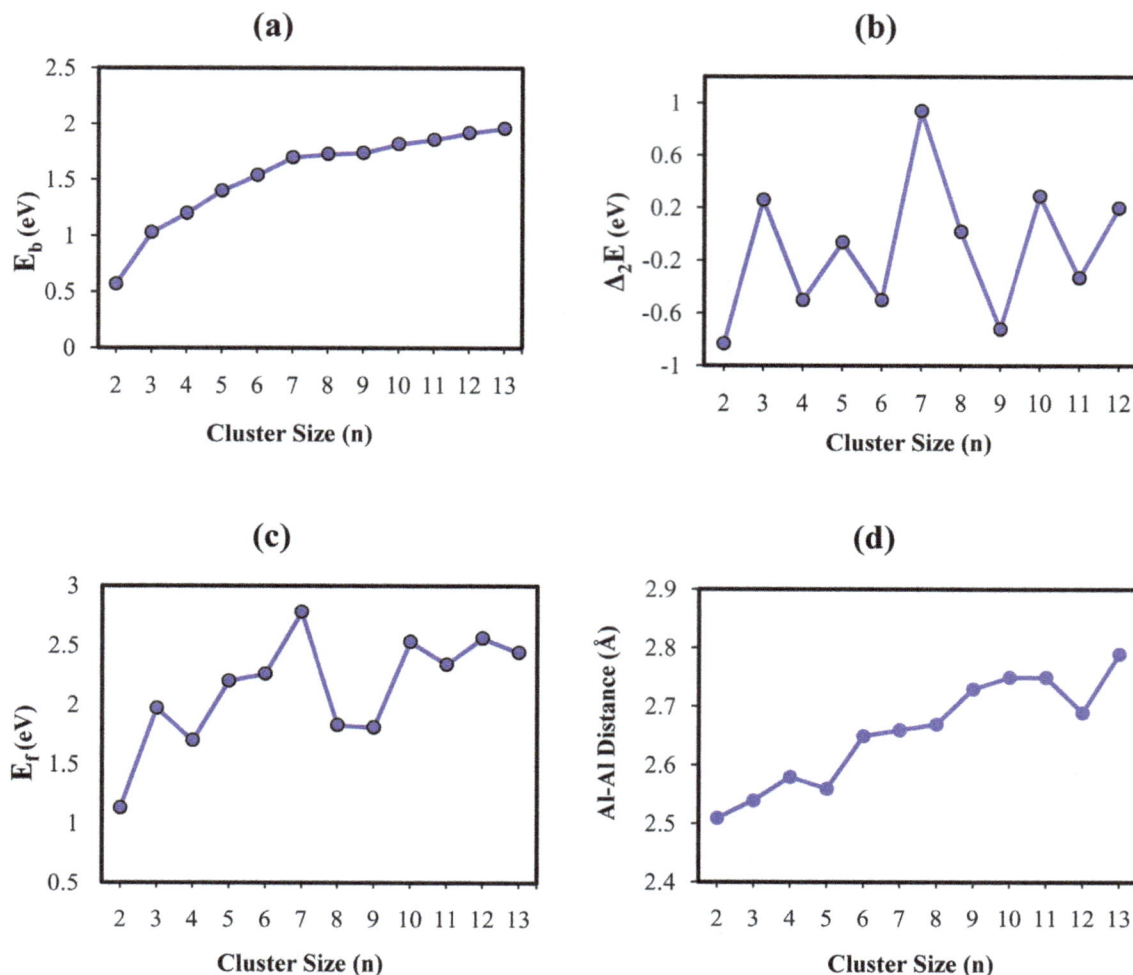

Fig. 2 Variation of binding energy per atom (**a**), second-order difference of cluster energies (**b**), fragmentation energy (**c**), and average Al–Al distance (**d**) for Al_n clusters as a function of cluster size computed at B3LYP/6–31 + G^* level of theory

energy (E_f) which were calculated according to the Eqs. 2 and 3, respectively [8].

$$\Delta_2E(Al_n) = E(Al_{n+1}) + E(Al_{n-1}) - 2E(Al_n) \qquad (2)$$

$$E_f(Al_n) = E(Al) + E(Al_{n-1}) - E(Al_n) \qquad (3)$$

where $E(Al_{n-1})$, and $E(Al_{n+1})$ are the energies of the most stable Al_{n-1}, and Al_{n+1} clusters, respectively.

Figure 2b presents variation of second-order difference of cluster energies (Δ_2E) as a function of cluster size where maxima are found at $n = 3$, 7, and 10 indicating more stability of these clusters compared to their neighbor clusters.

The fragmentation energy show the relative stability of clusters compared to the neighboring clusters. The higher value of fragmentation energy indicates stronger Al–Al interaction and, therefore, higher energy is required for evaporation of Al atom from the cluster (more stability of the cluster). According to the results of fragmentation energy (E_f) which are shown as a function of cluster size in

Fig. 2c, it is obvious that Al_3, Al_7, and Al_{10} clusters are more stable than their neighbor clusters. Overall, according to the binding energies per atom, second-order difference of cluster energies, and fragmentation energies it can be concluded that Al_7 is the most stable cluster among all Al_n clusters studied here.

Figure 2d shows variation of average Al–Al distance of the most stable clusters as a function of cluster size. As can be seen in this figure the average Al–Al distance gradually increases with cluster size while having local oscillation at Al_5, Al_8 and Al_{12} and finally approaches the bulk limit of 2.86 Å [2]. It seems that, unlike binding energy per atom, the average Al–Al distance of the most stable Al_n clusters approaches the bulk value rather rapidly with cluster size.

The reactivity of molecules can be investigated using global (chemical) hardness (η) [30]. Chemical hardness is defined as the second derivative of energy with respect to the number of electrons at constant external potential.

Using the finite difference approximation chemical hardness can be expressed as, $\eta = (IP - EA)/2$ [30], where IP and EA are the first ionization potential and electron affinity of the chemical system, respectively. Chemical hardness is resistance of a chemical entity to change in the number of electrons. Energetically speaking, hardness is one-half of the energy change for the disproportionation of a special chemical species according to the $S + S \rightarrow S^+ + S^-$ reaction [30]. Since always $IP_S \geq EA_S$, the minimum value of hardness is zero. Zero hardness indicates maximum softness and maximum softness means no energy change associated with the disproportionation reaction $(S + S \rightarrow S^+ + S^-)$. For example, a bulk metal has IP = EA $(\eta = 0)$ and maximum softness [30].

The chemical hardness was calculated approximately according to the Eq. 4 [31, 32].

$$\eta \cong \frac{E_{LUMO} - E_{HOMO}}{2} \tag{4}$$

where E_{LUMO} and E_{HOMO} are the energies of lowest unoccupied molecular orbital (LUMO) and highest occupied molecular orbital (HOMO) respectively.

Figure 3 presents variation of chemical hardness of Al_n clusters as a function of cluster size. The general trend of Fig. 3 is oscillating behavior of the chemical hardness as a function of cluster size. Such an oscillating behavior is also reported for $Zr_nO_{2n}H_2$ [33], Al–Au [34, 35], and Pd [1] clusters. According to this figure, Al_6 and Al_{13} have the maximum hardness while Al_2 and Al_9 have the minimum hardness. Therefore, it can be concluded that Al_6 and Al_{13} clusters have minimum tendency to exchange electrons (minimum reactivity) while Al_2 and Al_9 clusters have maximum tendency to exchange electrons (maximum reactivity).

Al_nSi_m clusters $(n = 4–6, m = 1–3, n + m = 5–7)$

Several possible structures for Al_nSi_m clusters as input file at different spin multiplicities (1, 3, 5 for clusters with even

electron numbers and 2, 4, 6 for clusters with odd electron numbers) were optimized at B3LYP level of theory using $6–31 + G^*$ basis set. Figure 4a–c presents the most stable structure of Al_nSi_m clusters along with the corresponding spin multiplicities. Other higher energy structures along with the corresponding spin multiplicities as well as relative energies are collected as supporting information in table S.3. The calculated vibrational frequencies of most stable structures of Al_nSi_m clusters computed at B3LYP/6-$31 + G^*$ level of theory are collected as supporting information in table S.4. Like Al_5 cluster, the Al_4Si, Al_3Si_2, and Al_2Si_3 clusters have planer structures. Therefore, it seems that substitution of Al atom(s) with Si atom(s) in Al_5 cluster do not change the geometry of theses clusters significantly. Unlike previous clusters, effect of Al atom(s) substitution with Si atom(s) on the geometry of Al_6 is significant as clearly shown in Fig. 4b. While the geometry of Al_6 and Al_5Si is trigonal prism, the geometry of Al_3Si_3 is tetragonal bipyramid. The tetragonal bipyramid geometry has been reported as the most stable structure of pure Si_6 cluster [10, 23]. Therefore, with increasing the amount of Si, the geometry of cluster changes from trigonal prism for pure Al_6 cluster toward tetragonal bipyramid for pure Si_6 cluster.

According to the Fig. 4c, substitution of Al atom(s) with Si atom(s) in Al_7 cluster has also small effect on the geometry of clusters.

Effect of cluster composition on the stability of Al_nSi_m clusters was discussed using binding energy per atom and fragmentation energy. The binding energy per atom of Al_nSi_m clusters was calculated according to the Eq. 5.

$$E_b = \frac{nE(Al) + mE(Si) - E(Al_nSi_m)}{n + m} \tag{5}$$

where in this equation m is the number of Si atoms in cluster, $E(Si)$ is the energy of Si atom in the most stable state and $E(Al_nSi_m)$ is the energy of the most stable Al_nSi_m cluster.

Variations of binding energy per atom of Al_nSi_m clusters as a function of atomic percentage of Si are shown in Fig. 5. The binding energies per atom increase linearly with adding Si to the pure Al_n clusters indicating that the stability of clusters can improve with increasing the amount of Si.

The fragmentation energy of Al_nSi_m clusters was calculated according to the Eq. 6

$$E_f(Al_nSi_m) = E(Al) + E(Al_{n-1}Si_n) - E(Al_nSi_m) \tag{6}$$

Variations of fragmentation energy of Al_nSi_m clusters as a function of atomic percentage of Si are shown in Fig. 6 where the most important feature is the oscillating behavior of E_f. The maxima appear in $m = 3$, 2, and 0 for clusters with $n + m = 5$, 6, and 7, respectively, indicating more

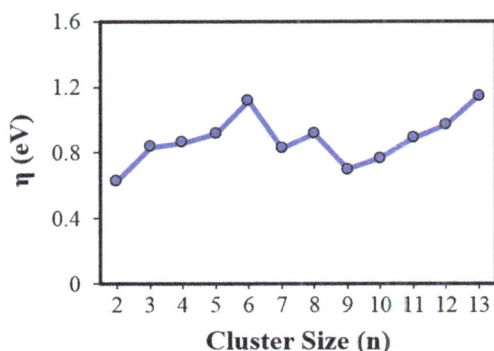

Fig. 3 Variation of chemical hardness for Al_n clusters as a function of cluster size computed at B3LYP/6–31 + G^* level of theory

Fig. 4 The most stable structure of Al_nSi_m clusters along with the corresponding spin multiplicities computed at B3LYP/6–31 + $G*$ level of theory

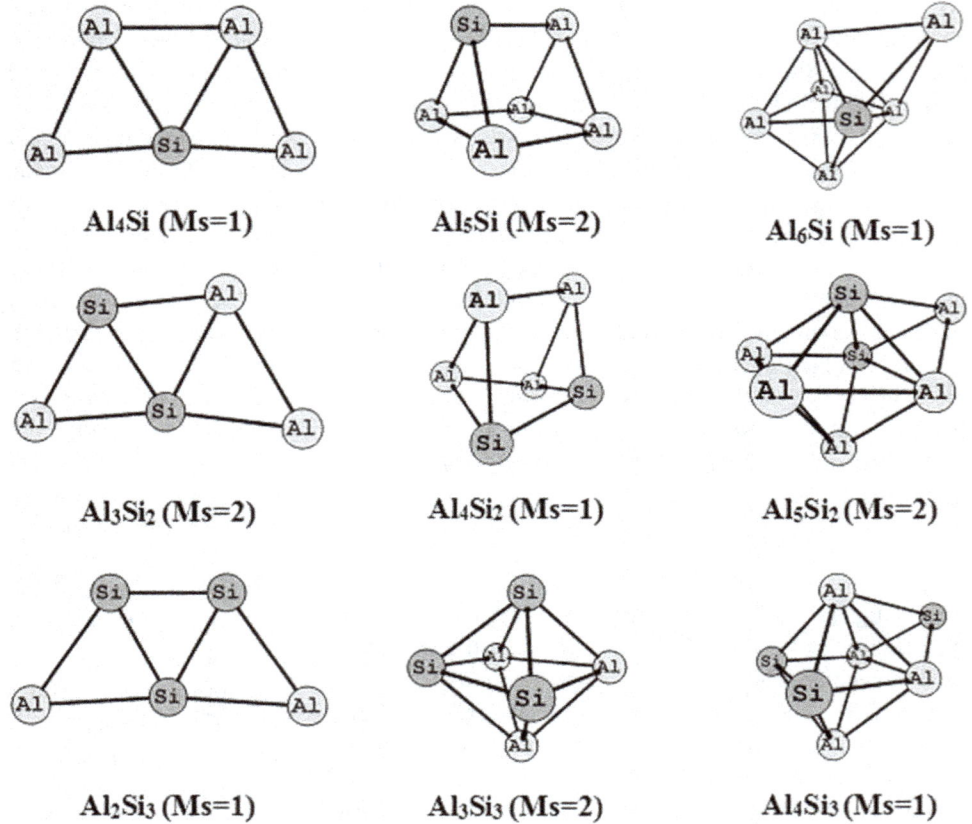

Al_4Si (Ms=1) Al_5Si (Ms=2) Al_6Si (Ms=1)

Al_3Si_2 (Ms=2) Al_4Si_2 (Ms=1) Al_5Si_2 (Ms=2)

Al_2Si_3 (Ms=1) Al_3Si_3 (Ms=2) Al_4Si_3 (Ms=1)

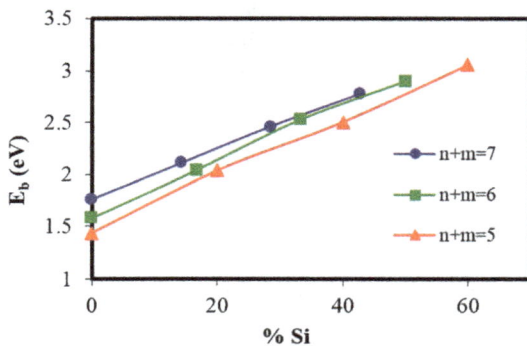

Fig. 5 Variation of binding energy per atom of Al_nSi_m clusters as a function of atomic percentage of Si computed at B3LYP/6–31 + $G*$ level of theory

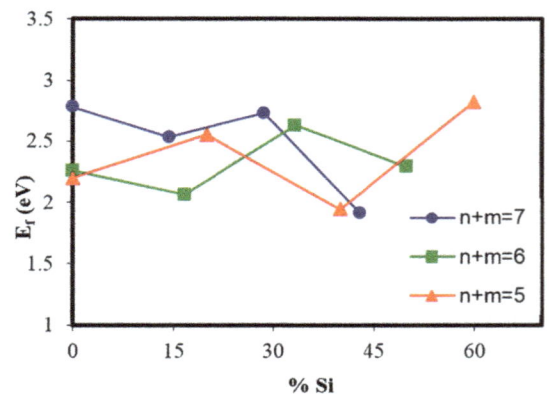

Fig. 6 Variation of fragmentation energy of Al_nSi_m clusters as a function of atomic percentage of Si computed at B3LYP/6–31 + $G*$ level of theory

stability of Al_2Si_3 (60 % Si), Al_4Si_2 (33.33 % Si), and Al_7 clusters among clusters with similar number ($n + m$) of atoms. It can be concluded that Al–Si interaction in Al_2Si_3 and Al_4Si_2 clusters is stronger than Al–Al interaction in Al_5 and Al_6 clusters, respectively, and evaporation of Al atom from Al_2Si_3 and Al_4Si_2 clusters requires more energy compared to the corresponding Al_5 and Al_6 clusters, respectively. For Al_7 cluster, on the other hand, Al–Al interaction is stronger than Al–Si interaction in Al_6Si, Al_5Si_2, and Al_4Si_3 clusters and evaporation of Al atom

from Al_7 cluster requires more energy than Al_6Si, Al_5Si_2, and Al_4Si_3 clusters.

The chemical hardness of Al_nSi_m clusters was calculated according to the Eq. 4 and results as a function of atomic percentage of Si are shown in Fig. 7. The most important feature of Fig. 7 is oscillating behavior of chemical hardness as a function of atomic percentage of Si. Therefore, some of Al_nSi_m clusters are softer than corresponding pure Al_n clusters while some others are harder. The Al_2Si_3 and

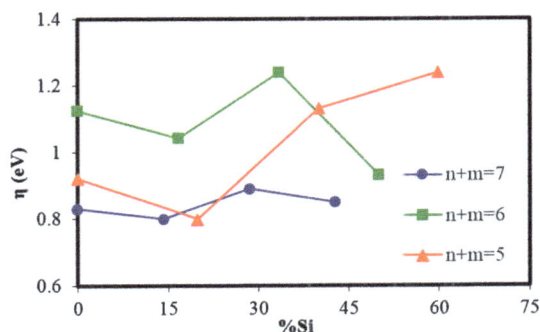

Fig. 7 Variation of chemical hardness of Al_nSi_m clusters as a function of atomic percentage of Si computed at B3LYP/6–31 + G^* level of theory

Al_4Si_2 clusters have the maximum values of chemical hardness among all clusters studied here.

Effect of basis set was investigated on the electronic properties of clusters considering 6–31 + G^*, 6–31 + G^{**} and 6–311 + G^* basis sets. The calculated results are shown in Tables 2 and 3 for Al_n and Al_nSi_m clusters, respectively. It is clear that basis set has no significant effect on the binding energy per atom, second-order difference of cluster energies and chemical hardness of clusters. Effect of basis set on the fragmentation energy of clusters is very little, but similar trend is observed for different basis set nevertheless.

Conclusions

The geometry, stability and properties of $Al_n(n = 1–13)$ and $Al_nSi_m(n + m = 5–7)$ nanoclusters at B3LYP level of theory using 6–31 + G^* basis set were investigated. The binding energy per atom of pure Al_n clusters increases monotonically to the maximum value of 2.01 eV for Al_{13} cluster while is smaller than the Al bulk cohesive energy of 3.39 eV. Analysis of binding energy per atom, second-order difference of cluster energies, and fragmentation energy reveals that Al_3, Al_7, and Al_{10} clusters are more stable than their neighbor clusters. It is observed that Al_7, Al_2Si_3 (60 % Si) and Al_4Si_2 (33.33 % Si) clusters have the maximum stability among pure and mixed clusters, respectively. Substituting of Al atom(s) with Si atom(s) change geometry of some clusters significantly while the geometry of some others remains without significant variation. Our results show that Al–Si interaction in Al_2Si_3 and Al_4Si_2 clusters is stronger than Al–Al interaction in Al_5 and Al_6 clusters, respectively, which means that evaporation of Al atom from Al_2Si_3 and Al_4Si_2 clusters requires more energy compared to the corresponding Al_5 and Al_6 clusters, respectively. For Al_7 cluster, on the other hand, Al–Al interaction is stronger than

Table 2 The calculated electronic properties of Al_n clusters at different basis sets

Cluster	E_b (eV)			E_f (eV)			Δ_2E (eV)			η (eV)		
	6-31 + G^*	6-31 + G^{**}	6-311 + G^*	6-31 + G^*	6-31 + G^{**}	6-311 + G^*	6-31 + G^*	6-31 + G^{**}	6-311 + G^*	6-31 + G^*	6-31 + G^{**}	6-311 + G^*
Al_2	0.58	0.59	0.59	1.34	1.17	1.18	-0.83	-0.84	-0.83	0.63	0.63	0.63
Al_3	1.06	1.06	1.06	1.97	2.01	2.01	0.26	0.26	0.26	0.84	0.84	0.83
Al_4	1.23	1.23	1.23	1.70	1.75	1.74	-0.50	-0.51	-0.51	0.86	0.86	0.86
Al_5	1.44	1.44	1.44	2.20	2.25	2.25	-0.06	-0.04	-0.05	0.92	0.91	0.90
Al_6	1.58	1.58	1.58	2.26	2.29	2.29	-0.51	-0.58	-0.57	1.12	1.12	1.12
Al_7	1.76	1.76	1.76	2.77	2.87	2.87	0.94	0.99	1.00	0.83	0.83	0.83
Al_8	1.78	1.78	1.78	1.83	1.87	1.89	0.02	0.01	0.01	0.92	0.92	0.92
Al_9	1.79	1.79	1.79	1.81	1.86	1.86	-0.72	-0.75	-0.74	0.68	0.68	0.69
Al_{10}	1.87	1.87	1.87	2.53	2.62	2.61	0.29	0.35	0.35	0.77	0.77	0.77
Al_{11}	1.91	1.91	1.90	2.24	2.27	2.26	-0.33	-0.39	-0.40	0.89	0.89	0.89
Al_{12}	1.97	1.97	1.97	2.57	2.66	2.66	0.12	0.16	0.14	0.97	0.97	0.97
Al_{13}	2.01	2.01	2.01	2.45	2.50	2.52	–	–	–	1.15	1.15	1.15

Table 3 The calculated electronic properties of Al$_n$Si$_m$ clusters at different basis sets

Cluster	E_b (eV)			E_f (eV)			η (eV)		
	6–31 + G*	6–31 + G**	6–311 + G*	6–31 + G*	6–31 + G**	6–311 + G*	6–31 + G*	6–31 + G**	6–311 + G*
Al$_4$Si	2.04	2.04	2.03	2.55	3.21	3.20	0.81	0.81	0.81
Al$_3$Si$_2$	2.50	2.50	2.50	1.94	1.94	1.92	1.13	1.13	1.12
Al$_2$Si$_3$	3.05	3.05	3.03	2.83	2.83	2.81	1.25	1.25	1.25
Al$_5$Si	2.04	2.04	2.04	2.06	2.06	2.07	1.04	1.04	1.04
Al$_4$Si$_2$	2.54	2.54	2.53	2.63	2.63	2.64	1.24	1.24	1.24
Al$_3$Si$_3$	2.90	2.93	2.92	2.29	2.29	2.31	0.93	0.93	0.93
Al$_6$Si	2.11	2.11	2.11	2.53	2.53	2.50	0.81	0.81	0.81
Al$_5$Si$_2$	2.46	2.46	2.45	2.73	1.96	1.95	0.89	0.89	0.89
Al$_4$Si$_3$	2.78	2.78	2.77	1.91	1.91	1.90	0.86	0.86	0.86

Al–Si interaction in Al$_6$Si, Al$_5$Si$_2$, and Al$_4$Si$_3$ clusters indicating that evaporation of Al atom from Al$_7$ cluster requires more energy than Al$_6$Si, Al$_5$Si$_2$, and Al$_4$Si$_3$ clusters. It is found that the chemical hardness of Al$_n$ and Al$_n$Si$_m$ clusters strongly depends on the cluster size as well as cluster composition.

Acknowledgments The Authors thank the Research Council of Semnan University for funding of this work.

References

1. Wen, J.Q., Xia, T., Zhou, H., Wang, J.F.: A density functional theory study of small bimetallic Pd$_n$Al (n = 1–8) clusters. J. Phys. Chem. Solids **75**, 528–534 (2014)
2. Rao, B.K., Jena, P.: Evolution of the electronic structure and properties of neutral and charged aluminum clusters: a comprehensive analysis. J. Chem. Phys. **111**, 1890–1904 (1999)
3. Kung, H.H., Kung, M.C.: Nanotechnology: applications and potentials for heterogeneous catalysis. Catal. Today **97**, 219–224 (2004)
4. Liu, Y., Hua, Y., Jiang, M., Jiang, G., Chen, J.: Theoretical study of the geometries and dissociation energies of molecular water on neutral aluminum clusters Al$_n$ (n = 2–25). J. Chem. Phys. **136**, 084703–084709 (2012)
5. Villanueva, M.S., Hernandez, A.B., Anota, E.C., Valdez, S., Cuchill, O.V.: Electronic and structural properties of Ti$_9$XO$_{20}$ (X = Ti, C, Si, Ge, Sn and Pb) clusters: a DFT study. Phys. E **65**, 120–124 (2015)
6. Arab, A., Gobal, F., Nahali, N., Nahali, M.: Electronic and structural properties of neutral, anionic, and cationic Rh$_x$Cu$_{4-x}$(x = 0–4) small clusters: a DFT study. J. Clust. Sci. **24**, 273–287 (2013)
7. Guo, L.: Density functional study of structural and electronic properties of Al$_n$As (1 ≤ n ≤ 15) clusters. J. Alloy. Compd. **527**, 197–203 (2012)
8. Feng, X.J., Luo, Y.H.: Structure and stability of Al-doped boron clusters by the density-functional theory. J. Phys. Chem. A **111**, 2420–2425 (2007)
9. Zhang, W., Han, Y., Yao, S., Sun, H.: Stability analysis and structural rules of titanium dioxide clusters (TiO$_2$)$_n$ with n = 1–9. Mater. Chem. Phys. **130**, 196–202 (2011)
10. Ding, W.F., Li, B.X.: A first-principles study of Al$_n$Si$_{m-n}$ clusters (m = 6, 9, 10; n ≤ m). J. Mol. Struct. Theochem. **897**, 129–138 (2009)
11. Akola, J., Hakkinen, H., Manninen, M.: Ionization potential of aluminum clusters. Phys. Rev. B. **58**, 3601–3604 (1998)
12. Viet Bac, P.T., Ogawa, H.: Hydrogen atom adsorption on aluminum icosahedral clusters: a DFT study. J. Alloy. Compd. **509**, S675–S678 (2011)
13. Guo, L., Yang, Y.: Theoretical investigation of molecular hydrogen adsorption and dissociation on Al$_n$V(n = 1–13) clusters. Int. J. Hydrogen Energ. **38**, 3640–3649 (2013)
14. Fuke, K., Nonose, S., Kikuchi, N., Kaya, K.: Reaction of aluminum clusters, Al$_n$ (n = 7–24), with oxygen and ammonia. Chem. Phys. Lett. **147**, 479–483 (1988)
15. Reber, A.C., Roach, P.J., Woodward, W.H., Khanna, S.N., Castleman, A.W.: Edge-induced active sites enhance the reactivity of large aluminum cluster anions with alcohols. J. Phys. Chem. A **116**, 8085–8091 (2012)
16. Das, S., Pal, S., Krishnamurty, S.: Understanding the site selectivity in small-sized neutral and charged Al$_n$ (4 ≤ n ≤ 7) clusters using density functional theory based reactivity descriptors: a validation study on water molecule adsorption. J. Phys. Chem. A **117**, 8691–8702 (2013)
17. Pino, I., Kroes, G.J., Van Hemert, M.C.: Hydrogen dissociation on small aluminum clusters. J. Chem. Phys. **133**, 184304–184312 (2010)
18. Yarovskya, I., Goldberg, A.: DFT study of hydrogen adsorption on Al$_{13}$ clusters. Mol. Simul. **31**, 475–481 (2005)
19. Xiang, H., Kang, J., Wei, S.H., Kim, Y.H., Curtis, C., Blake, D.: Shape control of Al nanoclusters by ligand size. J. Am. Chem. Soc. **131**, 8522–8526 (2009)
20. Mosch, C., Koukounas, C., Bacalis, N., Metropoulos, A., Gross, A., Mavridis, A.: Interaction of dioxygen with Al clusters and Al(111): a Comparative Theoretical Study. J. Phys. Chem. C **112**, 6924–6932 (2008)
21. Kumar, V., Bhattacharjee, S., Kawazoe, Y.: Silicon-doped icosahedral, cuboctahedral, and decahedral clusters of aluminum. Phys. Rev. B. **61**, 8541–8547 (2000)
22. Majumder, C., Kulshreshtha, S.K.: Influence of Al substitution on the atomic and electronic structure of Si clusters by density functional theory and molecular dynamics simulations. Phys. Rev. B. **69**, 115432–115439 (2004)

23. Li, B.X., Wang, G.Y., Ye, M.Y., Yang, G., Yao, C.H.: Geometric and energetic properties of Al-doped Si_n (n = 2–21) clusters: FP-LMTO-MD calculations. J. Mol. Struct. Theochem. **820**, 128–140 (2007)

24. Becke, A.D.: Density-functional exchange-energy approximation with correct asymptotic behavior. Phys. Rev. A **33**, 3098–3100 (1988)

25. Lee, C., Yang, W., Parr, R.G.: Development of the Colle-Salvetti correlation-energy formula into a functional of the electron density. Phys. Rev. B **37**, 785–789 (1988)

26. Frisch, M.J., et al.: Gaussian 03, Revision B 03. Gaussian, Inc., Pittsburgh, PA (2003)

27. Cox, D.M., Trevor, D.J., Whetten, R.L., Rohlfing, E.A., Kaldor, A.: Aluminum clusters: magnetic properties. J. Chem. Phys. **84**, 4651–4656 (1986)

28. Curotto, V.F., Diez, R.P.: Density functional study on the geometric features and growing pattern of Al_nN_m clusters (n = 1–4, m = 1–4, n + m ≤ 5). Comp. Mater. Sci. **50**, 3390–3396 (2011)

29. Zheng, X., Zhang, Y., Huang, S., Liu, H., Wang, P., Tian, H.: DFT study of structural, electronic and vibrational properties of pure $(Al_2O_3)_n$ (n = 9, 10, 12, 15) and Ni-doped $(Al_2O_3)_n$ (n = 9, 10) clusters. Appl. Surf. Sci. **257**, 6410–6417 (2011)

30. Parr, R.G., Pearson, R.G.: Absolute hardness: companion parameter to absolute electronegativity. J. Am. Chem. Soc. **105**, 7512–7516 (1983)

31. Tozer, D.J., Proft, F.D.: Computation of the hardness and the problem of negative electron affinities in density functional theory. J. Phys. Chem. A **109**, 8923–8929 (2005)

32. Chandra, A.K., Uchimaru, T.: Hardness profile: a Critical Study. J. Phys. Chem. A **105**, 3578–3582 (2001)

33. Jin, R., Zhang, S., Zhang, Y., Huang, S., Wang, P., Tian, H.: Theoretical investigation of adsorption and dissociation of H_2 on $(ZrO_2)_n$ (n = 1–6) clusters. Int. J. Hydrogen Energ. **36**, 9069–9078 (2011)

34. Paranthaman, S., Hong, K., Kim, J., Kim, D.E., Kim, T.K.: Density functional theory assessment of molecular structures and energies of neutral and anionic Al_n (n = 2–10) Clusters. J. Phys. Chem. A **117**, 9293–9303 (2013)

35. Wang, C., Kuang, X., Wang, H., Li, H., Gu, J., Liu, J.: Density-functional investigation of the geometries, stabilities, electronic, and magnetic properties of gold cluster anions doped with aluminum: Au_nAl^- (1 ≤ n ≤ 8). Comput. Theor. Chem. **1002**, 31–36 (2012)

19

Photobiological synthesis of noble metal nanoparticles using *Hydrocotyle asiatica* and application as catalyst for the photodegradation of cationic dyes

Thangavel Akkini Devi[1] · Narayanan Ananthi[1] · Thomas Peter Amaladhas[1]

Abstract Solar light induced photo catalysis by plasmonic nanoparticles such as Au and Ag is an important field in green chemistry. In this study an environmental benign method was investigated for the rapid synthesis of colloidal Ag and AuNPs using the extract of *Hydrocotyle asiatica*, as a reducing and stabilizing agent under sunlight irradiation. The nanoparticles were formed in few seconds and were characterized by UV–Vis., FT-IR, TEM, EDAX, XRD, DLS and Zetasizer. The nanoparticles were stable in aqueous solution for more than 6 months. TEM analysis established that the Ag and AuNPs were predominantly spherical with average size of 21 and 8 nm, respectively. The flavonoids and glycosides from the extract of *H. asiatica* were proved to be responsible for the reduction and capping through FT-IR analysis. The antimicrobial studies of AgNPs showed effective inhibitory activity against the clinical strains of gram-negative and positive bacteria. The localized surface plasmon resonance of AgNPs was used for the photo-driven degradation of cationic dyes (malachite green and methylene blue). Thus, this green technique can be used for bulk production of AgNPs, and thus prepared nanoparticles may be used for removal of dyes from effluent.

Graphical Abstract

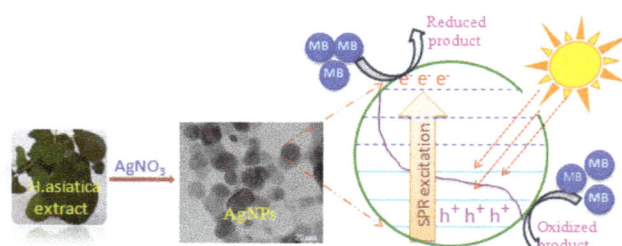

Keywords Biosynthesis · Silver and gold nanoparticles · *Hydrocotyle asiatica* · Malachite green · Methylene blue · Photobiological synthesis

Introduction

The essential of nanotechnology is to synthesize dispersed nanoparticles for potential applications in optics, biomedical sciences, drug delivery, catalysis and electronics. In the past, number of methods such as chemical [1–4], photochemical [5–8] and thermal [9, 10] have been developed to synthesize metallic nanoparticles. Recently an efficient method for fabrication of metallic nanoparticles from metallomicelles is reported [11]. Among these methods, photochemical method has gained considerable attention due to its convenience, but it employs toxic chemicals as reducing and stabilizing agent [12]. To avoid the utilization of highly toxic chemicals, the use of microorganisms and plant parts for nanoparticles synthesis has been explored [13]. Previous reports mention the use of microorganisms such as *Alternaria alternate* [14], and *Amylomyces rouxii* [15] for the synthesis of nanoparticles.

✉ Thomas Peter Amaladhas
  peteramaldhast@yahoo.co.in; peteramaladhas@gmail.com

  Thangavel Akkini Devi
  t.deviagni@gmail.com

  Narayanan Ananthi
  ananthi.red@gmail.com

[1] PG and Research Department of Chemistry, V.O. Chidambaram College, Tuticorin, Tamilnadu 628008, India

Recently, several plants extracts such as *Annona squamosa* [16], *Arbutus unedo* [17], *Cassia angustifolia* [18] and many others [19–27] have been used to synthesize silver (Ag) and gold (Au) nanoparticles (NPs). However, these natural sources, which have active ingredients, produce nanoparticles at slower rate. To overcome this, photo-assisted biosynthesis of Ag and AuNPs using many plants has been qualitatively investigated [28]. The sunlight irradiation technique has been formerly reported for the synthesis of nanoparticles using few plant materials such as *Allium sativum* [29], *Andrachnea chordifolia* [30], *Piper betle* [31], *Bacillus amyloliquefaciens* [32] and *Achyranthus aspera* [33]. Apart from plant materials, dendrimers and starch are also used as reducing/stabilizing agents in the sunlight-induced synthesis of nanoparticles [34, 35]. The nanoparticles produced by bio route have good stability due to the presence of capping agents like alkaloids, flavonoids and polyphenols, which are the major constituents present in the natural sources.

The localized surface plasmon resonance (LSPR) of noble metal nanoparticles makes them suitable for various applications. The surface plasmons live for few femto/picoseconds producing a shower of energetic electrons and holes and then dephase. By fabricating appropriately nanostructured materials that allow a reasonable fraction of these hot carriers to be harvested before they thermalize, the hot electrons can be transferred to appropriate catalyst systems and the material can be used to carry out light-enabled redox chemistry. Recently biosynthesized NPs have been reported to degrade organic dyes [36–42].

Hydrocotyle asiatica (Fig. 1) known as *Centella asiatica*, is a creeping perennial herbal plant with kidney-shaped leaves, found in India, Sri Lanka, Madagascar, South Africa, Australia, China, and Japan. *H. asiatica* is reported to contain triterpene acids (asiatic, madecassic acid, etc.), volatile and fatty oil (glycerides of palmitic acid, stearic, lignoceric, oleic, linoleic and linolenic acids), alkaloids (hydrocotylin), flavonoids (3-glucosylquercetin, 3-glucosylkaempferol and 7-glucosylkaempferol) and glycosides (asiaticoside A, asiaticoside B, madecassoside, etc.) [43]. The herb is well known for its high medicinal values; its

Fig. 1 *Hydrocotyle asiatica* leaves

antioxidant activity [44] and non-toxicity motivated us to carry out the present investigation.

Dyes are synthetic organic compounds released as effluent by many industries producing paper, plastic, leather, food, cosmetic, textile and medicine. The use of synthetic complex organic dyes as coloring materials in textile industries has increased significantly. Dyes in general, azo dyes in particular are carcinogenic, affecting reproductive organs and develop toxicity and neurotoxicity. Therefore, the dyes are to be necessarily removed from industrial effluent before discharging into the natural sources. The dye effluents are highly resistant to microorganisms so that their reduction using conventional biological treatment is generally ineffective and also resistant to destruction by physical–chemical treatments in a high effluent concentration. Methylene Blue (MB) is one of the phenothiazine cationic dyes, used in coloring paper, dyeing cottons, wools and so on. It makes very harmful impacts on living things causing difficulties in breathing, vomiting, diarrhea and nausea [45]. Malachite Green (MG), a triphenylmethane dye is extensively used in the leather, paper, silk, cotton, and jute dyeing processes and also used as biocide in the global aquaculture industry treating protozoal and fungal infections.

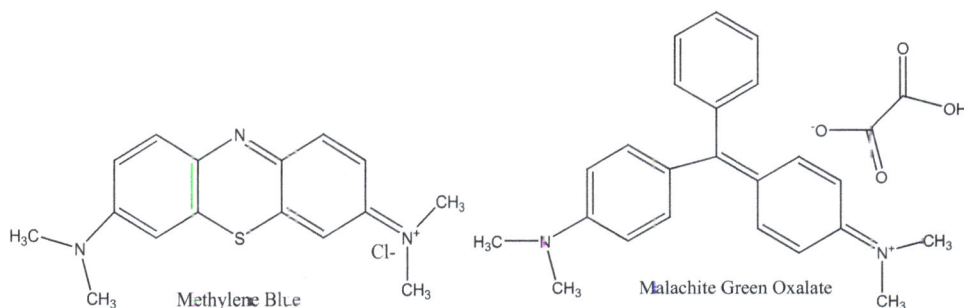

Methylene Blue

Malachite Green Oxalate

MG and its metabolites are known to cause mutagenic, carcinogenic, and teratogenic effects to living organisms [46]. Nanotechnology has been extended to the wastewater treatments in the recent years and due to high surface area, AgNPs exhibits an enhanced reactivity [47].

The present study deals with the biosynthesis of Ag and AuNPs in milligram quantities using abundantly available *H. asiatica* as a reducing and stabilizing agent. The kinetics of formation of Ag and AuNPs has been studied using Ultraviolet–Visible spectroscopy (UV–Vis) and the biosynthesized nanoparticles has been characterized using Transmission Electron Microscopy (TEM), Energy Dispersive X-ray Analysis (EDAX), X-ray Diffraction (XRD), Dynamic Light Scattering (DLS), Zetasizer and Fourier Transform Infra Red (FTIR) spectrometry. The potential applications of Ag and AuNPs as a catalyst for the photo degradation of two different cationic dyes, and as antibiotics have also been studied.

Materials and methods

Preparation of plant extract

The leaves of *H. asiatica* were collected from Eral, Tuticorin, Tamil Nadu, India. The plant was identified with the help of local flora and authenticated by botanical survey of India, Southern Circle, Coimbatore, Tamil Nadu, India. About 5 g of fresh leaves were washed thoroughly with tap water and finally with double distilled water and then cut into small pieces. These finely cut leaves were then boiled in 100 mL double distilled water in a 250 mL Erlenmeyer flask for a period of 5 min. After getting it cool, it was filtered through Whatman No. 41 filter paper. This clear filtrate was used for the synthesis of nanoparticles.

Biosynthesis of AgNPs

To make AgNPs, 5 mL of clear aqueous extract of *H. asiatica* was added to 10 mL of 10 mM AgNO$_3$ (Spectrochem, Mumbai, India, AR Grade) and 85 mL of double distilled water so as to make the final concentration to 1 mM. The solution mixture was then exposed to bright sunlight. The formation of AgNPs was observed by the appearance of yellowish brown color in 5 s and the periodic sampling was carried out to monitor the kinetics using UV–Vis. spectrophotometer. 1 mL of the aliquot was withdrawn from the reaction mixture and diluted to 10 mL with double distilled water before measuring the absorbance.

Biosynthesis of AuNPs

The aqueous extract of *H. asiatica* (2.0 mL) was added to 4.0 mL of 1 mM HAuCl$_4$.3H$_2$O (Sigma-Aldrich, USA,

99.9 % purity) and the solution mixture was then exposed to bright sunlight. The formation of AuNPs was observed by the appearance of pinkish purple color in 45 s. The periodic sampling was carried out to monitor the kinetics of formation of AuNPs. For the UV–Vis spectral analysis, 1 mL of the aliquot was withdrawn from the reaction mixture and diluted to 5 mL with double distilled water.

Analysis of photo-bio reduced Ag and AuNPs

The UV–Vis spectral analysis was performed on a JASCO, V-530 spectrophotometer. FT-IR spectra for the dry powder (obtained by evaporation of leaves extract) and nanoparticles were recorded in the range of 4000–400 cm^{-1} with Thermoscientific, Nicolet iS5 spectrometer. TEM analysis was done on a PHILIPS, CM 200 instrument operated at 200 kV, resolution 2.4 Å. XRD measurements were carried out using a Panalytical X'Pert Powder X'Celerator Diffractometer, with CuKα monochromatic filter. EDX (JEOL Model JED-2300) analysis was used for identifying the elemental composition of Ag and AuNPs. Electrokinetic property (zeta potential) of NPs was evaluated using Zetasizer 1000 HS (Malvern Instruments, UK) at Defence Food Research Laboratory (DFRL), Mysore.

Assessment of antimicrobial activity

For antimicrobial study, 50 mL of as-prepared colloidal suspension of nanoparticles was centrifuged at 13,000 rpm for 15 min and the nanoparticles were dispersed in 1 mL double distilled water and used. Antimicrobial activity of the synthesized Ag and AuNPs was determined using the disc diffusion assay method on pathogenic strains of *Salmonella paratyphi*, *Pseudomonas aeruginosa*, *Streptococcus pyogenes* and *Streptococcus faecalis*. The test microbial suspensions were spread on Petri dishes and freshly prepared Ag and AuNPs samples were introduced. The control samples lacking the precursor were used to assess the antimicrobial activity of the extract. The samples were initially incubated for 15 min at 4 °C (to allow diffusion) and later on at 37 °C for 24 h for the culture. Positive test results were scored when a zone of inhibition was observed around the well after the incubation period. The diameter of zones of inhibition was measured using a meter ruler, and the mean value for each organism was recorded and expressed in millimeter.

Evaluation of photo degradation of cationic dyes by AgNPs

To prepare the AgNPs in solid form, the colloidal solution was centrifuged at 13,000 rpm for 15 min and the pellet was washed with double distilled water to remove the excess biomolecules that were not capped and the washings were repeated for three times. The samples were dried at 60 °C, grounded and used for the degradation studies. 10 mg of AgNPs was added to 100 mL solution of 1×10^{-5} M methylene blue trihydrate (EMerck, Germany, Microscopy Grade) or malachite green oxalate (Spectrochem, Mumbai, Microscopy Grade) and the beaker was then exposed to sunlight with constant stirring using magnetic stirrer. To monitor the reaction, aliquots were withdrawn at predetermined time period and the suspension was centrifuged and the absorbance of the supernatant was subsequently measured using UV–Vis spectrophotometer.

Results and discussion

UV–Vis spectra of AgNPs

AgNPs are formed by the sunlight-induced reduction of Ag^+ ions by the addition of leaf extract. The colorless solution turned reddish brown, in 5 s of exposure to sunlight, indicating the formation of AgNPs. The formation of AgNPs is monitored using UV–Vis spectrophotometer in the range of 200–800 nm, where an intense peak was obtained at 436 nm, which is known as Localized Surface Plasmon Resonance (LSPR) band due to the excitation of free electrons in the nanoparticles. The peak is sharp and symmetrical in shape signifying the formation of mono dispersed AgNPs [33].

Figure 2a represents the UV–Vis spectra of synthesized AgNPs as a function of interaction time. From the figure it is evident that the formation of AgNPs mainly depends on the time of exposure to sunlight. It can be observed that the absorbance is increased with exposure time (Fig. 2b), indicating more Ag^+ ions are reduced with time. Initially the absorbance of the solution increases exponentially after that it tends to attain almost a constant value indicating the completion of the reaction. By monitoring the LSPR band from 0 to 55 min, it can be seen that there is a blue shift of λ_{max} from 441 to 411 nm with the consequent color changes from reddish brown to yellow. This kind of blue shift is being reported to be due to the formation of small-sized nanoparticles [48] and narrowing of peak with time implies monodispersed AgNPs are formed with time and about 30 min may be taken as appropriate to get uniform-sized nanoparticles.

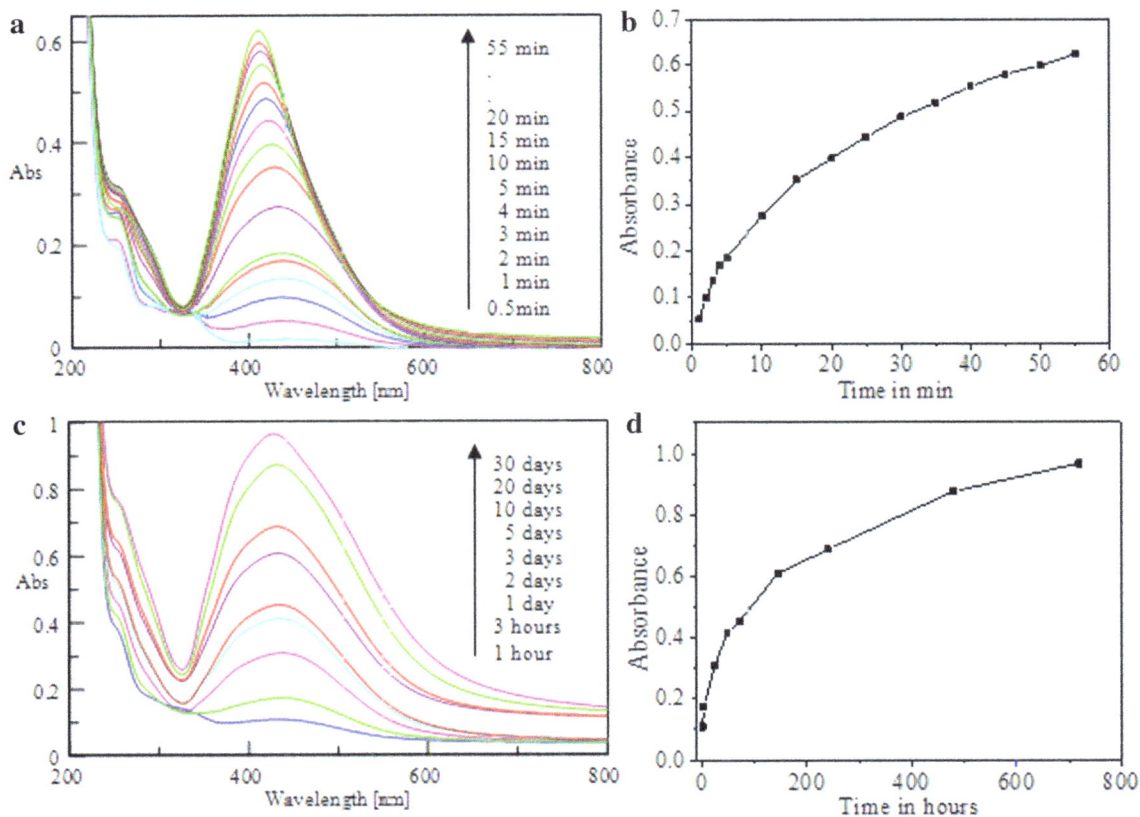

Fig. 2 UV–Vis spectra recorded as a function of time of reaction of *H. asiatica* extract with an aqueous solution of 1 mM AgNO₃: **a, c** presence and absence of sunlight, **b, d** variation of absorbance with time (presence and absence of sunlight)

For comparison, the title reaction is also studied at ambient conditions in the absence of sunlight. The reaction is slow and the appearance of yellowish brown color is noticed only after an hour, thereafter the color intensified slowly with time (Fig. 2c). The maximum absorption is observed at 437 nm and the LSPR band is broad signifying the poly-dispersed nature of nanoparticles. Initially the intensity increased almost linearly up to 6 days after that it tends to attain a maximum value (Fig. 2d) indicating the completion of reaction. It is observed that the λ_{max} is blue shifted from 437 nm to 428 nm with increase of reaction time from 1 to 30 days. The AgNPs are observed to be stable in solution for more than 1 month.

Effect of the synthesis variables on AgNPs

The effect of extract quantity, silver nitrate concentration, temperature and pH on the formation of AgNPs in the presence and absence of sunlight is also studied.

UV–Vis spectra of AgNPs synthesized in presence of sunlight

Effect of volume of extract

The volume of the extract was varied from 1 to 3 mL by keeping the AgNO$_3$ concentration constant. From Figure S1 (S—Supporting Materials), it is observed that initially the absorbance of LSPR band increases and then decreases. Minimum volume of leaf extract is preferred to have stable AgNPs.

Effect of AgNO$_3$ concentration

The AgNO$_3$ concentration was varied from 1 to 10 mM, while keeping the extract volume constant. On increasing the concentration of silver nitrate, the formation of NPs is also increased linearly up to 7 mM after that it tends to attain maximum value. There is no appreciable shift in the λ_{max} is observed (Figure S2).

Effect of pH

To study the effect of pH, extract pH was adjusted to 3–11 by adding 0.1 M sulphuric acid or 0.1 M sodium hydroxide. In all the pH ranges, the LSPR band appears between 411 and 429 nm indicating that there is no change in the shape of the NPs with change in pH (Figure S3). However, in alkaline pH (>9) the intensity of SPR band increases appreciably with a blue shift. This trend is reported earlier

also signifying the formation of large amount of AgNPs and the blue shift is attributed to formation of small-sized nanoparticles [17].

UV–Vis spectra of AgNPs synthesized in absence of sunlight

Effect of volume of extract

The volume of extract was varied from 0.5 to 2.5 mL with 25 mL of 1 mM AgNO$_3$ solution. The spectrum was recorded after 24 h of reaction. Figure S4a represents the UV–Vis spectra of AgNPs synthesized using *H. asiatica* at different extract volume. At higher volume of extract, SPR peaks are broad indicating that the AgNPs are poly-dispersed. The red shift in the absorption maximum shows the aggregation of nanoparticles leading to larger particle size. For the generation of small AgNPs, minimum quantity of leaf extract is preferred. It is observed that initially the absorbance of SPR band increases while increasing the quantity of extract; at higher volumes it tends to attain a constant value (Figure S4b).

Effect of AgNO$_3$ concentration

The AgNO$_3$ concentration was varied from 1 to 3 mM, while keeping the extract volume constant. Figure S5a represents UV–Vis spectra of AgNPs prepared using *H. asiatica* at different AgNO$_3$ concentration. On increasing the concentration of AgNO$_3$, the formation of NPs is also increased up to 2.5 mM after that there is no change in absorbance (Figure S5b). There is a shift in the λ_{max} from 449 to 443 nm when the concentration of AgNO$_3$ is increased from 1.5 to 3.0 mM indicating the aggregation of NPs. On adding extract to higher concentration of AgNO$_3$ (4 and 5 mM), a black colored suspension is formed, may be due to the formation of silver oxide.

Effect of pH

To study the effect of pH, extract pH was adjusted to 2, 3, 4, 6 and 9 and then the extract (5 mL) was added to 95 mL of 1 mM silver nitrate solution to make the total volume to 100 mL. The spectrum was recorded after 24 h of reaction. Figure S6 represents UV–Vis spectra of AgNPs synthesized using *H. asiatica* at different pH values. At lower pH 2, 3 and 4, rate of formation of AgNPs is very slow when compared to pH 6. The broad SPR bands observed at lower pH values are due to large anisotropic particles. At higher pH 9, rate of formation of AgNPs is high. This may be due to the large number of free functional groups available for silver binding, which facilitated higher number of silver ions to bind and subsequently form a large number of NPs with smaller diameters (λ_{max} 417 nm) [33].

Effect of temperature

To study the effect of temperature on AgNPs formation, the reaction was carried out at 35, 60, 80 and 100 °C in a water bath. The spectra were recorded after 1 h of reaction. As the temperature is increased, the rate of formation of AgNPs is also increased indicated by increase in the absorbance. Figure S7 represents UV–Vis spectra of AgNPs synthesized using *H. asiatica* at different temperature. At 35 °C, a broad SPR band indicates the formation of poly-dispersed AgNPs. With increase in the reaction temperature, the SPR band becomes narrow signifying the monodispersed nature of NPs. The λ_{max} of SPR band is shifted from 436 to 406 nm, due to the reduction in size of the NPs. With the increase of temperature from 35 to 80 °C more and more AgNPs are formed, after that not much increase is observed indicating the completion of the reaction.

All these results indicate the superiority of light induced method. More stable AgNPs are formed with uniform size distribution as implied by the sharp and narrow peaks in light reaction as compared to room temperature reaction. The rate of formation of nanoparticles is also very high in light reaction (min. vs days). As mentioned below the glycosides and flavonoids may act as reducing agents to reduce Ag^+ to AgNPs, the enhanced rate of reduction in the light irradiated reaction may be due to readily available electrons from the excitations of low energy transitions like π-π^* in the biomolecules.

UV–Vis spectra of AuNPs

UV–Vis spectra of AuNPs formed by *H. asiatica* extract in bright sunlight at different reaction time are given in Fig. 3. Initially the reaction mixture was colorless, after exposing to sunlight the color of the reaction mixture changed to dark purple in 45 s. The broad band observed initially at 556 nm is changed to sharp peaks with time with the increase of absorption intensity till 40 min with a noticeable blue shift in absorption band. This observation indicates an increase in percentage of nanoparticles in solution and decrease in size with time. However, the absorption intensity of solution slowly decreases after 40 min, which may be due to change in the dimension of anisotropic nanostructures [49]. Hence, 30 min is considered to be optimal time to get monodispersed AuNPs. For comparison formation of AuNPs is also studied in the absence of sunlight, but the reaction is very slow and the AuNPs got aggregated to give a blue precipitate with time hence the reaction is not studied further.

FT-IR studies

FT-IR measurements are carried out to identify the possible bio components in *H. asiatica* leaf extract, which are responsible for stabilization of Ag and AuNPs. The FT-IR spectra of leaf extract and Ag and AuNPs are represented in Fig. 4. The untreated leaves extract shows prominent absorption bands at 3413, 2927, 1630, 1450, 1383, 1272

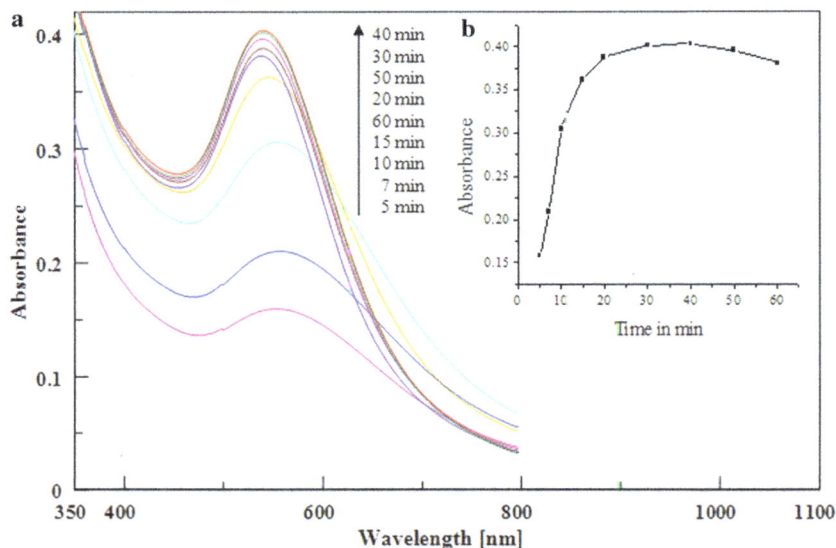

Fig. 3 a UV–Vis spectra recorded as a function of time of *H. asiatica* extract with an aqueous solution of 1 mM HAuCl$_4$ solution in sunlight, **b** variation of absorbance with time

and 1099 cm^{-1} (Fig. 4c). In the IR spectra, intense bands are observed at 3438, 2922, 1617, 1384 and 1097 cm^{-1} (Fig. 4b) and at 3451, 2923, 1629, 1384 and 1097 cm^{-1} (Fig. 4a) for Ag and AuNPs, respectively. Both spectra are almost identical indicating that the biomolecules responsible for stabilization are the same. The bands around 3400 cm^{-1} are due to alcoholic –OH stretching vibration, peak around 1630 cm^{-1} may be due to stretching vibrations of –C=C/–C=O, and the peak at 1384 cm^{-1} is probably due to –CH bending mode and the peak around 1099 cm^{-1} may be attributed to –C–O stretching vibration. The vibrational bands matching to the functional groups such as –OH, –C=O, –C=C and –C–O are probably derived from the water-soluble flavonoids and glycosides [50]. The presence of flavonoids in the extract was confirmed by Ferric chloride test (the extract when treated with few drops of Ferric chloride solution changed to blackish red color) and glycosides was confirmed by the Keller Killiani Test (the extract was treated with few drops of glacial acetic acid and Ferric chloride solution, mixed and concentrated sulphuric acid was added, lower reddish brown layer and upper acetic acid layer which turned bluish green indicated the presence of glycosides) [51]. The structures of important biomolecules present in *H. asiatica* leaves are given below.

TEM analysis

The morphology and size of the prepared Ag and AuNPs are analyzed by recording TEM images. TEM images of the Ag and AuNPs at different magnification are presented in Figs. 5 and 6 respectively. The TEM images clearly show that the nanoparticles are predominantly spherical in shape. The size is in the range of 6.6–33.6 and 8–35 nm for AgNPs in the presence, and absence of sunlight, respectively, and 5.2–10.3 nm for AuNPs in presence of sunlight. Figures 5i and 6f show the SAED pattern (circular ring) of Ag and AuNPs that revealed the polycrystalline nature of nanoparticles. The diffraction rings can be indexed on the basis of the fcc structure of Ag and AuNPs. Four rings arise due to reflections from (111), (200), (220) and (311) lattice planes of fcc of Ag and Au, respectively. This is further supported by broad Bragg's reflection observed in the XRD spectra.

EDAX analysis

EDAX analysis is carried out to confirm the elemental form of Ag and Au. The strong optical absorption peak is observed in the region of 3 keV (Fig. 7a) and 2 keV (Fig. 7b), which is typical for the absorption of metallic Ag

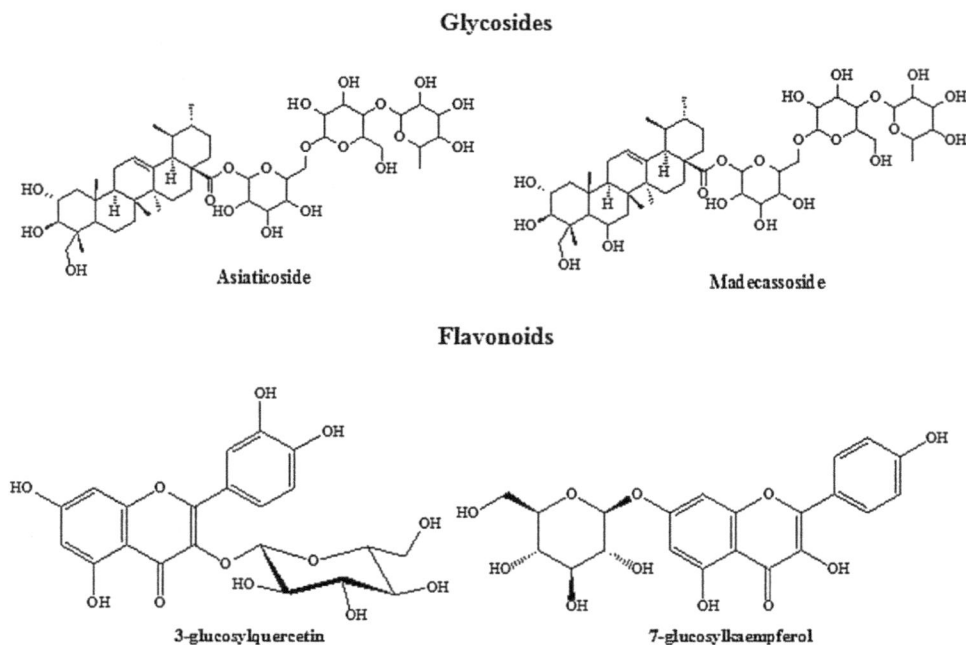

Glycosides

Asiaticoside

Madecassoside

Flavonoids

3-glucosylquercetin

7-glucosylkaempferol

The peak at 1630 cm^{-1} is shifted to 1617 cm^{-1} in the case of AgNPs indicating the involvement of flavonoids or glycosides containing C=O groups in the stabilization of AgNPs.

[52] and Au [53] nanocrystallites, respectively. In addition to Ag and Au, peaks due to carbon and oxygen are also present, which further support the stabilization of nanoparticles by biomolecules.

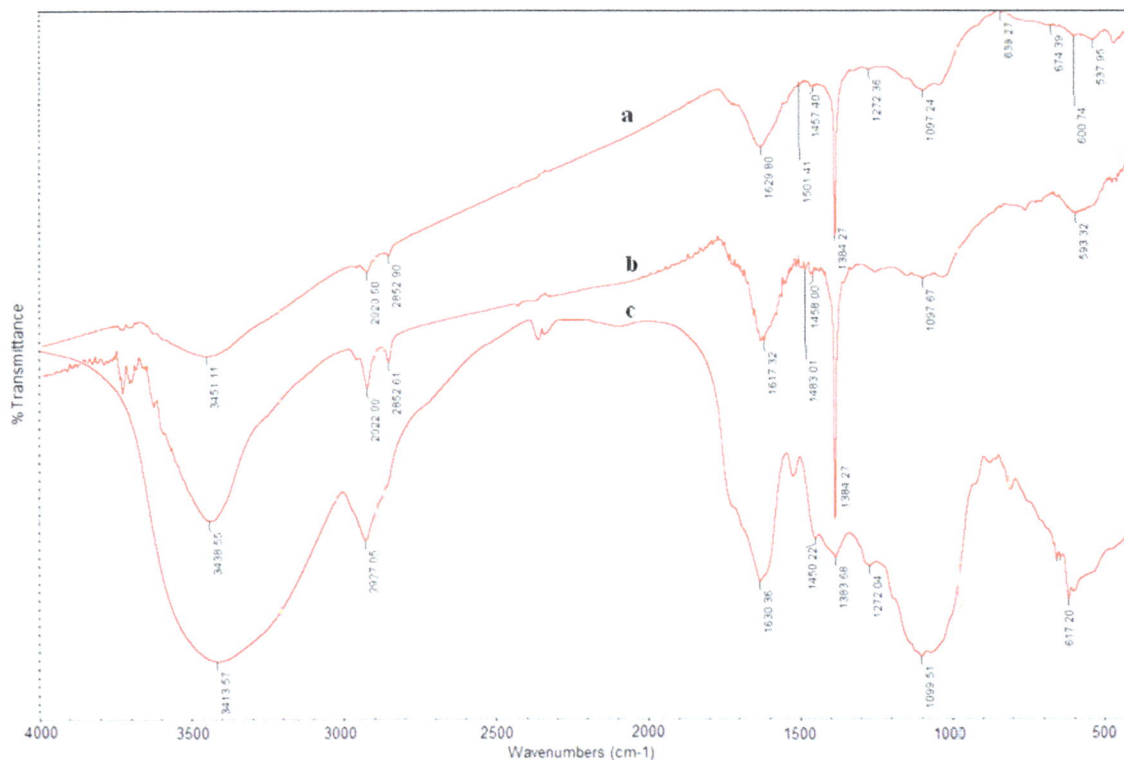

Fig. 4 FT-IR spectra of *a* Solid AuNPs, *b* Solid AgNPs and *c* dry powder obtained from the aqueous extract of *H. asiatica*

XRD studies

The powder X-ray diffraction patterns of the Ag and AuNPs are shown in Fig. 8a, b, respectively. XRD pattern of pure crystalline face-centered cubic (fcc) Ag has been published by the Joint Committee on Powder Diffraction Standards (JCPDS-2 4-0738). A comparison of XRD spectrum with the standard, confirm the crystalline nature of AgNPs, as evidenced by the peaks at 2θ values of 38.04°, 46.15°, 67.33° and 76.71°, corresponding to 111, 200, 220 and 311 planes, respectively. Similar peaks are obtained for the nanoparticles prepared in absence of light also.

The AuNPs show similar characteristic peaks of metallic fcc Au (JCPDS: 4-0784). The particle sizes of Ag and AuNPs are calculated by Debye–Scherrer Eq. (1) and the average sizes are found to be 17.26 and 12.84 nm, respectively.

$$D = k\lambda/\beta\cos\theta \tag{1}$$

where, k is the Scherrer constant, λ is the wavelength of the X-ray, β and θ are the half width of the peak and half of the Bragg angle, respectively.

In comparison to AgNPs, broad peaks are obtained in the XRD spectrum of AuNPs. Generally smaller-sized particles show peak broadening in the XRD pattern [54]. Broad peaks obtained in AuNPs are due to their smaller size. TEM results further augment this statement as smaller-sized AuNPs (8 nm) are produced in comparison to AgNPs (21 nm).

DLS and zeta potential studies

The data of DLS studies show that the average size of the AgNPs (Fig. 9a) is 53.64 nm and AuNPs (Fig. 9c) is 18.49 nm and the polydispersity index (PDI) is 0.245 and 0.330 for Ag and AuNPs, respectively, indicating moderate uniformity in the distribution of particles. The value of zeta potential of AgNPs (Fig. 9b) is −34.9 mV and AuNPs (Fig. 9d) is −20.7 with a single peak indicate the moderate repulsion between the nanoparticles. If the particles have a large negative zeta potential or large positive zeta potential, they will tend to repel each other and there will be no attraction to assemble together.

Antimicrobial studies

The antimicrobial activity of Ag and AuNPs is evaluated against various pathogenic strains including Gram-positive, Gram-negative bacteria. Generally, the antimicrobial

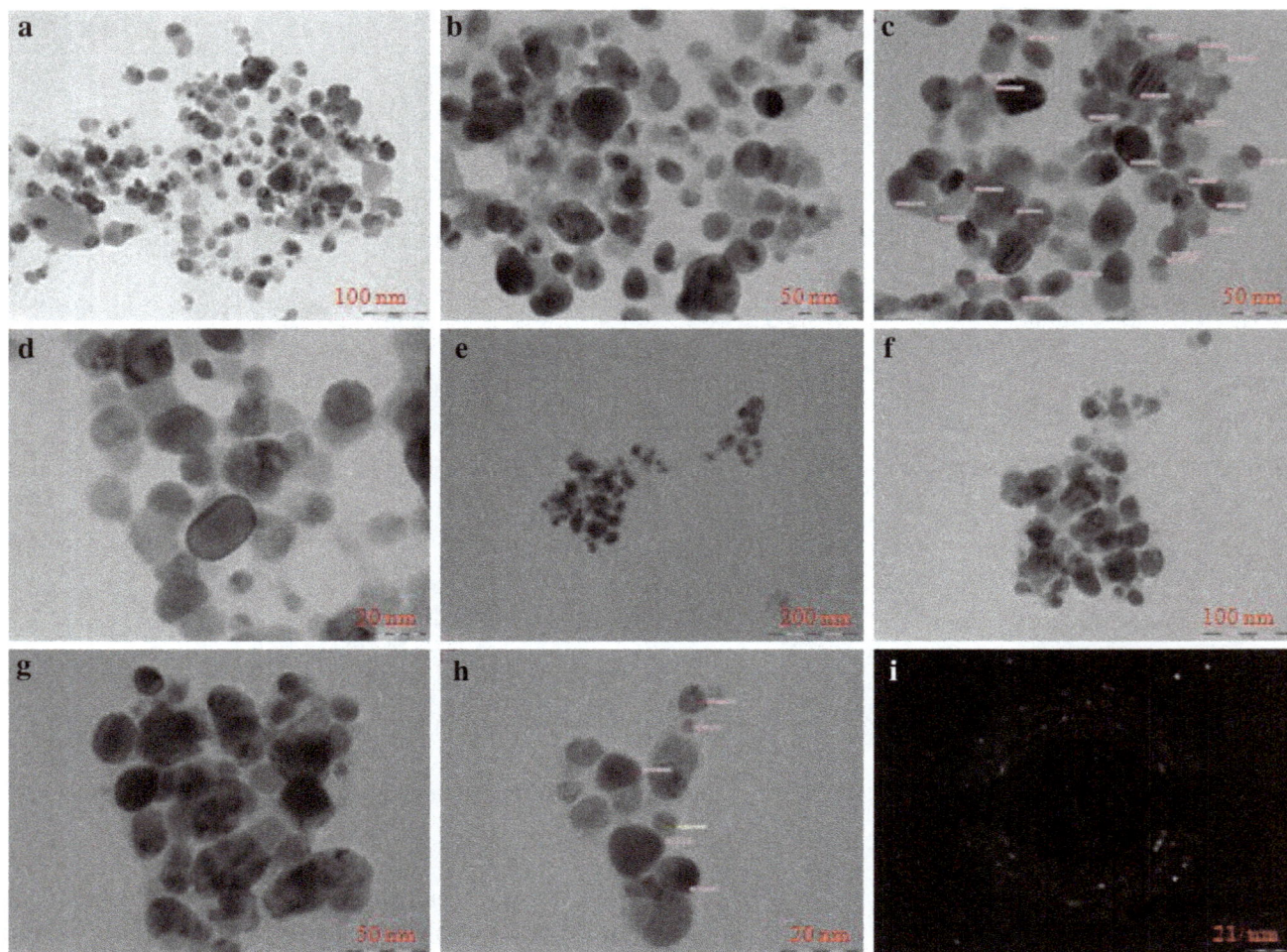

Fig. 5 TEM images of AgNPs at different magnification **a–d** in the presence of sunlight, **e–h** in the absence of sunlight, **i** SAED pattern of AgNPs in the presence of sunlight

behavior is due to the diffusion of nanoparticles into the bacteria resulting in the damage of cell membrane [55]. Another possibility recommended is the release of Ag^+ ions from the nanoparticles, which enhance the bactericidal properties of nanoparticles [56].

The extract and AuNPs have no activity against the tested bacteria, but AgNPs show desirable results. The maximum zone of inhibition for AgNPs is shown by Gram-negative bacteria, *S. paratyphi*, (17 mm), in comparison to Gram-positive bacteria such as *P. aeroginosa* (15 mm), *S. pyogenes* (11 mm) and *S. faecalis* (14 mm) (Figure S8), which may be explained by the fact that the cell wall of the gram-positive bacteria is made of thick peptidoglycan layer consisting of linear polysaccharide chains cross linked by short peptides, thus forming more rigid structure leading to difficult diffusion of AgNPs compared to the Gram-negative bacteria where the cell wall has been composed of thinner peptidoglycan layer [57].

Evaluation of photo degradation

To see the potential application of biosynthesized Ag and AuNPs for environmental remediation, the degradation of two different cationic dyes (MG and MB) is studied at neutral condition ($\sim$ pH 6). The experiments are carried out over period of 3 h both in the presence and absence of sunlight.

Degradation of malachite green

The change in the UV–Vis spectra of MG during the photodegradation process is given in Fig. 10b. Typically MG solutions display maximum absorbance at 315, 425, and 618 nm [58]. In the present study, the aqueous solution of MG solution has peaks at 316, 424 and 617 nm and the decrease in intensity of the major peak at 617 nm was monitored with time. In the absence of sunlight, the

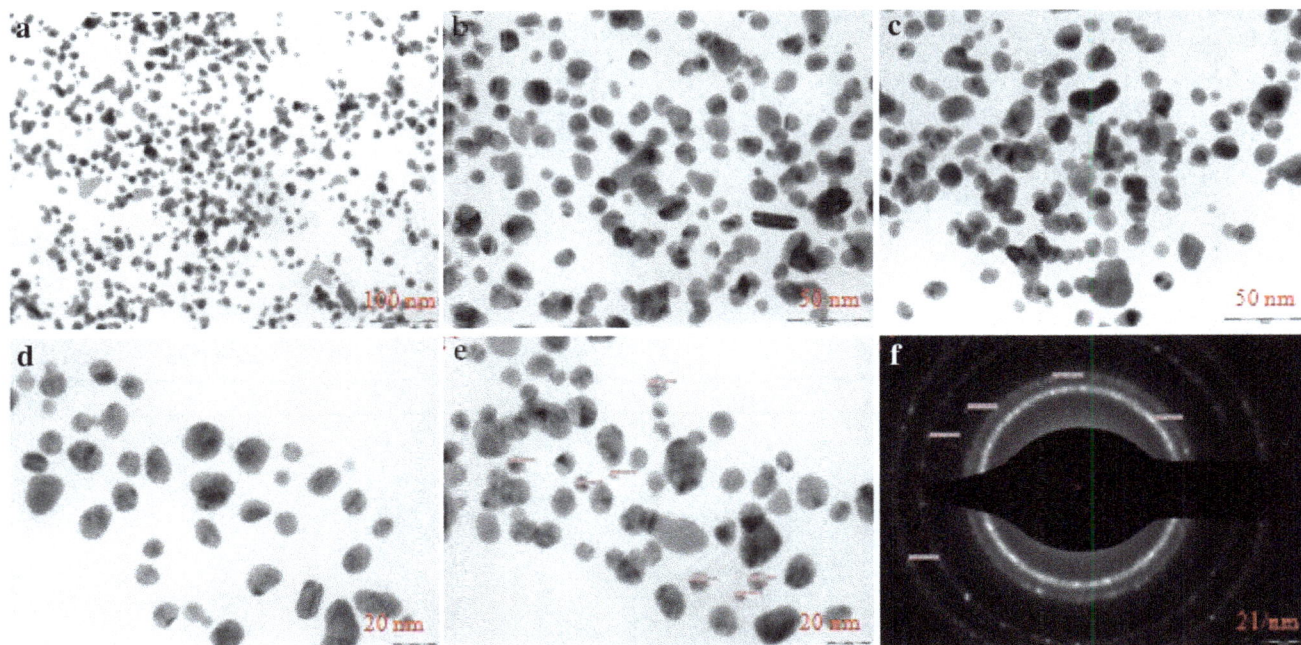

Fig. 6 **a–e** TEM images of AuNPs at different magnification and **f** SAED pattern

Fig. 7 EDAX spectra of
a AgNPs and **b** AuNPs

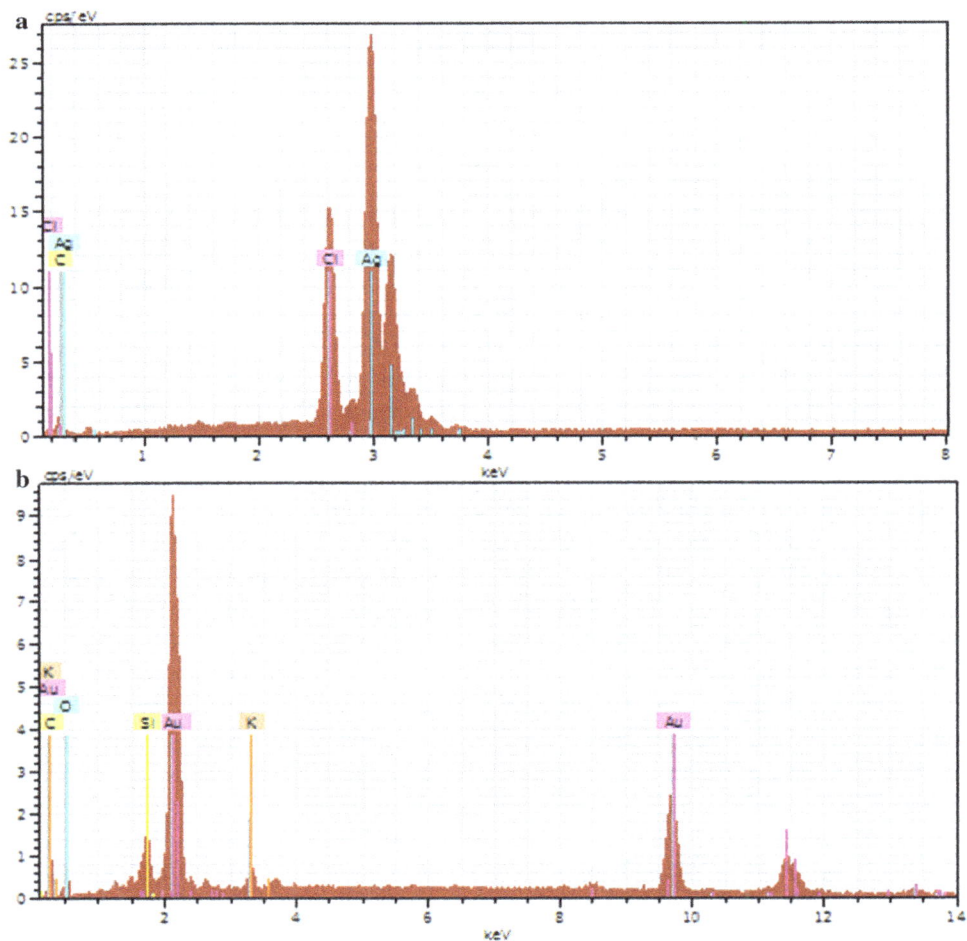

Fig. 8 XRD patterns of
a AgNPs and **b** AuNPs

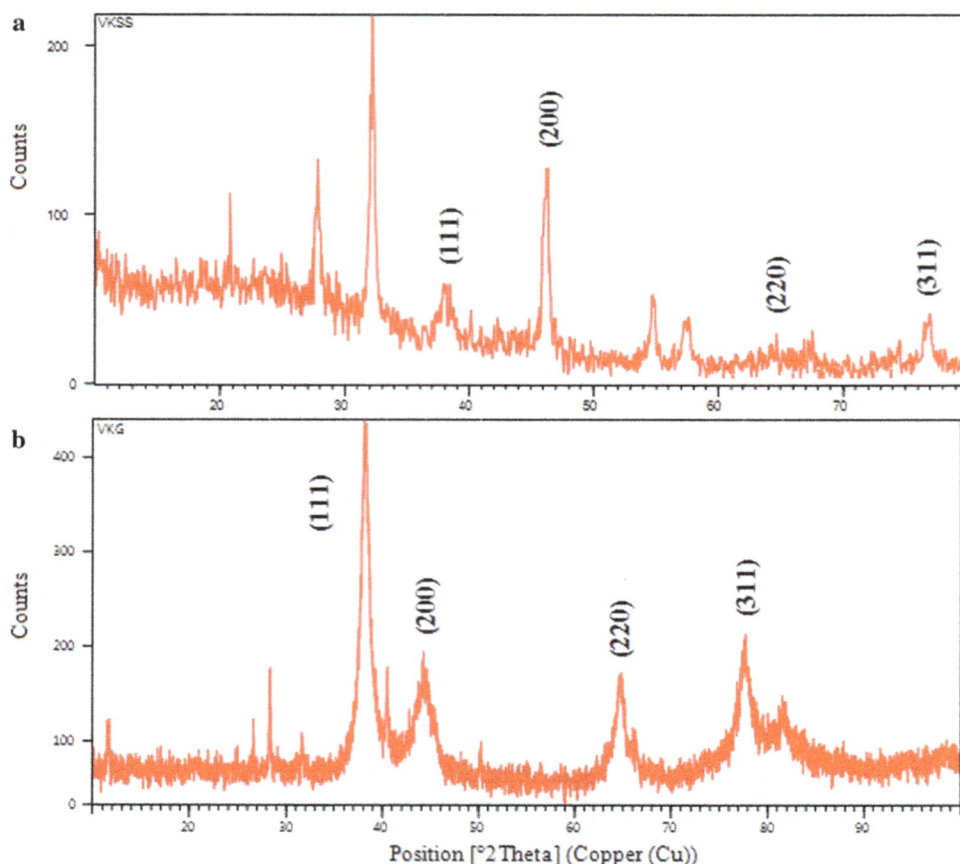

intensity of characteristic absorption band at 617 nm is decreased slowly (from 1.17 to 0.49 au in 180 min), corresponds to 57 % dye removal, but no shift in the λ_{max} is observed. During sunlight irradiation, the absorbance is decreased quickly (from 1.15 to 0.07 au in 180 min), which corresponds to 94 % removal of dye with a blue shift in λ_{max} (617–570 nm). No new absorption band is appeared around 255 nm indicating no leuco form of the dye is formed during the course of the reaction [59]. The plot of $-\ln(C/C_0)$ versus time is a straight line in both cases and the slope is equal to the rate of degradation, which is 0.38×10^{-2} and 1.51×10^{-2} min^{-1} in the absence and presence of sunlight, respectively. The fourfold increase in rate of the degradation in sunlight indicates the strong influence of sunlight on the reaction.

Another observation in the UV–vis spectra of MG is the initial decrease in the peak intensity at 617 nm (Fig. 10a, b). This may be due to the initial adsorption of MG over AgNPs [57]. From the decrease in initial peak intensity, the percentage adsorption of MG on the AgNPs is approximately 18 and 9.6 % in the absence and presence of sunlight, respectively. The less percentage of adsorption in the sunlight reaction may be due to the fact that the

degradation of MG starts as soon as the adsorption takes place and the mechanism of this reaction is different.

The chemical structure of MG varies with solution pH. Chromatic MG$^+$ exists between the pH of 3.5–5.0 ($\lambda_{max} = 617$ nm), the protonated species MGH$^+$ exists below pH 2 ($\lambda_{max} = 255$ nm) and at pH value above 8 it is colorless due to the formation of carbinol base [56]. To see the effect of pH on degradation of MG, the reaction is also studied at pH 9 (Figure S9). At pH 9 the intensity of the peak at 617 nm is decreased immediately after the addition of NaOH (1.14–0.71 au) and 0.71–0.30 within 5 min after that it decreased gradually. A new peak is also observed at 252 nm, which may be due to the alkali fading of dye (leuco form) [59]. The fact that, no additional peak is noticed around 255 nm and the λ_{max} is shifted from 617 to 570 nm in the sunlight irradiated reaction suggests that the dye undergo degradation in the presence of sunlight rather than simple adsorption [60].

Further to substantiate the degradation, the FT-IR spectra of the dye before and after treating with AgNPs were recorded. In the case of pure MG, the FT-IR spectrum (Figure S10b) exhibits peaks at 2924, 1585, 1371, 1170 and 723 cm^{-1}. The peaks at 2924 cm^{-1} is assigned to

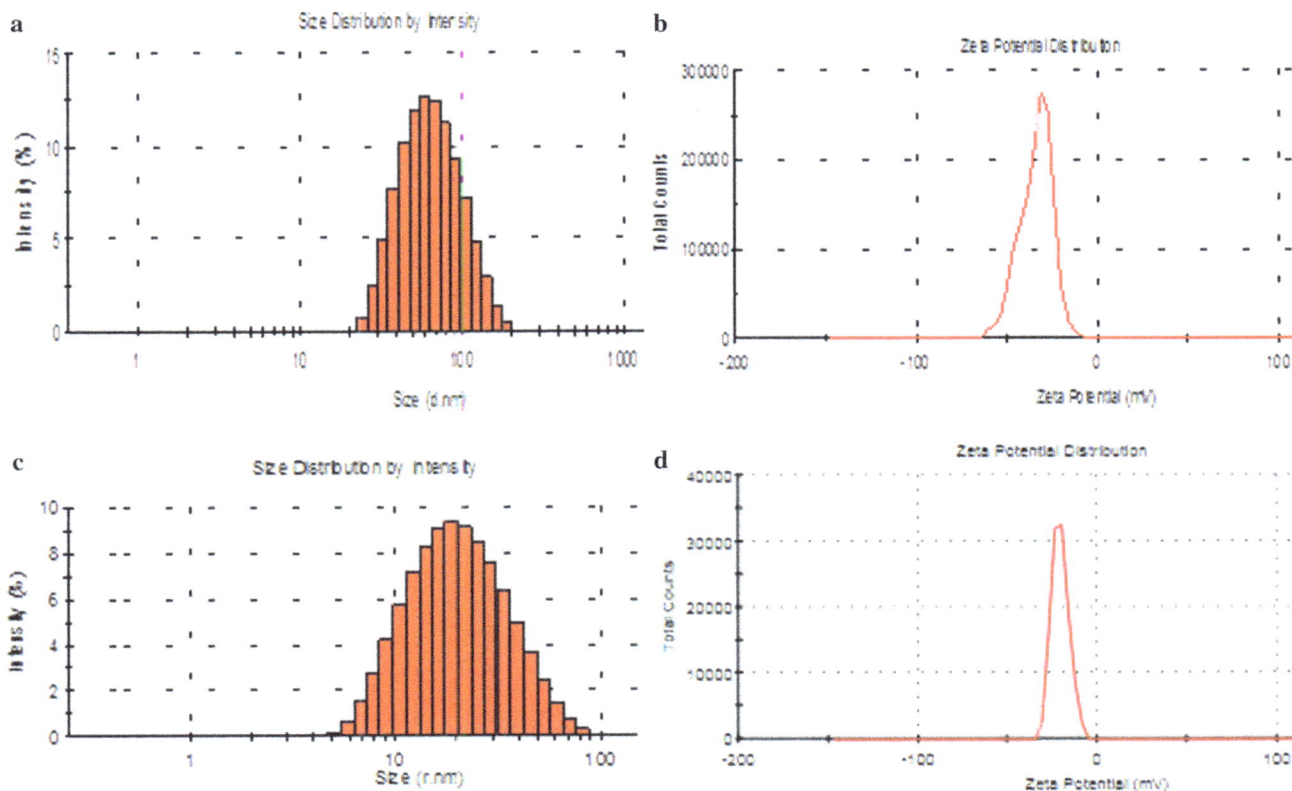

Fig. 9 Size distribution of **a** AgNPs; **c** AuNPs and Zeta potential of **b** AgNPs; **d** AuNPs

asymmetric C–H stretching of CH$_3$ group, the peak at 1585 cm^{-1} is assigned to C=C stretching of the benzene ring, the peak at 1170 cm^{-1} is due to C-N stretching vibration, the peak at 1371 cm^{-1} is assigned to stretching of −CH$_3$/−CH$_2$ and the peak at 725 cm^{-1} indicates symmetric out of plane bending of the ring hydrogen [61].

The FT-IR spectrum of solid AgNPs (Figure S10a) separated from the reaction mixture after the sunlight exposed degradation of MG shows that some peaks of MG have disappeared, specifically the peaks at 1371 and 1170 cm^{-1}, assigned to −CH$_3$/−CH$_2$ and C–N stretching vibrations, inferring the cleavage of C–N bond in N-methyl group. This is further supported by the shifting of λ_{max} in the UV–vis spectra in the presence of sunlight (Fig. 10b).

Degradation of methylene blue

Typically aqueous solutions of MB display maximum absorbance at 246, 292, 612 and 664 nm [62]. The decrease in intensity of peak at 663 nm was monitored with time and the change in the UV–Vis spectra during the photo degradation process is illustrated in Fig. 11. In the absence of sunlight, the intensity of characteristic absorption band of dye at 663 nm is decreased slowly, after 120 min 72 % of dye is removed without any shift in the λ_{max}. The

absorbance of dye is decreased from 0.71 to 0.19 au. During sunlight irradiation, the absorbance is decreased rapidly with a huge blue shift in λ_{max} (114 nm; 663–549 nm), and no new absorption band is observed. The absorbance is decreased quickly from 0.71 to 0.07 au corresponding to 90 % of removal of dye after 120 min of sunlight irradiation. The plot of $-\ln(C/C_0)$ versus time (Fig. 11d) is a straight line in both cases and the slope is equal to the rate of degradation, which is 0.75×10^{-2} and 1.72×10^{-2} min^{-1} in the absence and presence of sunlight, respectively, which is approximately 2.3-fold increase in sunlight reaction. No new peak is formed around 256 nm confirming the absence of leuco-methylene blue [63]. Another observation in the UV–Vis spectra is the sudden initial decrease in the peak intensity of MB at 663 nm (Fig. 11a, b). This may be due to the initial adsorption of MB over AgNPs. From the decrease in initial peak intensity, the percentage adsorption of MB on the AgNPs is approximately 28 and 33.8 % in the absence and presence of sunlight, respectively.

The FT-IR spectrum of AgNPs (Figure S11a) obtained after the sunlight exposed degradation of MB shows that some peaks have disappeared, specifically the peaks at 1489 and 1178 cm^{-1} both due to vibration of heterocyclic skeleton, indicating the cleavage of heterocyclic ring. Moreover, the peaks at 1350 cm^{-1} due to C–N stretching

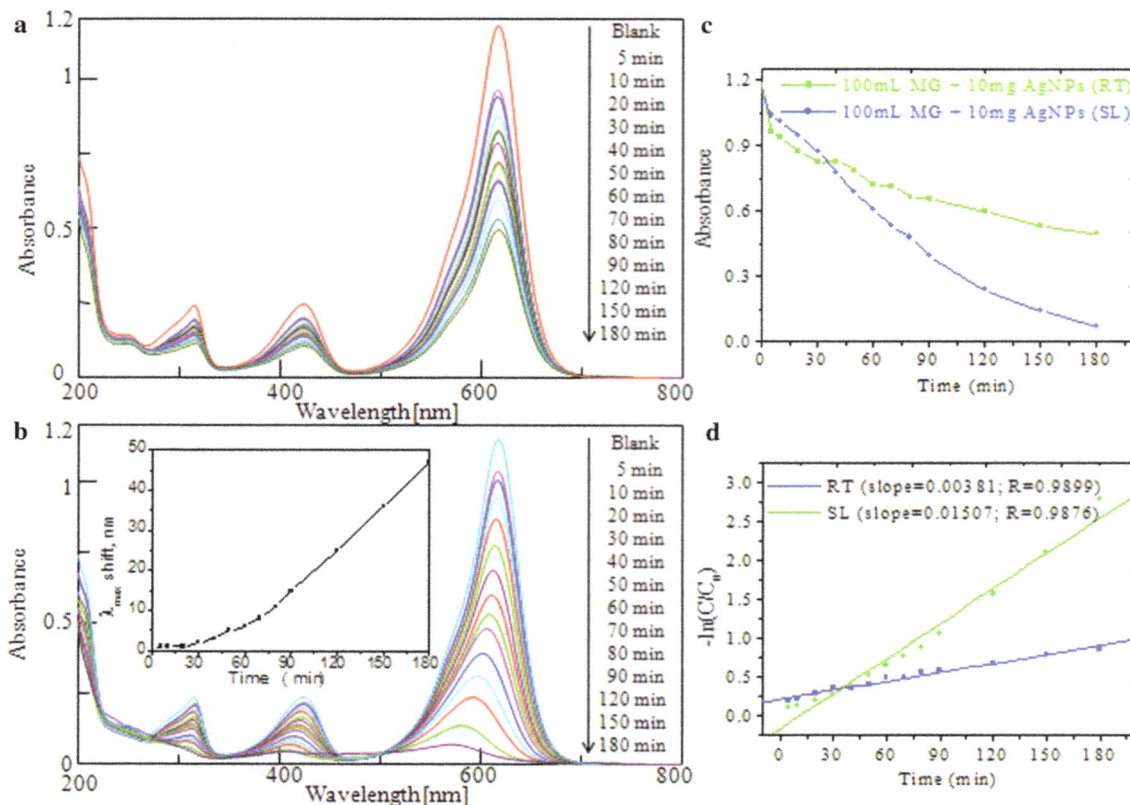

Fig. 10 UV–Vis spectra of MG as a function of time in the presence of AgNPs (MG—1 × 10^{-5} M, AgNPs-10 mg) **a** absence of sunlight, **b** presence of sunlight; *Inset*—shift in λ_{max} with time **c** plot of absorbance versus time and **d** plot of $-\ln(C/C_0)$ versus time

are also disappeared inferring the cleavage of C–N bond in N-methyl group. This is further supported by the large hypsochromic shift of λ_{max} in the UV–Vis spectrum in the presence of sunlight.

In the degradation of MB the intensity of LSPR band at 436 nm is increased with time both in the presence and absence of sunlight, but this trend is not observed in MG. After 90 min this peak spans entire visible region and no further decrease in intensity of peak at 663 nm is noticed even after 180 min although the blue color of MB is completely removed. This may be due to the interaction of AgNPs with dye molecules or any other intermediates formed during the reaction. The same trend was found in the recent articles and the authors attributed no reason [38, 39]. The intensity of peak at 246 nm decreased initially, after some time it started increasing in the reaction carried out in absence of light, but no such trend is observed in the light reaction. This may be due to interaction of AgNPs with adsorbed dye. The fact that this trend is not observed in light reaction means complete decomposition of intermediates containing phenothiazine moiety formed as a result of N-demethylation [64].

Parameters affecting dye degradation rate

Effect of dye concentration

The effect of concentration of MB (0.5, 1.0 and 2.0 × 10^{-5} M) was studied by keeping the catalyst dosage at 10 mg per 100 mL. While increasing the dye concentration from 0.5 to 1.0 × 10^{-5} M the rate of degradation also increases, but with further increase of concentration, the degradation rate decreases. This may be due to inability of light to reach the catalyst surface at high dye concentration [65]. The plot of $-\ln(C/C_0)$ versus time is a straight line (Figure S12). The same trend is observed in MG also.

Effect of catalyst dosage

The effect of catalyst dosage on the photocatalytic degradation of MB under sunlight was studied using different amounts of AgNPs (5, 10 and 15 mg per 100 mL) by keeping dye concentration constant (1 × 10^{-5} M). The degradation rate increases with increase of catalyst dosage (Figure S13). This may be due to an increased number of

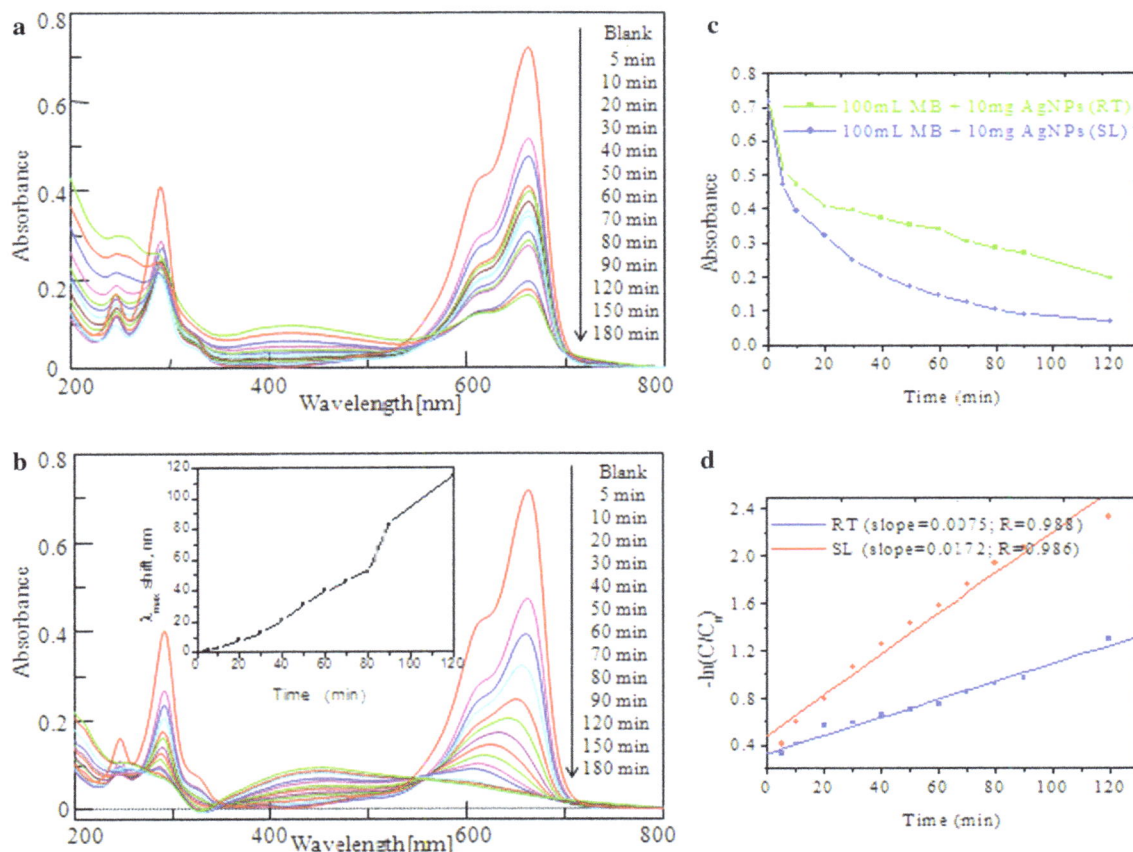

Fig. 11 UV–Vis spectra of MB as a function of time in the presence of AgNPs (MB—1×10^{-5} M, AgNPs—10 mg) **a** absence of sunlight, **b** presence of sunlight; *Inset*—shift in λ_{max} with time **c** plot of absorbance versus time and **d** plot of $-\ln(C/C_0)$ versus time

available adsorption and catalytic sites on the surface of the catalyst [66]. MG also shows the same trend.

Effect of pH

pH is one of the important parameters for study of dye degradation. The effect of pH on the degradation of dyes was studied at pH 4, 6 and 8 (pH 9 in the case of MG), which was maintained by adding 0.1 M H_2SO_4 or 0.1 M NaOH solution. At pH 4, degradation is less and at pH 8, degradation is high (Figures S14 and S15)

AuNPs could not be separated as solid by centrifugation and hence the degradation of MB and MG is studied in the presence of colloidal solutions of AuNPs. No appreciable degradation of the dye is noticed even after 3 h, may be due to the fact that the sunlight is insufficient to produce required electrons on the surface of AuNPs.

Mechanism for the degradation of dyes

Surfing chemical literature reveals that three possible mechanisms are proposed for the removal of the color of the dyes. 1. The dye may be converted to its leuco form by reduction or by increasing the pH [56]. 2. The dye may be adsorbed over the AgNPs, due to the high surface area of nanoparticles; large amount of dye can be removed. 3. The dye can be degraded on the surface of AgNPs due to hot electrons present on the surface created by the intraband transition of electrons (5sp) in the AgNPs due to the LSPR [67].

In this study the last mechanism seems to be operative (Fig. 12). The AgNPs prepared through biosynthetic route are negatively charged as supported by zeta potential (−34.9 mV). Initially the cationic dyes may get adsorbed over the AgNPs due to electrostatic interaction and due to proximity, the excited electrons may degrade the dye in the nanoparticles-adsorbate interface.

The blueshift in λ_{max} suggests the formation of N-demethylated intermediates in degradation pathway. The color of MB solutions becomes less intense (hypsochromic effect) when auxochromic groups (methyl or methylamine) are removed. For dyes containing auxochromic alkylamine groups, N-dealkylation plays an important role in photocatalytic degradation [68]. No new peaks for intermediates appeared during the degradation of both MG and MB and almost nil COD value after the reaction indicates the complete degradation of dyes in sunlight. Hence in the light reactions, it

Fig. 12 Schematic diagram of photodegradation of **a** MB and **b** MG under sunlight irradiation over AgNPs

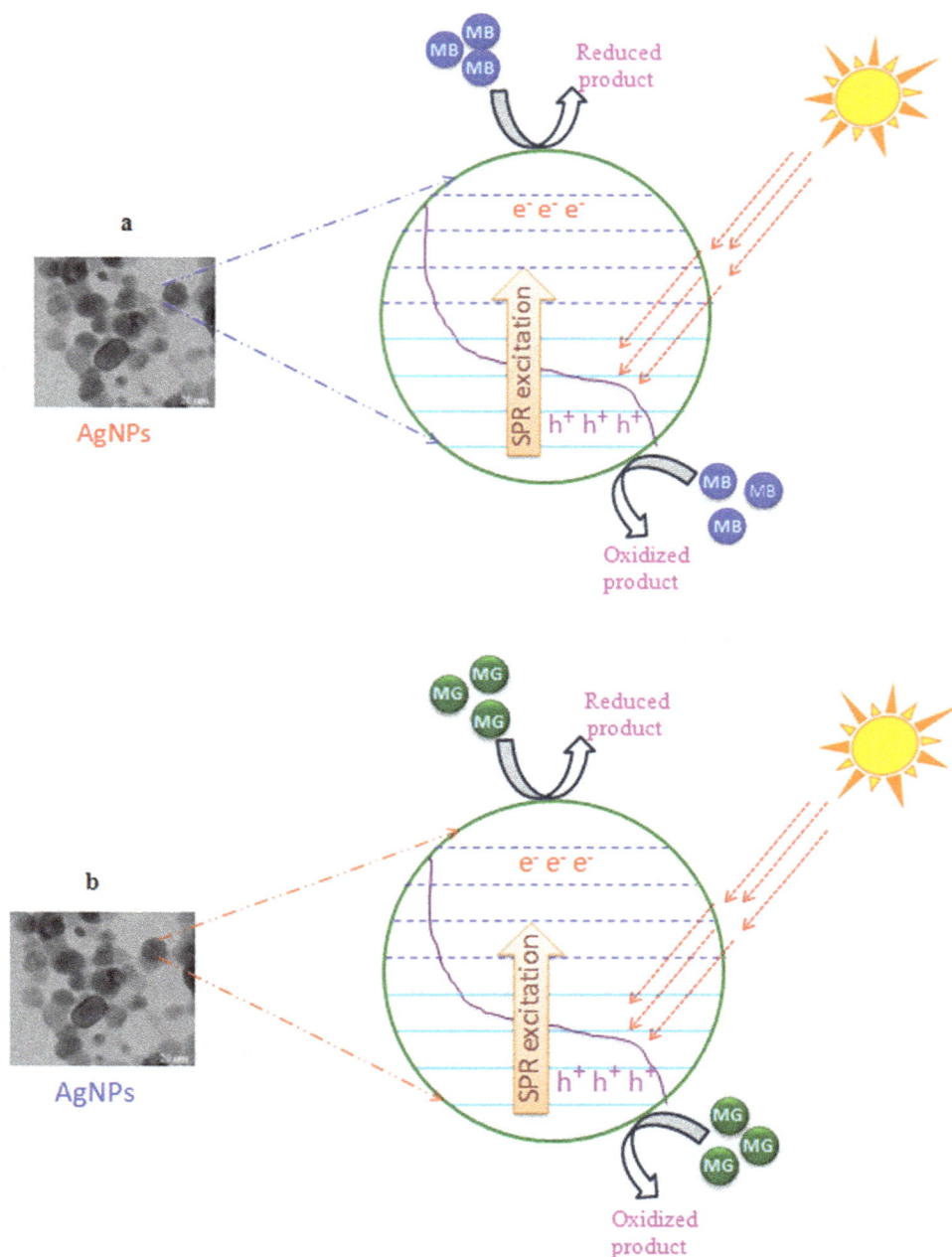

appears degradation takes place via demethylation process and in the absence of light the removal of color is due to adsorption. However, further research using other plasmonic nanoparticles is required to prove this mechanism.

Conclusions

H. asiatica leaves extract has been used as a reducing and capping agent for the photobiological synthesis of Ag and AuNPs. The formation of nanoparticles is observed by color change and confirmed by UV–Vis spectroscopy.

Active water-soluble biomolecules such as glycosides and flavonoids from the extract are responsible for the reduction and also act as stabilizing agent against agglomeration of nanoparticles. From TEM, the average sizes of Ag and AuNPs are found to be 21 and 8 nm, respectively. Small quantity of AgNPs (0.1 g/L) has potential capacity to remove cationic dyes in the presence of sunlight and have noticeable antibacterial activity. Utilization of surface plasmonic resonance excitation of noble metal nanoparticles for various catalytic processes is the current field in green chemistry. Since in this reaction a plant is used as a reductant and stabilizer for preparing noble metal nanoparticles and abundantly available sunlight is used to drive the photo degradation of dyes, this green route may be used for environmental remediation and in water treatment technology.

Acknowledgments This work is sponsored by UGC, New Delhi. One of the author, T. Akkini Devi thanks UGC for Project Fellowship. The authors express sincere thanks to the secretary, V.O.Chidambaram College, Tuticorin, India for providing necessary laboratory facilities. SAIF, IIT Bombay and DFRL, Mysore, India are acknowledged for TEM and DLS analysis.

References

1. Hiroki, H., Frank, E.O.: A simple large-scale synthesis of nearly monodisperse gold and silver nanoparticles with adjustable sizes and with exchangeable surfactants. Chem. Mater. **16**, 2509–2511 (2004)
2. Zhang, P., Chu, A.Y., Sham, T., Yao, Y., Lee, S.: Chemical synthesis and structural studies of thiol-capped gold nanoparticles. Can. J. Chem. **87**, 335–340 (2009)
3. Jack, B., Jeyanthinath, M., Annett, T., Erik, S.M., Umadevi, M.: Chemical synthesis of silver nanoparticles for solar cell applications. Phys. Status Solidi C **8**, 924–927 (2011)
4. Yasmeen, J., Baykal, A.: Sirajuddin.: green chemical synthesis of silver nanoparticles and its catalytic activity. J. Inorg. Organomet. Polym. **24**, 401–406 (2014)
5. Anjali, P.: Photochemical synthesis of gold nanoparticles via controlled nucleation using a bioactive molecule. Mater. Lett. **58**, 529–534 (2004)
6. Lilia, C.C., Flávia, R.O.S., Láercio, G.: A simple method to synthesize silver nanoparticles by photo-reduction. Colloids and surfaces A: physicochem. Eng. Aspects **305**, 54–57 (2007)
7. Huang, W., Chen, Y.: Photochemical synthesis of polygonal gold nanoparticles. J. Nanopart. Res. **10**, 697–702 (2008)
8. del Rodríguez-Torres, M.P., Luis, A.D., Pedro Salas, Claramaría, R.G., Martin O.: UV photochemical synthesis of heparin-coated gold nanoparticles. Gold Bull. **47**, 21–31 (2013)
9. Sun, X., Dong, S., Wang, E.: One-step synthesis and characterization of polyelectrolyte-protected gold nanoparticles through a thermal process. Polymer **45**, 2181–2184 (2004)
10. Jeevanandam, P., Srikanth, C.K.: Suchi Dixit.: synthesis of monodisperse silver nanoparticles and their self-assembly through simple thermal decomposition approach. Mater. Chem. Phys. **122**, 402–407 (2010)
11. Chaudhary, G.R., Singh, P., Kaur, G., Mehta, S.K., Kumar, S., Dilbaghi, N.: Multifaceted approach for the fabrication of metallomicelles and metallic nanoparticles using solvophobic bis-dodecylamine palladium (II) chloride as precursor. Inorg. Chem. **54**, 9002–9012 (2015)
12. Sakamoto, M., Fujistuka, M., Majima, T.: Light as a construction tool of metal nanoparticles: synthesis and mechanism. J. Photochem. Photobiol. C **10**, 33–56 (2009)
13. Morones, J.R., Elechiguerra, J.L., Camacho, A., Ramirez, J.T.: The bactericidal effect of silver nanoparticles. Nanotechnology **16**, 2346–2353 (2005)
14. Gajbhiye, M., Kesharwani, J., Avinash, I., Ariket, G., Rai, M.: Fungus-mediated synthesis of silver nanoparticles and their activity against pathogenic fungi in combination with fluconazole. Nanomed. Nanotech. Biol. Med. **5**, 382–386 (2009)
15. Javed, M., Dwivedi, S., Braj, R.S., Al-Khedhairy, A.A., Ameer, A., Alim, N.: Production of antimicrobial silver nanoparticles in water extracts of the fungus Amylomyces rouxii train KSU-09. Bioresour. Tech. **101**, 8772–8776 (2010)
16. Kumar, R., Roopan, S.M., Prabhakarn, A., Venkatesan, G.K., Chakroborty, S.: Agricultural waste Annona squamosa peel extract: biosynthesis of silver nanoparticles. Spectrochim. Acta Part A **90**, 173–176 (2012)
17. Kouvaris, P., Delimitis, A., Zaspalis, V., Papadopoulos, D., Tsipas, S.A., Nikolaos, M.: Green synthesis and characterization of silver nanoparticles produced using Arbutus unedo leaf extract. Mater. Lett. **76**, 18–20 (2012)
18. Amaladhas, T.P., Sivagami, S., Akkini Devi, T., Ananthi, N., Priya Velammal, S.: Biogenic synthesis of silver nanoparticles by leaf extract of Cassia angustifolia. Adv. Nat. Sci.: Nanosci. Nanotechnol. **3**, 1–7 (2012)
19. Gogoi, K., Saikia, J.P., Konwar, B.K.: Immobilizing silver nanoparticles (SNP) on Musa balbisiana cellulose. Colloids Surf. B **102**, 136–138 (2013)
20. Kathiresan, K., Alikunhi, N.M., Manickaswami, G., Nabikhan, A., Gopalakrishnan, A.: Synthesis of silver nanoparticles by coastal plant Prosopis chilensis (L.) and their efficacy in controlling vibriosis in shrimp Penaeus monodon. Appl. Nanosci. **3**, 65–73 (2013)
21. Edison, T.N.J.I., Sethuraman, M.G.: Electrocatalytic reduction of benzyl chloride by green synthesized silver nanoparticles using pod extract of Acacia nilotica. ACS Sustain. Chem. Eng. **1**, 1326–1332 (2013)
22. Shankar, S.S., Rai, A., Ahmad, A., Sastry, M.: Controlling the optical properties of lemongrass extract synthesized gold nanotriangles and potential application in infrared-absorbing optical coatings. Chem. Mater. **17**, 566–572 (2005)
23. Das, R.K., Sharma, P., Nahar, P., Bora, U.: Synthesis of gold nanoparticles using aqueous extract of Calotropis procera latex. Mater. Lett. **65**, 610–613 (2011)
24. Aswathy, A.S., Philip, D.: Green synthesis of gold nanoparticles using Trigonella foenum-graecum and its size-dependent catalytic activity. Spectrochim. Acta Part A Mol. Biomol. Spectrosc. **97**, 1–5 (2012)
25. Dash, S.S., Rakhi, M., Arun, K.S., Braja, G.B., Patra, B.K.: Saraca indica bark extract mediated green synthesis of poly shaped gold nanoparticles and its application in catalytic reduction. Appl. Nanosci. **4**, 485–490 (2014)
26. Sujitha, M.V., Kannan, S.: Green synthesis of gold nanoparticles using Citrus fruits (Citrus limon, Citrus reticulata and Citrus sinensis) aqueous extract and its characterization. Spectrochim. Acta Part A Mol. Biomol. Spectrosc. **102**, 15–23 (2013)
27. Nishant, S., Mausumi, M.: Biosynthesis and characterization of gold nanoparticles using zooglearanigera and assessment of its antibacterial property. J Clust Sci. **26**, 675–692 (2015)
28. Rajasekharreddy, P., Pathipati, U.R., Bojja, S.: Qualitative assessment of silver and gold nanoparticles synthesis in various plants: a photobiological approach. J. Nanopart. Res. **12**, 1711–1721 (2010)
29. Rastogi, L., Arunachalam, J.: Sunlight based irradiation strategy for rapid green synthesis of highly stable silver nanoparticles using aqueous garlic (Allium sativum) extract and their antibacterial potential. Mater. Chem. Phys. **129**, 558–563 (2011)
30. Zarchi, A.A.K., Mokhtari, N., Arfan, M., Rehman, T., Ali, M., Amini, M., Majidi, R.F., Shahverdi, A.R.: A sunlight-induced method for rapid biosynthesis of nanoparticles using an Andrachnea chordifolia ethanol extract. Appl. Phys. A **103**, 349–353 (2011)
31. Pathipati, U.R., Rajasekharreddy, P.: Green synthesis of silver-protein (core–shell) nanoparticles using Piper betle L. leaf extract and its ecotoxicological studies on Daphnia magna. Colloids Surf. A: Physicochem. Eng. Asp. **389**, 188–194 (2011)
32. Xuetuan, W., Mingfang, L., Li, W., Yang, L., Liang, X., Xu, L., Peng, K., Liu, H.: Synthesis of silver nanoparticles by solar irradiation of cell-free Bacillus amyloliquefaciens extracts and $AgNO_3$. Bioresour. Tech. **103**, 273–278 (2012)

33. Amaladhas, T.P., Usha, M., Naveen, S.: Sunlight induced rapid synthesis and kinetics of silver nanoparticles using leaf extract of *Achyranthes aspera* L. and their antimicrobial applications. Adv. Mat. Lett. **4**, 779–785 (2013)

34. Luo, Y.: Size-controlled preparation of dendrimer-protected gold nanoparticles: a sunlight irradiation-based strategy. Mater. Lett. **62**, 3770–3772 (2008)

35. Pienpinijtham, P., Han, X.X., Suzuki, T., Thammacharoen, C., Ekgasit, S., Ozaki, Y.: Micrometer-sized gold nanoplates: starch mediated photochemical reduction synthesis and possibility of application to tip-enhanced Raman scattering (TERS). Phys. Chem. Chem. Phys. **14**, 9636–9641 (2012)

36. Jebakumar, I.E.T., Sethuraman, M.G.: Instant green synthesis of silver nanoparticles using *Terminalia chebula* fruit extract and evaluation of their catalytic activity on reduction of methylene blue. Process Biochem. **47**, 1351–1357 (2012)

37. Vidhu, V.K., Philip, D.: Catalytic degradation of organic dyes using biosynthesized nanoparticles. Micron **56**, 54–62 (2014)

38. Vanaja, M., Paulkumar, K., Baburaja, M., Rajeshkumar, S., Gnanajobitha, G., Malarkodi, C., Sivakavinesan, M., Annadurai, G.: Degradation of methylene blue using biologically synthesized silver nanoparticles. Bioinorg. Chem. Appl. **2014**, 1–8 (2014)

39. Suvith, V.S.: Daizy Philip: catalytic degradation of methylene blue using biosynthesized gold and silver nanoparticles. Spectrochim. Acta Part A Mol. Biomol. Spectrosc. **118**, 526–532 (2014)

40. Hemant, P.B., Patil, C.D., Salunkhe, R.B., Suryawanshi, R.K., Salunke, B.K., Patil, S.V.: Transformation of aromatic dyes using green synthesized silver nanoparticles. Bioprocess Biosyst. Eng. **37**, 1695–1705 (2014)

41. Sinha, T., Ahmaruzzaman, M.: A novel green and template free approach for the synthesis of gold nanorice and its utilization as a catalyst for the degradation of hazardous dye. Spectrochim. Acta Part A Mol. Biomol. Spectrosc. **142**, 266–270 (2015)

42. Srinath, B.S, Ravishankar Rai, V.: Biosynthesis of gold nanoparticles using extracellular molecules produced by *Enterobacter aerogenes* and their catalytic study. J. Clust Sci. **26**, 1483–1494 (2015)

43. Matsuda, H., Morikawa, T., Ueda, H., Yoshikawa, M.: Medicinal Foodstuffs. XXVII. (1) Saponin Constituents of Gotu Kola (2): structures of New Ursane- and Oleanane-Type Triterpene Oligoglycosides, Centellasaponins B, C, and D, from *Centella asiatica* Cultivated in Sri Lanka. Chem. Pharm. Bull. **49**, 1368–1371 (2001)

44. Subathra, M., Shila, S., Devi, M.A., Panneerselvam, C.: Emerging role of *Centella asiatica* in improving age-related neurological antioxidant status. Exp. Geront. **40**, 707–715 (2005)

45. Kavitha, D., Namasivayam, C.: Experimental and kinetic studies on Methylene blue adsorption by coir pith carbon. Bioresour. Technol. **98**, 14–21 (2007)

46. Srivastava, S., Sinha, R., Roy, D.: Toxicological effects of malachite green. Aquat. Toxicol. **66**, 319–329 (2004)

47. Kang, S.F., Liao, C.H., Po, S.T.: Decolorization of textile wastewater by photo-fenton oxidation technology. Chemosphere **41**, 1287–1294 (2000)

48. Sheny, D.S., Joseph, M., Daizy, P.: Phytosynthesis of Au, Ag and Au–Ag bimetallic nanoparticles using aqueous extract and dried leaf of *Anacardium occidentale*. Spectrochim. Acta Part A **79**, 254–262 (2011)

49. Venkatesh, G., Kalpana, U., Prasad, M.N.V., Rao, N.V.S.: Green synthesis of gold and silver nanoparticles using *Achyranthes aspera* L. leaf extract. Adv. Sci. Eng. Med. **4**, 1–6 (2012)

50. ShahirJamil, S., Nizami, Q., Salam, M.: *Centella asiatica* (Linn.) óA Review. Natural Product Radiance **6**, 158–170 (2007)

51. Satheesh Kumar, B., Suchethakumari, N., Vadisha, S.B., Sharmila, K.P., Mahesh Prasad, B.: Preliminary phytochemical screening of various extracts of Punica granatum peel, whole fruit and seeds. Nitte Univ. J. Health Sci. **4**, 34–38 (2012)

52. Magudapathy, P., Gangopadhyay, P., Panigrahi, B.K., Nair, K.G.M., Dhara, S.: Electrical transport studies of Ag nanoclusters embedded in glass matrix. Phys. B **299**, 142–146 (2001)

53. Narayanan, K.B., Sakthivel, N.: Facile green synthesis of gold nanostructures by NADPH-dependent enzyme from the extract of *Sclerotium rolfsii*. Colloids Surf. A Physicochem. Eng. Asp. **380**, 156–161 (2011)

54. Wani, I.A., Aparna, G., Ahmed, J., Ahmad, T.: Silver nanoparticles: ultrasonic wave assisted synthesis, optical characterization and surface area studies. Mater. Lett. **65**, 520–522 (2011)

55. Panacek, A., Kvitek, L., Prucek, R., Kolar, M., Vecerova, R., Pizurova, N.: Silver colloid nanoparticles: synthesis, characterization, and their antibacterial activity. J. Phys. Chem. B **110**, 16248–16253 (2006)

56. Kim, K.J., Sung, W.S., Suh, B.K., Moon, S.K., Choi, J.S., Kim, J.G.: Antifungal activity and mode of action of silver nanoparticles on *Candida albicans*. Biometals **22**, 235–242 (2009)

57. Shrivastava, S., Bera, T., Roy, A., Singh, G., Rao, P.R., Dash, D.: Characterization of enhanced antibacterial effects of novel silver nanoparticles. Nanotechnology **18**, 225103–225111 (2007)

58. Ouarda, M., Oualid, H., Christian, P.: Sonochemical degradation of malachite green in water. Chem. Eng. Prog. **62**, 47–53 (2012)

59. Lee, Y.C., Kim, J.Y., Shin, H.J.: Removal of malachite green (MG) from aqueous solutions by adsorption, precipitation, and alkaline fading using talc. Sep. Sci. Technol. **48**, 1093–1101 (2013)

60. Han, H.T., Khan, M.M., Kalathil, S., Lee, J., Cho, M.H.: Simultaneous enhancement of methylene blue degradation and power generation in microbial fuel cell by gold nanoparticles. Ind. Eng. Chem. Res. **52**, 8174–8181 (2013)

61. Ganesh, P., Satish, K., Ganesh, S., Sanjay, G.: Biodegradation of malachite green by *Kocuria rosea* MTCC 1532. Acta Chim. Slov. **53**, 492–498 (2006)

62. Sonal, S., Rimi, S., Charanjit, S., Bansa, S.: Enhanced photocatalytic degradation of methylene blue using $ZnFe_2O_4$/MWCNT composite synthesized by hydrothermal method. Indian J. Mater. Sci. **2013**, 1–6 (2013)

63. Impert, O., Katafias, A., Kita, P., Mills, A., Pietkiewicz-Graczyk, A., Wrzeszcz, G.: Kinetic and mechanism of fast leuco-methylene blue oxidation by copper (II)-halide species in acidic aqueous media. Dalton Trans. (2003). doi:10.1039/B205786G

64. Zhang, T., Oyama, T., Aoshima, A., Hidaka, H., Zhao, J., Serpone, N.: Photooxidative demethylation of methylene blue in aqueous TiO_2 dispersions under UV irradiation. J. Photochem. Photobiol. A Chem. **140**, 163–172 (2001)

65. Sivakumar, P., Gaurav Kumar, G.K., Sivakumar, P., Renganathan, S.: Synthesis and characterization of ZnS-Ag nanoballs and its application in photocatalytic dye degradation under visible light. J. Nanostruct. Chem. (2014). doi:10.1007/s40097-014-0107-0

66. Hashemzadeh, F., Rahimin, R., Ghaffarinejad, A.: Mesoporous nanostructures of Nb2O5 obtained by an EISA route for the treatment of malachite green dye-contaminated aqueous solution under UV and visible light irradiation. Ceram. Int. **40**, 9817–9829 (2014)

67. Sarina, S., Waclawik, E.R., Zhu, H.: Photocatalysis on supported gold and silver nanoparticles under ultraviolet and visible light irradiation. Green Chem. **15**, 1814–1833 (2013)

68. Liu, Y., Yu, H.B., Zhan, S.H., Li, Y.L.Z.N., Yang, X.Q.: Fast degradation of Methylene Blue with electrospun hierarchical-Fe_2O_3 nanostructured fibers. J. Sol–Gel. Sci. Technol. **58**, 716–723 (2011)

20

SnO$_2$ nanoparticles as an efficient heterogeneous catalyst for the synthesis of 2H-indazolo[2,1-b]phthalazine-triones

Cinnathambi Subramani Maheswari[1] · Chandrabose Shanmugapriya[2] · Krishnan Revathy[2] · Appaswami Lalitha[1]

Abstract A new protocol for the synthesis of 2H-indazolo[2,1-b]phthalazine-trione derivatives is described via a one-pot three-component condensation reaction of phthalhydrazide, dimedone or 1,3-cyclohexanedione and aromatic aldehydes catalyzed by SnO$_2$ nanoparticles as a heterogeneous catalyst under solvent-free conditions. The SnO$_2$ nanoparticles (NPs) were characterized by FT-IR, X-ray diffraction (XRD), field emission scanning electron microscopy (FE-SEM) and energy dispersive X-ray analyzer (EDAX). The advantages of the protocol are the shorter reaction time, simple work-up procedure and reusable catalyst.

Graphical Abstract

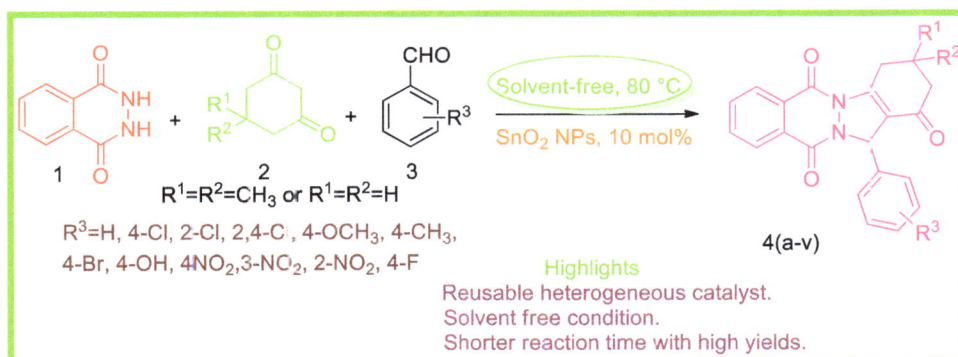

Keywords New protocol · 2H-indazolo[21-b]phthalazine-trione · SnO$_2$ nanoparticles · Solvent-free · Heterogeneous catalyst · Reusable catalyst

Introduction

Generally, nitrogen containing heterocyclic compounds have a wide range of pharmacological and clinical applications [1–6]. Among them, nitrogen incorporated phthalazine containing bridgehead hydrazine heterocyclic compounds are highly desirable drugs with anticonvulsant [7], cardiotonic [8], vasorelaxant [9], antimicrobial [10],

✉ Appaswami Lalitha
  lalitha2531@yahoo.co.in

[1] Department of Chemistry, Periyar University, Periyar Palkalai Nagar, Salem, Tamil Nadu 636011, India

[2] Department of Chemistry, Sri Sarada College For Women (Autonomous), Salem, Tamil Nadu 636016, India

antifungal [11], anticancer [12] and anti-inflammatory [13] activities. In addition, phthalazine derivatives also have excellent optical properties [14, 15]. Due to the notable significances, we would like to establish a new facile protocol for the synthesis of phthalazine scaffolds.

In recent years, the literature reports reveal that only a limited number of solvent-free protocols are available for the synthesis of phthalazine derivatives using the condensation reaction of phthalhydrazide, dimedone, and aromatic aldehydes with various catalysts [16–25]. However, the reported methods have some limitations such as non-biodegradable and expensive metal catalysts, corrosive reaction media and longer reaction times due to which new synthetic protocols may be suggested for the synthesis of phthalazine derivatives via one-pot multicomponent reactions. Multicomponent reactions (MCRs) are more essential to construct the complex organic molecule in a single step without isolation of any intermediates [29, 30]. MCRs with heterogeneous catalyst and solvent-free process are more attractive to the researchers, because they will minimize the waste and usage of hazardous chemicals [31, 32]. Moreover, the heterogeneous catalysts have more enrapture to the researchers because of their selectivity and reactivity in organic transformations [33–35] and they provide more advantages over homogeneous catalysts due to the more active sites. In addition, they are environmental friendly, cost effective, insoluble in solvents and reusable leading to easy work-up procedures. In general, the nanomaterials exhibit superior mechanical, thermal, chemical, electrical, optical and catalytic properties. Therefore, we extend our present research work by employing SnO_2 nanoparticles as a heterogeneous catalyst for the synthesis of phthalazine derivatives. Tin oxide (SnO_2) has a broad spectrum of applications which includes its use in electrochemical [36] and photo-catalytical processes [37], gas sensors [38, 39], optical electronic devices [40] and solar cells [41]. Additionally, the SnO_2 NPs have an excellent catalytic behavior in the synthesis of various heterocyclic compounds [42–45].

The present research work utilizes SnO_2 NPs as a reusable catalyst for the synthesis of 2H-indazolo[2,1-b]phthalazine-trione derivatives. To the best of our knowledge, there is no literature report available for the SnO_2 NPs-catalyzed synthesis of 2H-indazolo[2,1-b]phthalazine-triones.

Experimental Methods

General

All reagents and common solvents were purchased from Loba chemie and Sigma-Aldrich and used without further purification. The products were isolated and characterized by physical and spectral analysis. ^{1}H NMR and ^{13}C NMR spectra were recorded on Bruker (DRX Avance-400) spectrometer in the presence of tetramethylsilane as an internal standard. The IR spectra were recorded on Bruker (ATR FT-IR) apparatus using Zn-Se technique. Reactions were monitored by TLC using precoated silica gel plates and melting points were determined on an electro thermal apparatus, and are uncorrected. Powder X-ray diffraction (XRD) was carried out on a Philips diffractometer of X'pert Company with monochromatized Cu Ka radiation ($k = 1.5406$ Å). Field emission scanning electron microscopy (FE-SEM) and the X-ray dispersive analysis were recorded by FE-SEM (LEO 1455VP), operated at a 15 kV accelerating voltage.

General procedure for the preparation of the SnO_2 nanoparticles

The SnO_2 NPs were prepared by taking 2 g (0.1 M) of stannous chloride dihydrate ($SnCl_2 \cdot 2H_2O$) in 50 ml of distilled water and dissolved. Simultaneously, an aqueous ammonia solution was added drop by drop under continuous stirring for 15–20 min while maintaining the pH of the solution between 9 and 11. The resulting gel was washed with deionized water until the pH reached 7, filtered and dried at 80 °C for 24 h to remove water. Then, the powdered sample was calcinated at 600 °C for 2 h and the finally obtained powder sample was characterized with FT-IR, powder X-ray diffraction (XRD), field emission scanning electron microscopy (FE-SEM) and X-ray dispersive analyses.

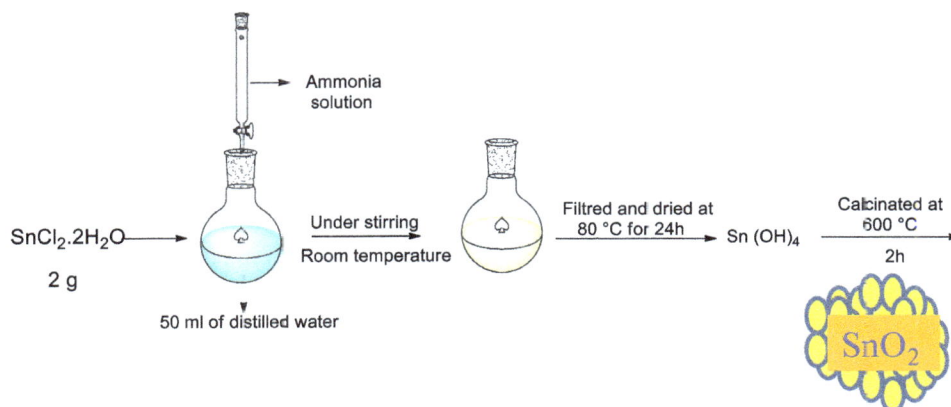

General procedure for the synthesis of 2H-indazolo[2,1-b]phthalazine-trione derivatives

An equimolar mixture of phthalhydrazide (1.0 mmol), aromatic aldehyde (1.0 mmol), 5,5-dimethyl-1,3-cyclo-hexanedione or 1,3-cyclohexanedione (1.0 mmol) and 10 mol % of SnO$_2$ NPs was magnetically stirred over an oil bath at 80 °C under solvent-free condition for an appropriate time. The completion of the reaction was monitored by TLC using ethyl acetate–hexane (1:1) as the eluent. After completion of the reaction, the reaction mass was cooled to room temperature and 3 ml of chloroform was added and stirred well. The product was soluble in chloroform and the catalyst was recovered by simple fil-tration. The catalyst was washed with 3 ml of acetone and dried at 100 °C. Finally, the recovered catalyst was sub-jected to the next run of the reaction. The solvent was removed under vacuum to get the crude product, which on further recrystallization with ethanol afforded the pure product (compared with literatures Ref. [16–18, 26–28]). The pure products were characterized with FT-IR, ^{1}H and ^{13}C NMR spectroscopic techniques.

Results and discussion

Characterization of SnO$_2$ NPs

The SnO$_2$ NPs were synthesized using an already reported procedure [38]. The synthesized SnO$_2$ NPs were charac-terized by Fourier transform-infrared spectroscopy, powder X-ray diffraction, field emission scanning electron micro-scopy and energy dispersive X-ray spectroscopy.

FT-IR and XRD analysis

We have compared the FT-IR spectra of as-synthesized form and calcinated samples of SnO$_2$ NPs (Fig. 1).

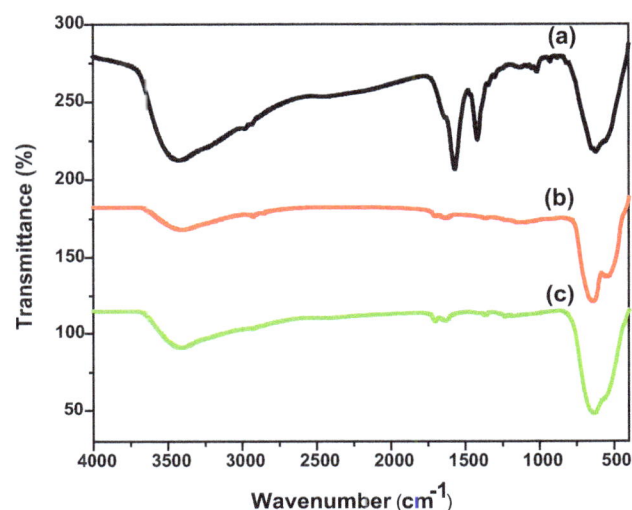

Fig. 1 The comparative FT-IR spectra of: **a** Before calcinations of SnO$_2$ NPs, **b** After calcinations of SnO$_2$ NPs and **c** SnO$_2$ NPs after the sixth run of the reaction

Figure 1a shows a broad band at 3460.78 cm^{-1} which represents the presence of water molecule bound to Sn surface. The band at 1573.15 cm^{-1} is due to the bending vibration of water molecules and the peak at 1423.15 cm^{-1} represents the unreactive SnCl$_2$. The band at 625 cm^{-1} indicates the Sn–OH. Furthermore, the calci-nated sample has shown a band at 646.31 cm^{-1} corre-sponds to Sn–O–Sn and the band present at 553.94 cm^{-1} is due to the stretching vibration of Sn–O. The intensity of peaks due to water molecules is very much decreased in the case of calcinated sample. The observed results reveal that the Sn–O band is not present in the fresh sample. Figure 1c shows the IR spectrum of the SnO$_2$ catalyst after the sixth run of the reaction in which there is no change in the peak position and intensity when compared to that of the calci-nated catalyst supporting its reusability results. From the X-ray diffraction spectrum, the formations of SnO$_2$ NPs are confirmed (Fig. 2). All the diffraction peaks are in good

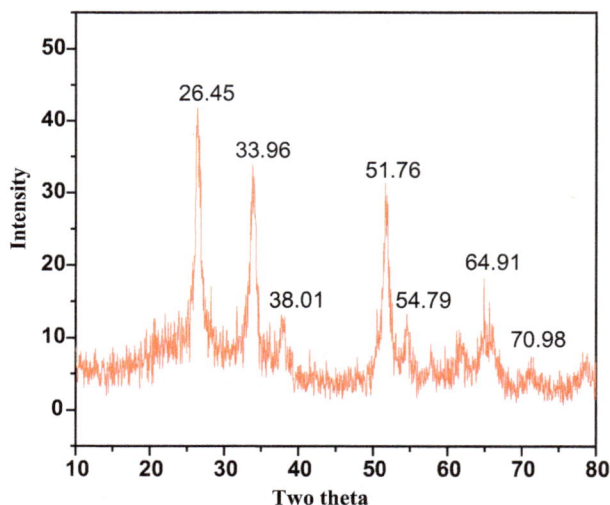

Fig. 2 XRD spectra of SnO$_2$ NPs

agreement with the standard (JCPDS No- 41-1445) which can be indexed as the tetragonal structure of SnO$_2$ with lattice constant of $a = 4.78$, $c = 3.17$ and unit cell volume $d = 72.66$. No other diffraction peaks are observed in the XRD pattern that clearly indicates the purity of SnO$_2$ NPs. The average crystallite size of the SnO$_2$ NPs was calculated using Scherrer formula as, $(D = k\lambda/\beta\cos\theta)$, where D is the average crystallite size, k is the Scherrer constant, λ is the wavelength of the X-ray beam, β is the $(\pi/2)$ FWHM (Full width at half maximum) and θ is the diffraction angle. The average crystallite size of the SnO$_2$ NPs was found to be 23.4 nm. Meanwhile, these results were in good agreement with the literature report [46].

FE-SEM and EDAX analysis

The field emission scanning electron microscopy (FE-SEM) that shows the morphology and size of the SnO$_2$ NPs is presented in Fig. 3. The morphology and size of the individual particles could not be measured due to their extremely small size and heavy aggregation. In general, the formation of agglomerated nanoparticles can be explained by nucleation–aggregation–agglomeration–growth mechanism proposed by Rodriguez-Clemente et al. [47]. The particle size of the SnO$_2$ NPs is in the range of 44–71.79 nm approximately and also the SnO$_2$ NPs have been aggregated in spherical morphology. Finally, the energy dispersive X-ray spectrum exhibits the atomic weight % of tin 33.33 and 66.67% of oxygen present in the SnO$_2$ NPs (Fig. 4).

Catalytic activity

The formation of SnO$_2$ NPs made us to utilize the NPs as a catalyst for the one-pot synthesis of 2H-indazolo[2,1-b]phthalazine-triones under various reaction conditions.

We have optimized the amount of catalyst required to catalyze the reaction where the temperature and solvent have also been screened. Initially, we have performed the model reaction with a mixture of 4-chlorobenzaldehyde (1 mmol), dimedone (1 mmol) and phthalhydrazide (1 mmol) without catalyst and solvent and found that there was no reaction even after five hours (Table 1, Entry 1) revealing the need of a catalyst to activate the reaction. By employing 20 mol % of various Lewis acid catalysts such as NiCl$_2$.6H$_2$O, CuCl$_2$.2H$_2$O, CaCl$_2$.2H$_2$O and MnCl$_2$.4-H$_2$O for the reaction, the yield of the expected product was very low even after 60 min of the reaction (Table 1, Entries 2-5). Next, we investigated the reaction with 20 mol % of SnCl$_2$·2H$_2$O, where the moderate yield of the product was observed (Table 1, Entry 6). The model reaction when conducted at 80 °C with 10 mol % of SnO$_2$ bulk led to prolonged reaction time with 72% yield (Table 1, Entry 7) which made us to standardize the reaction using SnO$_2$ NPs. Using 5 mol % SnO$_2$ NPs, the reaction provided 70% of the product with lesser reaction time when compared with other metal chlorides (Table 1, Entry 8) and 10 mol % of the SnO$_2$ NPs catalyzed the reaction to an excellent yield of the product (Table 1, Entry 9) with shorter reaction time. Further increase in the mol % of SnO$_2$ NPs did not improve the yield of the product (Table 1, Entries 10 and 11). Meanwhile, the model reaction was studied with 10 mol % of SnO$_2$ NPs under various temperatures. At 80 and 90 °C, there was no change in the yield and reaction time, (Table 1, Entry 12) but the yield decreased at 70 or 60 °C (Table 1, entries 13 and 14).

With the optimized catalyst, solvent-mediated reaction did not succeed well which may be due to the interruption of solvent molecules for the direct contact between the reactants and the catalyst (Table 1, entries 15–17). We have compared the yield of the product and the reaction time with previous literature in Table 2, entries 1–10. Apparently, some of the reported catalysts afforded good yields but most of the catalysts were hazardous to the environment and the catalyst quantity needed was also more in addition to longer reaction time. The catalytic activity of SnO$_2$ NPs may be due to the presence of more acidic (Sn^{4+}) and the basic (O^{2-}) surface sites [37] that could readily induce the organic transformations. We have concluded that 10 mol % of SnO$_2$ NPs at 80 °C was more favorable under solvent-free condition to provide excellent yield of the target product with shorter reaction time (Table 1, entry 9) and the optimization results are presented in Table 1. The reusability of the catalyst was also successfully checked up to seven runs.

Using the above optimized conditions, we have taken various aromatic aldehydes having different electron withdrawing and donating substituents in the ortho, meta

Fig. 3 FE-SEM images of the SnO$_2$ NPs

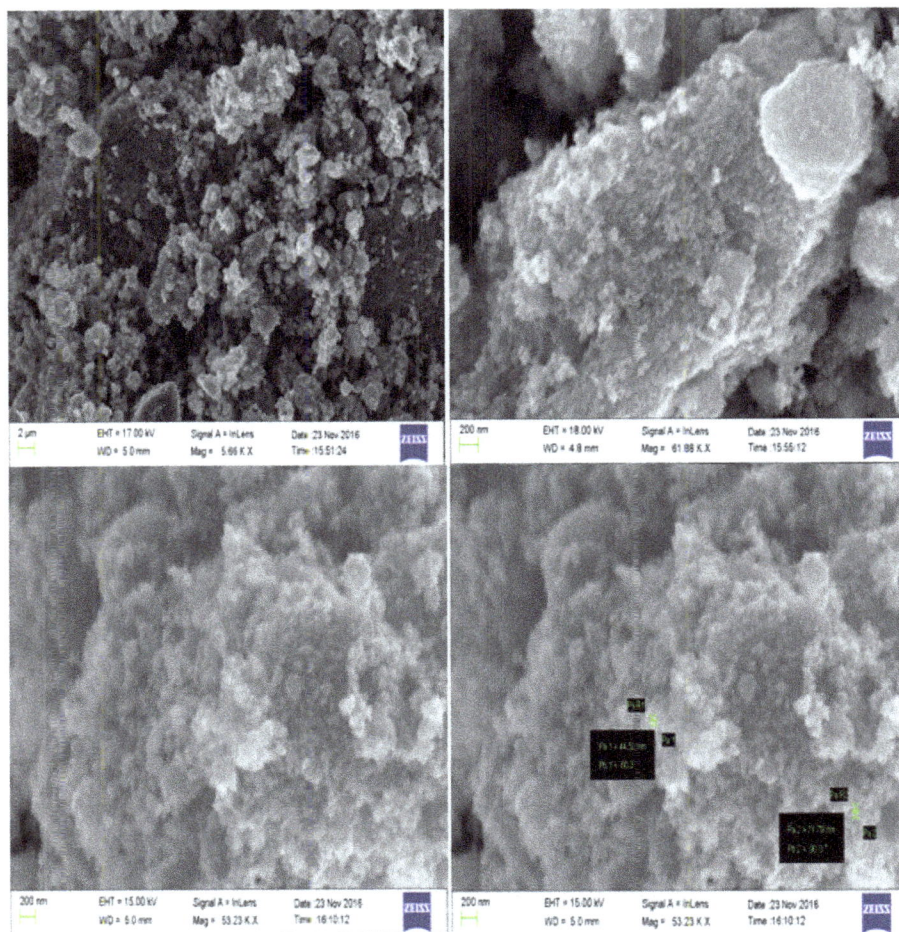

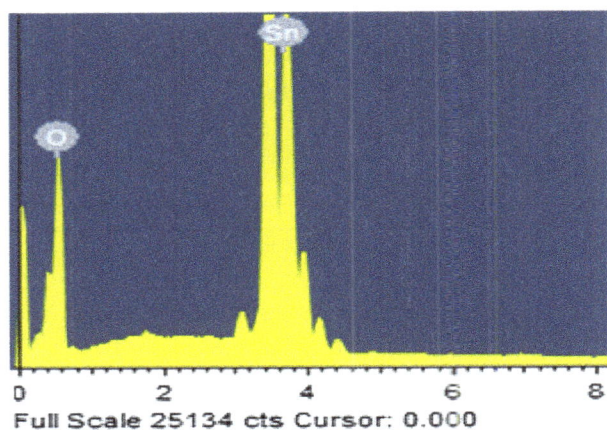

Fig. 4 EDAX spectrum of SnO$_2$ NPs

and para positions and cyclic diketones for the synthesis of 2H-indazolo[2,1-b]phthalazine-trione derivatives (Table 3). Almost all aldehydes provided good to excellent yields of the products irrespective of the position as well as electron donating and withdrawing nature of the group in the aryl aldehydes (Scheme 1).

The synthesized 2H-indazolo[2,1-b]phthalazine-trione derivatives were confirmed by ^{1}H-NMR and ^{13}C-NMR spectroscopic techniques. The ^{1}H-NMR spectrum of the compound **4p** exhibited a peak at δ 2.24–3.59 ppm which could confirm the presence of six methylene protons in the cyclohexenone ring and another singlet appeared at δ 3.75 ppm which indicates the presence of 4-OCH$_3$ on the aromatic ring. The methine (C–H) proton has given a single peak at δ 6.42 ppm and the remaining aromatic protons exhibited peaks at δ 6.84–8.35 ppm. The ^{13}C-NMR spectrum of the compound (**4p**) exhibited peak at δ 22.35–64.60 ppm which corresponds to the presence of aliphatic carbon in the cyclohexenone ring and the peak appeared at δ 114.11 ppm due to the presence of methine carbon. The characteristic peak of carbonyl carbon appeared at δ 192.62 ppm and the rest of the aromatic carbons appeared at δ 159.74–119.68 ppm.

The plausible reaction mechanism proposed for the synthesis of 2H-indazolo[2,1-b]phthalazine-trionederivatives is represented in Scheme 2. Initially, the SnO$_2$ NPs catalyzed the Knoevenagel condensation reaction between dimedone (**I**) and aromatic aldehyde (**II**) providing **III**.

Table 1 Optimization of reaction condition for the synthesis of 2*H*-indazolo[2,1-*b*]phthalazine-triones

Entry	Catalyst (mol%)	Reaction conditions	Time (min)	Yield[a] (%)
1	No catalyst	Solvent-free, 80 °C	>300	No reaction
2	NiCl$_2$·6H$_2$O (20)	Solvent-free, 80 °C	60	10
3	CuCl$_2$·2H$_2$O (20)	Solvent-free, 80 °C	60	17
4	CaCl$_2$·2H$_2$O (20)	Solvent-free, 80 °C	60	12
5	MnCl$_2$·4H$_2$O (20)	Solvent-free, 80 °C	60	20
6	SnCl$_2$·2H$_2$O (20)	Solvent-free, 80 °C	60	48
7	SnO$_2$ (10) bulk	Solvent-free, 80 °C	44	72
8	SnO$_2$ (5)	Solvent-free, 80 °C	26	80
9	SnO$_2$ (10)	Solvent-free, 80 °C	12	96
10	SnO$_2$ (15)	Solvent-free, 80 °C	12	96
11	SnO$_2$ (20)	Solvent-free, 80 °C	12	96
12	SnO$_2$ (10)	Solvent-free, 90 °C	12	96
13	SnO$_2$ (10)	Solvent-free, 70 °C	12	81
14	SnO$_2$ (10)	Solvent-free, 60 °C	12	66
15	SnO$_2$ (10)	Ethanol, Reflux	12	60
16	SnO$_2$ (10)	Methanol, Reflux	12	48
17	SnO$_2$ (10)	Water, Reflux	>60	20

Reaction condition: 4-chlorobenzaldehyde (1 mmol), dimedone or 1,3-cyclohexanedione (1 mmol) and phthalhydrazide (1 mmol) and 10 mol % SnO$_2$ NPs, solvent-free condition

[a] Isolated yield

Table 2 Comparative study of the present protocol with previous reports for the synthesis of 2*H*-indazolo[2,1-*b*]phthalazine-triones

Entry	Catalyst	Reaction conditions	Time (min)	Yield %	Ref
1	MNPs-guanidine (0.030 g)	Solvent-free, 70 °C	34	94	[16]
2	Nano-ASA (0.04 g)	Solvent-free, 110 °C	10	100	[17]
3	*p*-TSA (0.1 g)	Solvent-free, 80 °C	30	93	[18]
4	Iodine (0.1 g)	Solvent-free, 25–30 °C	10	92	[19]
5	PPA-SiO$_2$(0.1 g)	Solvent-free, 100 °C	30	92	[20]
6	Et$_3$N–SO$_3$HCl (0.043 g)	Solvent-free, 20 °C	40	93	[21]
7	PEG-SO$_3$H (0.61 g)	Solvent-free, 80 °C	40	91	[22]
8	SO$_3$H-FMSM (0.020 g)	Solvent-free, 110 °C	18	95	[23]
9	H$_4$SiW$_{12}$O$_{40}$ (0.2 g)	Solvent-free, 100 °C	60	75	[24]
10	ZrOCl$_2$.8H$_2$O (0.096 g)	Solvent-free, 80 °C	60	83	[25]
11	SnO$_2$ (0.015 g)	Solvent-free, 80 °C	12	96	Present work

This step is facilitated by the activation of carbonyl group in aromatic aldehyde by SnO$_2$ NPs. Michael addition of phthalhydrazide to **III** provides **IV** which on cyclization afforded the target product (**V**).

Reusability of the catalyst

Next, we have also checked the reusability of SnO$_2$ NPs. The reusability of the catalyst was investigated with the model reaction using an equimolar mixture of 4-chlorobenzaldehyde (1 mmol), dimedone (1 mmol) and phthalhydrazide (1 mmol) performed with 10 mol % of SnO$_2$ NPs under solvent-free condition. After completion

of the reaction, the reaction mass was cooled to room temperature and 3 ml of chloroform was added and stirred well. After that the product soluble in chloroform was separated by filtration. The recovered catalyst was washed with 3 ml of acetone to remove any organic residues present in the catalyst. The recovered catalyst was dried at 100 °C and subjected to the consecutive runs with same experimental condition. The yield of the product and the reaction time were not much affected up to fifth run after that a slight decrement in the yield of the product was observed. From these results, it has been concluded that 10 mol % of SnO$_2$ NPs can be efficiently reused for more than five to six consecutive runs without loss in its catalytic

Table 3 Synthesis of 2H-indazolo[2,1-b]phthalazine-trione derivatives catalyzed by SnO$_2$ NPs under solvent-free condition

Entry	ArCHO	R^1, R^2	Product (4a-4v)	Time (min)	Yield^a (%)	Mp (° C)	Lit. Mp (°C)[Ref]
1	C$_6$H$_5$	CH$_3$	4a	14	93	206–208	[17]
2	4-ClC$_6$H$_4$	CH$_3$	4b	12	96	259–261	[17]
3	2-ClC$_6$H$_4$	CH$_3$	4c	13	85	261–263	[17]
4	2,4-ClC$_6$H$_3$	CH$_3$	4d	12	88	220–222	[16]
5	4-OCH$_3$C$_6$H$_4$	CH$_3$	4e	15	91	216–218	[17]
6	4-CH$_3$C$_6$H$_4$	CH$_3$	4f	14	90	227–229	[17]
7	4-BrC$_6$H$_4$	CH$_3$	4 g	13	93	260–262	[17]
8	4-OHC$_6$H$_4$	CH$_3$	4 h	13	90	290–292	[26]
9	4-NO$_2$C$_6$H$_4$	CH$_3$	4i	12	92	218–221	[17]
10	3-NO$_2$C$_6$H$_4$	CH$_3$	4j	12	87	267–269	[18]
11	2-NO$_2$C$_6$H$_4$	CH$_3$	4 k	13	84	236–238	[16]
12	4-FC$_6$H$_4$	CH$_3$	4 l	14	94	218–220	[18]
13	C$_6$H$_5$	H	4 m	13	86	229–230	[16]
14	4-ClC$_6$H$_4$	H	4n	15	93	268–270	[27]
15	2,4-ClC$_6$H$_3$	H	4o	13	86	273–274	[27]
16	4-OCH$_3$C$_6$H$_4$	H	4p	17	90	252–253	[16]
17	4-CH$_3$C$_6$H$_4$	H	4q	19	87	242–244	[16]
18	4-BrC$_6$H$_4$	H	4r	17	88	279–280	[27]
19	4-OHC$_6$H$_4$	H	4 s	22	86	262–263	[28]
20	4-NO$_2$C$_6$H$_4$	H	4t	20	86	251–253	[16]
21	3-NO$_2$C$_6$H$_4$	H	4u	22	84	226–227	[26]
22	4-FC$_6$H$_4$	H	4v	17	90	259–261	[16]

^{a} Reaction condition: aromatic aldehydes (1 mmol), dimedone or 1,3-cyclohexanedione (1 mmol) and phthalhydrazide (1 mmol) and 10 mol % of SnO$_2$ NPs, solvent-free condition at 80 °C

^{b} Isolated yield

Scheme 1 Synthesis of 2H-indazolo[2,1-b]phthalazine-triones

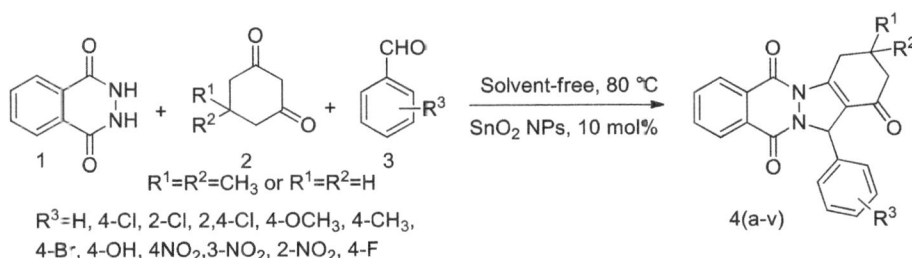

Scheme 2 A plausible mechanism for the one-pot three-component synthesis of 2H-indazolo[2,1-b]phthalazine-triones catalyzed by 10 mol % of SnO$_2$ under solvent-free condition

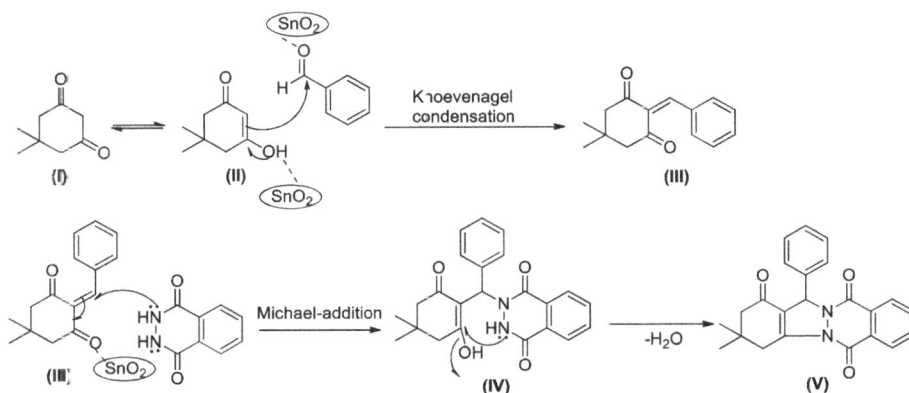

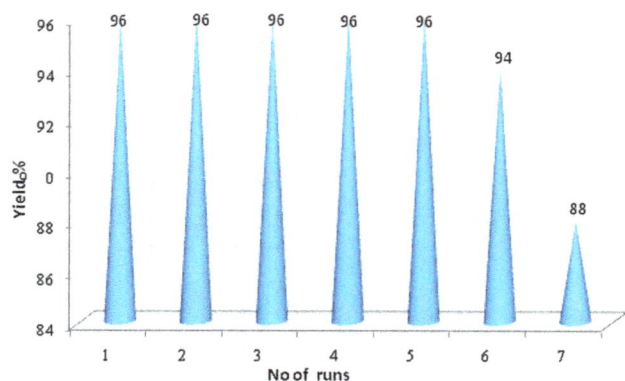

Fig. 5 Reusability of the catalytic system

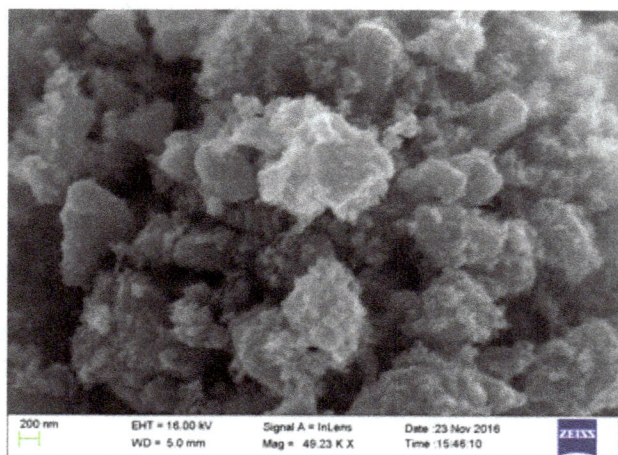

Fig. 6 FE-SEM image of SnO$_2$ NPs after sixth run

activity (Fig. 5). After the sixth run, the recovered SnO$_2$ NPs were further subjected to FE-SEM analysis which was shown in Fig. 6.

Conclusion

We have demonstrated the SnO$_2$ NPs as an effective catalyst for the synthesis of 2H-indazolo[2,1-b]phthalazine-triones derivatives via one-pot three-component condensation reaction of the aromatic aldehydes, dimedone/1,3-cyclohexanedione and phthalhydrazide under solvent-free condition at 80 °C. The SnO$_2$ NPs have remarkable advantages such as cheap and efficient catalyst. The excellent yield of the product without using column chromatography, shorter reaction time and easy work-up procedure and the reusability of the catalyst are the added advantages of this method. The present protocol seems to be a better alternative compared to other reported protocols.

References

1. Taj, T., Kamble, R.R., Gireesh, T.M., Hunnur, R.K., Margankop, S.B.: One-pot synthesis of pyrazoline derivatised carbazoles as antitubercular, anticancer agents, their DNA cleavage and antioxidant activities. Eur. J. Med. Chem. **46**, 4366–4373 (2011)
2. Bergstrom, F.W.: Heterocyclic nitrogen compounds. Chem. Rev. **35**, 77–277 (1944)
3. Padmaja, A., Payani, T., Dinneswara Reddy, G., Padmavathi, V.: Synthesis, antimicrobial and antioxidant activities of substituted pyrazoles, isoxazoles, pyrimidine and thioxopyrimidine derivatives. Eur. J. Med. Chem. **44**, 4557–4566 (2009)
4. Litvinov, V.P.: Multicomponent cascade heterocyclisation as a promising route to targeted synthesis of polyfunctional pyridines. Russ. Chem. Rev. **72**, 69–85 (2003)
5. Xu, Y., Guo, Q.-X.: Syntheses of heterocyclic compounds under microwave irradiation. Heterocycles **63**, 903–974 (2004)
6. Clark, M.P., Laughlin, S.K., Laufersweiler, M.J., Bookland, R.G., Brugel, T.A., Golebiowski, A., Sabat, M.P., Townes, J.A., VanRens, J.C., Djung, J.F., Natchus, M.G., De, B., Hsieh, L.C., Xu, S.C., Walter, R.L., Mekel, M.J., Heitmeyer, S.A., Brown, K.K., Juergens, K., Taiwo, Y.O., Janusz, M.J.: Development of orally bioavailable bicyclic pyrazolones as inhibitors of tumor necrosis factor-α production. J. Med. Chem. **47**, 2724–2727 (2004)
7. Zhang, L., Guan, L.P., Sun, X.Y., Wei, C.X., Chai, K.Y., Quan, Z.S.: Synthesis and anticonvulsant activity of 6-alkoxy-[1, 2, 4]triazolo[3,4-a]phthalazines. Chem. Bio. Drug Des. **73**, 313–319 (2009)
8. Nomoto, Y., Obase, H., Takai, H., Teranishi, M., Nakamura, J., Kubo, K.: Studies on cardiotonic agents. III.: synthesis of 1-[1-(6,7-dimethoxy-4-quinazolinyl)-4-piperidinyl]-3-substituted 2-imidazolidinone and 2-imidazolidinethione derivatives. Chem. Pharm. Bull. **38**, 2467–2471 (1990)
9. Del Olmo, E., Barboza, B., Ybarra, M.I., Lopez-Perez, J.L., Carron, R., Sevilla, M.A., Bosellid, C., San Felicianoa, A.: Vasorelaxant activity of phthalazinones and related compounds. Bioorg. Med. Chem. Lett. **16**, 2786–2790 (2006)
10. Barbuceanu, S.-F., Saramet, G., Almajan, G.L., Draghici, C., Barbuceanu, F., Bancescu, G.: New heterocyclic compounds from 1,2,4-triazole and 1,3,4-thiadiazole class bearing diphenylsulfone moieties. Synthesis, characterization and antimicrobial activity evaluation. Eur. J. Med. Chem. **49**, 417–423 (2012)
11. Ryu, C.K., Park, R.E., Ma, M.Y., Nho, J.H.: Synthesis and antifungal activity of 6-arylamino-phthalazine-5,8-diones and 6,7-bis(arylthio)-phthalazine-5,8-diones. Bioorg. Med. Chem. Lett. **17**, 2577–2580 (2007)
12. De, P., Baltas, M., Lamoral-Theys, D., Bruyere, C., Kiss, R., Bedos-Belval, F., Saffon, N.: Synthesis and anticancer activity evaluation of 2(4-alkoxyphenyl)cyclopropylhydrazides and triazolophthalazines. Bioorg. Med. Chem. **18**, 2537–2548 (2010)
13. Abdalla, M.S.M., Hegab, M.I., Abo Taleb, N.A., Hasabelnaby, S.M., Goudah, A.: Synthesis and anti-inflammatory evaluation of some condensed [4-(3,4-dimethylphenyl)-1(2H)-oxo-phthalazin-2-yl]acetic acid hydrazide. Eur. J. Med. Chem. **45**, 1267–1277 (2010)
14. Cheng, Y., Ma, B., Wudl, F.: Synthesis and optical properties of a series of pyrrolopyridazine derivatives: deep blue organic luminophors for electroluminescent devices. J. Mater. Chem. **9**, 2183–2188 (1999)
15. Raghuvanshi, D.S., Singh, K.N.: A highly efficient green synthesis of 1H-pyrazolo[1,2-b]phthalazine-5,10-dione derivatives and their photophysical studies. Tetrahedron Lett. **52**, 5702–5705 (2011)

16. Atashkar, B., Rostami, A., Gholami, H., Tahmasbi, B.: Magnetic nanoparticles Fe3O4-supported guanidine as an efficient nanocatalyst for the synthesis of 2H-indazolo[2,1-b]phthalazine-triones under solvent-free conditions. Res. Chem. Intermed. **41**, 3675–3681 (2015)

17. Kiasat, A.R., Noorizadeh, S., Ghahremani, M., Saghanejad, S.J.: Experimental and theoretical study on one-pot, three-component route to 2H-indazolo[2,1-b]phthalazine-triones catalyzed by nano-alumina sulforic acid. J. Mol. Struct. **1036**, 216–225 (2013)

18. Sayyafi, M., Soorki, A.A., Bazgir, A.: One-pot synthesis and antibacterial activities of novel 1H-pyridazino[1,2-a]indazole-1,6,9(2H,11H)-triones. Chem. Pharm. Bull. **56**, 1289–1291 (2008)

19. Varghese, A., Nizam, A., Kulkarni, R., George, L.: Solvent-free synthesis of 2H-indazolo[2,1-b] phthalazine-triones promoted by cavitational phenomenon using iodine as catalyst. Eur. J. Chem. **4**, 132–137 (2013)

20. Shaterian, H.R., Hosseinian, A., Ghashang, M.: Reusable silica supported poly phosphoric acid catalyzed three-component synthesis of 2H-indazolo [2, 1-b] phthalazine-trione derivatives. Arkivoc **II**, 59–67 (2009)

21. Pouramiri, B., TavakolinejadKermani, E : One-pot, four-component synthesis of new 3,4,7,8-tetrahydro-3,3-dimethyl-11-aryl-2H-pyridazino[1,2-a]indazole-1,6,9(11H) triones and 2H-indazolo[2,1-b]phthalazine-1,6,11(13H)-triones using an acidic ionic liquid N, N-diethyl-N-sulfoethanammorium chloride ([Et3 N–SO3H]Cl) as a highly efficient and recyclable catalyst. Tetrahedron Lett. **57**, 1006–1010 (2016)

22. Hasaninejad, A., Zare, A., Shekouhy, M.: Highly efficient synthesis of triazolo [1,2-a] indazole-triones and novel spirotriazolo [1,2-a] indazole-tetraones under solvent-free conditions. Tetrahedron **67**, 390–400 (2011)

23. AlinasabAmiri, A., Javanshir, S., Dolatkhah, Z., Dekamin, M.G.: SO3H-functionalized mesoporous silica materials as solid acid catalyst for facile and solvent-free synthesis of 2H-indazolo[2,1-b]phthalazine-1,6,11-trione derivatives. New J. Chem. **39**, 9665–9671 (2015)

24. Hassankhani, A., Mosaddegh, E., Yousef Ebrahimipour, S.: Tungstosilicic acid as an efficient catalyst for the one-pot multicomponent synthesis of triazolo[1,2-a]indazole-1,3,8-trione derivatives under solvent-free conditions. Arab. J. Chem. **9**, S936–S939 (2016)

25. Tavakoli, H.R., Moosavi, S.M., Bazgir, A.: ZrOCl2·8H2O as an efficient catalyst for the three-component synthesis of triazoloindazoles and indazolophthalazines. J. Korean Chem. Soc. **57**, 472–475 (2013)

26. Davarpanah, J., Rezaee, P., Elahi, S.: Synthesis and characterization of a porous acidic catalyst functionalized with an imidazole ionic liquid, and its use for synthesis of phthalazinedione and phthalazinetrione heterocyclic compounds. Res. Chem. Intermed. **41**, 9903–9915 (2015)

27. Nagarapu, L., Bantu, R., Mereyala, H.B.: TMSCl-mediated one-pot, three-Component synthesis of 2H-indazolo[2,1-b]phthalazine-triones. J. Heterocyclic Chem. **46**, 728–731 (2009)

28. Khurana, J.M., Magoo, D.: Efficient one-pot syntheses of 2H-indazolo[2,1-b] phthalazine-triones by catalytic H2SO4 in water–ethanol or ionic liquid. Tetrahedron Lett. **50**, 7300–7303 (2009)

29. Toure, B.B., Hall, D.G.: Natural product Synthesis using multi-component reaction strategies. Chem. Rev. **109**, 4439–4486 (2009)

30. Razvan, C.C., Ruijter, E., Orru, R.V.A.: Multicomponent reactions: advanced tools for sustainable organic synthesis. Green Chem. **16**, 2958–2975 (2014)

31. Bamoniri, A., Moshtael-Arani, N.: Nano-Fe3O4 encapsulated-silica supported boron trifluoride as a novel heterogeneous solid acid for solvent-free synthesis of arylazo-1-naphthol derivatives. RSC Adv. **5**, 16911–16920 (2015)

32. Maddila, S.N., Maddila, S., vanZyl, W.E., Jonnalagadda, S.B.: Mn doped ZrO2 as a green, efficient and reusable heterogeneous catalyst for the multicomponent synthesis of pyrano[2,3-d]pyrimidine derivatives. RSC Adv. **5**, 37360–37366 (2015)

33. Tarannum, S., Siddiqui, Z.N.: Fe(OTs)3/SiO2: a novel catalyst for the multicomponent synthesis of dibenzodiazepines under solvent-free conditions. RSC Adv. **5**, 74242–74250 (2015)

34. Karthikeyan, G., Pandurangan, A.: Heteropolyacid (H3PW12O40) supported MCM-41: an efficient solid acid catalyst for the green synthesis of xanthenedione derivatives. J. Mol. Catal. A-Chem. **311**, 36–45 (2009)

35. Adharvana Chari, M., Syamasundar, K.: Silicagel supported sodium hydrogensulfate as a heterogenous catalyst for high yield synthesis of 3, 4-dihydropyrimidin-2 (1H)-ones. J. Mol. Catal. A-Chem. **221**, 137–139 (2004)

36. Lu, Y.C., Ma, C., Alvarado, J., Kidera, T., Dimov, N., Meng, Y.S., Okada, S.: Electrochemical properties of tin oxide anodes for sodium-ion batteries. J. Power Sources **284**, 287–295 (2015)

37. Bayal, N., Jeevanandam, P.: Sol-gel synthesis of SnO2-MgO nanoparticles and their photocatalytic activity towards methylene blue degradation. Mater. Res. Bull. **48**, 3790–3799 (2013)

38. Sharp, S.L., Kumar, G., Vicenzi, E.P., Bocarsly, A.B., Heibel, M.: Formation and structure of a tin-iron oxide solid-state system with potential applications in carbon monoxide sensing through the use of cyanogelchemistry. Chem. Mater. **10**, 880–885 (1998)

39. Yang, G., Haibo, Z., Biying, Z.: Monolayer dispersion of oxide additives on SnO2 and their promoting effects on thermal stability of SnO2 ultrafine particles. J. Mater. Sci. **35**, 917–923 (2000)

40. Chopra, K.L., Major, S., Pandya, D.K.: Transparent conductors-A status review. Thin Solid Films **102**, 1–46 (1983)

41. Ferrere, S, Zaban, A., Gregg, B.A.: Dye sensitization of nanocrystalline tin oxide by perylene derivatives. J. Phys. Chem. B. **101**, 4490–4493 (1997)

42. Sharghi, H., Ebrahimpourmoghaddam, S., Memarzadeh, R., Javadpour, S.: Tin oxide nanoparticles (NP-SnO2): preparation, characterization and their catalytic application in the Knoevenagel condensation. J. Iran. Chem. Soc. **10**, 141–149 (2013)

43. Fallah, N.S., Mokhtary, M.: Tin oxide nanoparticles (SnO2-NPs): an efficient catalyst for the one-pot synthesis of highly substituted imidazole derivatives. J. Taibah Univ. Sci. **9**, 531–537 (2015)

44. Dehbashi, M., Aliahmad, M., Shafiee, M.R.M., Ghashang, M.: SnO2 nanoparticles: preparation and evaluation of their catalytic activity in the oxidation of aldehyde derivatives to their carboxylic acid and sulfides to sulfoxide analogs. Phosphorus Sulfur **188**, 864–872 (2013)

45. Yelwande, A.A., Navgire, M.E., Tayde, D.T., Arbad, B.R., Lande, M.K.: SnO2/SiO2 nanocomposite catalyzed one-pot synthesis of 2-arylbenzothiazole derivatives. Bull. Korean Chem. Soc. **33**, 1856–1860 (2012)

46. Zolfigol, M.A., Baghery, S., Moosavi-Zare, A.R., Vahdat, S.M., Alinezhad, H., Norouzi, M.: Design of 1-methylimidazolium tricyanomethanide as the first nanostructured molten salt and its catalytic application in the condensation reaction of various aromatic aldehydes, amides and β-naphthol compared with tin dioxide nanoparticles. RSC Adv. **5**, 45027–45037 (2015)

47. Lopez-Macipe, A., Gomez-Morales, J., Rodriguez-Clemente, R.: Nanosized hydroxyapatite precipitation from homogeneous calcium/citrate/phosphate solutions using microwave and conventional heating. Adv. Mater. **10**, 49–53 (2009)

21

Characterization and optical studies of PVP-capped silver nanoparticles

Ali Mirzaei[1] · Kamal Janghorban[1] · Babak Hashemi[1] · Maryam Bonyani[1] · Salvatore Gianluca Leonardi[2] · Giovanni Neri[2]

Abstract In this study, the size-controlled synthesis of silver nanoparticles (Ag NPs) via chemical reduction method by $NaBH_4$ as a reducing agent and poly(vinyl pyrrolidone) or PVP as a stabilizing agent is reported. Changing of ratios between reducing agent and stabilizing agent relative to $AgNO_3$-optimized conditions for synthesis of stable Ag NPs was studied. The formation of Ag NPs was tracked by UV–Vis spectroscopy, X-ray diffraction (XRD), X-ray photoelectron spectroscopy (XPS), and photoluminescence (PL) spectroscopy. Particle size distribution was studied by particle size analyzer, and the morphology was examined by scanning electron microscopy (SEM) and transmission electron microscopy (TEM). The optical properties of the synthesized Ag NPs were also investigated. The optimized Ag NPs were very stable even after 1 month that was due to effective stabilization by PVP molecules. The mechanism of Ag NPs formation and stabilization is discussed in detail.

Keywords Ag NPs · PVP · Chemical reduction · Optical properties

Introduction

Ag nanoparticle (Ag NPs) is clusters of silver atoms of 1–100 nm at least in one dimension [1, 2]. Owing to their unique physical and chemical properties, they are widely used in various areas of science and technology, such as surface-enhanced Raman spectroscopy [3], catalysts [4], anti-bacterial materials [5], sensors [6], lubricating materials [7], and so on, with exponentially increasing production. As the properties of Ag NPs depend on their sizes, up to now, various methods, such as spray pyrolysis synthesis [8], microwave irradiation synthesis [9], DC arc thermal plasma synthesis [10], chemical synthesis [11], hydrothermal synthesis [12], UV irradiation synthesis [13], sonochemical synthesis [14], laser ablation synthesis [15], thermal decomposition synthesis [16], atom beam-sputtering synthesis [17], and so forth, have been employed to prepare Ag NPs with different sizes and shapes.

Among the various methods for synthesis of Ag NPs, the wet chemical reduction method is the most popular synthesis route, thanks to its simplicity, low cost and ability to produce large quantities of Ag NPs [1]. In a typical chemical reduction process, a reducing agent, such as $NaBH_4$ [18], hydrazine [19], triethanolamine [20], L-ascorbic acid [21], glucose [22], and aniline [11], and a protective agent, such as poly(vinyl alcohol) (PVA) [23], poly(vinyl pyrrolidone) (PVP), cellulose [24], alkanethiols [25], alkylamines [26], dodecanethiol [27], and carboxyliccacids [28], help to obtain Ag NPs with various sizes [29].

However, almost all of above studies mostly studied the effect of different reducing agents and stabilizing agents on the final morphology of Ag NPs [30, 31], while the control of particle size using the chemical reduction route is rarely reported. For example, Bastús et al. [32] reported the possibility to synthesize highly monodisperse Ag NPs from 10 to

✉ Ali Mirzaei
alisonmirzaee@yahoo.com

[1] Department of Materials Science and Engineering, Shiraz University, Shiraz, Iran

[2] Department of Engineering, University of Messina, Contrada di Dio, 98166 Messina, Italy

200 nm by the precise kinetic control of the reaction, in particular the use of two reducing agents, i.e., sodium citrate and tannic acid, the temperature of the solution, and the Ag seeds to Ag precursor ratio. Li et al. [33] reported synthesis of monodisperse, quasi-spherical silver nanocrystals directly in water via adding the aqueous solution of a mixture of AgNO₃, sodium citrate, and potassium iodide into the boiling aqueous solutions of ascorbic acid. In addition, Liu et al. [34] demonstrated a successful one-step seeded growth route to Ag quasi-nanoparticles of uniform sizes in the range of 19–140 nm, by carefully tune the kinetics of the seeded growth reaction to an appropriate level by the means of coordinating acetonitrile to an Ag (I) salt and adjusting the reaction temperature.

In this paper, a simple and efficient chemical reduction method was explored to synthesize the size-controlled Ag NPs. The formation of Ag NPs and their electrical and optical properties are investigated and discussed in detail.

Experiment

Ag NP synthesis

The starting materials were silver nitrate (>99.9%, AgNO₃) sodium borohydride (>98%, NaBH₄) and PVP (M.W. > 1.3×10^5), purchased from Merck (Germany). All the chemicals were used without further purification, and redistilled water was used to prepare all the solutions.

Figure 1 schematically shows the synthesis procedure, which refers to Creighton's method that uses sodium borohydride as the reducing agent [35]. The solutions of PVP and AgNO₃ were prepared separately by dissolving appropriate amounts of AgNO₃ (10 and 25 mM AgNO₃ in 10 ml) and PVP in distilled water and in well-cleaned dry beakers at room temperature. With the aim of preparing stable Ag NPs with different sizes, several PVP (based on the weights of a PVP unit 111) to AgNO₃ ratios (S) were

tested to determine the optimal conditions (Table 1). Then, two solutions were mixed together and stirred for 30 min, at an ice-cold bath. In a separate flask, appropriate amount of NaBH₄ was dissolved in the distilled water at zero temperature. The ratio of NaBH₄ to AgNO₃ (R) was varied to find the best results (Table 1). Then, reducing agent solution was added to Ag⁺–PVP solution dropwise to reduce the Ag ions to Ag NPs. A transparent bright yellow color was observed immediately due to the formation of the silver colloids. The solutions were kept in a dry place study their stabilities over months. In addition, the necessary characterizations were done for successfully synthesized Ag NPs.

Characterization

Characterization of Ag NPs was achieved by different techniques. X-ray diffraction analysis (XRD) was performed using a Philips X-Pert diffractometer operating with monochromatic CuKα ($\lambda = 1.54056$ Å) radiation at 40 kV and 30 mA. The XRD pattern was recorded at room temperature at a scan rate 0.05°/s and 2θ from 30° to 90°.

Morphological analysis was carried out using scanning electron microscopy (SEM) using a JEOL 5600 LV instrument operating with a 20 kV accelerating voltage. Crystallinity and morphological studies of synthesized powders were performed with a JEOL JEM 2010 electron microscope (LaB₆ electron gun) operating at 200 kV.

X-ray photoelectron spectroscopy (XPS) was employed to determine the composition of Ag NPs. XPS was performed on an ESCALAB MKII X-ray photoelectron spectrometer (VG Instruments, CA, USA), using non-monochromatized Mg–Kα X-rays as the excitation source. The binding energies for the sample were calibrated by setting the measured binding energy of C1s to 284.60 eV.

The photoluminescence (PL) spectrum of the Ag NPs was measured on a Hitachi F-4500 FL spectrophotometer with an Xe lamp with the excitation wavelength of 355 nm

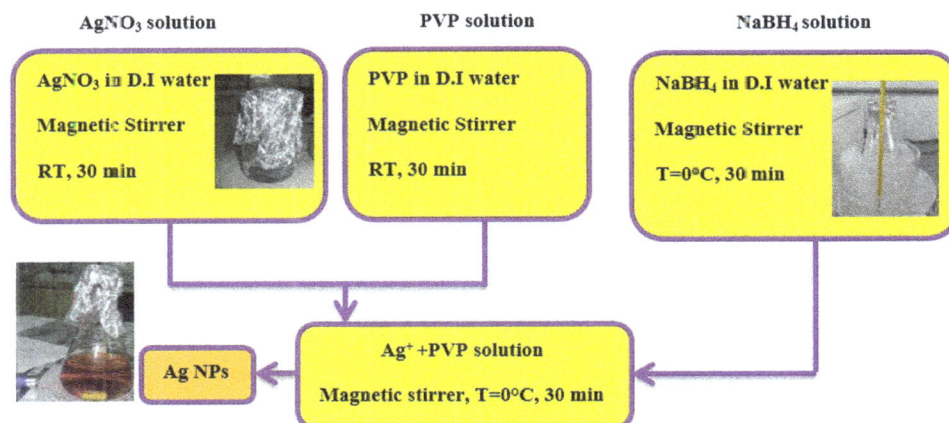

Fig. 1 Schematic representation of different steps of Ag NPs synthesis

AgNO₃ solution

AgNO₃ in D.I water
Magnetic Stirrer
RT, 30 min

PVP solution

PVP in D.I water
Magnetic Stirrer
RT, 30 min

NaBH₄ solution

NaBH₄ in D.I water
Magnetic Stirrer
T=0°C, 30 min

Ag⁺ +PVP solution
Magnetic stirrer, T=0°C, 30 min

Ag NPs

Table 1 Overview of the different experiment parameters performed for the synthesis of Ag NPs during this work

No	[AgNO$_3$] mM in 10 ml	[SBH]/[AgNO$_3$]	[PVP]/[AgNO$_3$]	T (°C)	Appearance and stability
1	10	0.5	0	Ice-cold	Black, unstable
2	10	0.7	0	Ice-cold	Black, unstable
3	10	1	0	Ice-cold	Black, unstable
4	10	1(aged NaBH$_4$)	0	Ice-cold	Black, unstable
5	10	1.3	0	Ice-cold	Black, unstable
6	10	0.7	0	Ice-cold	Black, unstable
7	10	0.7	0.5	Ice-cold	Yellow, stable
8	10	0.7	1	Ice-cold	Yellow, stable
9	25	0.7 (aged NaBH$_4$)	0.5	Ice-cold	Gray, unstable
10	25	0.7	0.5	Ice-cold	Yellow, stable
11	25	0.7	1.5	Ice-cold	Yellow, stable
12	25	0.7	3	Ice-cold	Yellow, stable
13	25	0.7	0	RT	Gray, unstable
14	25	0.7	1.5	RT	Gray, unstable
15	25	0.7	3	RT	Gray, unstable

and with the wavelength-scanning mode at room temperature in the air.

Thermal analysis was carried out on a "Mettler Toledo" thermal gravimetric analyzer. Analysis was performed in the presence of N$_2$, in the temperature range between 25 and 600 °C, with a heating rate of 10 °C/min.

Optical absorbance of Ag NPs was recorded with a 1 cm path length quartz cell using a UV–Vis spectrophotometer (Perkin–Elmer Lambda 2 spectrophotometer) as a function of wavelength in the range from 200 to 800 nm with a 1 nm resolution.

Results and discussions

Parameters' optimization for the synthesis of Ag NPs

The reduction of Ag ions (Ag$^+$) to Ag NPs was roughly monitored by visual inspection of the solution. Solutions of colloidal Ag NPs have distinctive yellow color arising from their tiny dimensions. Since Ag$^+$ does not have any color, therefore, the formation of Ag NPs has been observed by a change in their color. The addition of AgNO$_3$ solution to distilled water produces a colorless precursor solution. Its UV–Vis spectrum shows a sharp peak at 300 nm (Fig. 2), which is associated with Ag$^+$ ions in the solution. This is the first stage of the AgNO$_3$ reduction reaction, which is dissociated into Ag$^+$ and NO$_3^-$ ions.

The borohydrides, such as NaBH$_4$, are very strong reducing agents, and can reduce most metal salts to elemental metals. The chemical reaction for reduction of AgNO$_3$ in the presence of NaBH$_4$ is as follows [36]:

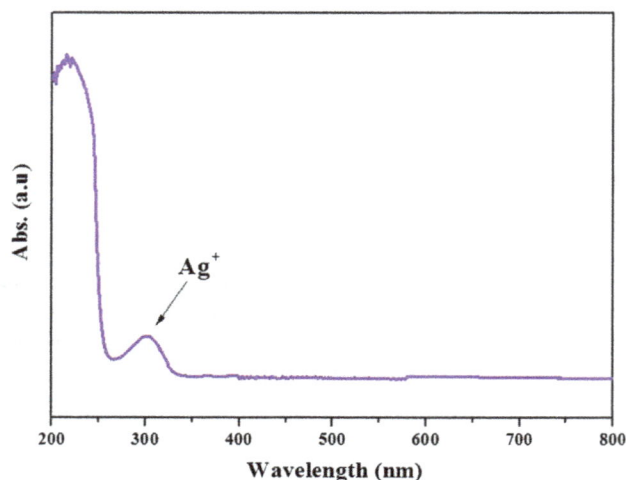

Fig. 2 UV–Vis spectrum of AgNO$_3$ in water

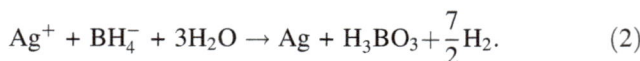

$$AgNO_3 \rightarrow Ag^+ + NO_3^- \tag{1}$$

$$Ag^+ + BH_4^- + 3H_2O \rightarrow Ag + H_3BO_3 + \frac{7}{2}H_2. \tag{2}$$

Hydrogen gas is produced by the reduction of silver ions as well as the slow reduction of water by the sodium borohydride at zero temperature:

$$BH_4^- + 3H_2O \rightarrow H_2BO_3^- + 4H_2(g). \tag{3}$$

Cloutier et al. [37] reported that the solutions of NaBH$_4$ contain certain amount of sodium hydroxide as a result of the NaBH$_4$ hydrolysis followed by the hydrolysis of sodium metaborate. Furthermore, NaBH$_4$ hydrolysis rate slows down with time because of the increasing concentration of the sodium hydroxide:

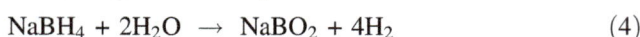

$$NaBH_4 + 2H_2O \rightarrow NaBO_2 + 4H_2 \tag{4}$$

$$NaBO_2 + H_2O \rightarrow NaOH + H_3BO_3. \tag{5}$$

Therefore, ageing time of $NaBH_4$ solutions is a crucial problem, so in all cases, before each set of experimental runs, fresh ice-cold $NaBH_4$ solutions were prepared.

Particle formation in the silver-borohydride system follows three distinct stages. First, upon mixing, the reaction between borohydride and silver occurs rapidly, resulting in the formation of small particles (2–3 nm). In the second stage, these particles grow to achieve sizes of 8–20 nm. In the final stage, the borohydride is consumed by reaction with water. This results in the loss of BH_4^- and the solution passing from a reducing to an oxidizing environment. The resulting changes in pair potential can drive the particle to aggregates [38].

When sodium borohydride was used for reduction of silver, the reducing reaction was very intense, and in the absence of a protective dispersing agent, the resulting particle sizes would increase as a result of agglomeration effects; therefore, PVP as a surfactant was used to prevent growth of Ag NPs.

A summary of the Ag NPs synthesis experiments performed in this work is listed in Table 1. Without PVP, all

Ag colloids at zero temperature and room temperature were unstable and, therefore, not appropriate for further studies. In addition, Ag NPs prepared with aged $NaBH_4$ resulted in the formation of unstable colloids that the reason for this is formation of sodium hydroxide according to equations of (4) and (5). However, when PVP was added as stabilizer, it resulted in formation of stable Ag NPs. Zero temperature was chosen for synthesis, because at room temperature, the particles were grown more rapidly, and aggregation took place at a higher rate, resulting in formation of unstable Ag NPs. According to obtained results and based on stability of Ag NPs over time, these parameters were chosen to prepare Ag NPs: concentration of $AgNO_3 = 25$ mM, R ($[NaBH_4]/[AgNO_3]) = 0.7$, S ($[PVP]/[AgNO_3]) = 0.5$, 1.5, and 3, and $T = 0$ °C.

Characterization of synthesized Ag NPs

SEM micrographs of synthesized Ag NPs are shown in Fig. 3a and b for Ag NPs with $S = 2.5$ and $S = 3$, respectively, where almost spherical morphology of Ag NPs can be observed. The particle size in these figures can

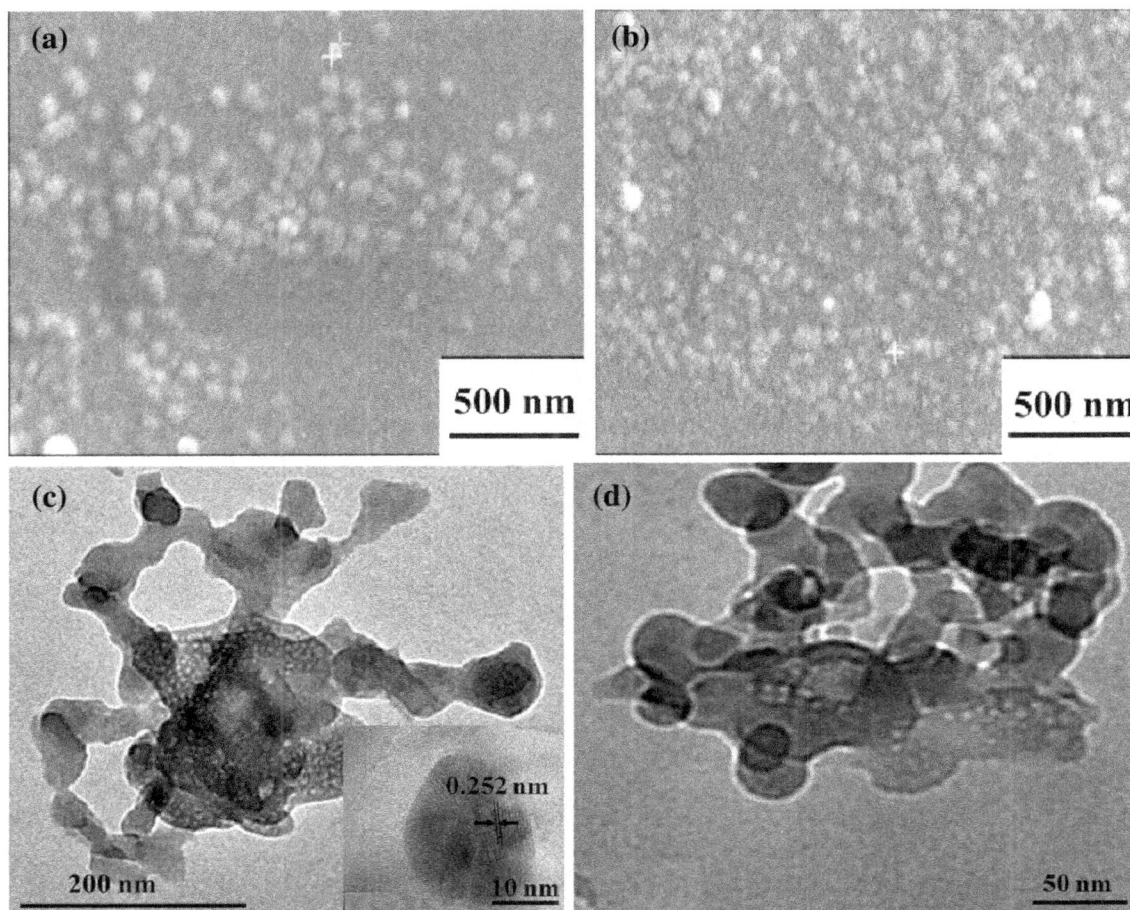

Fig. 3 SEM images of Ag NPs after drying **a** $S = 1.5$, **b** $S = 3$ TEM micrographs, **c** S = 1.5 (*inset*: HRTEM image), and **d** S = 3

Fig. 4 Particle size distribution of Ag NPs with $R = 0.7$ and different values of S

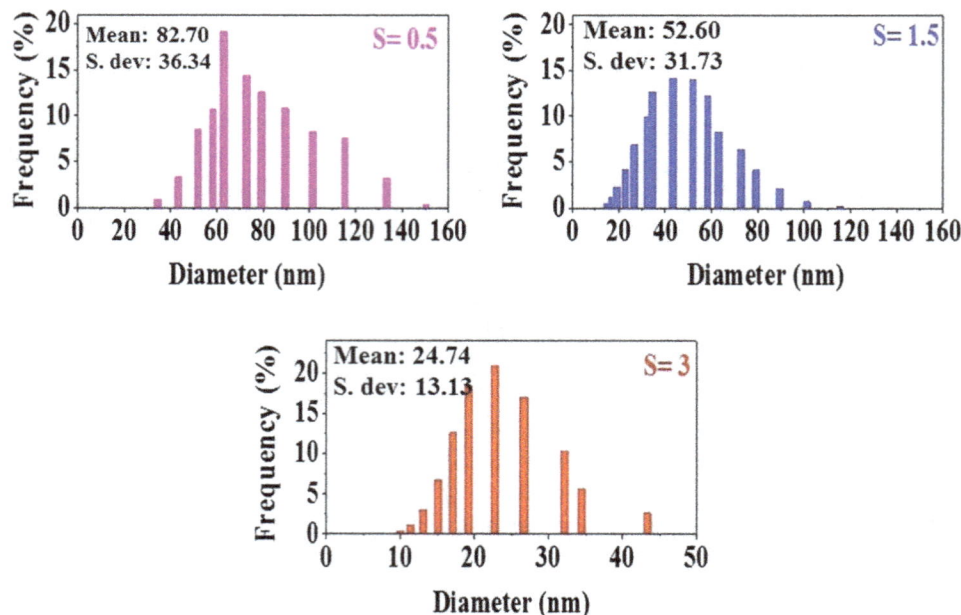

be roughly estimated 75–100 and 40–80 nm. To see if the PVP is presented around Ag NPs, TEM analysis was performed and TEM micrographs are presented in Fig. 3c and d for $S = 2.5$ and $S = 3$, respectively, where PVP around the Ag NPs is clearly visible. The HRTEM micrograph in the inset of Fig. 3c confirms the crystalline nature of synthesized Ag NPs, and lattice spacing of 0.252 nm can be attributed to the (111) planes of Ag NPs. Figure 4 shows particle size distribution of Ag NPs at fixed R, (0.7) and different values of S. It is worthy to note that the synthesis procedure was very sensitive to change of amount of reducing agent, and in case of $R < 0.7$, the reduction of Ag NPs was not completed, and in case of $R > 0.7$, the synthesized Ag NPs agglomerated instantly. Therefore, $R = 0.7$ was chosen for synthesis procedure and amount of S was changed. As it can be seen, in general, with increase of S, particle sizes decrease. In case of Ag NPs with $S = 0.5$, broad particle size distribution (~ 30 to 150 nm) was observed, which was due to insufficient capping of PVP on the surfaces of Ag NPs. However, with increase of S to 1.5, particle size distribution was spanned from 10 nm to about 120 nm. The smallest particles were obtained in Ag NPs with $S = 3$, where particle size distribution was narrow and spanned from 10 to 45 nm. As shown in Fig. 4c, the mean of particles for this sample is ~ 25 which is smaller than other samples. Even though there are some particles with sizes around 10 nm in the samples with $S = 3$, the relatively smaller particle sizes reported in the literatures [32–34], can be related to the use of PVP with different molecular weights as well as use of different reducing agents. To obtain more smaller Ag NPs, higher amount of PVP with different molecular weights (to

prevent high viscosity of solution) can be used. When S was above 3, the viscosity of the solution dramatically increased, and thus, S did not increase above 3. The crystallinity of prepared Ag NPs with $S = 3$ and $R = 0.7$ was investigated by XRD (Fig. 5). It is clearly observed that all the diffraction peaks with 2θ values of 38.2°, 44.35°, 64.5°, 77.4°, and 81.57° can be indexed as the (111), (200), (220), (311), and (222) crystal planes of the face-centered-cubic (fcc) phase of the silver and no trace of Ag_2O or AgO was found. Therefore, as-synthesized Ag NPs have high purity, and neither silver compound nor impurity has been intermixed in. At the same time, diffraction peaks indicate good crystallinity of ultrafine silvers. The lattice constant 'a' calculated from the XRD pattern for dried Ag NPs is 4.0845 Å, a value, which is in a close agreement with previously reported data in the literature for silver ($a = 4.086$ Å, JCPDS 04- 0783). The relatively broad

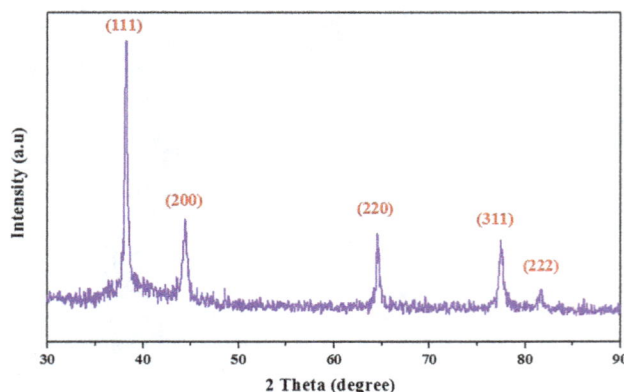

Fig. 5 XRD pattern of Ag NPs

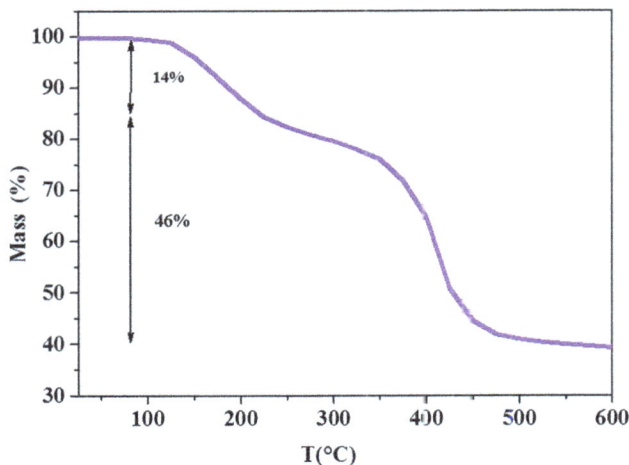

Fig. 6 TGA curve of PVP-capped Ag NPs

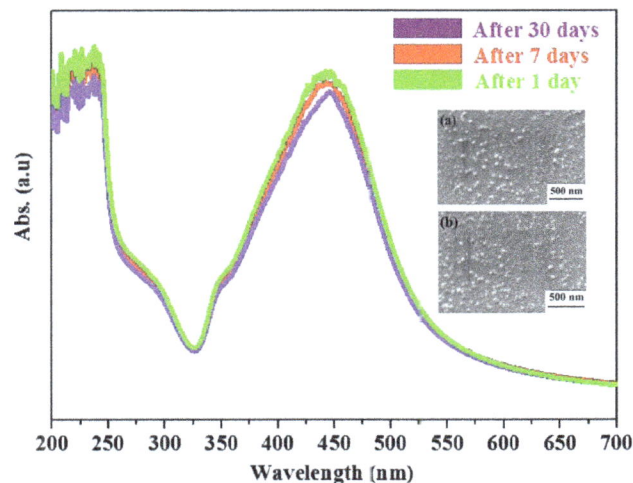

Fig. 7 UV–Vis spectra of Ag NPs after 1, 7, and 30 days. *Inset* **a** SEM micrograph after 7 days and **b** SEM micrograph after 30 days

peaks indicate the ultrafine average size of products. From the full-width at half-maximum of diffraction peaks, the average crystallite size of Ag NPs has been calculated using Debye–Scherrer equation [39]:

$$D = \frac{C\lambda}{\beta \cos\theta} \qquad (6)$$

where D is crystal size (Å), C is the shape factor that was taken equal to 0.94, λ is the wavelength of X-ray used (1.5418 Å), β is full-width at half-maximum (FWHM) in radians, and θ is Bragg's diffraction angle for the peak in degrees. The grain size of Ag NPs has been found in the range of ~18 nm.

Figure 6 shows the TGA curve of PVP-capped Ag NPs with $S = 3$ and $R = 0.7$. Very small mass loss before 100 °C is due to desorption of adsorbed water on the surface of Ag NPs. The mass loss from 100 to ~200 °C is attributed to the evaporation of low-molecular weight molecules which were covered Ag NPs. Mass loss observed from ~200 to 430 °C corresponds to the burning of organic species from capping layer (PVP). Above 460 °C, the TGA curve tends to be slowly smooth, reaching an almost constant weight. The total mass loss was about 60%, which confirms that the Ag NPs are coated with PVP.

Stability of Ag NPs

The ability to store chemically synthesized NPs for later use is beneficial to many applications. The stability of Ag NPs was investigated by monitoring the color of the reaction mixture and measuring the absorption spectra. The results obtained indicate that there is no obvious difference in position and symmetry of absorption peak during the initial 7 days (Fig. 7). After 1 month, the position of the peak has a small shift (from 443 to 448 nm), suggesting the

formation of slightly larger particles without any aggregation. Insets (a) and (b) in Fig. 7 show the SEM images, where the particle sizes did not changes in comparison with fresh colloids (Fig. 3d). Thus, colloidal silver can remain stable at room temperature for as long as several weeks or months. The reason for the stability of the Ag NPs is effective stabilization of Ag NPs due to the presence of PVP. Figure 8 shows stability of Ag NPs during 90 days, where after 90 days, color of Ag NPs was gradually turned into gray, indicating formation of bigger Ag particles and breakdown of stability. As the particles destabilize, the original extinction peak will decrease in intensity (due to the depletion of stable nanoparticles), and the peak will broaden due to the formation of aggregates (Fig. 9).

High-resolution XPS spectrum of the Ag 3d region presented in Fig. 10 shows two deconvoluted peaks located at 368.08 and 374.08 eV (with a spin–orbit separation of 6.0 eV) that can be attributed to Ag $3d_{5/2}$ and Ag $3d_{3/2}$, respectively [40]. For metal silver, 368.2 eV for Ag 3d5/2 and 374.2 eV for Ag 3d3/2 are reported. These results suggest a strong interaction between the carboxyl oxygen atoms in the PVP chain and Ag NPs. Furthermore, no peak corresponding to Ag_2O (367.8 eV) or AgO (367.4 eV) is observed in the XPS spectrum of Ag NPs, which indicates that the Ag NPs are all in form of metallic Ag [41].

A primary purpose of the introducing PVP is to protect the Ag NPs from growing and agglomerating. The structures of the monomer, vinylpyrrolidone, and the repeat unit of PVP are shown in Fig. 11a and b, respectively. According to the literature [42], the main reason of PVP protecting Ag NPs is N in PVP coordinate with silver and forms the protection layer. The reactions would be like Fig. 11c. PVP preventing the Ag NPs from aggregation is

Fig. 8 Stability of Ag NPs during 90 days

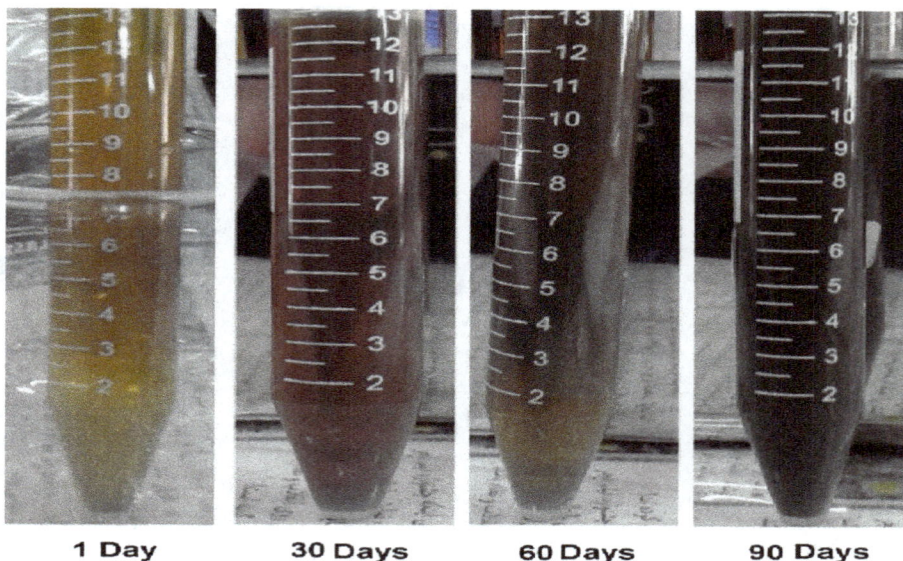

1 Day   30 Days   60 Days   90 Days

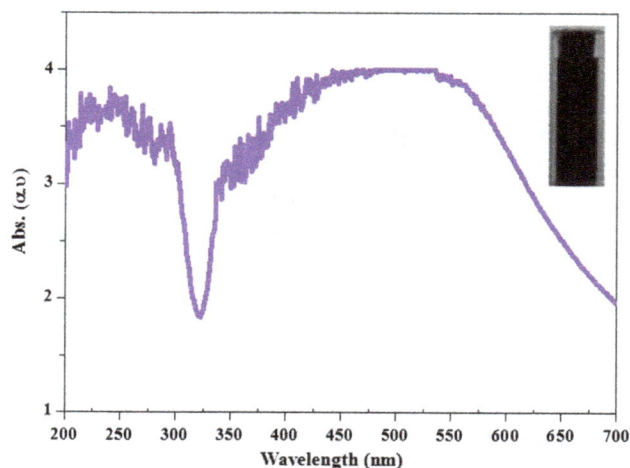

Fig. 9 UV–Vis of unstable Ag NPs after 90 days

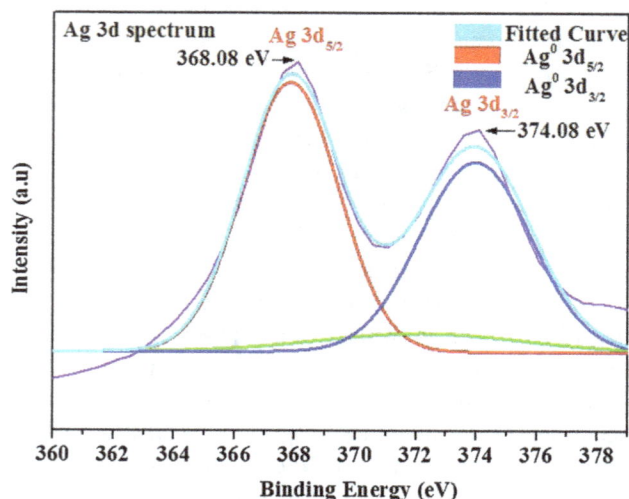

Fig. 10 XPS spectrum of Ag NPs

the main role of the surfactant. In fact, the steric effect arising from the long polyvinyl chain of PVP on the surface of Ag NPs may contribute to the anti-agglomeration of them.

Optical studies

Optical absorption spectra of Ag NPs measured as a function of the wavelength of incident photons are shown in Fig. 12. It seems that samples with $S = 0.5$ and 1 are not totally converted to Ag NPs, as maximum peak of their adsorption is lower than other samples. In case of Ag NPs with $S = 0.5$ and $S = 1$, it seems that there is not enough PVP + NaBH$_4$ to complete reduction of silver ions. For samples with $S = 1.5$, 2, 2.5, and 3, chemical reduction is complete. The absorption maximum is almost the same, with a little shift. These results are consistent with results by Zou et al. [43], where with the increase of stabilizer amount, the position of maximum absorption wavelength slightly shifted towards red.

In the high absorption region (where $\alpha > 10^4$ cm^{-1}), involving interband optical transitions between valence and conduction bands, the absorption coefficient data as a function of wavelength are, determined using the Tauc formula, given by the following equation [44]:

$$\alpha h v = B^* \left(h v - E_g^{\text{Opt}} \right)^m \tag{7}$$

where B^* is the edge width parameter representing the materials' quality and is calculated from the linear part of this relation, hv is the photon energy (=hc/λ, where hc is 1239.83 eV), E_g^{Opt} is the optical energy gap of the material, and "m" is a number which characterized the mechanism of transition process. $m = 1/2$, 3/2, for direct transition and

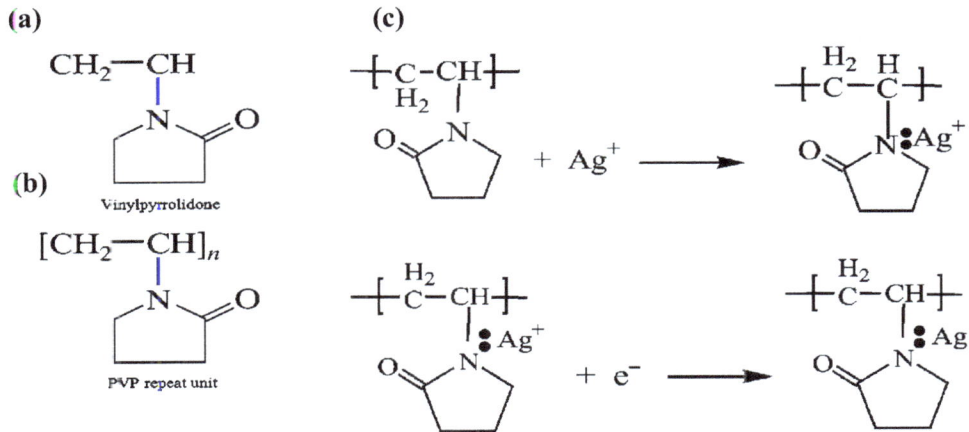

Fig. 11 a Structure of the monomer, vinylpyrrolidone. **b** Repeat unit of PVP. **c** Reaction between PVP and Ag

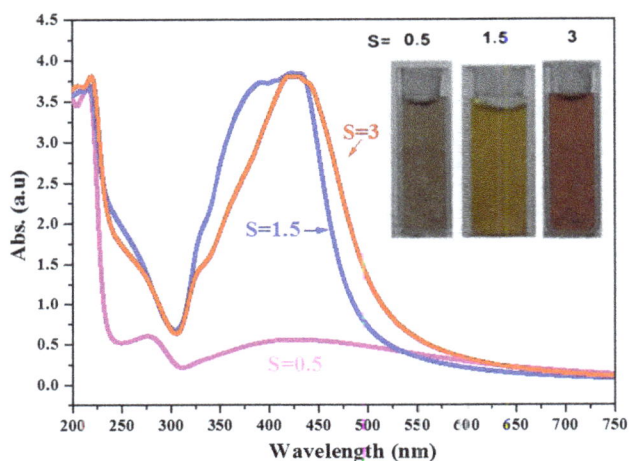

Fig. 12 Optical absorption spectra of Ag NPs with different values of S and $R = 0.7$

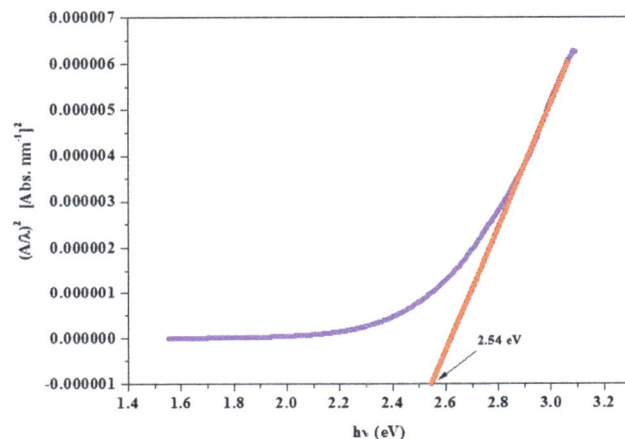

Fig. 13 Tauc plot for Ag NPs ($S = 3$, $R = 0.7$)

$m = 1, 2, 3$, for indirect transitions. $\alpha(v)$ is the absorption coefficient defined by the Beer–Lambert's law as follows:

$$\alpha = \frac{2.303 \text{Abs.}}{d} \tag{8}$$

where d and Abs. are the optical path (1 cm) and absorbance of Ag colloids, respectively. Because of $d = 1$ cm in optical measurements of Ag NPs, we could simply use the following formula instead of Eq. (8):

$$\alpha = 2.303 \text{ Abs.} \tag{9}$$

Using equations of (7) and (9), we can plot $\left(\frac{\text{Abs}}{\lambda}\right)^2$ vs. hv. E_g of the Ag NPs were estimated by extrapolating the linear portion of $\left(\frac{\text{Abs}}{\lambda}\right)^2$ vs. hv curves to $\left(\frac{\text{Abs}}{\lambda}\right)^2 = 0$, as shown in Fig. 13. The band gap value (2.54 eV) of Ag NPs is very close to the interband (d to sp) threshold energy of bulk Ag, and the determined n and k values are also close to those of bulk Ag. The bandgap is due to the d to sp interband gap, instead of any quantized energy gap, since there is no quantization in Ag NPs [45].

Figure 14 shows the variations of PL emission Ag NPs ($S = 3$, $R = 0.7$) dispersed in water, under excitation wavelength of 355 nm. Apparently, it can be observed that

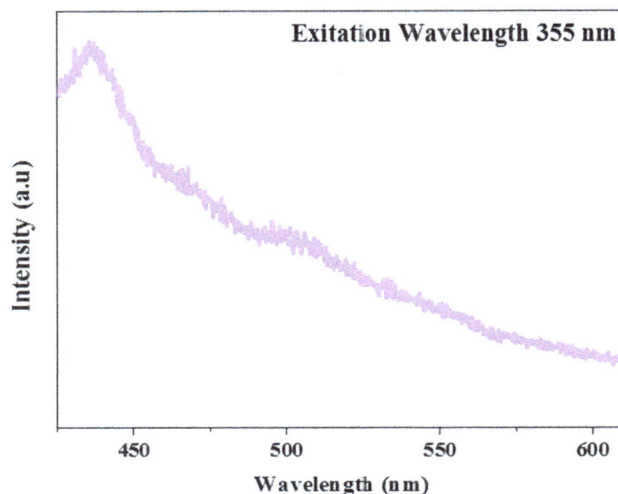

Fig. 14 Room-temperature photoluminescence emission spectrum of Ag NPs dispersed in water (Excitation wavelength was 355 nm)

the strongest emission peaks are appeared at around 440 nm. In general, the UV emission band is originated from the direct recombination of the free excitons through an exciton–exciton collision process. In analogy with PL spectra of other noble metals, this PL peak can be assigned to radiative recombination of Fermi-level electrons and *sp*- or *d*-band holes [46].

Conclusion

In brief, we successfully prepared colloidal Ag NPs stabilized with PVP by a simple chemical reduction route using sodium borohydride as a reducing agent. Different characterization techniques, such as SEM, TEM, XRD, XPS, PSA, UV–Vis, and PL, were used to characterize synthesized Ag NPs. It was found that the average particle size can be controlled with a narrow size distribution by controlling the amount of PVP added. The particles size decreased with increasing PVP concentration. The synthesized NPs were found to be stable for 1 month, so they can be used for a long time after preparation. The size-controlled, well-dispersed Ag NPs prepared are capable of being dispersed in water with potential applications in dielectric and biological fields.

Acknowledgements This work was partially supported by the Iran nanotechnology council.

References

1. Manikam, V.R., Cheong, K.Y., Razak, K.A.: Chemical reduction methods for synthesizing Ag and Al nanoparticles and their respective nanoalloys. Mat. Sci. Eng. B Solid. **176**(3), 187–203 (2011)
2. Liu, J.F., Yu, S.J., Yin, Y.G., Chao, J.B.: Methods for separation, identification, characterization and quantification of silver nanoparticles. Trends Anal. Chem. **33**, 95–106 (2012)
3. Setua, P., Chakraborty, A., Seth, D., Bhatta, M.U., Satyam, P.V., Sarkar, N.: Synthesis, optical properties, and surface enhanced Raman scattering of silver nanoparticles in nonaqueous methanol reverse micelles. J. Phys. Chem. C **111**(10), 3901–3907 (2007)
4. Jiangmei, Y.A.N., Huiwang, T.A.O., Muling, Z.E.N.G., Jun, T.A.O., Zhang, S., Zhiying, Y.A.N., Wei, W.A.N.G., Jiaqiang, W.A.N.G.: PVP-capped silver nanoparticles as catalyst for oxidative coupling of thiols to disulfides. Chin. J. Catal. **30**(9), 856–858 (2009)
5. Kong, H.Y., Jang, Y.S.: Antibacterial properties of novel poly(-methyl methacrylate nanofiber containing silver nanoparticles. Langmuir **24**, 2051–2056 (2008)
6. Mirzaei, A., Janghorban, K., Hashemi, B., Bonavita, A., Bonyani, M., Leonardi, S.G., Neri, G.: Synthesis, characterization and gas sensing properties of Ag@ α-Fe$_2$O$_3$ core-shell nanocomposites. Nanomaterials **5**(2), 737–749 (2015)
7. Sun, L., Zhang, Z.J., Wu, Z.S., Dang, H.X.: Synthesis and characterization of DDP coated Ag nanoparticles. Mater. Sci. Eng., A **379**, 378–383 (2004)
8. Pingali, K.C., Rockstraw, D.A., Deng, S.: Silver nanoparticles from ultrasonic spray pyrolysis of aqueous silver nitrate. Aerosol Sci. Tech. **39**(10), 1010–1014 (2005)
9. Joseph, S., Mathew, B.: Microwave-assisted facile synthesis of silver nanoparticles in aqueous medium and investigation of their catalytic and antibacterial activities. J. Mol. Liq. **197**, 346–352 (2014)
10. Shinde, M., Pawar, A., Karmakar, S., Seth, T., Raut, V., Rane, S., Bhoraska, S., Amalnerkar, D.: Uncapped silver nanoparticles synthesized by DC arc thermal plasma technique for conductor paste formulation. J. Nano. Res. **11**, 2043–2047 (2009)
11. Khan, Z., Al-Thabaiti, S.A., Obaid, A.Y., Al-Youbi, A.O.: Preparation and characterization of silver nanoparticles by chemical reduction method. Colloids Surf. B **82**, 513–517 (2011)
12. Yang, J., Pan, J.: Hydrothermal synthesis of silver nanoparticles by sodium alginate and their applications in surface-enhanced Raman scattering and catalysis. Acta Mater. **60**, 4753–4758 (2012)
13. Lu, Y., Yu, M., Schrinner, M., Ballauff, M., Möller, M.W., Breu, J.: In situ formation of Ag nanoparticles in spherical polyacrylic acid brushes by UV irradiation. J. Phys. Chem. C **111**, 7676–7681 (2007)
14. Zhang, Z., Li, V.J.: Synthesis and characterization of silver nanoparticles by a sonochemical method. Rare Metal Mater. Eng. **41**, 1700–1705 (2012)
15. Nikolov, A.S., Nedyalkov, N.N., Nikov, R.G., Atanasov, P.A., Alexandrov, M.T.: Characterization of Ag and Au nanoparticles created by nanosecond pulsed laser ablation in double distilled water. Appl. Surf. Sci. **257**, 5278–5282 (2011)
16. Abu-Zied, B.M., Asiri, A.M.: An investigation of the thermal decomposition of silver acetate as a precursor for nano-sized Ag-catalyst. Thermochim. Acta **581**, 110–117 (2014)
17. Mishra, Y.K., Mohapatra, S., Kabiraj, D., Mohanta, B., Lalla, N.P., Pivin, J.C., Avasthi, D.K.: Synthesis and characterization of Ag nanoparticles in silica matrix by atom beam sputtering. Scr. Mater. **56**, 629–632 (2007)
18. Sun, Y., Liu, Y., Guizhe, Z., Zhang, Q.: Effects of hyperbranched poly(amido-amine)s structures on synthesis of Ag particles. J. Appl. Polym. Sci. **107**, 9–13 (2008)
19. Kim, K.D., Nam Han, D., Kim, H.T.: Optimization of experimental conditions based on the Taguchi robust design for the formation of nano-sized silver particles by chemical reduction method. Chem. Eng. J. **104**, 55–61 (2004)
20. Jia, Z., Sun, H., Gu, Q.: Preparation of Ag nanoparticles with triethanolamine as reducing agent and their antibacterial property. Colloids Surf. A Physicochem. Eng. Asp. **419**, 174–179 (2013)
21. Zeng, J., Zheng, Y., Ryceng, M., Tao, J., Li, Z., Zhang, Q., Zhu, Y., Xi, Y.: Controlling the shapes of silver nanocrystals with different capping agents. J. Am. Chem. Soc. **132**, 8552–8553 (2010)
22. Chen, Z., Gang, T., Yan, X., Li, X., Zhang, J., Wang, Y., Chen, X., Sun, Z., Zhang, K., Zhao, B., Yang, B.: Ordered silica microspheres unsymmetrically coated with Ag nanoparticles and Ag-nanoparticles-doped polymer voids fabricated by microcontact printing and chemical reduction. Adv. Mater. **18**, 924–929 (2006)
23. Chou, K.S., Ren, C.Y.: Synthesis of nanosized silver particles by chemical reduction method. Mater. Chem. Phys. **64**, 241–246 (2000)
24. Magdassi, S., Bassa, A., Vinetsky, Y., Kamyshny, A.: Silver nanoparticles as pigments for water-based ink-jet inks. Chem. Mater. **15**, 2208–2217 (2003)
25. Murthy, S., Bigioni, T.P., Wang, Z.L., Khoury, J.T., Whetten, R.L.: Liquid-phase synthesis of thiol-derivatized silver nanocrystals. Mater. Lett. **30**, 321–325 (1997)
26. Li, Y., Wu, Y., Ong, B.S.: Facile synthesis of silver nanoparticles

useful for fabrication of high-conductivity elements for printed electronics. J. Am. Chem. Soc. **127**, 3266–3267 (2005)

27. Seoudi, R., Shabaka, A., Elsayed, Z.A., Anis, B.: Effect of stabilizing agent on the morphology and optical properties of silver nanoparticles. Phys. E **44**, 440–447 (2011)

28. Lee, K.J., Jun, B.H., Choi, J., Lee, Y.I., Joung, J., Oh, Y.S.: Environmentally friendly synthesis of organic-soluble silver nanoparticles for printedelectronics. Nanotechnology **18**, 335601–335606 (2007)

29. Lu, Y.C., Chou, K.S.: A simple and effective route for the synthesis of nano-silver colloidal dispersions. J. Chin. Inst. Chem. Eng **39**, 673–678 (2008)

30. Guo, G., Gan, W., Luo, J., Xiang, F., Zhang, J., Zhou, H., Liu, H.: Preparation and dispersive mechanism of highly dispersive ultrafine silver powder. Appl. Surf. Sci. **256**. 6683–6687 (2010)

31. Wang, D., Song, C., Hu, Z., Zhou, X.: Synthesis of silver nanoparticles with flake-like shapes. Mater. Lett. **59**, 1760–1763 (2005)

32. Bastús, N.G., Merkoçi, F., Piella, J., Puntes, V.: Synthesis of highly monodisperse citrate-stabilized silver nanoparticles of up to 200 nm: kinetic control and catalytic properties. Chem. Mater. **26**(9), 2836–2846 (2014)

33. Li, H., Xia, H., Wang, D., Tao, X.: Simple synthesis of monodisperse, quasi-spherical, citrate-stabilized silver nanocrystals in water. Langmuir **29**(16), 5074–5079 (2013)

34. Liu, X., Yin, Y., Gao, C.: Size-tailored synthesis of silver quasi-nanospheres by kinetically controlled seeded growth. Langmuir **29**(33), 10559–10565 (2013)

35. Creighton, J.A., Blatchford, C.G., Albrecht, M.G.: Plasma resonance enhancement of Raman scattering by pyridine adsorbed on silver or gold sol particles of size comparable to the excitation wavelength. J. Chem. Soc. Faraday Trans. **75**, 790–798 (1979)

36. Pradhan, N., Pal, A., Pal, T.: Silver nanoparticle catalyzed reduction of aromatic nitro compounds. Colloids Surf. A Physicochem. Eng. Asp. **196**, 247–257 (2002)

37. Cloutier, C.R., Gyenge, E., Alfantazi, A.: Physicochemical properties of alkaline aqueous sodium metaborate solutions. J. Fuel Cell Sci. Tech. **4**, 88–98 (2007)

38. Van Hyning, D.L., Klemperer, W.G., Zukoski, C.F.: Silver nanoparticle formation: predictions and verification of the aggregative growth model. Langmuir **17**, 3128–3135 (2001)

39. Mirzaei, A., Janghorban, K., Hashemi, B., Bonyani, M., Leonardi, S.G., Neri, G.: Highly stable and selective ethanol sensor based on α-Fe$_2$O$_3$ nanoparticles prepared by Pechini sol–gel method. Ceram. Int. **42**(5), 6136–6144 (2016)

40. Ghodselahi, T., Neishaboorynejad, T., Arsalani, S.: Fabrication LSPR sensor chip of Ag NPs and their biosensor application based on interparticle coupling. Appl. Surf. Sci. **343**, 194–201 (2015)

41. Li, H.J., Zhang, A.Q., Hu, Y., Sui, L., Qian, D.J., Chen, M.: Large-scale synthesis and self-organization of silver nanoparticles with Tween 80 as a reductant and stabilizer. Nanoscale Res. Lett. **7**, 612–623 (2012)

42. Wang, H., Qiao, X., Chen, J., Wang, X., Ding, S.: Mechanisms of PVP in the preparation of silver nanoparticles. Mater. Chem. Phys. **94**, 449–453 (2005)

43. Zou, J., Xu, Y., Hou, B., Wu, D., Sun, Y.: Controlled growth of silver nanoparticles in a hydrothermal process. China Particuol. **5**(3), 206–212 (2007)

44. Ghobadi, N.: Band gap determination using absorption spectrum fitting procedure. Int. Nano Lett. **3**, 1–4 (2013)

45. Khana, M.A., Kumar, S., Ahameda, M., Alrokayana, S.A., Alsalhi, M.S., Alhoshana, M., Alcwayyana, A.S.: Structural and spectroscopic studies of thin film of silver nanoparticles. Appl. Surf. Sci. **257**, 10607–10612 (2011)

46. Sarkar, R., Kumbhakar, P., Mitra, A.K., Ganeev, R.A.: Synthesis and photoluminescence properties of silver nanowires. Curr. Appl. Phys. **10**, 853–857 (2010)

Permissions

All chapters in this book were first published in JNSC, by Springer International Publishing AG.; hereby published with permission under the Creative Commons Attribution License or equivalent. Every chapter published in this book has been scrutinized by our experts. Their significance has been extensively debated. The topics covered herein carry significant findings which will fuel the growth of the discipline. They may even be implemented as practical applications or may be referred to as a beginning point for another development.

The contributors of this book come from diverse backgrounds, making this book a truly international effort. This book will bring forth new frontiers with its revolutionizing research information and detailed analysis of the nascent developments around the world.

We would like to thank all the contributing authors for lending their expertise to make the book truly unique. They have played a crucial role in the development of this book. Without their invaluable contributions this book wouldn't have been possible. They have made vital efforts to compile up to date information on the varied aspects of this subject to make this book a valuable addition to the collection of many professionals and students.

This book was conceptualized with the vision of imparting up-to-date information and advanced data in this field. To ensure the same, a matchless editorial board was set up. Every individual on the board went through rigorous rounds of assessment to prove their worth. After which they invested a large part of their time researching and compiling the most relevant data for our readers.

The editorial board has been involved in producing this book since its inception. They have spent rigorous hours researching and exploring the diverse topics which have resulted in the successful publishing of this book. They have passed on their knowledge of decades through this book. To expedite this challenging task, the publisher supported the team at every step. A small team of assistant editors was also appointed to further simplify the editing procedure and attain best results for the readers.

Apart from the editorial board, the designing team has also invested a significant amount of their time in understanding the subject and creating the most relevant covers. They scrutinized every image to scout for the most suitable representation of the subject and create an appropriate cover for the book.

The publishing team has been an ardent support to the editorial, designing and production team. Their endless efforts to recruit the best for this project, has resulted in the accomplishment of this book. They are a veteran in the field of academics and their pool of knowledge is as vast as their experience in printing. Their expertise and guidance has proved useful at every step. Their uncompromising quality standards have made this book an exceptional effort. Their encouragement from time to time has been an inspiration for everyone.

The publisher and the editorial board hope that this book will prove to be a valuable piece of knowledge for researchers, students, practitioners and scholars across the globe.

List of Contributors

M. Ramakrishna, G. Nageswara Rao and Dandamudi Rajesh Babu
Department of Chemistry, Sri Sathya Sai Institute of Higher Learning, Prasanthinilayam, Puttaparthi 515134, Andhra Pradesh, India

R. M. Gengan
Chemistry Department, Durban University of Technology, Durban 4001, South Africa

S. Chandra
LN Government College, Ponneri 601204, Tamil Nadu, India

Agbaje Lateef, Akeem Akinboro, Musibau A. Azeez, Taofeek A. Yekeen and Sunday A. Ojo
Department of Pure and Applied Biology, Ladoke Akintola University of Technology, PMB 4000, Ogbomoso, Nigeria

Tesleem B. Asafa
Department of Mechanical Engineering, Ladoke Akintola University of Technology, PMB 4000, Ogbomoso, Nigeria

Iyabo C. Oladipo
Department of Science Laboratory Technology, Ladoke Akintola University of Technology, PMB 4000, Ogbomoso, Nigeria

Luqmon Azeez
Department of Chemical Sciences, Osun State University, Osogbo, Nigeria

Evariste B. Gueguim-Kana
Department of Microbiology, University of KwaZulu-Natal, Private Bag X01, Scottsville, Pietermaritzburg 3209, South Africa

Lorika S. Beukes
Microscopy and Microanalysis Unit, School of Life Sciences, University of KwaZulu-Natal, Private Bag X01, Scottsville, Pietermaritzburg 3209, South Africa

Munisamy Manjunathan
Department of Chemistry, BWDA Arts and Science College, Tindivanam 604304, India

Department of Chemistry, Pondicherry University, Pondicherry 605014, India

Yesuvadimai Jerlin Jose
Department of Chemistry, BWDA Arts and Science College, Tindivanam 604304, India
Department of Chemistry, St. Josephs College, Tiruchirappalli 620002, India

Savariraj Joseph Selvaraj
Department of Chemistry, St. Josephs College, Tiruchirappalli 620002, India

Pankaj Madkikar, Michele Piana, Thomas Mittermeier, Christoph Denk and Hubert A. Gasteiger
Chair of Technical Electrochemistry, Department of Chemistry and Catalysis Research Center, Technische Universitä't Mü'nchen, 85748 Garching, Germany

Xiaodong Wang
Johnson Matthey Catalysts (Germany) GmbH, Bahnhofstr. 43, 96257 Redwitz, Germany

Alessandro H. A. Monteverde Videla and Stefania Specchia
Department of Applied Science and Technology, Politecnico di Torino, Corso Duca degli Abruzzi 24, 10129 Turin, Italy

C. Ashajyothi and R. Kelmani Chandrakanth
Department of Biotechnology, Gulbarga University, Gulbarga 585106, Karnataka, India

K. Handral Harish and Nileshkumar Dubey
Oral Sciences Disciplines, Faculty of Dentistry, National University of Singapore, Singapore 117510, Singapore

Amirali Abbasi and Jaber Jahanbin Sardroodi
Molecular Simulation Laboratory (MSL), Azarbaijan Shahid Madani University, Tabriz, Iran
Department of Chemistry, Faculty of Basic Sciences, Azarbaijan Shahid Madani University, Tabriz, Iran
Computational Nanomaterials Research Group (CNRG), Azarbaijan Shahid Madani University, Tabriz, Iran

Somayeh Firoozi and Mina Jamzad
Department of Chemistry, Shahr-e-Qods Branch, Islamic Azad University, Tehran, Iran

Mohammad Yari
Department of Chemistry, Islamshahr Branch, Islamic Azad University, P.O. Box: 33135-369, Islamshahr, Iran

P. Senthilkumar, D. S. Ranjith Santhosh Kumar, B. Sudhagar, M. Vanthana, M. Hajistha Parveen, S. Sarathkumar, Jeslin Cheriyan Thomas, A. Sandhiya Mary and Chandramouleeswaran Kannan
PG and Research Department of Biotechnology, Kongunadu Arts and Science College, Coimbatore 640 029, Tamilnadu, India

Md Niharul Alam, Sreeparna Das, Shaikh Batuta and Naznin Ara Begum
Department of Chemistry, Visva-Bharati (Central University), Santiniketan 731 235, India

Debabrata Mandal
Department of Chemistry, University College of Science and Technology, University of Calcutta, 92, Acharya Prafulla Chandra Road, Kolkata 700 009, India

Zahra Khaghanpour and Sanaz Naghibi
Department of Materials Engineering, Shahreza Branch, Islamic Azad University, Pasdaran St., PO Box 86145-311, Shahreza, Isfahan Province, Iran

Arumugam Sengottaiyan, Chinnapan Sudhakar, Thangaswamy Selvankumar, Kandasamy Selvam and Palanisamy Srinivasan
Department of Biotechnology, Mahendra Arts and Science College (Autonomous), Kalippatti, Namakkal, Tamil Nadu 637501, India

Adithan Aravinthan
College of Veterinary Medicine, Biosafety Research Institute, Chonbuk National University, Iksan 570-752, South Korea

Muthusamy Govarthanan
Department of Biotechnology, Mahendra Arts and Science College (Autonomous), Kalippatti, Namakkal, Tamil Nadu 637501, India
Division of Biotechnology, Advanced Institute of Environment and Bioscience, College of Environmental and Bioresource Sciences, Chonbuk National University, Iksan 570-752, South Korea

Koildhasan Manoharan
Raja Duraisingam Government Arts and Science College, Sivagangai, Tamil Nadu, India

S. Kalaiarasi and M. Jose
Department of Physics, Sacred Heart College (Autonomous), Tirupattur 635601, India

Neda Rohani and Azam Marjani
Department of Chemistry, Arak Branch, Islamic Azad University, Arak 38361-1-9131, Iran

Fatemeh F. Bamoharram
Research Center for Animal Development Applied Biology - Department of Nanobiotechnology, Mashhad Branch, Islamic Azad University, Mashhad 91865-397, Iran

Majid M. Heravi
Department of Chemistry, Alzahra University, Tehran 1993891176, Iran

S. Priya Velammal, T. Akkini Devi and T. Peter Amaladhas
PG and Research Department of Chemistry, V.O. Chidambaram College, Tuticorin 628008, Tamil Nadu, India

A. R. Allafchian and S. S. Hashemi
Nanotechnology and Advanced Materials Institute, Isfahan University of Technology, Isfahan 84156–83111, Iran

S. Z. Mirahmadi-Zare
Department of Molecular Biotechnology at Cell Science Research Center, Royan Institute for Biotechnology, ACECR, Isfahan, Iran

S. A. H. Jalali
Institute of Biotechnology and Bioengineering, Isfahan University of Technology, Isfahan 84156–83111, Iran

Department of Natural Resources, Isfahan University of Technology, Isfahan 84156–83111, Iran

M. R. Vahabi
Department of Natural Resources, Isfahan University of Technology, Isfahan 84156–83111, Iran

Javad Safaei-Ghomi and Seyed Hadi Nazemzadeh
Department of Organic Chemistry, Faculty of Chemistry, University of Kashan, P.O. Box 87317-51167, Kashan, Islamic Republic of Iran

Hossein Shahbazi-Alavi
Department of Organic Chemistry, Faculty of Chemistry, University of Kashan, P.O. Box 87317-51167, Kashan, Islamic Republic of Iran

Young Researchers and Elite Club, Islamic Azad University, Kashan Branch, Kashan, Islamic Republic of Iran

Ramin Mostafalu, Akbar Heydari and Marzban Arefi
Chemistry Department, Tarbiat Modares University, P.O. Box 14155-4838, Tehran, Iran

Abbas Banaei and Fatemeh Ghorbani
Research and Development, Padideh Shimi Jam Co., Eshtehard Industrial Town, Karaj, Iran

Indrajit Shown and Abhijit Ganguly
Institute of Atomic and Molecular Sciences, Academia Sinica, Taipei, Taiwan

Thangavel Akkini Devi, Narayanan Ananthi and Thomas Peter Amaladhas
PG and Research Department of Chemistry, V.O. Chidambaram College, Tuticorin, Tamilnadu 628008, India

Cinnathambi Subramani Maheswari and Appaswami Lalitha
Department of Chemistry, Periyar University, Periyar Palkalai Nagar, Salem, Tamil Nadu 636011, India

Chandrabose Shanmugapriya and Krishnan Revathy
Department of Chemistry, Sri Sarada College For Women (Autonomous), Salem, Tamil Nadu 636016, India

Ali Mirzaei, Kamal Janghorban, Babak Hashemi and Maryam Bonyani
Department of Materials Science and Engineering, Shiraz University, Shiraz, Iran

Salvatore Gianluca Leonardi and Giovanni Neri
Department of Engineering, University of Messina, Contrada di Dio, 98166 Messina, Italy

Index

www.ingramcontent.com/pod-product-compliance
Lightning Source LLC
Chambersburg PA
CBHW082045190326
41458CB00010B/3466